Plant Science: Methods for Growth and Development

Plant Science: Methods for Growth and Development

Edited by Agatha Wilson

SYRAWOOD
PUBLISHING HOUSE

New York

Published by Syrawood Publishing House,
750 Third Avenue, 9th Floor,
New York, NY 10017, USA
www.syrawoodpublishinghouse.com

Plant Science: Methods for Growth and Development
Edited by Agatha Wilson

International Standard Book Number: 978-1-68286-672-6 (Hardback)

Cataloging-in-Publication Data

Plant science : methods for growth and development / edited by Agatha Wilson.
    p. cm.
Includes bibliographical references and index.
ISBN 978-1-68286-672-6
1. Botany. 2. Growth (Plants). 3. Plants--Development. I. Wilson, Agatha.
QK46 .P53 2019
580--dc23

TABLE OF CONTENTS

Permissions

List of Contributors

Index

PREFACE

In my initial years as a student, I used to run to the library at every possible instance to grab a book and learn something new. Books were my primary source of knowledge and I would not have come such a long way without all that I learnt from them. Thus, when I was approached to edit this book; I became understandably nostalgic. It was an absolute honor to be considered worthy of guiding the current generation as well as those to come. I put all my knowledge and hard work into making this book most beneficial for its readers.

Plant science is the in-depth study of plants, their growth, life processes, genetics and other related aspects. It is a sub-field of biology and also known as botany. It encompasses all aspects of plant reproduction, evolution, pathology, physiology, genetics and taxonomy. It branches out into sub-disciplines of plant biochemistry, plant ecology, plant anatomy, etc. This field plays a crucial role in agriculture, biodiversity conservation, horticulture, environmental management, etc. The objective of this book is to give a general view of the different areas of plant science, and their applications. It aims to shed light on some of the unexplored aspects and recent researches in this field. This book includes contributions of experts and scientists which will provide innovative insights into plant science. It is appropriate for students seeking detailed information in this area as well as for botanists, agricultural engineers, ecologists, professionals and experts involved in this area at various levels.

I wish to thank my publisher for supporting me at every step. I would also like to thank all the authors who have contributed their researches in this book. I hope this book will be a valuable contribution to the progress of the field.

Editor

A high-throughput stereo-imaging system for quantifying rape leaf traits during the seedling stage

Xiong Xiong[1†], Lejun Yu[1†], Wanneng Yang[2,3], Meng Liu[2], Ni Jiang[1], Di Wu[3], Guoxing Chen[4], Lizhong Xiong[2], Kede Liu[2] and Qian Liu[1*]

Abstract

Background: The fitness of the rape leaf is closely related to its biomass and photosynthesis. The study of leaf traits is significant for improving rape leaf production and optimizing crop management. Canopy structure and individual leaf traits are the major indicators of quality during the rape seedling stage. Differences in canopy structure reflect the influence of environmental factors such as water, sunlight and nutrient supply. The traits of individual rape leaves traits indicate the growth period of the rape as well as its canopy shape.

Results: We established a high-throughput stereo-imaging system for the reconstruction of the three-dimensional canopy structure of rape seedlings from which leaf area and plant height can be extracted. To evaluate the measurement accuracy of leaf area and plant height, 66 rape seedlings were randomly selected for automatic and destructive measurements. Compared with the manual measurements, the mean absolute percentage error of automatic leaf area and plant height measurements was 3.68 and 6.18%, respectively, and the squares of the correlation coefficients (R^2) were 0.984 and 0.845, respectively. Compared with the two-dimensional projective imaging method, the leaf area extracted using stereo-imaging was more accurate. In addition, a semi-automatic image analysis pipeline was developed to extract 19 individual leaf shape traits, including 11 scale-invariant traits, 3 inner cavity related traits, and 5 margin-related traits, from the images acquired by the stereo-imaging system. We used these quantified traits to classify rapes according to three different leaf shapes: mosaic-leaf, semi-mosaic-leaf, and round-leaf. Based on testing of 801 seedling rape samples, we found that the leave-one-out cross validation classification accuracy was 94.4, 95.6, and 94.8% for stepwise discriminant analysis, the support vector machine method and the random forest method, respectively.

Conclusions: In this study, a nondestructive and high-throughput stereo-imaging system was developed to quantify canopy three-dimensional structure and individual leaf shape traits with improved accuracy, with implications for rape phenotyping, functional genomics, and breeding.

Keywords: Stereo-imaging system, Canopy three-dimensional structure, Individual leaf traits, Morphological classification

*Correspondence: qianliu@mail.hust.edu.cn
†Xiong Xiong and Lejun Yu contributed equally to this work
[1] Britton Chance Center for Biomedical Photonics, Wuhan National Laboratory for Optoelectronics, Huazhong University of Science and Technology, 1037 Luoyu Rd., Wuhan 430074, People's Republic of China
Full list of author information is available at the end of the article

Background

Oilseed rape (*Brassica napus*) is an important species that is cultivated in many countries for its valuable oil and protein [1–5]. The area planted with oilseed rape has rapidly increased in recent decades [6]. The leaf is of fundamental importance to the rape, acting as the power generator and aerial environmental sensor of the plant [7, 8]. Leaves are primarily involved in photosynthesis and transpiration, thereby influencing crop yield [9, 10]. The size, shape, area and number of leaves are of great significance to plant science, allowing scientists to distinguish between different species and even to model climate change [11]. Moreover, plant canopy architecture is of major interest for plant phenotyping. Variations in canopy structure have been linked to canopy function and have been attributed to genetic variability as well as a reaction to environmental factors such as light, water, and nutrient supplies as well as stress [12]. Thus, canopy structure is an essential variable for plant's adaptation to its environment [13, 14]. It is therefore important to study the oilseed rape phenotypic traits of both individual leaf shape and plant canopy structure.

Many researchers have carried out studies of individual leaf traits [15–19]. In many cases, rapeseed species can be distinguished by aspects of leaf shape, flower shape, or branching structure. Shape is, of course, important in many other disciplines [11]. To characterize these properties, O'Neal et al. [20] applied a desk-top scanner and public domain software to extract individual leaf shape traits, including leaf height, leaf width. However, the efficiency of this process is problematic: each leaf must be removed from the plant and scanned into a digital format. In addition, some complex traits such as leaf serration and leaf margin can't be assessed by this method. In an attempt to measure leaf area easier and more accurate, the new software namely "Compu Eye, Leaf & Symptom Area" was developed by Bakr et al. [21], etc. The purpose of this software is to obtain the symptom area for each leaf. But, this method offers no method to quantify leaf serration and morphology traits. Thus, this software has some limitations in practical. Igathinathane et al. [22] designed software that uses the computer monitor as the working surface to trace leaf outline and determines leaf area, perimeter, length, and width. This software offers no method to quantify leaf serration and inner cavity-related traits. Also, this is a semi-automatic program and the interactive processes are complex and tedious. Weight et al. [23] reported the development of LeafAnalyser, which is an excellent tool to facilitate PCA analysis of leaf shape parameters. However, the leaf petiole region did not remove when analyzing the leaf traits and the software was not released as open source, negating the possibility of further development by the community. Bylesjö

et al. [7] designed a tool to extract classical indicators of blade dimensions and leaf area, as well as measurements that indicate asymmetry in leaf shape and leaf serration traits. This software not only obtains object boundaries but also analyzes serration traits. However, it requires the user to analyze leaves in vitro and to correctly characterize the blade azimuth for subsequent image analysis, which limits the throughput of the analysis. Dengkui et al. [24] designed a tool to acquire plant growth information by abstracting the plant morphological characters, size and color of leaves, etc. The morphological operation has been used to remove petiole, which will influence the accuracy of leaf margin information extraction. But, the method offers no method to quantify leaf serration and inner cavity-related traits. Yang et al. [25] designed a device "HLS" for assessing leaf number, area, and shape. The device is efficient and can process multiple blades in parallel. However, all blades must be cut from the plant before insertion into the HLS device. Furthermore, the present equipment for extracting serrated blade edge traits is insufficient.

The work described above focuses on the morphological traits of individual leaves. However, three-dimensional canopy structure also plays an important role in sustaining plant function. The canopy structure contains useful information regarding developmental stage during the vegetation period as well as yield-forming parameters [26, 27]. Three-dimensional imaging methods can be broadly classified into two types: active and passive [14]. Commonly used active light projection technologies include laser scanning and structured light. Light detection and ranging (LIDAR) laser scanners have emerged as a powerful active sensing tool for direct three-dimensional measurement of plant height, canopy structure, plant growth, and shape responses [28]. The precision of laser scanner systems is very high, but the scanning time is very long, reducing the system's throughput. In structure light systems, the Kinect Microsoft RGB-depth camera [29] is used as a depth camera to shine light onto the object scene. The light reflected from the scene is used to build the depth image by measuring the deformation of the spatially structured lighting pattern [30]. The system produces 640×480 pixels RGB-depth images coded with a 16-bits dynamic that are acquired at a rate of 30 frames per second [13]. The imaging speed of these systems is very high, nearly satisfying the demands of real-time measurement. However, the measurement accuracy exhibits low spatial resolution in comparison with a standard RGB camera. One problem with laser-based and structured light systems is that they do not work well with reflective objects, and it is often necessary to coat the surface with a non-reflective layer that can lead to the collection of unsatisfactory texture data [31]. In addition, the method used for volumetric reconstruction from multiple images has been proposed to be a passive three-dimensional imaging technology [32–34]. It works by obtaining multiple images from different fixed angles. Here, a rotated plate is used to achieve multi-angle imaging, which will result in time-consuming rotations. Moreover, this method requires a significant amount of post-processing. Also, binocular/multi-view stereo imaging approach is another major passive three-dimensional imaging technology [35, 36]. There are some applications of using binocular/multi-view stereo vision for plant sensing. For automatic robot or vehicle-mounted system, the binocular stereo system is a common component for obtaining distance depth information or field plant 3D structure [37–40]. Moreover, the binocular stereo is also used in small- to medium-sized plant canopies reconstruction. Ivanov et al. [41] applied film-based stereo photogrammetry to reconstruct the maize canopy, where the plant canopy geometrical structure was analyzed and different simulation procedures were carried out to analyze leaf position and orientation and leaf area distribution. Andersen et al. [42] designed simulated annealing (SA) binocular stereo match algorithm for young wheat plants and analyzed height and total leaf area for single wheat plant. Biskup et al. [43] designed a stereo vision system with two cameras to build 3D models of soybean canopy and analyzed the angle of inclination of the leaves. Also, for isolated leaf, Biskup established a stereoscopic imaging system, which quantifies surface growth of isolated leaf discs floating on nutrient solution in wells of microtiter plates [44]. Müller-Linow et al. [12] developed a software package, which provides tools for the quantification of leaf surface properties within natural canopies via 3-D reconstruction from binocular stereo images. Furthermore, the multi-view stereo 3D reconstruction for plant phenotyping is also widely used combining with SfM- and MVS-based photogrammetric method. Lou et al. [45] described an accurate multi-view stereo (MVS) 3D reconstruction method of plants using multi-view images, which takes both accuracy and efficiency into account. Several plants, including arabidopsis, wheat and maize, are used to evaluate the performance of reconstruction algorithm. Rose et al. [46] developed a multi-view stereo system to evaluate the potential measuring accuracy of a SfM- and MVS-based photogrammetric method for the task of organ-level tomato plant phenotyping. The leaf area, main stem height and convex hull of the complete tomato plant are analyzed. Miller et al. [47] applied a low-cost hand-held camera to accurately extract height, crown spread, crown depth, stem diameter and volume of small potted trees. The multi-view stereo-photogrammetry was used to generate 3D point clouds. From the literatures above, the binocular stereo is usually used in small-sized

plant canopies reconstruction by using two top-view cameras and the multi-view stereo 3D reconstruction method is applied for organ-level plant 3D phenotyping.

In this study, we attempt to create a three-dimensional surface model of the rape canopy from images taken by double top-view cameras, and we estimate geometric attributes such as plant height and canopy leaf area. For RGB images collected using a stereo-imaging system, a novel image analysis pipeline for the accurate quantification seedling rape leaf traits was developed. We are thus able to perform leaf shape analysis, including contour signatures and shape features.

Results and discussion
Development of a stereo-imaging system
In order to extract canopy leaf area, plant height and canopy three-dimensional structure, we developed a stereo-imaging system consisting of three major units: an imaging unit, a transportation unit and a control unit (Fig. 1a). For the imaging unit, we utilized two identical RGB cameras [AVT Stingray F-504B/F-504C, Allied Vision Technologies Corporation, 2452 (H) × 2056 (V) resolution] with 8 mm fixed focal lenses (M1214-MP, Computar Corporation), two LED lamps and a lifting platform. The RGB cameras are fixed to ensure that the two main optical axes are parallel and that the two imaging planes are located at the same horizontal level. An automatic trigger acquires image pairs, and software was developed to obtain the color image pairs simultaneously. The lifting platform can be used to adjust the imaging region [537.5 mm (H) × 449.9 mm (V)] and spatial resolution (0.2188 mm/pixel). In addition, the computer workstation (HP xw6400, Hewlett-Packard Development Company, USA) plays the role of central control unit, and utilizes software developed by LabVIEW 8.6 (Nation Instruments, USA) to communicate with the two RGB cameras. In order to achieve high-throughput measurements, the stereo-imaging system was integrated into an automated high-throughput phenotyping facility developed in our previous work [48]. Two optional processing modes were developed to reconstruct the three-dimensional structure of the seedling rape canopy (Fig. 1b) and to extract individual rape leaf traits (Fig. 1c).

Three-dimensional structure of the seedling rape canopy
Plant canopy structure can be described by a range of complex and variable phenotypic traits that dictate the function of plant [49]. Here, canopy three–dimensional point cloud data were extracted from pairs of digital color images obtained under a constant light environment. The point cloud size for each canopy reconstruction is nearly 5.5 Mb. The user-friendly software interface for three-dimensional reconstruction is shown in Additional file 1:

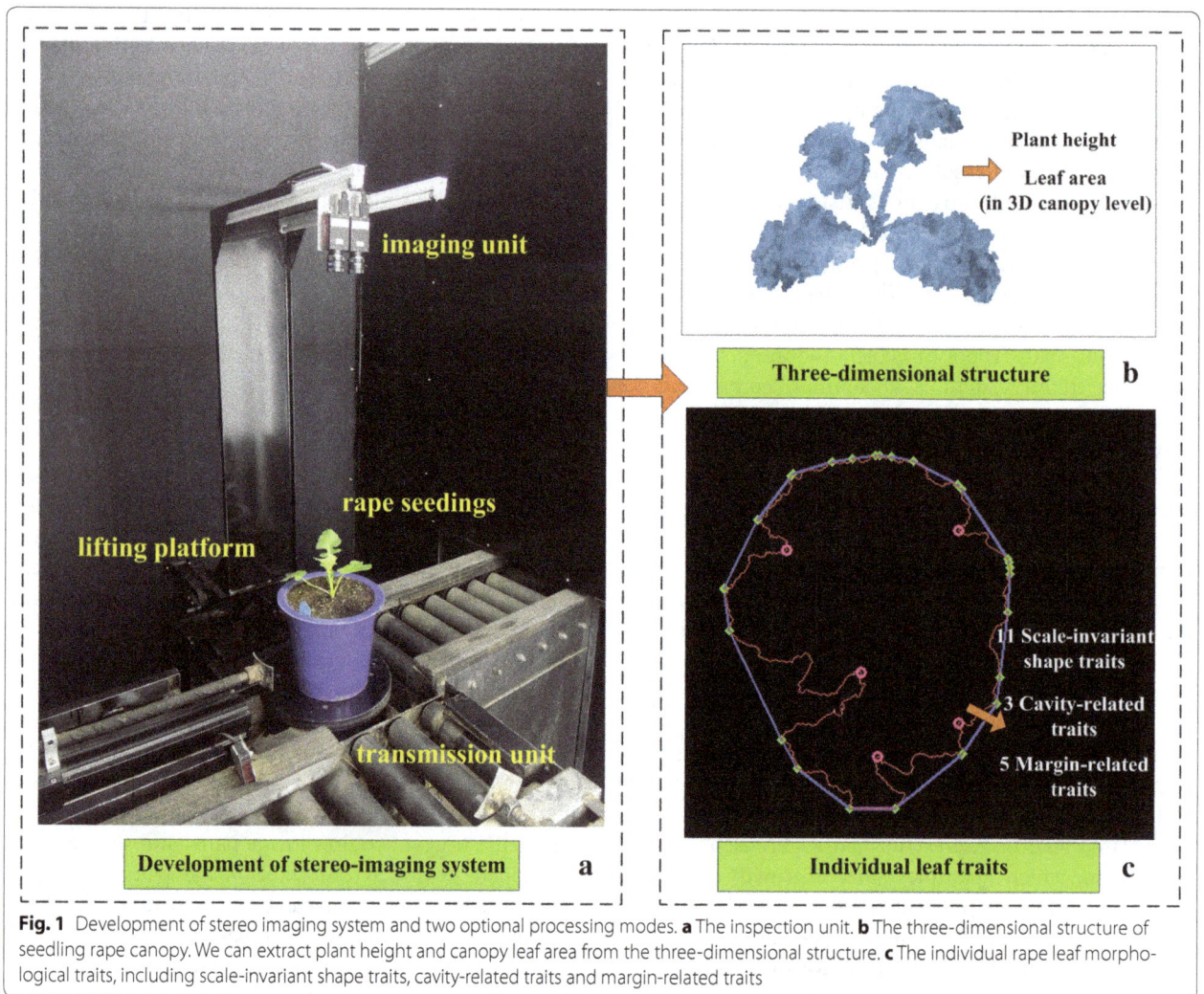

Fig. 1 Development of stereo imaging system and two optional processing modes. **a** The inspection unit. **b** The three-dimensional structure of seedling rape canopy. We can extract plant height and canopy leaf area from the three-dimensional structure. **c** The individual rape leaf morphological traits, including scale-invariant shape traits, cavity-related traits and margin-related traits

Figure S3, and the final reconstructed seedling rape canopy shown in Fig. 2 from three different perspectives.

Leaf area and plant height

From the generated three-dimensional structure, we were able to extract two important parameters: leaf area and plant height. To evaluate leaf area based on canopy level, the Delaunay algorithm was used in the process of three-dimensional mesh generation. Figure 3 shows the detailed process for leaf triangular patches generation. After the stereo-imaging for seedling rape canopy, a set of 3D point cloud can be obtained (Fig. 3a). Color differences in point cloud represents different rape leaf. The matlab functions "trimesh" and "delaunay" based on Lifting Method [50] are applied to achieve Delaunay algorithm. The "delaunay" function produces an isolated triangulation, which is useful for applications like plotting surfaces via the "trimesh" function. The stack of triangular patches forms the 3D leaf region and a smoothing mechanism is used to extract smooth triangular

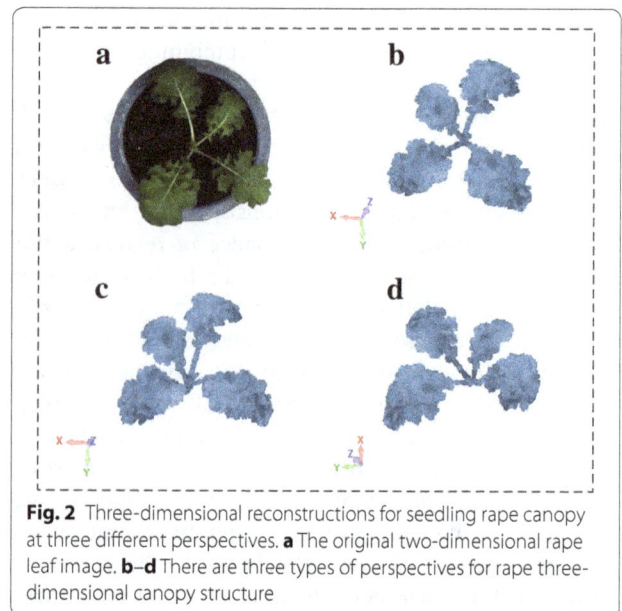

Fig. 2 Three-dimensional reconstructions for seedling rape canopy at three different perspectives. **a** The original two-dimensional rape leaf image. **b–d** There are three types of perspectives for rape three-dimensional canopy structure

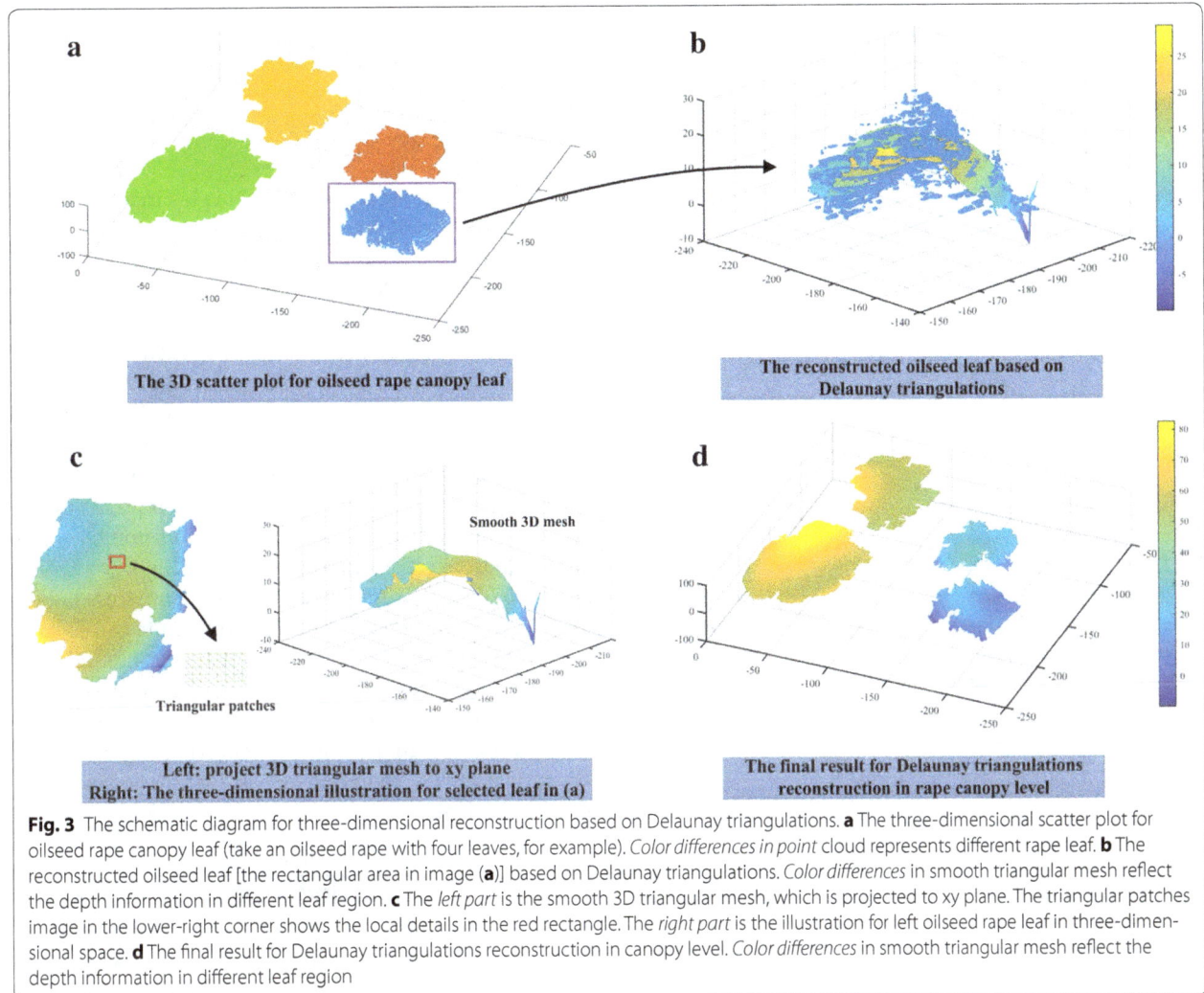

Fig. 3 The schematic diagram for three-dimensional reconstruction based on Delaunay triangulations. **a** The three-dimensional scatter plot for oilseed rape canopy leaf (take an oilseed rape with four leaves, for example). *Color differences in point* cloud represents different rape leaf. **b** The reconstructed oilseed leaf [the rectangular area in image (**a**)] based on Delaunay triangulations. *Color differences* in smooth triangular mesh reflect the depth information in different leaf region. **c** The *left part* is the smooth 3D triangular mesh, which is projected to xy plane. The triangular patches image in the lower-right corner shows the local details in the red rectangle. The *right part* is the illustration for left oilseed rape leaf in three-dimensional space. **d** The final result for Delaunay triangulations reconstruction in canopy level. *Color differences* in smooth triangular mesh reflect the depth information in different leaf region

mesh (Fig. 3b, c). Finally, we can obtain the Delaunay triangulations in rape canopy level and the sum area of all smooth triangular meshes is the canopy leaf area (Fig. 3d).

To evaluate the measurement accuracy of leaf area and plant height (vertical distance from the edge of a plastic pot to the tip of longest leaf), 66 rape seedling plants were randomly selected for manual measurement. Figure 4 shows the results of manual observation versus automatic observation. The *MAPE* values were 3.68% for leaf area and 6.18% for plant height, and the squares of the correlation coefficients (R^2) were 0.984 and 0.845, respectively. Detailed experimental data are presented in Additional file 2. The result shows that the stereo-vision method has a good potential for accurate measurement.

Individual leaf traits

Shape-based individual leaf traits, such as leaf size, vein network and leaf margin, are currently used for plant species identification and quantitative trait loci mapping [11,

16, 51, 52]. These shape-based morphological traits can be extracted using our image analysis pipeline. However, these traits alone do not reflect differences of leave shape due to variation in leaf size. We therefore must consider several new characteristic parameters that are not influenced by leaf size. Here, 19 shape related traits, including 11 scale-invariant traits, 3 inner cavity-related traits, and 5 margin-related traits, are proposed (Additional file 3). The definitions of all shape-related traits are shown in Fig. 5 and Table 1, and the computational formulas are provided in Eqs. 1–9.

$$Aspect\ Ratio = \frac{Length_{bounding\ box}}{Width_{bounding\ box}} \tag{1}$$

$$Rectangularity = \frac{Area_{object}}{Area_{bounding\ box}} \tag{2}$$

$$Area\ Convexity = \frac{Area_{object}}{Area_{convex\ hull}} \tag{3}$$

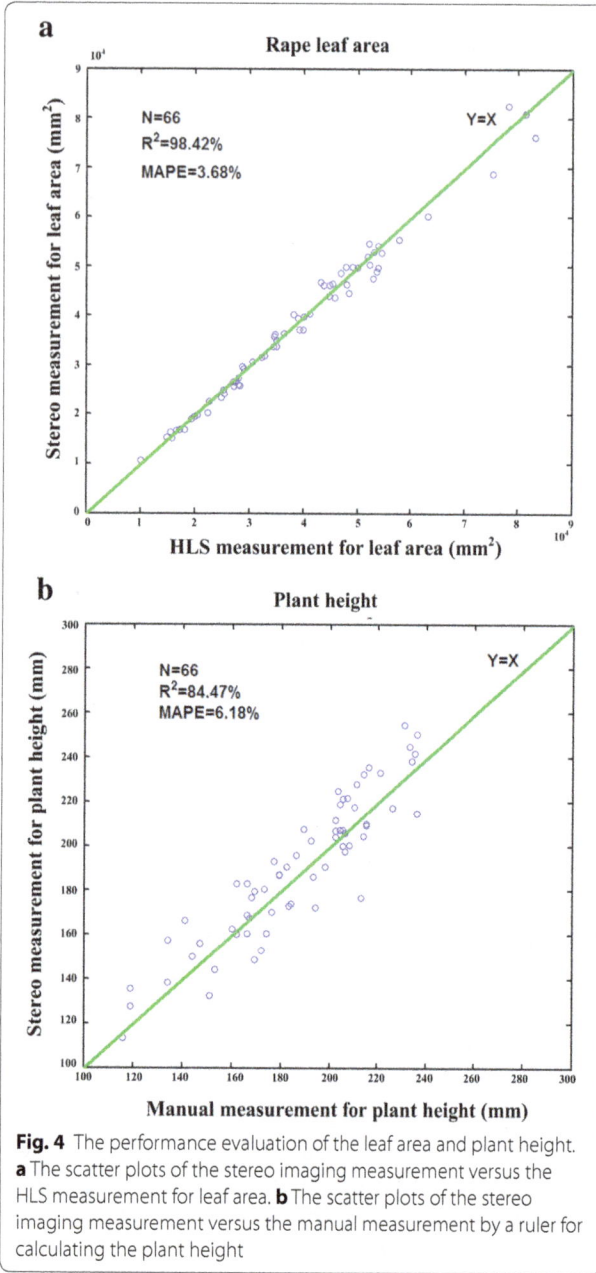

Fig. 4 The performance evaluation of the leaf area and plant height. **a** The scatter plots of the stereo imaging measurement versus the HLS measurement for leaf area. **b** The scatter plots of the stereo imaging measurement versus the manual measurement by a ruler for calculating the plant height

$$Perimeter\ Convexity = \frac{Perimeter_{object}}{Perimeter_{convex\ hull}} \qquad (4)$$

$$Sphericity = \frac{4\pi \times Area_{object}}{Perimeter_{convex\ hull}^2} \qquad (5)$$

$$Eccentricity = \frac{Axis\ Length_{long}}{Axis\ Length_{short}} \qquad (6)$$

$$Form\ Factor = \frac{4\pi \times Area_{object}}{Perimeter_{object}^2} \qquad (7)$$

$$P/A = \frac{Perimeter_{object}}{Area_{object}} \qquad (8)$$

$$Circularity = \frac{Radius_{inscribed\ circle}}{Radius_{excircle}} \qquad (9)$$

As seen in Fig. 5a, the red and purple circles represent the inscribed circle and circumscribed circle, respectively. The yellow ellipse indicates the result of elliptical fitting for a rape leaf. The green rectangle represents the minimum circumscribed box of the blade region. The inner cavities are by definition surrounded by a boundary region that is not connected to the outer boundary of the object. As seen in Fig. 5b, the green regions delineate inner cavities. In addition, the red lines (Fig. 5c), indicate the outer boundary of the individual leaf, while the blue lines demarcate the convex hull. The green points on the convex hull lines represent the serration points, which reflect to the vertices in all directions. The intermediate region between two serration points defines an indent. For each indent region, two serration points can be connected by a straight line, and the depth of the indent is measured as the longest distance from an indent point to the corresponding straight line. The effectiveness of the indents is calculated using the following Eq. 10. When the ratio is greater than 0.3, we consider the indents to be effective indents.

$$iseffective = \frac{depth}{min(height, width)/2} \qquad (10)$$

where, *height* and *width* represent the number of minimum circumscribed box rows and cols, respectively. In addition, *depth* represents the distance from the indent point to the corresponding convex hull straight line.

The software interface for extracting individual leaf traits is shown in Additional file 4: Figure S4.

Stepwise discriminant analysis
801 samples were randomly selected and divided into two groups: 402 samples with three different leaf shapes (mosaic-leaf, semi-mosaic-leaf, and round-leaf) comprising the training group and 399 samples with three different leaf shapes comprising the testing group. According to our classification, eleven of nineteen significant traits were selected by stepwise discriminant analysis as independent variables to construct two decision functions. The final classification using the two decision functions is shown in Fig. 6.

The black square blocks represent the center of the three different leaf shapes, which can be calculated using two decision functions. In Fig. 6 (red points mean round-leaf, green points mean semi-mosaic-leaf and blue points mean mosaic-leaf) and Table 2, the classification

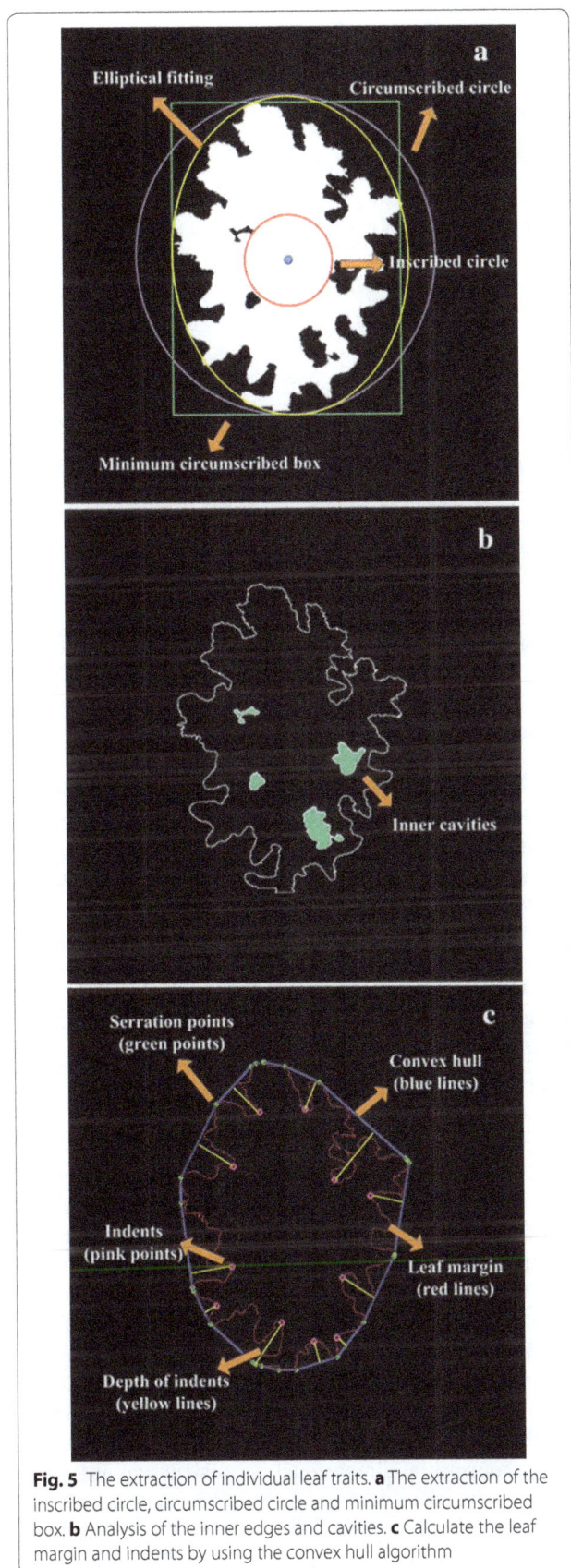

Fig. 5 The extraction of individual leaf traits. **a** The extraction of the inscribed circle, circumscribed circle and minimum circumscribed box. **b** Analysis of the inner edges and cavities. **c** Calculate the leaf margin and indents by using the convex hull algorithm

accuracy is given. We found that 92.7% of the tested grouped cases were correctly classified. In addition, we applied the leave-one-out cross-validation (LOO-CV) [54] to assess the accuracy of the classification model and found that 94.4% of the cross-validated grouped cases were correctly classified.

Stepwise discriminant analysis (SDA) [55] has been proven to effectively classify different rape leaf shapes. Moreover, in stepwise discriminant analysis, the first selected variable carries more weight in classification. In this study, the first four selected variables—form factor (FF), area convexity (AC), the average depth of effective indents (ADEI), and perimeter convexity (PC)—represent nearly 97% of the classification ability, as shown in Fig. 7. As we can see from Fig. 7, introducing a new variable will have little impact on the classification results after the selection of the first four variables. For FF, the round-leaf always has small perimeter for a given area, so the FF for the round-leaf shape is smaller than that of the mosaic-leaf and semi-mosaic-leaf. AC represents the ratio of the leaf area to the leaf convex hull area, which is an important parameter that reflects leaf morphology. For mosaic-leaf, the serrate border feature increases the convex area of the leaf. Thus, the duty ratio relative to its convex hull decreases markedly in comparison with the two other leaf shapes. Due to its serrated edge, the mosaic-leaf exhibits deeper indents compared with other two shapes.

Support vector machine (SVM)

Support vector machine (SVM) is a standard classification technique that has been shown to produce state-of-the-art results in many classification problems [54, 56]. To apply SVM to our rape seedling leaf classification, nearly half of the 801 rape samples (402 samples) were used as training parameters and the other samples (399 samples) without labels were divided into a testing group for comparison of the results. We found that 94.7% of the tested grouped cases were correctly classified. In addition, the leave-one-out (LOO-CV) accuracy is 95.6% for the cross-validated group. The specific classification accuracy is shown in Table 3.

Random forest

In essence, the random forest (RF) model is a multiple decision trees classifier, and it is widely used in regression analysis and multi-classification [57, 58]. In this study, 402 samples with three different shapes were used to construct random forest model. The other 399 samples with category labels as testing group were applied to evaluate the performance of classification. The final classification accuracy for a test group is 91.7%, and the leave-one-out cross-validation accuracy is 94.8%. The specific classification accuracy is shown in Table 4.

Table 1 Definitions of nineteen leaf shape-related traits

Classification	Variable	Definition
Scale-invariant traits	AA	The aspect ratio of leaf minimum circumscribed box
	R	The ratio of leaf area to minimum circumscribed box area
	AC	The ratio of leaf area to leaf convex hull area
	PC	The ratio of leaf circumference to leaf convex hull perimeter
	S	The ratio of leaf area to the square of leaf convex hull perimeter
	E	The ratio of long axis of ellipse to short axis of ellipse
	FF	The ratio of leaf area to the square of leaf perimeter
	PAR	The ratio of leaf circumference to leaf area
	SFD	Reflect the effectiveness of the occupies space without image cropping [53]
	IFD	Reflect the effectiveness of the occupies space with image cropping [53]
	C	The ratio of inscribed circle radius to circumscribed circle radius
Cavity traits	NIC	The number of inner cavities
	APIC	The average perimeter of inner cavities
	AAIC	The average area of inner cavities
Margin related traits	TNI	Total number of indents
	ENI	Effective number of indents
	ADI	The average depth of indents
	ADEI	The average depth of effective indents
	AVE	The average calculated value of effectiveness

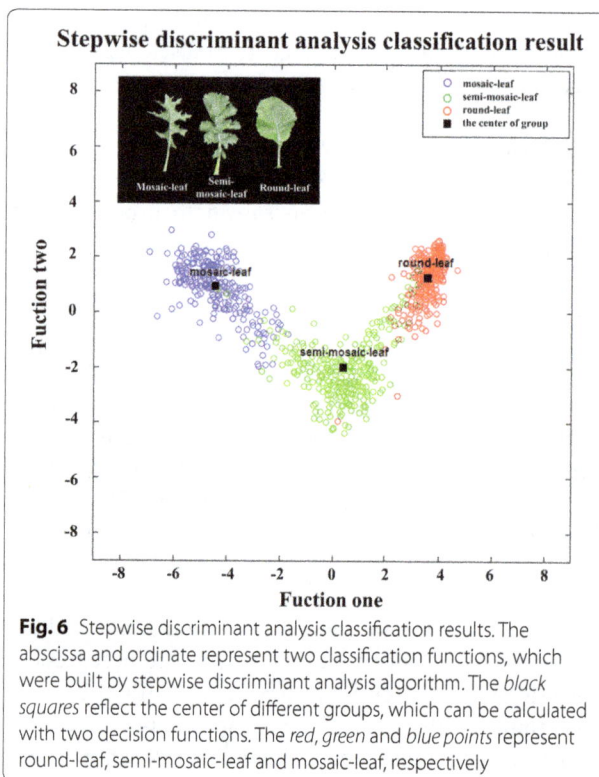

Fig. 6 Stepwise discriminant analysis classification results. The abscissa and ordinate represent two classification functions, which were built by stepwise discriminant analysis algorithm. The *black squares* reflect the center of different groups, which can be calculated with two decision functions. The *red, green* and *blue points* represent round-leaf, semi-mosaic-leaf and mosaic-leaf, respectively

satisfactorily classify leaves. The final classification accuracy for test group (399 samples) is 92.7, 94.7, 91.7% for stepwise discriminant analysis (SDA), support vector machine (SVM) and random forests (RF), respectively. In addition, the leave-one-out cross validation classification accuracy is 94.4% for SDA, 95.6% for SVM and 94.8% for RF algorithm. Among them, the stepwise discriminant analysis has a better prediction effect on round-leaf, while the support vector machine classifier is the most sensitive to mosaic-leaf. From the perspective of predicated group accuracy and the leave-one-out cross-validated results, the most reliable forecasting model was established by SVM algorithm.

The performance of efficiency and accuracy

In this work, the stereo-imaging system is integrated to the high-throughput phenotyping facility. Each pot-grown rape would be transported by the conveyor, and the image pairs were acquired by the two top-view cameras at the same time. The inspection procedure is fully automated and highly efficient (45 s per plant) [48]. All the image processing works are carried out after the completion of image acquiring. Here, the time for image processing consists of two parts: canopy 3D reconstruction and individual leaf traits extraction. Usually, the time for canopy three-dimensional reconstruction is closely linked to the size of oilseed rape. After evaluated with 10 different size seedling rape samples, the average processing time for each canopy 3D reconstruction and data

A comparison of the performance of the three methods of classification

As shown in Table 5, all three methods were able to

Table 2 Stepwise discriminant analysis classification results

	Leaf type	Predicted group membership			Total
		Mosaic	Semi-mosaic	Round	
Trained group	% Mosaic-leaf	98.3	1.7	0	100.0
	Semi-mosaic-leaf	1.4	90.3	8.3	100.0
	Round-leaf	0	0	100.0	100.0
Tested group	% Mosaic-leaf	97.5	2.5	0	100.0
	Semi-mosaic-leaf	7.6	88.9	3.5	100.0
	Round-leaf	0	7.4	92.6	100.0
Cross-validated[a]	% Mosaic-leaf	95.9	4.1	0	100.0
	Semi-mosaic-leaf	2.1	89.3	8.7	100.0
	Round-leaf	0	1.5	98.5	100.0

96.0% of trained grouped cases correctly classified

92.7% of tested grouped cases correctly classified

94.4% of cross-validated grouped cases correctly classified

[a] Cross validation is conducted only for those cases in the analysis. In cross validation, each case is classified by the functions derived from all cases other than that case

Table 3 SVM classification results

	Leaf type	Predicted group membership			Total
		Mosaic	Semi-mosaic	Round	
Trained group	% Mosaic-leaf	100.0	0	0	100.0
	Semi-mosaic-leaf	0	99.3	0.7	100.0
	Round-leaf	0	0	100.0	100.0
Tested group	% Mosaic-leaf	100.0	0	0	100.0
	Semi-mosaic-leaf	6.2	92.4	1.4	100.0
	Round-leaf	0	7.4	92.6	100.0
Cross-validated[a]	% Mosaic-leaf	96.7	3.3	0	100.0
	Semi-mosaic-leaf	3.1	93.1	3.8	100.0
	Round-leaf	0	2.6	97.4	100.0

99.8% of trained grouped cases correctly classified

94.7% of tested grouped cases correctly classified

95.6% of cross-validated grouped cases correctly classified

[a] Cross validation is conducted only for those cases in the analysis. In cross validation, each case is classified by the functions derived from all cases other than that case

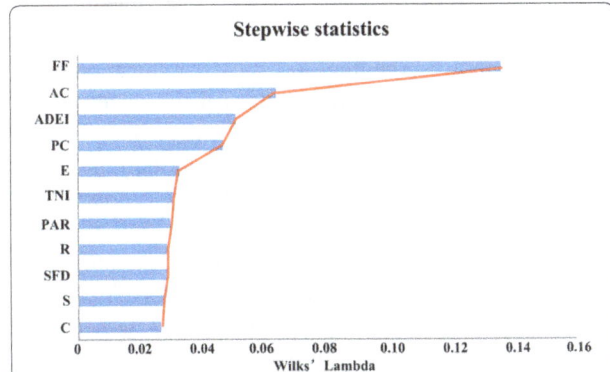

Fig. 7 The performance analysis of traits in stepwise discriminant analysis. The screening results of the stepwise discriminant analysis. The traits from top to bottom reflect the order of characteristics screening. The value of Wilks' Lambda statistic represents the discriminant ability after entering current traits

extraction of leaf area and plant height is about 43.46 s; for manual interaction, the time for each individual leaf extraction is about 8 s. The detailed description for manual part is shown in video (Additional file 5). Moreover, the two independent parts could run in parallel to save time. In this way, the total processing time depends on the longer part. So, the manual interaction does not lag the efficiency of the high-throughput platform. In addition, the manual interaction method can extract individual leaf with more accuracy compared with automatic segmentation, and the efficiency is also satisfactory.

Assuming that the system can work 8 h a day, then, about 660 pots can handle just one day, which is an acceptable number for high-throughput.

From the view of measurement accuracy for rape seedling leaf area, a comparison result of two different methods is shown in Fig. 8a. The red scatter points represent the leaf area result of two-dimensional projective method by only use one top-view image. After the rape was segmented from the background in the top-view image, the area of each rape is calculated by multiplying the pixel area and average spatial resolution, while blue scatter points indicate the result of three-dimensional stereo measurement by analyzing canopy structure. The $MAPE$ values were 3.68% for three-dimensional stereo measurement and 11.44% for two-dimensional projective measurement, and the square of correlation coefficients (R^2) for three- and two-dimensional measurements was 0.984 and 0.938, respectively. Obviously, compared with two-dimensional imaging, the stereo measurement considering more spatial information of rape leaf can indicate more accuracy of leaf area in real world. Moreover, the area errors are almost below eight percent by using the stereo measurement method (Fig. 8b).

The overlapping situation in stereo-imaging

The overlap of oilseed rape leaves is surely a difficult issue for binocular stereo-imaging. In this study, the oilseed rapes are at the seedling stage, which have less overlapping situation. Actually, for some situation (round-leaf), the overlap can be solved and recovered. The following

Table 4 Random forest classification results

	Leaf type	Predicted group membership			Total
		Mosaic	Semi-mosaic	Round	
Trained group	% Mosaic-leaf	100.0	0	0	100.0
	Semi-mosaic-leaf	0	100.0	0	100.0
	Round-leaf	0	0	100.0	100.0
Tested group	% Mosaic-leaf	99.2	0.8	0	100.0
	Semi-mosaic-leaf	8.3	87.5	4.2	100.0
	Round-leaf	0	10.4	89.6	100.0
Cross-validated[a]	% Mosaic-leaf	96.7	3.3	0	100.0
	Semi-mosaic-leaf	3.5	91.0	5.5	100.0
	Round-leaf	0	3.0	97.0	100.0

100.0% of trained grouped cases correctly classified

91.7% of tested grouped cases correctly classified

94.8% of cross-validated grouped cases correctly classified

[a] Cross validation is conducted only for those cases in the analysis. In cross validation, each case is classified by the functions derived from all cases other than that case

part only considers the round-leaf (Fig. 9a). The detailed implementation steps are as follows: In the first step, we need to segment the overlapped leaf binary image. Secondly, the contour of overlapped leaf is extracted by using the front binary image. Then, the polygonal approximation [59] is used to represent the overlapped contour. This is an important step to trim away the small-scale rough fluctuations. Next, we need to detect the concave points [60] and segment the polygonal contour (Fig. 9b). Finally, the ellipse fitting [61] is chosen to recover the overlapped leaf region for round-leaf (Fig. 9c). The detailed algorithm description can refer to Additional files 6 and Additional File 7: Figure S5. The key for above algorithm is based on a priori knowledge: the oilseed rape leaf is approximate circle. Thus, for mosaic-leaf and semi-mosaic-leaf, the above method is useless.

Conclusion

In this study, we establish a nondestructive and high-throughput stereo-imaging system for screening leaf canopy three-dimensional structure and individual leaf

phenotypic traits. Compared with manual measurements, the squares of the correlation coefficients (R^2) for leaf area and plant height are 0.984 and 0.845, respectively. Moreover, 19 morphological traits were applied in morphology classification of three different rape leaf shapes. Three classifiers (SDA, SVM, and RF) were used and compared, and the better classification accuracy with SVM is 94.7% for 399 test samples. In conclusion, we developed a high-throughput stereo-imaging system to quantify leaf area, plant height, and leaf shape with more accuracy, which will benefit rape phenotyping, functional genomics, and breeding.

Methods
Plant materials and measurements

In total, 801 *Brassica napus* with three different shapes, including mosaic-leaf, semi-mosaic-leaf and round-leaf (Additional file 8: Figure S2), were analyzed in this study. Seeds were sown and germinated, and plants were grown up to the seedling stage. All plants were cultivated in plastic pots of 23.5 cm diameter with approximately 6 L of experimental soil. All pots were randomly distributed over a glasshouse compartment to control the growth conditions. Approximately 30 days after sowing, three experienced agronomists recorded the leaf shape using the visual method. The final statistical classification result would abide by the majority rule. All the experimental samples were measured with our stereo-imaging system to obtain image pairs. Among them, 66 rape plants were randomly selected to reconstruct the canopy three-dimensional structure, extract leaf area and calculate plant height. To estimate the accuracy of measurement, the plant leaf area was measured with the HLS [25] and plant height was measured manually by well-trained worker. In order to evaluate the extraction of individual leaf traits, a biological classification for three different leaf shapes was proposed. All samples were divided into two groups: one group consists of 402 samples with three labels (mosaic-leaf, semi-mosaic-leaf and round-leaf) as the training group. The other group consists of 399 samples without labels as the testing group. All the training group samples are selected randomly to balance the number of different shapes.

Table 5 A comparison of three classification methods

Methods of classification	Predicted group accuracy				Leave-one-out cross-validated (LOO-CV)
	Mosaic	Semi-mosaic	Round	Total (n = 399)	
Stepwise discriminant analysis	98.3%	90.3%	100.0%	92.7%	94.4%
Support vector machine (SVM)	100%	92.4%	92.6%	94.7%	95.6%
Random forests classifier	99.2%	87.5%	89.6%	91.7%	94.8%

Fixed half selected for training (402), and the other half (399) for testing

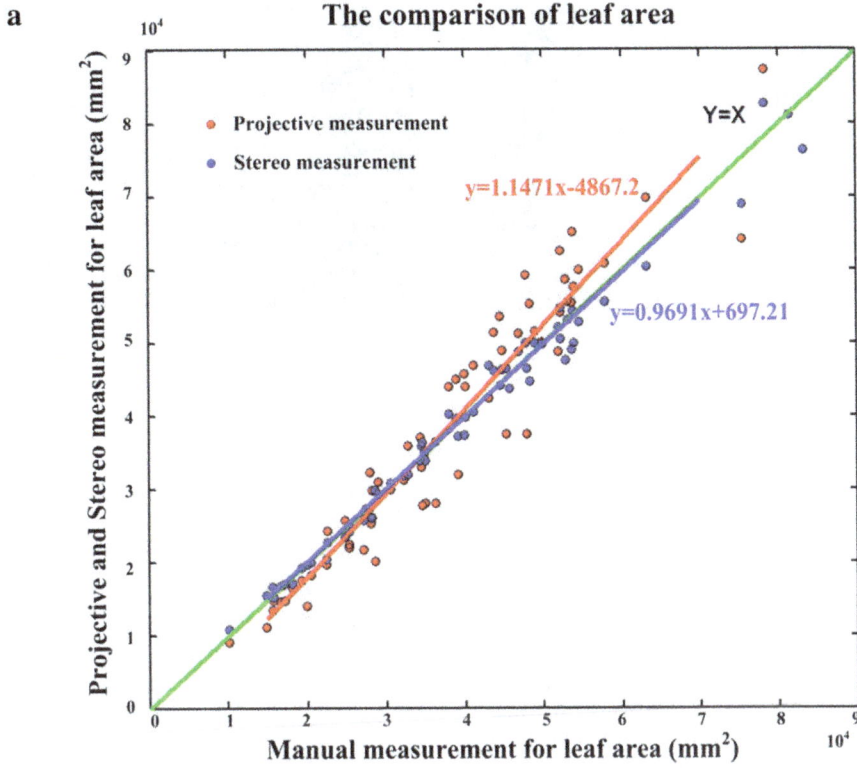

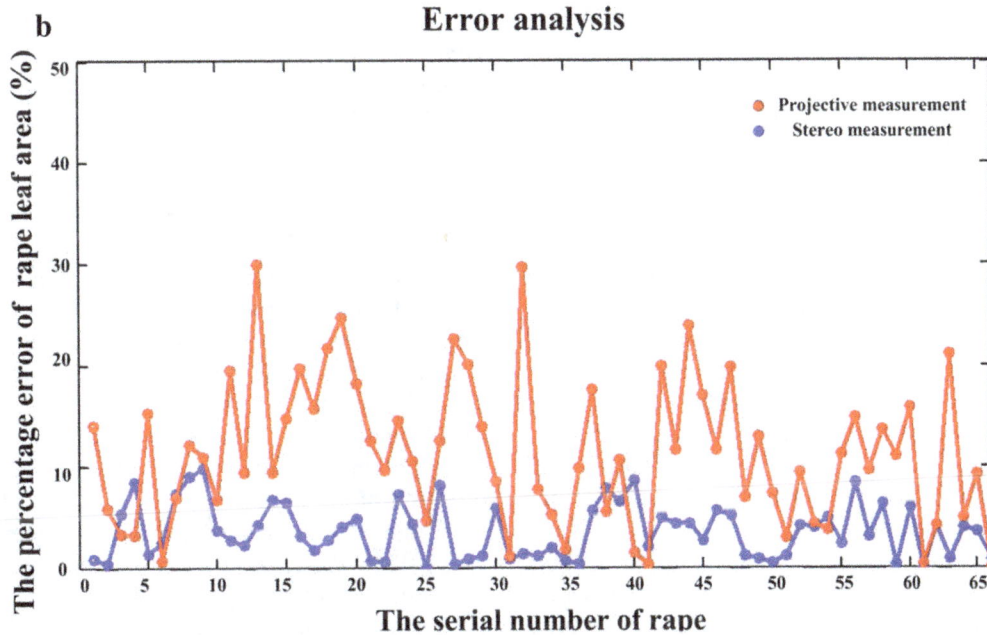

Fig. 8 The comparison of two different methods for leaf area. **a** The *red scatter points* represent the leaf area result of the two-dimensional projective method by only using one top image. The *blue scatter points* indicate the result of three-dimensional measurement by using stereo imaging algorithm for leaf area. **b** The distribution of percentage error with two different methods

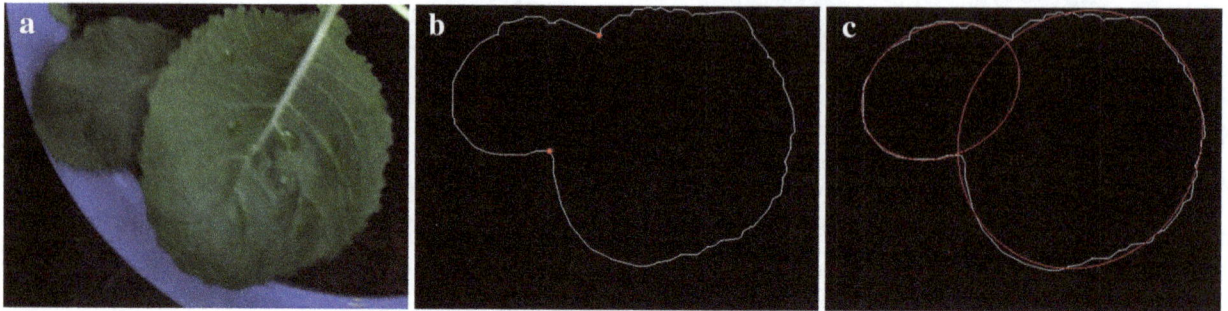

Fig. 9 Recovering the overlapped leaf region for round-leaf. **a** The original overlapped round-leaf. **b** The concave points (*red*) and polygonal boundary (*white*) of overlapped round-leaf. **c** The ellipse fitting is chosen to recover the overlapped leaf region for round-leaf

Image analysis for canopy three-dimensional reconstruction

The main content of this part is to describe the specific image processing steps of canopy 3D reconstruction for seedling rape. The first step is camera stereo calibration. To achieve this step, a black and white calibration pattern [31] pasted on a plastic plate was used to obtain 20–25 image pairs. To ensure the accuracy of the calibration, the imaging angles should have obvious differences. For original image pairs (Fig. 10a), the corresponding feature in the left and right original image is not on the same horizontal baseline. Here, Bouguet algorithm [62] was used to rectify two original images. The final rectified image pairs were shown in Fig. 10b. Considering the influence of environmental light, an automatically segmenting method [63], adopts normalized RGB component to get binary image of blade region (Fig. 10c). Considering the slender characteristics of the stem, the morphological opening operation is used to remove the stem, and connected component mark technology is used to distinguish different leaves (Fig. 10d). The next work is to match the corresponding feature points in the left and right rectified images. Here, the library for efficient large-scale stereo matching [64] is used to compute the left and right disparity map. In the actual situation, two thorny situations might happen [65, 66]. The first one is mismatch, which means that there are no matching pixels or wrong matching pixels. The second thorny situation is occlusion, which means that some pixels appears only in an image, and can't see in another image (Additional file 9: Figure S6). If we don't take some special measures to focus on region where the mismatched and occluded situations are serious, there will have some wrong in the process of 3D point clouds extraction. So, it is important to rectify the disparity map. The specific process is described in Additional file 10 and the rectified left disparity image is shown in Fig. 10e. According to the principle of triangular range (Additional file 11: Figure S1), we can

extract the three-dimensional point cloud data of the canopy leaves. After removing the isolate points, triangle patches are used as the surface of canopy leaves by using Delaunay triangulation algorithm [50]. The final result of canopy reconstruction is shown in Fig. 2b–d. Detailed processing for canopy 3D reconstruction and triangle patches generation has been described in Additional file 10 and Fig. 3. The code for Delaunay algorithm is shown in Additional file 12.

Image analysis for individual leaf

The main content of this part is to describe the specific image processing steps for individual leaf extraction. Firstly, the user needs to click the left mouse button and drag it to choose a rectangular box. In this rectangular box, the individual leaf must be typical and representative (Fig. 11a). All selected rectangular images are saved in PNG format for subsequent analysis. Usually, the selected individual leaf has a long petiole part, the existence of which will seriously impact the analysis of blade traits. So, the next step is to remove petiole. The difference between blade and petiole in color and texture is tiny. But from the view of shape, petiole region is more slender than blade. With that mechanism we can remove petiole. The detailed operating processing includes the following steps: (1) Marking two points on the petiole (Fig. 11b) and rotating the rectangular image so that the direction of the petiole is downward (Fig. 11c). (2) Segmenting rotated rectangular image to obtain binary leaf image (Fig. 11d). Here, the excess green vegetation (*ExG*) [67] and excess red vegetation (*ExR*) indices [68] were used to extract binary leaf image. (3) From the bottom to top search binary image to remove the pixel width less than a specified threshold area (Fig. 11e). Here, the area threshold is set to 25, which is an appropriate value determined by lots of preliminary experiments. After removing the petiole, the next step is to remove connected components that were erroneously selected. The situation that other partial leaf region might be chosen in

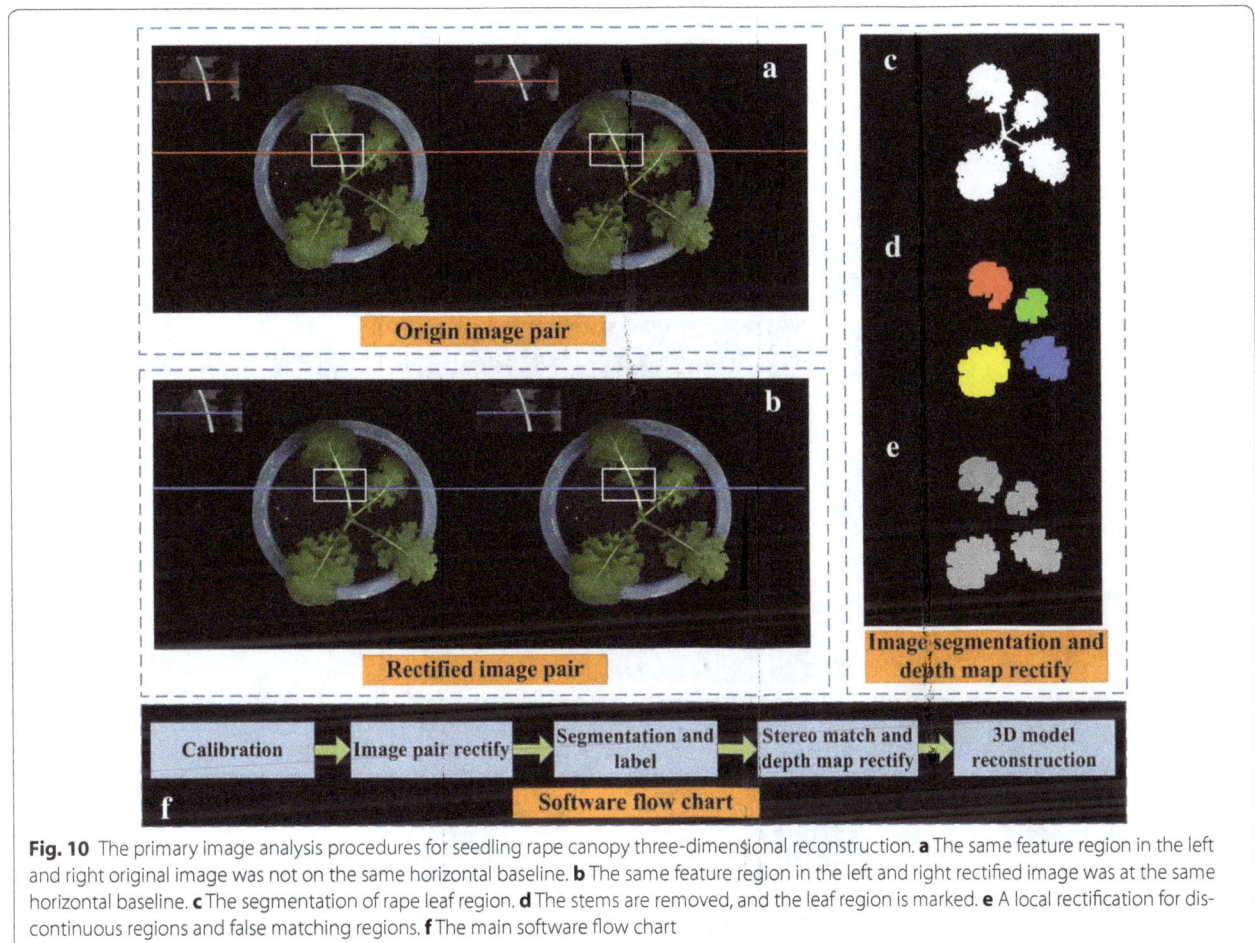

Fig. 10 The primary image analysis procedures for seedling rape canopy three-dimensional reconstruction. **a** The same feature region in the left and right original image was not on the same horizontal baseline. **b** The same feature region in the left and right rectified image was at the same horizontal baseline. **c** The segmentation of rape leaf region. **d** The stems are removed, and the leaf region is marked. **e** A local rectification for discontinuous regions and false matching regions. **f** The main software flow chart

the rectangular region was always happened. Usually, the target leaf region had the largest area. So, only thing we need to do is to keep the largest connected component as the target individual leaf. In addition, because of the binarization segmentation error, small holes might appear on the target blade region (the red square in Fig. 11e). Usually, these holes have a small number of pixels compared with other real holes. An area threshold is used to fill the small area holes. The final individual leaf is shown in Fig. 11f. The detailed processing flow and computational formulas are in Additional file 13.

Three different classification methods
Stepwise discriminant analysis statistical method
For stepwise discriminant analysis (SDA), the specific operating approaches are as follows: All traits are selected as the input variables of the algorithm. Then, the SDA algorithm will select a variable that has the most significant discriminant ability. Next, the selecting for second variable based on the first variable, which indicates that combining the first and second variables will have the most significant discriminant ability. By that

analogy, the third variable will be selected. Because of the mutual relationship between different variables, the previous variable may lose significant discriminant ability after inputting the new variable. Then, we will inspect the discriminant ability of all previous selected variables to find the disabled variables, remove them, and go on to find new variables until no significant variables can be removed. In this study, stepwise discriminant analysis training was achieved using (SPSS v.22 software), which is a proven technique for meaningfully classifying different shapes [69]. The detailed description is shown in Additional file 14.

Support vector machine statistical method
The support vector machine (SVM) is a common supervised learning algorithm that has been shown to provide state-of-the-art performance in many classification problems. The main thinking of the SVM is to establish a classification hyperplane as the decision curved surface, which maximizes the gap of positive samples and negative samples. The LIBSVM-matlab toolkit [70] was used here to conduct SVM model. All the 801 samples with

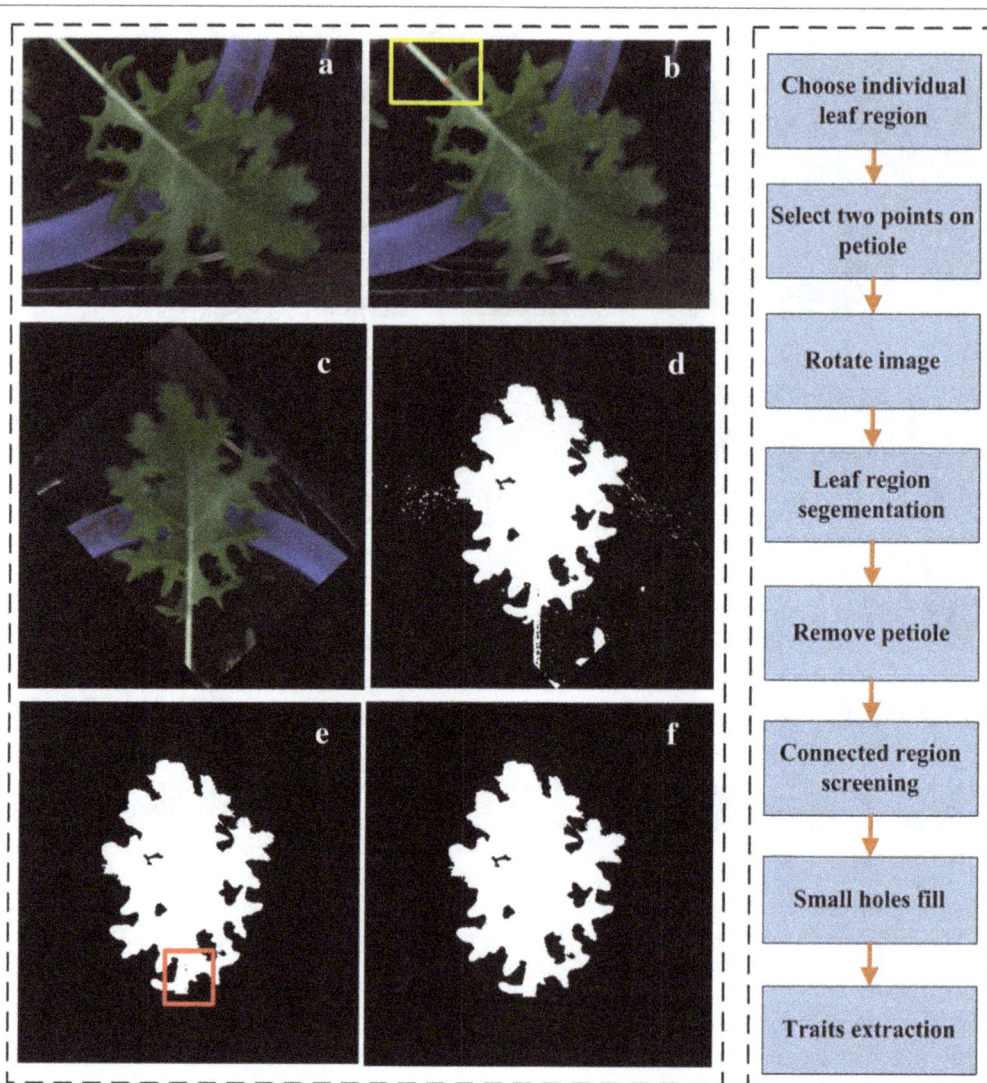

Fig. 11 The primary procedure for individual leaf traits extract. **a** The original rape leaf image. **b** The selected single rape leaf region. **c** Select two points on the image. The first point is the intersective region of stem and leaf and the second point is used to decide the direction of rotation. **d** The rotated oilseed rape leaf image. **e** Using normalized EG and ER to segment the leaf region. **f** Remove the petiole region and connected region screening

three different leaf shapes were randomly divided into two groups (402 samples for training and 399 samples for testing). Firstly, we should limit all data into a certain range. Here, the interval from 0 to 1. The purpose for data normalization [71] is to ensure the convergence of the SVM algorithm. At the same time, it will improve the accuracy of classification. Next, we can train the model. Here, the kernel function is generated to use polynomials and the kernel parameter is set to 1.5. Also, the penalty parameter is set to 2. The genetic algorithm (GA) was used to choose the best value of kernel parameter and penalty parameter. The detailed algorithm process refers to Additional file 14 and the code is shown in Additional file 15.

Random forest statistical method

The random forest (RF) classifier is a combination of multiple decision trees. In this study, the open source randomforest-matlab toolkit [72] was adopted to build random forest classifier. Abhishek Jaiantilal, of the University of Colorado, Boulder, is the primary developer. Here, the number of decision trees in my random forest is 1000 and the other parameters adopt the default value. 801 samples were randomly selected and divided into two groups: 402 samples comprising the training group and 399 samples for testing group. When the test samples enter into the random forest, every decision tree will independently classify the category it belongs to. The final statistical classification result will abide by the majority rule. The detailed algorithm process refers to Additional file 14 and the code is shown in Additional file 16.

Additional files

Additional file 1: Figure S3. The software interface for seedling rape canopy three-dimensional reconstruction.

Additional file 2. Original data of leaf area and canopy height.

Additional file 3. Original data of individual leaf traits.

Additional file 4: Figure S4. The software interface for extracting individual leaf traits.

Additional file 5. Supplementary document for manual interaction.

Additional file 6. Supplementary document for overlap recovery algorithm.

Additional file 7: Figure S5. Overlap recovery algorithm flow.

Additional file 8: Figure S2. Three different shapes of rape leaf in seedling stage.

Additional file 8: Figure S6. The occluded situation in binocular stereo-imaging system.

Additional file 10. Supplementary document for three-dimensional reconstruction.

Additional file 11: Figure S1. The triangle range finding for optical path.

Additional file 12. Supplementary document for Delaunay code.

Additional file 13. Supplementary document for individual leaf traits.

Additional file 14. Supplementary document for three classification methods.

Additional file 15. Supplementary document for SVM code.

Additional file 16. Supplementary document for RF code.

Abbreviations
AA: the aspect ratio of leaf minimum circumscribed box; R: the ratio of leaf area to minimum circumscribed box area; AC: the ratio of leaf area to leaf convex hull area; PC: the ratio of leaf circumference to leaf convex hull perimeter; S: the ratio of leaf area to the square of leaf convex hull perimeter; E: the ratio of long axis of ellipse to short axis of ellipse; FF: the ratio of leaf area to the square of leaf perimeter; PAR: the ratio of leaf circumference to leaf area; SFD: reflect the effectiveness of the occupies space without image cropping; IFD: reflect the effectiveness of the occupies space with image cropping; C: the ratio of inscribed circle radius to circumscribed circle radius; NIC: the number of inner cavities; APIC: the average perimeter of inner cavities; AAIC: the average area of inner cavities; TNI: total number of indents; ENI: effective number of indents; ADI: the average depth of indents; ADEI: the average depth of effective indents; AVE: the average calculated value of effectiveness; SDA: stepwise discriminant analysis; SVM: support vector machine ; RF: random forest; GA: genetic algorithm.

Authors' contributions
XX developed the image analysis pipeline, performed the evaluation experiment, analyzed the data, and drafted the manuscript. LY and WY designed and built the hardware, analyzed the data, performed the evaluation experiment, and contributed in writing the manuscript. ML planted and managed the experimental material. NJ contributed in developed image analysis pipeline. DW provided support with hardware development and implementation of the evaluation experiment. GC planted and managed the experimental material and guided experiment. LX contributed in writing the manuscript. KL contributed in writing the manuscript. QL supervised the study and contributed in writing the manuscript. All authors read and approved the final manuscript.

Author details
[1] Britton Chance Center for Biomedical Photonics, Wuhan National Laboratory for Optoelectronics, Huazhong University of Science and Technology, 1037 Luoyu Rd., Wuhan 430074, People's Republic of China. [2] National Key Laboratory of Crop Genetic Improvement and National Center of Plant Gene Research, Huazhong Agricultural University, Wuhan 430070, People's Republic of China. [3] College of Engineering, Huazhong Agricultural University, Wuhan 430070, People's Republic of China. [4] MOA Key Laboratory of Crop Eco-physiology and Farming System in the Middle Reaches of the Yangtze River, Huazhong Agricultural University, Wuhan 430070, People's Republic of China.

Acknowledgements
This work was supported by Grants from the National Program on High Technology Development (2013AA102403), and the Scientific Conditions and Resources Research Program of Hubei Province of China (2015BCE044). We would like to thank Dr. Jiang for technical assistance.

Competing interests
The authors declare that they have no competing interests.

References
1. Paulauskas A, Jodinskienė M, Griciuvienė L, et al. Morphological traits and genetic diversity of differently overwintered oilseed rape (*Brassica napus* L.) cultivars. Zemdirb Agric. 2013;100(4):409–16.
2. Lorin M, Jeuffroy MH, Butier A, et al. Undersowing winter oilseed rape with frost-sensitive legume living mulch: Consequences for cash crop nitrogen nutrition. Field Crops Res. 2016;193:24–33.
3. Pospišil M, Brčic M, Husnjak S. Suitability of soil and climate for oilseed rape production in the Republic of Croatia. Agric Conspec Sci. 2011;76(1):35–9.
4. Pessel D, Lecomte J, Emeriau V, et al. Persistence of oilseed rape (*Brassica napus* L.) outside of cultivated fields. Theor Appl Genet. 2001;102(6–7):841–6.
5. Rathke GW, Diepenbrock W. Energy balance of winter oilseed rape (*Brassica napus* L.) cropping as related to nitrogen supply and preceding crop. Eur J Agron. 2006;24(1):35–44.
6. Powers S, Pirie E, Latunde-Dada A, et al. Analysis of leaf appearance, leaf death and phoma leaf spot, caused by Leptosphaeria maculans, on oilseed rape (*Brassica napus*) cultivars. Ann Appl Biol. 2010;157(1):55–70.
7. Bylesjö M, Segura V, Soolanayakanahally RY, Rae AM, Trygg J, Gustafsson P, Jansson S, Street NR. LAMINA: a tool for rapid quantification of leaf size and shape parameters. BMC Plant Biol. 2008;8(1):82.
8. Efroni I, Eshed Y, Lifschitz E. Morphogenesis of simple and compound leaves: a critical review. Plant Cell. 2010;22(4):1019–32.
9. Yang W, Duan L, Chen G, et al. Plant phenomics and high-throughput phenotyping: accelerating rice functional genomics using multidisciplinary technologies. Curr Opin Plant Biol. 2013;16(2):180–7.
10. Pérez-Pérez JM, Esteve-Bruna D, Micol JL. QTL analysis of leaf architecture. J Plant Res. 2010;123(1):15–23.
11. Cope JS, Corney D, Clark JY, et al. Plant species identification using digital morphometrics: a review. Expert Syst Appl. 2012;39(8):7562–73.
12. Muller-Linow M, Pinto-Espinosa F, Scharr H, et al. The leaf angle distribution of natural plant populations: assessing the canopy with a novel software tool. Plant Methods. 2015;11:11.
13. Chéné Y, Rousseau D, Lucidarme P, et al. On the use of depth camera for 3D phenotyping of entire plants. Comput Electron Agric. 2012;82:122–7.
14. Omasa K, Hosoi F, Konishi A. 3D lidar imaging for detecting and understanding plant responses and canopy structure. J Exp Bot. 2007;58(4):881–98.

15. Shibayama M, Sakamoto T, Takada E, et al. Estimating paddy rice leaf area index with fixed point continuous observation of near infrared reflectance using a calibrated digital camera. Plant Prod Sci. 2011;14(1):30–46.

16. Tsukaya H. Organ shape and size: a lesson from studies of leaf morphogenesis. Curr Opin Plant Biol. 2003;6(1):57–62.

17. Neto JC, Meyer GE, Jones DD. Individual leaf extractions from young canopy images using Gustafson–Kessel clustering and a genetic algorithm. Comput Electron Agric. 2006;51(1–2):66–85.

18. Bama BS, Valli S M, Raju S, et al. Content based leaf image retrieval (CBLIR) using shape, color and texture features. Indian J Comput Sci Eng. 2011;2(2): 202–11.

19. Nam Y, Hwang E, Kim D. A similarity-based leaf image retrieval scheme: joining shape and venation features. Comput Vis Image Underst. 2008;110(2):245–59.

20. O'Neal ME, Landis DA, Isaacs R. An inexpensive, accurate method for measuring leaf area and defoliation through digital image analysis. J Econ Entomol. 2002;95(6):1190–4.

21. Bakr EM. A new software for measuring leaf area, and area damaged by *Tetranychus urticae* Koch. J Appl Entomol. 2005;129(3):173–5.

22. Igathinathane C, Prakash VSS, Padma U, et al. Interactive computer software development for leaf area measurement. Comput Electron Agric. 2006;51(1–2):1–16.

23. Weight C, Parnham D, Waites R. Technical advance: leafAnalyser: a computational method for rapid and large-scale analyses of leaf shape variation. Plant J. 2008;53(3):578–86.

24. Dengkui A, Minzan L, Li Z. Measurement of tomato leaf area using computer image processing technology. Sensor Lett. 2010;8(1):56–60.

25. Yang W, Guo Z, Huang C, et al. Genome-wide association study of rice (*Oryza sativa* L.) leaf traits with a high-throughput leaf scorer. J Exp Bot. 2015;66(18):5605–15.

26. Paulus S, Behmann J, Mahlein AK, et al. Low-cost 3D systems: suitable tools for plant phenotyping. Sensors (Basel). 2014;14(2):3001–18.

27. Frasson RPM, Krajewski WF. Three-dimensional digital model of a maize plant. Agric For Meteorol. 2010;150(3):478–88.

28. Dornbusch T, Lorrain S, Kuznetsov D, et al. Measuring the diurnal pattern of leaf hyponasty and growth in Arabidopsis—a novel phenotyping approach using laser scanning. Funct Plant Biol. 2012;39(11):860.

29. Chen S, Li YF, Zhang J, et al. Active sensor planning for multiview vision tasks. Berlin: Springer; 2008.

30. Henry P, Krainin M, Herbst E, et al. RGB-D mapping: using Kinect-style depth cameras for dense 3D modeling of indoor environments. Int J Robot Res. 2012;31(5):647–63.

31. Baumberg A, Lyons A, Taylor R. 3D S.O.M.—a commercial software solution to 3D scanning. Graph Models. 2005;67(6):476–95.

32. Zhou Z. 3D reconstruction from multiple images using single moving camera [D]. 2015.

33. Park J-S. Interactive 3D reconstruction from multiple images: a primitive-based approach. Pattern Recogn Lett. 2005;26(16):2558–71.

34. Herman M, Kanade T. Incremental reconstruction of 3D scenes from multiple, complex images. Artif Intell. 1986;30(3):289–341.

35. Seitz S M, Curless B, Diebel J, et al. A comparison and evaluation of multi-view stereo reconstruction algorithms. In: IEEE computer society conference on computer vision and pattern recognition; 2006. p. 519–528.

36. Scharstein D, Szeliski R. A taxonomy and evaluation of dense two-frame stereo correspondence algorithms: stereo and multi-baseline vision. New York: IEEE; 2002. p. 131–40.

37. Kise M, Zhang Q. Creating a panoramic field image using multi-spectral stereovision system. Comput Electron Agric. 2008;60(1):67–75.

38. Xia C, Li Y, Chon TS et al. A stereo vision based method for autonomous spray of pesticides to plant leaves. In: IEEE international symposium on industrial electronics. IEEE; 2009. p. 909–914.

39. Xiang R, Jiang H, Ying Y. Recognition of clustered tomatoes based on binocular stereo vision. Comput Electron Agric. 2014;106:75–90.

40. Wang C, Zou X, Tang Y, et al. Localisation of litchi in an unstructured environment using binocular stereo vision. Biosyst Eng. 2016;145:39–51.

41. Ivanov N, Boissard P, Chapron M, et al. Computer stereo plotting for 3-D reconstruction of a maize canopy. Agric For Meteorol. 1995;75(1–3):85–102.

42. Andersen HJ, Reng L, Kirk K. Geometric plant properties by relaxed stereo vision using simulated annealing. Comput Electron Agric. 2005;49(2):219–32.

43. Biskup B, Scharr H, Schurr U, et al. A stereo imaging system for measuring structural parameters of plant canopies. Plant Cell Environ. 2007;30(10):1299–308.

44. Biskup B, Scharr H, Fischbach A, et al. Diel growth cycle of isolated leaf discs analyzed with a novel, high-throughput three-dimensional imaging method is identical to that of intact leaves. Plant Physiol. 2009;149(3):1452–61.

45. Lou L, Liu Y, Han J, et al. Accurate multi-view stereo 3D reconstruction for cost-effective plant phenotyping. In: International Conference, Iciar; 2014. p. 349–356.

46. Rose JC, Paulus S, Kuhlmann H. Accuracy analysis of a multi-view stereo approach for phenotyping of tomato plants at the organ level. Sensors. 2015;15(5):9651–65.

47. Miller J, Morgenroth J, Gomez C. 3D modelling of individual trees using a handheld camera: accuracy of height, diameter and volume estimates. Urban For Urban Green. 2015;14(4):932–40.

48. Yang W, Guo Z, Huang C, et al. Combining high-throughput phenotyping and genome-wide association studies to reveal natural genetic variation in rice. Nat Commun. 2014;5:5087.

49. Song Y, Wilson R, Edmondson R, et al. Surface modelling of plants from stereo images. 2007;312–319:312–9.

50. Edelsbrunner H, Seidel R. Voronoi diagrams and arrangements. Discr Comput Geom. 1986;1(1):25–44.

51. Farooq M, Tagle AG, Santos RE, et al. Quantitative trait loci mapping for leaf length and leaf width in rice cv. IR64 derived lines. J Integr Plant Biol. 2010;52(6):578–84.

52. Duan L, Yang W, Huang C, et al. A novel machine-vision-based facility for the automatic evaluation of yield-related traits in rice. Plant Methods. 2011;7(1):1–13.

53. Duan L, Huang C, Chen G, et al. High-throughput estimation of yield for individual rice plant using multi-angle RGB imaging. In: Computer and computing technologies in agriculture VIII. Springer, Berlin; 2015. p. 1–12.

54. Iosifidis A, Gabbouj M. Multi-class support vector machine classifiers using intrinsic and penalty graphs. Pattern Recogn. 2016;55:231–46.

55. Shi Z, Cheng J-L, Huang M-X, et al. Assessing reclamation levels of coastal saline lands with integrated stepwise discriminant analysis and laboratory hyperspectral data. Pedosphere. 2006;16(2):154–60.

56. Fung GM, Mangasarian OL. Multicategory proximal support vector machine classifiers. Mach Learn. 2005;59(1–2):77–97.

57. Breiman L. Random forests. Mach Learn. 2001;45(1):5–32.

58. Pardo M, Sberveglieri G. Random forests and nearest shrunken centroids for the classification of sensor array data. Sensors Actuat B Chem. 2008;131(1):93–9.

59. Bai X, Sun C, Zhou F. Splitting touching cells based on concave points and ellipse fitting. Pattern Recogn. 2009;42(11):2434–46.

60. Awrangjeb M, Lu G, Fraser C S, et al. **A** fast corner detector based on the chord-to-point distance accumulation technique. In: Digital image computing: techniques and applications. IEEE Computer Society; 2009. p. 519–525

61. White NDG, Jayas DS, Gong Z. Separation of touching grain kernels in an image by ellipse fitting algorithm. Biosyst Eng. 2005;92(2):135–42.

62. Jean-Yves B. Camera calibration toolbox for matlab [EB/OL]. http://www.vision.caltech.edu/. 12, 2011.

63. Jiang N, Yang W, Duan L, et al. A nondestructive method for estimating the total green leaf area of individual rice plants using multi-angle color images. J Innov Opt Health Sci. 2015;08(02):1550002.

64. Geiger A, Roser M, Urtasun R. Efficient large-scale stereo matching. In: Computer vision—ACCV 2010. Springer, Berlin; 2010. p. 25–38.

65. Cochran SD, Medioni G. 3-D surface description from binocular stereo. IEEE Trans Pattern Anal Mach Intell. 1992;14(10):981–94.

66. Fua P. A parallel stereo algorithm that produces dense depth maps and preserves image features. Mach Vis Appl. 1993;6(1):35–49.

67. Woebbecke D, Meyer G, Von Bargen K, et al. Color indices for weed identification under various soil, residue, and lighting conditions. Trans ASAE. 1995;38(1):259–69.

68. Meyer G, Mehta T, Kocher M, et al. Textural imaging and discriminant analysis for distinguishingweeds for spot spraying. Trans ASAE. 1998;41(4):1189.

69. Petalas C, Anagnostopoulos K. Application of Stepwise discriminant analysis for the identification of salinity sources of groundwater. Water Resour Manage. 2006;20(5):681–700.

70. Chang C-C, Lin C-J. LIBSVM: a library for support vector machines. ACM Trans Intell Syst Technol (TIST). 2011;2(3):27.

71. Navarro PJ, Fernando P, Julia W, et al. Machine learning and computer vision system for phenotype data acquisition and analysis in plants. Sensors 2016;16(5):641.

72. Liaw A, Wiener M. Classification and regression by random forest. R News 2001; 23(23).

2

Development of high-throughput methods to screen disease caused by *Rhizoctonia solani* AG 2-1 in oilseed rape

Fryni Drizou[1]*[iD], Neil S. Graham[1], Toby J. A. Bruce[2] and Rumiana V. Ray[1]

Abstract

Background: *Rhizoctonia solani* (Kühn) is a soil-borne, necrotrophic fungus causing damping off, root rot and stem canker in many cultivated plants worldwide. Oilseed rape (OSR, *Brassica napus*) is the primary host for anastomosis group (AG) 2-1 of *R. solani* causing pre- and post-emergence damping-off resulting in death of seedlings and impaired crop establishment. Presently, there are no known resistant OSR genotypes and the main methods for disease control are fungicide seed treatments and cultural practices. The identification of sources of resistance for crop breeding is essential for sustainable management of the disease. However, a high-throughput, reliable screening method for resistance traits is required. The aim of this work was to develop a low cost, rapid screening method for disease phenotyping and identification of resistance traits.

Results: Four growth systems were developed and tested: (1) nutrient media plates, (2) compost trays, (3) light expanded clay aggregate (LECA) trays, and (4) a hydroponic pouch and wick system. Seedlings were inoculated with virulent AG 2-1 to cause damping-off disease and grown for a period of 4–10 days. Visual disease assessments were carried out or disease was estimated through image analysis using ImageJ.

Conclusion: Inoculation of LECA was the most suitable method for phenotyping disease caused by *R. solani* AG 2-1 as it enabled the detection of differences in disease severity among OSR genotypes within a short time period whilst allowing measurements to be conducted on whole plants. This system is expected to facilitate identification of resistant germplasm.

Keywords: *Rhizoctonia solani*, Oilseed rape, High-throughput phenotyping, Disease, Plant characteristics

Background

Rhizoctonia solani (Kühn) [teleomorph *Thanatephorus cucumeris* (Donk)] is a necrotrophic soil-borne fungus belonging to the phylum Basidiomycota. The species is sub-divided into anastomosis groups (AG) based on genetic and biological characteristics, as well as host-specific pathogenicity [1, 2]. Among the groups, AG 2-1 is the most destructive to oilseed rape (OSR, *Brassica napus*) and other members of the Brassicaceae [3, 4]. Under favourable temperatures, ranging from 18 to 20 °C, moist soil conditions and in the presence of the host, the growing hyphae infect young OSR seedlings causing pre- and post-emergence damping-off and root rot [4–6]. Damping-off is characterised by the formation of brown lesions and eventually rotting of the hypocotyl [7]. The infection can also result in root rot and stem rot in older plants [7, 8]. *B. napus* is a widely cultivated crop for oil production for human consumption and biodiesel, as well as for animal fodder. It is an amphiploid species derived from the crossing of *Brassica rapa* and *Brassica oleracea* and has undergone breeding for the optimisation of oil production and yields [9]. Although many studies have attempted to identify resistant or tolerant genotypes of *B. napus* and related species, currently there are no known resistant OSR genotypes to AG 2-1 [3, 5].

*Correspondence: fryni.drizou@nottingham.ac.uk
[1] Division of Plant and Crop Sciences, School of Biosciences, University of Nottingham, Sutton Bonington Campus, Loughborough, Leicestershire, UK
Full list of author information is available at the end of the article

Babiker et al. [3] assessed the survival of 85 genotypes of *B. napus* and other *Brassica* species 4 weeks after sowing in inoculated soil. Their results showed that all genotypes were susceptible, the majority of seedlings died and only 18 genotypes survived with survival rates ranging from 8.3 to 88.3% [3].

The pathogen can be partially controlled using seed treatments prior to sowing [10] and via cultural practises [4, 8]. However, these control measures only reduce the inoculum in the soil and thus delay the infection. The use of biofumigation and seed meals, from Brassicaceous plants, that usually suppress soil-borne pathogens [11, 12] or the application of beneficial biological control organisms such as *Trichoderma* and binucleate *Rhizoctonia* [8], are not effective against *R. solani* AG 2-1. Consequently, the identification of traits and genes associated with resistance to *R. solani* AG 2-1 is an essential step towards the development of sustainable integrative control strategies for this pathogen.

An important factor in developing a method is to consider the epidemiology of the pathogen and the specificity of the pathosystem. In the case of *R. solani* and *B. napus* seed germination, emergence and survival under inoculated conditions can potentially reveal phenotypic differences among genotypes that play a role in susceptibility or resistance towards AG 2-1. The developmental rate of genotypes is likely to influence disease outcome [6, 8], therefore plants that emerge faster are expected to perform better. Additionally, plant characteristics such as hypocotyl length and root architecture may explain the ability of certain genotypes to escape infection. Furthermore, the progress of disease as well as its severity in different plant organs could potentially indicate genetic differences among different genotypes. At present the most popular method to assess disease severity and classify different genotypes and plant species to their susceptibility to *R. solani* is using pots with soil or soil-free media [3, 4, 7]. Although screening in soil is realistic and provides an ideal environment for the fungi, it is time consuming, labour intensive and requires extensive controlled environment space. This limits the number of plants that can be screened quickly and cheaply. Another major bottleneck in identification of resistance to soil borne pathogens, apart from the time and space required when using inoculated soil or compost to cause disease, is the uncertainty and/or reproducibility of moderate disease on which to detect consistent differences between genotypes.

The aim of the present work was to develop a low cost, rapid and high-throughput method to enable the screening of OSR genotypes for identification of *R. solani* AG 2-1 resistance. Four different methods were tested: media nutrient plates, hydroponic growth in pouches and growth in trays with compost or light expanded clay aggregate (LECA). The methods were evaluated to screen disease and/or assess plant physiological characteristics within a short period of time during the early stages of infection among different OSR genotypes.

Methods

Inoculum and seeds

Rhizoctonia solani AG 2-1 (#1934 from the University of Nottingham isolate collection), originally isolated from OSR plants, was used to produce inoculum. The pathogenicity of this isolate to OSR was previously confirmed by Sturrock et al. [13]. The inoculum was grown on Potato Glucose Agar (PGA; Sigma-Aldrich, UK) at room temperature (18–20 °C) for a period of 10–14 days prior to the inoculation. In order to exclude contamination by other pathogens and ensure their germination, seeds were surface sterilised with 4% sodium hypochlorite (Parazone, Jeyes Limited, UK) for 5 min followed by three rinses with distilled autoclaved water and then pre-germinated on round filter paper (diameter 85 mm, GE Healthcare Whatman, UK) with 3 ml of sterile water and kept in dark at room temperature (18–20 °C) for 2 days. A group of eight *B. napus* genotypes, not previously tested for AG 2-1 resistance, was used for the evaluation of the methods to evaluate their performance against AG 2-1. The group consisted of seven commercial winter oilseed cultivars 'Temple'(conventional), 'Abaco'(conventional), 'Lioness'(conventional), 'Grizzly'(conventional), 'Galileo'(conventional), 'Sequoia'(semi-dwarf hybrid) and 'ES Betty'(restored hybrid) and one fodder type ('Canard').

Nutrient media plates

Square petri dishes-plates (120 × 120 × 17 mm Greiner Bio-One International) were filled with sterile 50% Hoagland No. 2 Basal Salt Mixture (Sigma-Aldrich, UK), pH 5.8 and 1% w/v agar (Agar–Agar granular powder, Fisher Scientific, UK). On each plate 3 seedlings of each genotype were placed 2 cm from the top of the plate with equal distances between them. For the inoculation, 1 plug (5 × 5 mm) of *R. solani* AG 2-1 from a colony growing on PGA was placed below each seed and 1 cm above the bottom of the plate. The control plates were not inoculated. Inoculated and control plates were sealed with parafilm and kept in an upright position in a controlled environment room at 18 °C and 12 h light:12 h dark. Photosynthetically active radiation (PAR) was 218.5 μmol s^{-1} m^{-2} at a height of 4 cm (LI-250A light meter, LI-COR Biosciences).

Hydroponic growth in pouch and wick system

A method previously developed for high-throughput phenotyping of roots in tanks [14, 15] was modified for screening disease caused by *R. solani* AG 2-1. The construction of the tank consisted of a metal frame with 9 drip trays and 192 growth-pouch positions. Each pouch was made of an acrylic bar, onto which 2 filter papers (Anchor Paper Company, St Paul, MN, USA) were placed on each side and covered with a black polythene sheet (Cransford, Polyethylene Ltd, Suffolk, UK). The filter papers and the sheets were held on the bars with fold-back clips (19 mm). Prior to sowing, pouches were left to soak overnight in nutrient solution (25% Hoagland's in 2 L of purified water per tray). During the experiment filter papers on growth pouches remained soaked by adding purified water in the trays in equal amounts. Filter papers and clips were autoclaved and acrylic bars were bleached and sprayed with 70% ethanol prior to their use, to eliminate contamination. One seedling was placed in each side of the growth pouch, in the middle and approximately 3 cm from the top of the filter paper and left to grow for 3 days in a controlled environment room (18 °C, 12 h light:12 h dark). Then the seedlings were inoculated by adding 1 mycelia PGA plug (5 × 5 mm) 3 cm below the tip of the primary root and another 2 plugs diametrically opposite to each other and 3 cm away from the top of the primary root. For the control seedlings PGA plugs (5 × 5 mm) without inoculum were used.

Growth in compost trays

Plastic trays (6143, Beekenkamp Verpakkingen, Netherlands) with 308 wells (3 × 3 cm) were filled with compost (Levington F2s, Everris Limited, UK) up to 2 cm and then each well was inoculated with 1 mycelia PGA plug (5 × 5 mm) of *R. solani* AG 2-1. A layer (0.5 cm) of compost was added above the inoculum and 3 pre-germinated surface sterilised seeds of OSR were placed in each well and covered with compost in order to fill up the well (1.5 cm layer). For the control wells 1 PGA plug without inoculum was added in each well. The trays were left in a controlled environment room (18 °C, 12 h light:12 h dark).

Growth in light expanded clay aggregate (LECA) trays

Light expanded clay aggregate (LECA) was used to develop a screening method that kept the roots of young seedlings intact. Each compartment of a plastic tray (6143, Beekenkamp Verpakkingen, Netherlands) with 308 wells (3 × 3 cm) was filled with approximately 3 LECA particles (size 4–10 mm; Saint-Gobain Weber Limited, UK) enough to block the bottom and then 1 mycelia PGA plug (5 × 5 mm) of AG 2-1 was added for the inoculated treatment or 1 PGA plug for the control treatment. LECA

particles were added to fill each compartment up to the ¾ of the well volume and then 2 pre-germinated seeds were added. Another layer of LECA was used to fill the wells to the top. An equal amount of 25% Hoagland's in 0.5 L of purified water was supplemented in each well of the tray.

Assessments on disease and plant characteristics

In nutrient media plates and in hydroponic pouches disease as well as plant characteristics (hypocotyl, primary root and lateral root length, lateral root and leaf number) were assessed using the same method but at different time points; Nutrient media plates were assessed at 4, 7 and 10 days post inoculation (dpi) while the seedlings in the hydroponic pouches only at 4 dpi. Disease assessment was made with disease severity categories modified from Khangura et al. [7]; for hypocotyl rot the seedlings were categorised on a scale of 0–3 (0 = no lesions, 1 = lesions on hypocotyls affecting <25% of the length of the hypocotyl, 2 = lesions covering 26–75% of the length of the hypocotyl, 3 = lesions covering >75% of the length of the hypocotyl), for primary root rot on a 0–6 scale (0 = no lesions, 1 = small lesions on primary root, 2 = discoloration up to 50% of primary root, 3 = discoloration 51–75% of the primary root, 4 = discoloration >75% and necrosis covering up to 30% of primary root, 5 = necrosis covering 31–60% of primary root, 6 = necrosis covering >61% or dead root) and for leaf disease on a 0–3 scale (0 = no lesions, 1 = disease affecting up to 25% of total leaf area, 2 = disease affecting 25–50% of total leaf area, 3 = disease affecting 51–75% of total leaf area, 4 = completely necrotic leaves of total leaf area). Disease index (DI %) was calculated as: [S (no. plants in disease category) × numerical value of disease category) × 100]/[(no. plants in all categories) × (maximum value on rating scale)]. Plant images were taken from the plates using a digital SLR camera (Canon 1100D, EOS Utility software, Canon Inc., Tokyo, Japan) and analysed with ImageJ (version 1.4.7, Schneider et al. [16]) software and used for the assessment of plant characteristics.

In compost trays, emergence and survival were assessed daily, 2 days after planting and for a period of 5 days. Final counts of emergence and survival were taken on the 10 dpi and then seedlings were removed from the wells, washed and assessed for disease. For non-emerged seedlings, soil was removed and examined to ensure that control seedlings (or seeds) were healthy while the inoculated were heavily infected (dead). For the disease assessments, the above disease scale was modified by including another level for seedlings suffering from pre-emergence damping-off (not emerged) and those that they did not survive due to post-emergence damping-off. Thus for hypocotyl rot, seedlings were rated on a 0–4 scale (4 = completely dead or/and not emerged), for primary root rot on a 0–7 scale (7 = completely dead or/and not emerged) and for leaf disease on a 0–5 scale (5 = not emerged). The percentage of disease index was calculated as described before. Control seedlings that did not emerge were scaled as healthy, as they were found in the compost without any disease symptoms.

Survival of seedlings in trays with LECA was estimated 5 dpi, then the seedlings were removed and images were taken to estimate disease (Canon 1300D, EOS Utility software, Canon Inc., Tokyo, Japan) and analysed with ImageJ (version 1.4.7, software). Seedlings that had not emerged in the control treatment, were assessed in order to ensure that they were viable and not infected, contrary with seedlings that had not emerged in the inoculated treatment which were heavily infected. In contrast to the other methods, disease was estimated as a percentage of the infected plant area to the total plant area for hypocotyls and for roots.

Experimental design and statistical analysis
All statistical analysis was performed using GenStat (15th Edition, VSN International Ltd, Hemel Hempstead, UK). The experiments for each method were designed as randomized blocks with two factors; genotype and inoculum. Where appropriate disease development, seedling emergence, survival and plant characteristics were analysed using analysis of variance (ANOVA) for repeated measures. General ANOVA was used for variables assessed

less than three times. Each method consisted of two replicated experiments, analysed as replicates when there were no significant interactions detected. Disease progress on the genotypes was analysed by excluding the non-inoculated controls in each of the four methods.

Results

Nutrient media plates
Disease development on the roots of inoculated seedlings in nutrient media plates revealed significant differences during the 10 days of the experiment ($P = 0.006$; Fig. 1). Disease developed slower on the genotype 'Grizzly', which had consistently less disease compared to the other genotypes. 'Abaco' followed 'Grizzly' but did not have significantly different disease severity compared to the other genotypes (Fig. 1). Disease on hypocotyl and roots was inconsistent between the two replicate experiments (results not shown).

Over time, AG 2-1 significantly reduced the length or the number of assessed plant characteristics apart from hypocotyl length (Table 1). Inoculated seedlings had significantly fewer leaves, smaller and fewer lateral roots, shorter primary roots and as a result total length of roots was also reduced (Table 1). However, hypocotyl growth was not different between inoculated and control seedlings ($P = 0.216$). There were no interactions between inoculum and genotype and in both inoculated and un-inoculated seedlings consistent differences were

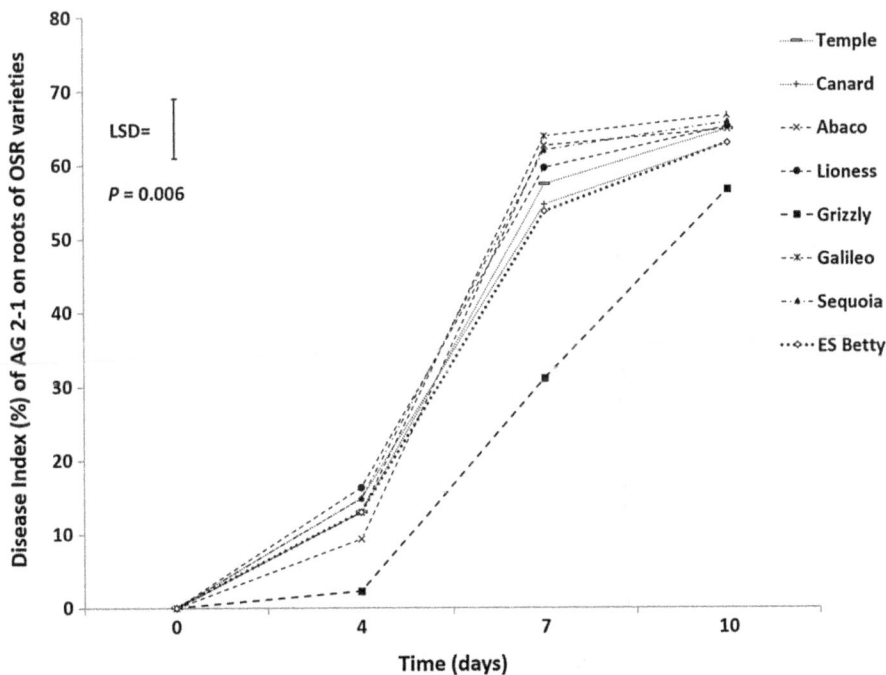

Fig. 1 Progress of disease caused by AG 2-1 on roots of seedlings of the eight varieties growing in media plates

Table 1 Plant characteristics under inoculated (AG 2-1) and un-inoculated (control) conditions during the 10 days of the experiment in nutrient media plates

Treatment	Hypocotyl length			Leaves number			Lateral RL			Lateral root number			Primary RL			Total RL		
	4 days	7 days	10 days	4 days	7 days	10 days	4 days	7 days	10 days	4 days	7 days	10 days	4 days	7 days	10 days	4 days	7 days	10 days
AG 2-1	1.99	2.30	2.37	2	3.14	1.97	1.27	2.26	2.63	12.21	21.64	23.1	7.09	7.57	7.48	8.37	9.83	10.07
Control	1.78	2.06	2.37	1.99	3.12	3.84	1.16	2.37	3.30	10.79	22.42	26.72	7.98	10.60	11.51	9.13	12.97	14.81
$P_{(time*inoculum)}$	0.216			<0.001			<0.001			0.021			<0.001			<0.001		
$LSD_{(time*inoculum)}$	0.26			0.23			0.44			3.15			1.07			1.20		
$P_{(time)}$	<0.001			<0.001			<0.001			<0.001			<0.001			<0.001		
$LSD_{(time)}$	0.15			0.15			0.18			1.76			0.30			0.35		

Lengths are expressed as cm. $P_{(time*inoculum)}$ values and $LSD_{(time*inoculum)}$ (ANOVA) were used for the comparison between the two treatments and $P_{(time)}$ values and $LSD_{(time)}$ for the comparison among different days

RL root lengths

observed in the growth of each of these plant characteristic between the different days (Table 2). Lateral root length ($P < 0.001$) and total root length ($P = 0.001$) were significantly different between the different genotypes over the 10 days. Hypocotyl length was different among the varieties for each of the 3 days, with 'Grizzly' always having shorter hypocotyl and longest lateral roots (Table 2). Additionally, the number of lateral roots was also significantly different between the genotypes with 'Canard' always having more lateral roots. Significant differences for primary root length between varieties were observed for day 4 and 7 but not on day 10. Seedlings of 'Grizzly' had consistently shorter primary roots (Table 2). Significant differences between genotypes in total length of the roots and number of leaves were observed only on the 4th and 7th day, respectively.

Hydroponic growth in pouch and wick system

Infection of seedlings with AG 2-1 did not result in significant differences in disease severity between the genotypes for any of the examined plant organs (Table 3). Inoculated seedling characteristics were all significantly affected by disease 4 dpi compared to their controls except for lateral root number ($P = 0.066$; Table 4). Additionally, significant variation was observed between genotypes for some of their morphological characteristics (Table 4): hypocotyl length ($P < 0.001$), lateral root length ($P = 0.011$) and lateral root number ($P = 0.011$) were significantly different. The length of the hypocotyl was significantly reduced in infected seedlings with 'Grizzly', 'Galileo' and 'Sequoia' being most affected. 'Canard' had the least reduction and 'ES Betty' had no reduction in hypocotyl length despite the disease (Table 4). In general, 'Canard' had shorter hypocotyls compared to the rest while 'Abaco' and 'Sequoia' had longer ones. The number of leaves of inoculated seedlings was significantly reduced compared to controls for all genotypes but no differences were observed among the genotypes. Lateral roots of genotypes were significantly shorter under inoculation with 'ES Betty' and 'Grizzly' being more affected with reduction of length of 72.2 and 88.1% respectively. Although lateral root length was significantly reduced in infected seedlings, lateral root number was not affected. Nevertheless, genotypes differed in the number of lateral roots with 'Canard' having more lateral roots. The length of the primary roots was significantly reduced due to infection of AG 2-1 in all genotypes with more pronounced reduction in 'Grizzly' (61.8%), 'Sequoia' (55.9%) and 'ES Betty' (48.5%). The total length of roots was also significantly reduced due to the infection with AG 2-1 with 'ES Betty', 'Sequoia' and 'Grizzly' having the greatest reduction of length. Despite the effect of AG 2-1

infection the genotypes did not significantly differ in primary and total root lengths (Table 4).

Growth in compost trays

Inoculation of seedlings in compost trays with AG 2-1 resulted in significant differences on disease severity between the genotypes on hypocotyls ($P = 0.003$) and leaves ($P < 0.001$) but not in roots ($P = 0.073$; Fig. 2). 'ES Betty' and 'Canard' were consistently least affected, followed by 'Abaco' and 'Sequoia', 'Lioness' and 'Grizzly' (Fig. 2). 'Galileo' and 'Temple' were the genotypes with significantly more disease (Fig. 2).

Emergence of seedlings was significantly different between genotypes ($P < 0.001$) and inoculation with AG 2-1 reduced seedling emergence in almost all varieties apart from 'Canard', 'Grizzly' and 'ES Betty' ($P < 0.001$). However, there was no significant interaction between genotypes and treatment ($P = 0.186$) (Table 5). Infection of seedlings with AG 2-1 enabled us to detect differences in survival between inoculated and non-inoculated control seedlings ($P < 0.001$) and there were significant differences between genotypes in seedling survival ($P = 0.004$; Fig. 3). 'Canard' was the genotype with significantly greater survival and the only one with no significant differences between inoculated and control seedlings (Fig. 3). 'Sequoia', 'Abaco', 'ES Betty' and 'Grizzly' followed, with the first two not being significantly different from 'Canard'. The poorest survival was observed for 'Galileo', 'Temple' and 'Lioness' (Fig. 3).

Growth in LECA trays

AG 2-1 was able to grow and infect seedlings grown in trays filled with LECA. The inoculation resulted in disease symptoms 5 days post inoculation ($P < 0.001$) and enabled assessment through image analysis. Screening for disease revealed significant differences between the tested genotypes for both disease on hypocotyls ($P = 0.002$) and on roots ($P = 0.006$). 'Sequoia' was the genotype with consistently less disease on both roots and hypocotyls followed by 'ES Betty' (Fig. 4). 'Canard' and 'Lioness' ranked in the middle and had significantly lower disease than 'Grizzly' ($P = 0.002$). 'Galileo', 'Temple', 'Abaco' and 'Grizzly' were the genotypes with the highest disease levels (Fig. 4). Disease severity on roots indicated that genotypes had similar responses to AG 2-1 infection: 'Sequoia' was the genotype with the least disease followed by 'ES Betty' and 'Lioness'; 'Canard' ranked in the middle, and 'Temple' was the genotype with the most severe disease symptoms on roots ($P = 0.006$; Fig. 4).

Inoculation with AG 2-1 reduced seedling survival ($P < 0.001$) 5 dpi but survival was not significantly different between genotypes ($P = 0.107$) and no significant

Table 2 Plant characteristics of the tested genotypes, in nutrient media plates

Genotype	Hypocotyl length			Leave number			Lateral RL			Lateral root number			Primary RL			Total RL		
	4 days	7 days	10 days	4 days	7 days	10 days	4 days	7 days	10 days	4 days	7 days	10 days	4 days	7 days	10 days	4 days	7 days	10 days
Temple	2.25	2.34	2.87	2.00	3.25	3.14	1.11	1.61	2.76	9.28	18.94	20.92	7.59	9.46	9.03	8.71	11.06	11.79
Canard	2.11	2.68	2.81	2.00	3.61	3.06	1.34	2.23	2.87	18.19	29.31	30.31	8.79	10.22	10.33	10.13	12.45	13.20
Abaco	1.77	2.08	2.10	2.01	3.24	3.07	1.30	2.09	2.53	14.12	22.70	26.67	6.83	8.19	8.93	8.14	10.28	11.45
Lioness	2.23	1.27	1.35	2.00	3.18	2.81	1.25	2.25	2.72	12.49	20.89	23.65	7.68	8.82	9.23	8.93	11.06	11.95
Grizzly	1.18	1.27	1.35	2.00	2.97	3.07	1.16	3.11	4.23	6.71	18.32	24.00	4.65	7.16	7.92	5.81	10.26	12.16
Galileo	1.73	1.99	2.27	2.00	2.78	2.49	1.04	2.34	2.76	11.78	25.00	29.94	8.08	9.91	10.43	9.12	12.24	13.19
Sequoia	1.79	2.01	2.03	2.00	3.14	2.93	1.29	2.38	3.14	10.75	18.14	23.03	8.91	9.72	10.08	10.20	12.10	13.22
ES Betty	1.99	2.31	2.52	2.00	2.86	2.69	1.23	2.51	2.73	8.69	22.97	20.78	7.74	9.22	9.85	8.97	11.73	12.58
P	<0.001	<0.001	<0.001	0.99	<0.001	0.254	0.781	0.041	0.043	<0.001	0.105	0.001	<0.001	0.021	0.186	<0.001	0.129	0.49
LSD	0.39	0.39	0.59	0.07	0.32	0.57	0.41	1.15	1.02	3.26	8.04	5.35	1.35	1.59	1.86	1.40	1.57	1.88

Lengths are expressed in cm. Comparisons for each plant characteristic among genotypes were made by using P values and LSD (ANOVA)

RL root length

Table 3 Disease index on hypocotyls, roots and leaves of the tested genotypes after inoculation with AG 2-1 for 4 days on the hydroponic growth pouches

Genotype	Disease index (%)		
	Hypocotyl	Root	Leaves
Temple	61.1	54.2	22.9
Canard	69.4	68.1	35.4
Abaco	66.7	54.2	18.8
Lioness	69.4	54.2	18.8
Grizzly	72.2	72.2	47.9
Galileo	77.8	45.8	37.5
Sequoia	75.0	52.8	35.4
ES Betty	58.3	58.3	16.7
P	0.935	0.663	0.533
LSD	32.88	29.28	34.88

For the comparison of disease severity among genotypes within each plant part P values and LSD were used (ANOVA)

interaction was observed between genotypes and treatment ($P = 0.716$).

Discussion

The primary aim of this study was to develop a high throughput method for evaluation of OSR resistance to disease caused by *R. solani* AG 2-1, as a first step towards the identification of traits that could be used in future breeding programs. Early infection of OSR by *R. solani* AG 2-1 leads to pre- and post-emergence damping off which reduces crop establishment, but infection in later stages towards the maturity of plant is less damaging [8]. Therefore, our objective was to develop methods to enable assessment of the early stages of disease progression. A key aspect of our work was to develop a low cost, rapid method that would enable screening of a large number of different OSR genotypes. The four developed methods here (nutrient media plates, hydroponic growth in pouches, trays with compost or LECA) lasted no more than 10 days and enabled the screening of up to 240 seedlings. We used a simple and cheap inoculation technique with mycelial plugs, which allows the induction of disease symptoms and minimises the time for inoculum production to 7 days.

Plant growth in media plates is a commonly used method for the evaluation of seedling growth and root architecture phenotyping. We aimed to further test this for the assessment of initial infection and disease development. Our results indicated that nutrient media plates are a good method for disease phenotyping of roots: both fungal hyphae and root systems grew successfully on the surface of the media. All the steps of infection and disease development could be observed and differences in disease severity amongst different genotypes were detected. Also, due to the horizontal growth of the root system, root architecture was easily measured. Unfortunately, in contrast to roots, this method is not suitable for assessing disease in hypocotyls and leaves. There was no consistency in disease severity among genotypes between the two replicate experiments with hypocotyls. In many

Table 4 Comparison of plant characteristics between inoculated (AG 2-1) and un-inoculated (Control) seedlings of different OSR genotypes 4 days after inoculation on hydroponic growth pouches

Genotype	Hypocotyl length		Leaves number		Lateral RL		Lateral root number		Primary RL		Total RL	
	AG 2-1	Control	AG 2-1	Control	AG 2-1	Control	AG 2-1	Control	AG 2-1	Control	AG 2-1	Control
Temple	1.87	2.19	1.63	1.99	0.52	1.29	2.00	4.85	1.40	2.44	1.92	3.73
Canard	1.31	1.40	1.50	2.08	1.15	1.62	4.67	5.83	2.09	2.83	3.24	4.45
Abaco	2.79	2.93	2.00	2.00	0.49	0.91	2.33	3.17	1.85	2.40	2.34	3.30
Lioness	2.13	2.62	1.33	2.08	0.54	0.93	3.92	3.58	1.94	2.98	2.48	3.91
Grizzly	1.48	2.41	1.46	1.99	0.15	1.22	0.42	1.65	0.68	1.78	0.82	3.00
Galileo	1.81	2.54	1.54	2.00	0.32	0.49	1.33	1.08	1.39	2.55	1.70	3.04
Sequoia	2.50	3.16	1.54	2.00	0.33	1.18	2.33	3.25	1.50	3.40	1.83	4.58
ES Betty	2.50	2.34	1.71	2.00	0.47	1.69	2.75	5.08	1.74	3.38	2.21	5.07
$P_{(genotype)}$	<0.001		0.721		0.011		0.011		0.299		0.115	
$LSD_{(genotype)}$	0.56		0.32		0.49		2.34		1.13		1.37	
$P_{(inoculum)}$	0.005		<0.001		<0.001		0.066		<0.001		<0.001	
$LSD_{(inoculum)}$	0.28		0.16		0.25		1.17		0.57		0.68	

Lengths are expressed in cm. $P_{(genotype)}$ and $LSD_{(genotype)}$ were used for the comparison among genotypes and $P_{(inoculum)}$ and $LSD_{(inoculum)}$ for the comparison between treatments (ANOVA)

RL root length

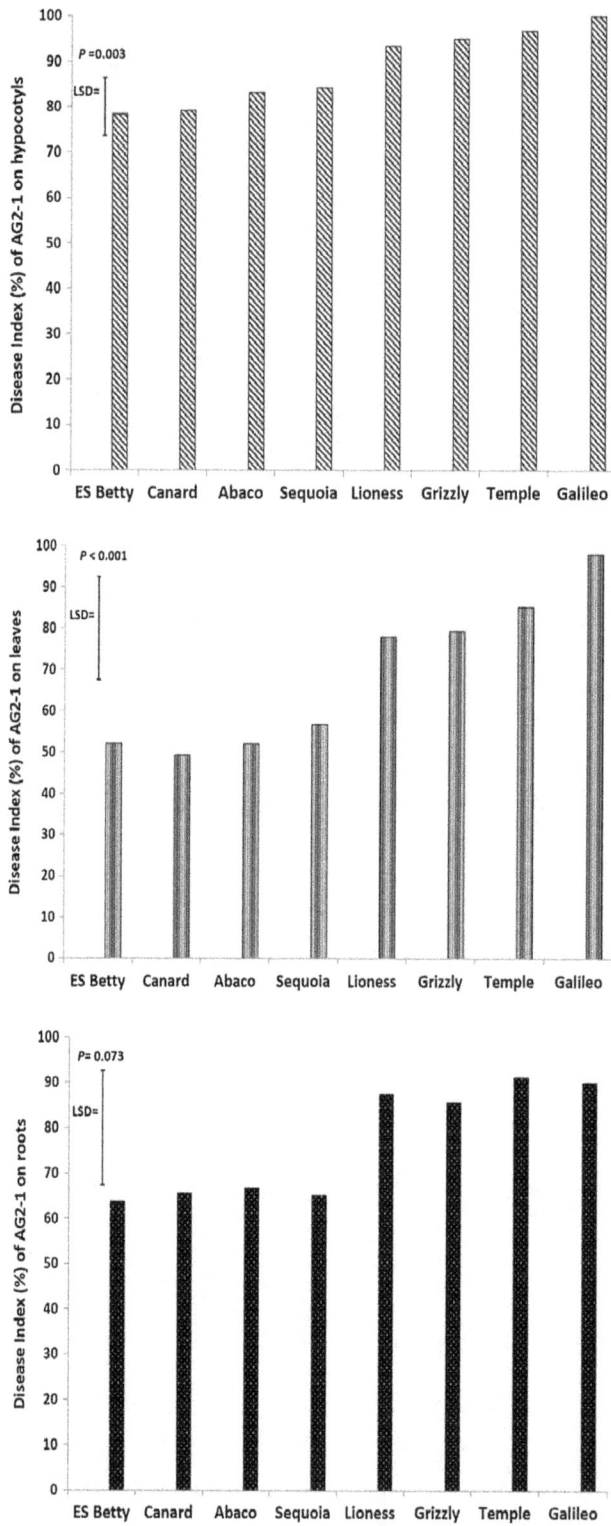

Fig. 2 Disease on hypocotyls, leaves and roots of the tested genotypes 10 days after inoculation in compost trays

Table 5 Comparison of emergence between inoculated (AG 2-1) and un-inoculated (Control) seedlings of different OSR genotypes 10 dpi in compost trays

Genotype	Emergence (%)	
	AG 2-1	Control
Temple	30.0	67.2
Canard	83.9	88.3
Abaco	60.0	98.3
Lioness	43.3	88.9
Grizzly	42.8	57.8
Galileo	11.7	64.4
Sequoia	63.3	98.9
ES Betty	53.3	77.2
$P_{(genotype)}$	<0.001	
$LSD_{(genotype)}$	18.586	
$P_{(Inoculum)}$	<0.001	
$LSD_{(inoculum)}$	9.293	
$P_{(inoculum*genotype)}$	0.186	
$LSD_{(inoculum*genotype)}$	26.284	

$P_{(inoculum)}$ and $LSD_{(inoculum)}$ were used for the comparison between treatments and $P_{(inoculum*genotype)}$ and $LSD_{(inoculum*genotype)}$ for the interaction between genotypes and treatments (ANOVA)

both leaves and roots of inoculated seedlings compared to controls, with reduction of healthy leaf area, root length (both primary and lateral) and lateral root number. The results are in agreement with a recent study showing that AG 2-1 causes severe disease by significantly reducing root length and density of inoculated OSR plants and is capable of killing the seedling within 6 dpi [13]. The analysis of plant characteristics showed that genotypes differ in lateral root and total root length as well as their growth rates. Among the genotypes, 'Grizzly' was the only one that consistently had significantly lower disease but also shorter hypocotyl and primary root compared to other genotypes. Therefore, it might be that the slower growth rate contributed to delay in infection and thus resulted in lower disease levels observed on plates. 'Grizzly' is a winter hybrid known to carry genes for stem canker resistance and for that reason is included in breeding programs [17], however in our tests with 56.6% of root disease 'Grizzly' was susceptible to AG 2-1.

Advanced high-throughput methods have been developed to screen the root system [18] and to quantify traits and identify Quantitative Trait Loci (QTLs) [14]. Atkinson et al. [14] screened a mapping population of wheat seedlings aiming to identify QTLs linked with root traits in hydroponic pouch and wick system. Also Thomas et al. [15] used this approach for screening a range of OSR genotypes under control environment and field conditions. Here we modified the method for screening disease caused by AG 2-1 in OSR. Our results showed that *R. solani* was able to grow on filter paper and infect young OSR seedlings causing disease symptoms 4 dpi.

cases, hypocotyls escaped hyphae and tended to grow towards the lids of the plates. In the same way the leaves of these plants were also escaping the pathogen. Consequently, this variation in growth led to the uneven and inconsistent infection among genotypes and between experiments. Nonetheless, disease significantly affected

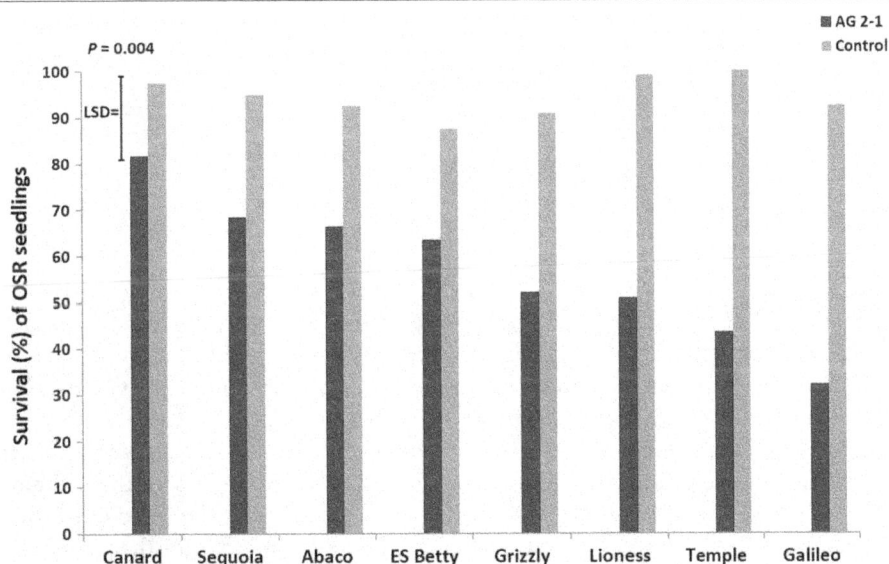

Fig. 3 Percentage of survival of different OSR genotypes 10 days post inoculation in compost trays. Comparisons for the interaction between treatment and genotype were made with *P* values and LSD (ANOVA)

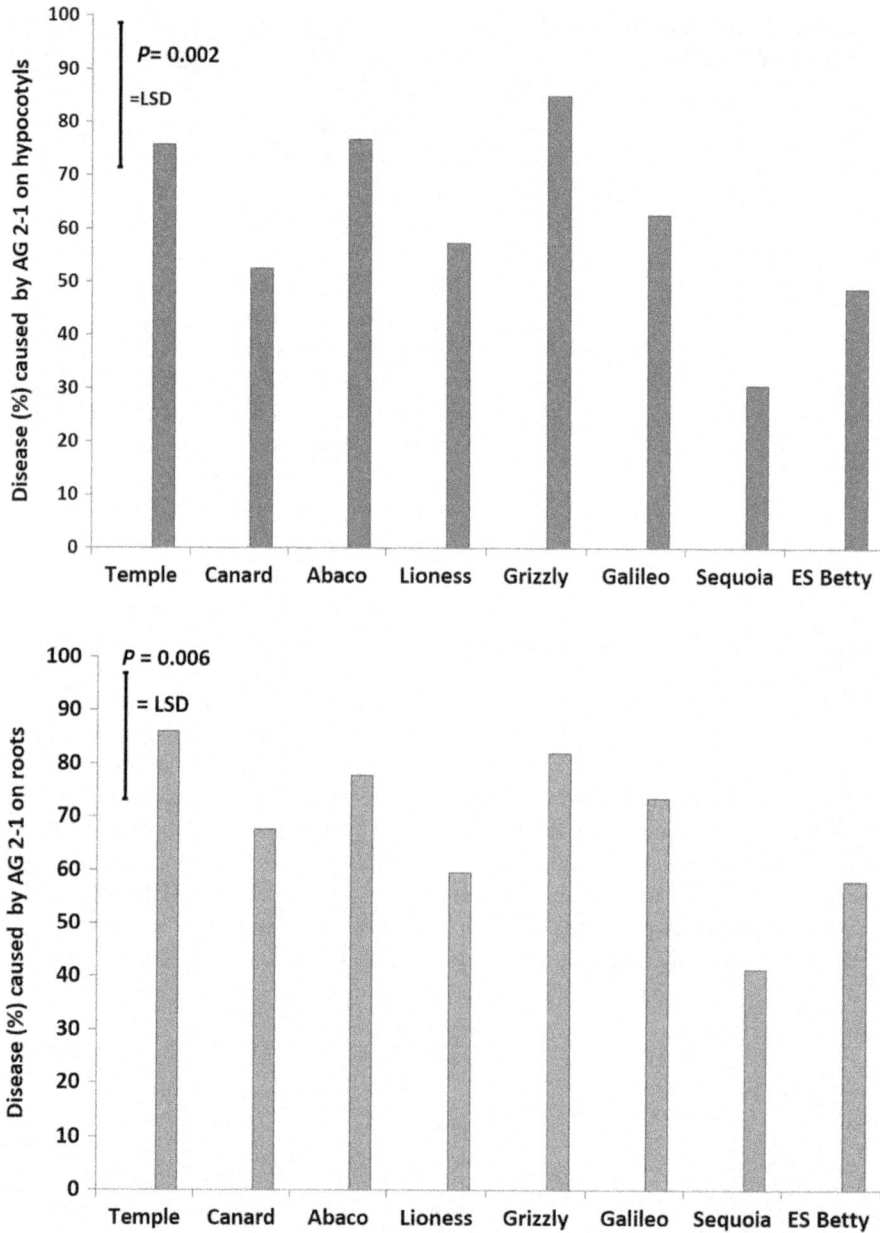

Fig. 4 Disease on hypocotyls and roots of the tested genotypes 5 days after inoculation in trays with LECA

Within this time, disease developed on hypocotyls, roots and leaves and resulted in their reduction in inoculated plants compared to controls. However, no differences were detected between genotypes for disease and all were observed to be highly susceptible under this method of inoculation. It is likely that the tested genotypes are characterized by only minor differences and the present screening method could not detect them under the tested conditions. However, this is in contrast with the results of the other two methods, where significant differences on

disease severity were observed. Different inoculum densities and length of inoculation periods were tested (results not shown) prior to the present experimental procedure, which appeared to be the most consistent. Possibly the moist environment of the filter paper and the polythene sheet as well as the lack of the soil environment altered hyphal growth and the infection process. *R. solani* is a soil-borne pathogen, thus the presence of soil with nutrients, organic matter and aeration play a pivotal role in its epidemiology. In this growing system the polythene sheet was attached to the filter paper but in the position of the seedlings, small aerate cavities were formed possibly enabling the pathogen to grow better. As a result, pathogen hyphae were denser close to the seedling and eventually led to greater disease on plants, whilst in the other methods pathogen growth was more even. Nonetheless, this method enabled us to detect differences in plant characteristics between inoculated and un-inoculated control seedlings, as well as differences among genotypes in a short period of time.

Soil and compost are most commonly used for the evaluation of plant resistance against soil-borne pathogens. In the case of *R. solani*, the vast majority of studies focussing on plant responses to pathogen exposure, have used soil [3, 5], soil free media [4] or a combination of both [10]. In this way, the experiments simulate more realistic conditions that occur in the field and a better evaluation of the plants response to the pathogen is observed. Therefore we decided as a suitable alternative that the third method should be developed with compost. In contrast to other studies, we used multiple cell-trays which save space and time by enabling us to screen more than 100 different genotypes per tray in a single experiment. The trays were also ideal to assess the early stages of infection in young seedlings that are less than 10 days old. An additional benefit of this method is that it enabled the recording of emergence and survival of seedlings and hence record pre- and post-emergence damping off. Low emergence of inoculated seedlings compared to controls, indicated susceptibility of those cultivars to pre-emergence damping off and confirmed the detrimental effect of AG 2-1 to OSR during early growth stages.

Disease screening on hypocotyls and leaves was easily conducted, but in contrast the extraction and assessment of the delicate roots of seedlings damaged by root rot was difficult and time consuming. Despite meticulous work, it was hard to keep the roots intact. We were unable to detect significant differences in root disease between cultivars in this method but we were able to detect differences in disease severity of hypocotyls and leaves. 'ES Betty' and 'Canard' were consistently the two genotypes with the lowest disease while 'Temple' and 'Galileo' were the most susceptible. This is in agreement with emergence and survival data and it can be an indication that these genotypes may carry both quantitative and qualitative traits allowing them to perform better against AG 2-1. In this research all genotypes were pre-germinated in order to standardise our methods, and therefore their germination rates under inoculated conditions were not assessed. However, it is possible that some genotypes are able to germinate and emerge faster and therefore escape and/or be less affected by the infection. Indeed, Sturrock

et al. [13], suggested that rapid germination of OSR seedlings may enable the early establishment of a strong root system allowing better nutrient uptake and growth and consequent recovery from AG 2-1 infection.

We aimed to improve the method by eliminating high inoculum pressure and most importantly by reducing damage to roots to be able to better discriminate the genotypes in our disease assessments. Therefore we decided first to reduce the time that the seedlings were exposed to the pathogen from 10 to 5 dpi. Secondly we used a medium that would not affect seedling growth but would minimise the damage to the root system upon removal. In this respect, LECA particles with the addition of nutrient solution appeared to be an appropriate medium. LECA has been receiving a growing acceptance as an environmental friendly natural material with great benefits in civil engineering and gardening. Currently there is a limited number of published studies examining the use of LECA as a growing medium [19–21] and to the best of our knowledge only one study has examined the growth of a fungi in LECA [22]. In this study the authors showed that arbuscular mycorrhizal fungi (AMF) were not able to colonise their tested plant, *Paspalum notatum*, when grown in LECA and consequently concluded that LECA was not colonised effectively by AMF [22]. However, the results of the current study show that the necrotrophic pathogen *R. solani* AG 2-1 was able to grow on the surface of LECA particles, observed as hyphal mass and infect OSR seedlings. The inoculation period of 5 days was sufficient to induce disease symptoms without killing the seedlings. At the same time differences in disease severity of the tested genotypes were detected for both hypocotyls and roots. The use of LECA preserved the roots intact during their collection from the trays and therefore allowed more accurate disease assessments. Taking images of the seedlings and analysing them with ImageJ not only allowed us to complete the experiments faster but also to estimate the disease more objectively compared to more subjective visual assessments which are not taking into account differences in growth and development of the seedlings. The OSR genotypes had different responses to AG 2-1 infection: 'Sequoia' was the least affected for both damping off and root rot, followed by 'ES Betty'. Although disease affected the survival of inoculated OSR seedlings compared to the controls, we were not able to detect significant differences in survival of seedlings between the different genotypes at 5 dpi.

Comparison of different methods

Assessing the severity of disease caused by AG 2-1 on hypocotyls and/or roots of young seedlings is the most important measure for the identification of active genetic resistance. Nonetheless, other traits related to rapid development and growth for crop establishment such as root architecture and emergence or survival are important for the identification of disease escape. Each of the four methods we developed has positive and negative aspects: Nutrient media plates enabled the recording of the infection progress and the collection of data on root traits but were not suitable for disease screening of hypocotyls and leaves. Growth in hydroponic pouches can be high-throughput, fast screening method but the moist environment altered *R. solani* growth and we could not detect any difference in disease severity among the tested OSR genotypes. Screening on trays with compost was more realistic approach that makes available holistic disease screens for the plant as well as measurements of emergence and survival. Nevertheless, damage to the root system prevented accurate disease assessment and measurements of root architecture traits and a longer time was required to detect differences. However, the use of LECA holds the benefits of screening in compost trays but also enables the roots to be intact and detect differences between genotypes in root rot disease. We were unable to detect differences in survival most likely due to short infection period of 5 dpi. Most importantly 5 dpi screening in LECA resulted in moderate disease of seedlings compared to screening in compost and this might be the reason that we have small differences in in the ranking of genotypes between the two methods. Considering the severity of disease 5 dpi and the lack of resistance in the tested genotypes, further screening for a longer period for detection of differences in survival using this method was not pursued here. In Table 6 we provide a basic estimation of the cost of screening 100 genotypes by each method, based on the cost of consumables and equipment used; the hydroponic pouch and wick system was the most expensive method as the requirements for building the system were high compared to the other methods that use petri dishes and well trays. As mentioned previously, the choice of method should be based on the scientific aim; in the present study we aimed to identify a low cost high-throughput screening method which would enable the detection of potential resistant OSR genotypes to root diseases such as AG 2-1. Therefore, we required a method

that allowed the detection of differences in disease severity and resultant changes to plant morphological characteristics. Screening in trays with LECA fulfilled these criteria it enables fast and high-throughput screening with the assessment of early infection stages. Therefore it is an applicable method for the detection of resistant OSR cultivars to AG 2-1.

Conclusion

The present study provides a new low cost, high-throughput screening method for the identification of potential OSR cultivars that are resistant to root diseases such as *R. solani* AG 2-1. This method can be used as an early step for the evaluation of germplasm prior to testing under field conditions. Additionally, it confirms that AG 2-1 is an extremely pathogenic isolate to OSR [3–6]; the inoculum density used resulted in low survival of young seedlings 10 dpi in compost trays and high disease levels ranked from 30 to 85% 5 dpi in trays with LECA. None of the genotypes tested in the current study were resistant. Future screening of diverse populations of *B. napus* and *Brassica* "species is essential to elucidate if there is any resistance against this destructive pathogen.

Abbreviations

OSR: oilseed rape; AG: anastomosis group; LECA: light expanded clay aggregate; PGA: potato glucose agar; dpi: days post inoculation; DI: disease index; LSD: least significant difference of means; RL: root length; QTL: quantitative trait loci; AMF: arbuscular mycorrhizal fungi.

Authors' contributions

The experiments were designed by FD with the contribution of RVR and NSG. Experiments were performed by FD. Data analysed by FD with the contribution of RR, production of figures and tables was performed by FD with the contribution of RVR and NSG. The manuscript was written by FD with the contribution of RVR, NSG and TJAB. All authors read and approved the final manuscript.

Author details

[1] Division of Plant and Crop Sciences, School of Biosciences, University of Nottingham, Sutton Bonington Campus, Loughborough, Leicestershire, UK. [2] School of Life Sciences, Keele University, Keele, Staffordshire, UK.

Acknowledgements

Not applicable.

Competing interests

The authors declare that they have no competing interests.

Table 6 Estimation of cost for the screen of 100 genotypes in the developed methods

Method	Cost (£) for 100 genotypes
Hydroponic pouch and wick system	348
Nutrient media plates	27.3
Trays with compost	1.05
Trays with LECA	2.14

The estimation excludes the cost for the camera that was used in the hydroponic pouch and wick system, on nutrient media plates and trays with LECA

Funding

This work was supported by the Biotechnology and Biological Sciences Research Council (Grant Number BB/M008770/1), through the Nottingham-Rothamsted Doctoral Training Programme.

References

1. Anderson NA. The genetics and pathology of *Rhizoctonia solani*. Annu Rev Phytopathol. 1982;20(1):329–47.
2. Ogoshi A. Ecology and pathogenicity of anastomosis groups of *Rhizoctonia solani* Kühn. Annu Rev Phytopathol. 1987;25:125–43.
3. Babiker E, et al. Evaluation of Brassica species for resistance to *Rhizoctonia solani* and binucleate Rhizoctonia (*Ceratobasidum* spp.) under controlled environment conditions. Eur J Plant Pathol. 2013;136(4):763–72.
4. Yang J, Verma PR. Screening genotypes for resistance to pre-emergence damping-off and postemergence seedling root rot of oilseed rape and canola caused by *Rhizoctonia solani* AG-2-1. Crop Prot. 1992;11(5):443–8.
5. Acharya SN, et al. Screening rapeseed/canola for resistance to damping-off and seedling root rot caused by *Rhizoctonia solani*. Can J Plant Pathol. 1984;6(4):325–8.
6. Kataria HR, Verma PR. *Rhizoctonia solani* damping-off and root rot in oilseed rape and canola. Crop Prot. 1992;11(1):8–13.
7. Khangura RK, Barbetti MJ, Sweetingham MW. Characterization and pathogenicity of Rhizoctonia species on canola. Plant Dis. 1999;83(8):714–21.
8. Verma PR. Biology and control of *Rhizoctonia solani* on rapeseed: a review. Phytoprotection. 1996;77(3):99–111.
9. Allender CJ, King GJ. Origins of the amphiploid species *Brassica napus* L. investigated by chloroplast and nuclear molecular markers. BMC Plant Biol. 2010;10(1):1–9.
10. Lamprecht SC, et al. Evaluation of strategies for the control of canola and lupin seedling diseases caused by Rhizoctonia anastomosis groups. Eur J Plant Pathol. 2011;130(3):427–39.
11. Cohen MF, Yamasaki H, Mazzola M. *Brassica napus* seed meal soil amendment modifies microbial community structure, nitric oxide production and incidence of Rhizoctonia root rot. Soil Biol Biochem. 2005;37(7):1215–27.
12. Handiseni M, et al. Effect of Brassicaceae seed meals with different glucosinolate profiles on Rhizoctonia root rot in wheat. Crop Prot. 2013;48:1–5.
13. Sturrock CJ, et al. Effects of damping-off caused by *Rhizoctonia solani* anastomosis group 2-1 on roots of wheat and oil seed rape quantified using X-ray computed tomography and real-time PCR. Front Plant Sci. 2015;6:461.
14. Atkinson JA, et al. Phenotyping pipeline reveals major seedling root growth QTL in hexaploid wheat. J Exp Bot. 2015;66:2283–92.
15. Thomas CL, et al. High-throughput phenotyping (HTP) identifies seedling root traits linked to variation in seed yield and nutrient capture in field-grown oilseed rape (*Brassica napus* L.). Ann Bot. 2016;118(4):655–65.
16. Schneider CA, Rasband WS, Eliceiri KW. NIH image to ImageJ: 25 years of image analysis. Nat Methods. 2012;9(7):671–5.
17. Jestin C, et al. Connected populations for detecting quantitative resistance factors to phoma stem canker in oilseed rape (*Brassica napus* L.). Mol Breed. 2015;35(8):16.
18. Hund A, Trachsel S, Stamp P. Growth of axile and lateral roots of maize: I development of a phenotying platform. Plant Soil. 2009;325(1–2):335–49.
19. Laznik Z, Znidarcic D, Trdan S. Control of *Trialeurodes vaporariorum* (Westwood) adults on glasshouse-grown cucumbers in four different growth substrates: an efficacy comparison of foliar application of *Steinernema feltiae* (Filipjev) and spraying with thiamethoxamn. Turk J Agric For. 2011;35(6):631–40.
20. Graber A, Junge R. Aquaponic systems: nutrient recycling from fish wastewater by vegetable production. Desalination. 2009;246(1–3):147–56.
21. Trdan S, Znidarcic D, Vidrih M. Control of *Frankliniella occidentalis* on glasshouse-grown cucumbers: an efficacy comparison of foliar application of *Steinernema feltiae* and spraying with abamectin. Russ J Nematol. 2007;15(1):25–34.
22. Douds DD, et al. Pelletized biochar as a carrier for AM fungi in the on-farm system of inoculum production in compost and vermiculite mixtures. Compost Sci Util. 2014;22(4):253–62.

3

A nondestructive method to estimate the chlorophyll content of *Arabidopsis* seedlings

Ying Liang[1,2], Daisuke Urano[1], Kang-Ling Liao[1], Tyson L. Hedrick[1], Yajun Gao[2*] and Alan M. Jones[1,3*] ⓘ

Abstract

Background: Chlorophyll content decreases in plants under stress conditions, therefore it is used commonly as an indicator of plant health. *Arabidopsis thaliana* offers a convenient and fast way to test physiological phenotypes of mutations and treatments. However, chlorophyll measurements with conventional solvent extraction are not applicable to *Arabidopsis* leaves due to their small size, especially when grown on culture dishes.

Results: We provide a nondestructive method for chlorophyll measurement whereby the red, green and blue (RGB) values of a color leaf image is used to estimate the chlorophyll content from *Arabidopsis* leaves. The method accommodates different profiles of digital cameras by incorporating the ColorChecker chart to make the digital negative profiles, to adjust the white balance, and to calibrate the exposure rate differences caused by the environment so that this method is applicable in any environment. We chose an exponential function model to estimate chlorophyll content from the RGB values, and fitted the model parameters with physical measurements of chlorophyll contents. As proof of utility, this method was used to estimate chlorophyll content of G protein mutants grown on different sugar to nitrogen ratios.

Conclusion: This method is a simple, fast, inexpensive, and nondestructive estimation of chlorophyll content of *Arabidopsis* seedlings. This method lead to the discovery that G proteins are important in sensing the C/N balance to control chlorophyll content in *Arabidopsis*.

Keywords: *Arabidopsis thaliana*, C/N sensing, Chlorophyll content, ColorChecker chart, Heterotrimeric G protein complex, Stress assay

Background

The chlorophyll content of leaves is an indirect indicator of the health and nutritional status of the plant [1]. Traditional methods to calculate the chlorophyll content include a destructive chemical extraction and a nondestructive measurement of chlorophyll fluorescence. The former method, while direct, is tedious and unsuitable for continuous monitoring individual plants because of its destructive manner. The latter method needs expensive instruments of which none are presently suitable for small leaves such as the commonly used *Arabidopsis*

cotyledons. It is important to develop a non-destructive method to estimate chlorophyll content for *Arabidopsis* because it is a genetic model plant, however traditional chlorophyll extraction is not useful due to the small size of the *Arabidopsis* leaves grown on agar plates. Recently, digital photographic imaging showed great promise for quantitating plant phenotypes [2]. Indirect methods are available but none are yet suitable for *Arabidopsis*. Sass et al. [3] developed a protocol to convert the RGB values of a color image into a hue saturation value (HSV), and showed that the hue value was correlated to the chlorophyll content estimated by a destructive method. A similar color-image method was used to assess the nitrogen status of rice under natural light condition [4]. Riccardi et al. [5] found that an exponential function model displays the best correlation between the RGB values and the chlorophyll content through single and multiple

*Correspondence: yajungao@nwsuaf.edu.cn; alan_jones@unc.edu
[1] Department of Biology, The University of North Carolina at Chapel Hill, Coker Hall, CB#3280, Chapel Hill, NC 27599-3280, USA
[2] College of Natural Resources and Environment, Northwest A&F University, Yangling 712100, Shaanxi, China
Full list of author information is available at the end of the article

regression in quinoa and amaranth leaves. No similar color-image methods have been adapted for *Arabidopsis* chlorophyll content, in particular *Arabidopsis* seedlings grown on agar plates. The lack of a quantitative method for measuring chlorophyll of plate-grown *Arabidopsis* restricted previous studies on stress-induced phenotypes to subjective assessment without quantitation [6, 7].

Chlorophyll content in leaves is affected by the carbon (C) and nitrogen (N) balance. Genetics studies using *Arabidopsis* revealed that the C/N balance is regulated through multiple signaling cascades of *abscisic acid-insensitive 1* (*ABI1*), *hexokinase 1* (*HXK1*), nitrate transporters, glutamate receptor (*AtGLR1.1*) and heterotrimeric G proteins [6, 8–12]. Many mutant alleles in these pathways confer altered chlorophyll content regulated by the C/N balance in *Arabidopsis* [6, 9, 12], therefore these mutants are useful for validating the utility of our digital image method. Heterotrimeric G proteins consist of one canonical Gα subunit (*GPA1*), one Gβ subunit (*AGB1*), three Gγ subunits (*AGGs*) and three atypical extra-large Gα proteins (*XLGs*) [13–15] in *Arabidopsis*. G protein signaling pathway senses glucose levels in the environment [16, 17], and is also involved in nitrogen use efficiency in rice [10]. *Regulator of G protein Signaling 1* protein (AtRGS1) is a component of the glucose sensor [11, 12, 18]. This protein modulates the activation state of G signaling.

In this study, we describe a convenient and nondestructive method to estimate leaf chlorophyll of *Arabidopsis* seedlings grown on agar plates using calibrated-RGB images. We also provide instructions how to adapt it to other small leave samples. We quantitated chlorophyll content in small *Arabidopsis* seedlings grown on different C: N ratios in the agar medium. The results indicated that G proteins play important roles in sensing and/or responding to the C/N balance in this chlorophyll response.

Materials needed

1. *Arabidopsis* seedlings grown on agar plates. In this study, we used the following T-DNA insertion mutant alleles: *agb1-2* [19], *rgs1-2* [20], *gpa1-3* [21], *xlg1xlg2xlg3* [13] which combines these alleles *xlg1-1* (SAIL_760H08) [13], *xlg2-2* (SALK_062645), *xlg3-2* (SAIL_107656) [13], and *xlg/gpa1* which combines the *xlg1-1, xlg2-1, xlg3-2* and *gpa1-3 alleles above* [22] (in press).
2. A digital camera that captures images in RAW format.
3. X-rite ColorChecker classic chart (http://xritephoto.com/colorchecker-classic).
4. Software: ImageJ https://imagej.nih.gov/ij/ (public source; imageJ 1.50i is recommended),

DNG converter http://www.adobe.com (public source),
DNG profile editor http://www.adobe.com (public source) and
PhotoShop or Lightroom http://www.adobe.com (license required).
Optional software: Matlab http://www.mathworks.com/(license required).

5. Plug-in programs for ImageJ are provided here in Additional file 1: S1, Additional file 2: S2, Additional file 3: S3 and Additional file 4: S4. Additional file 1: S1 and Additional file 2: S2 are for chlorophyll calculation and Additional file 3: S3 and Additional file 4: S4 are for reading the RGB values of the images. Additional file 1: S1 and Additional file 3: S3 are the .java source code used to generate the .class files (Additional file 2: S2 and Additional file 4: S4). In most cases, only the .class files are needed, however the corresponding .java files are provided for those wanting to examine or improve the programs.

Note 1: This plugin was successfully tested in both ImageJ 1.48v and ImageJ 1.50i.

6. Optional: Materials for chlorophyll extraction: 80% acetone in water, spectrophotometer

Plant growth

1. Grow *Arabidopsis* seedlings in the light on square 8 cm × 8 cm plastic Petri plates (VWR; Cat No. 60872-310) with a 40-mL layer of agar and media suitable for plant growth.
2. Thirty-six seedlings are placed individually within the 36-gridded area of the plate; one seedling per grid. There should be no overlap of seedlings. This spacing is important for the software to automatically detect seedlings for chlorophyll calculations. There is also an option described later to create a different grid then the default 6 × 6 grid (step 8 of the protocol).

Note 2: It is not necessary to use a square petri dish on which to arrange the seedlings; any rectangle background with samples in a matrix format will work. It is also possible to treat seedlings in liquid culture or on some matrix other than agar and then transfer them to the square agar plates for photography.

Note 3: For this study, seedlings were grown on the indicated media arranged on square plates as described above and photographed as will be described below. Specifically, Murashige and Skoog (MS) Modified Medium w/o Nitrogen (Plantmedia; Cat No. 30630200-1) supplied with 0.8% phytoagar and 1 g/L MES was used. The pH was adjusted to 5.7 with KOH. Filter-sterilized D-glucose was added to the medium to adjust the glucose concentration as indicated. A stock solution of 1 M KNO_3 was used as the nitrogen resource. The agar plates of sterilized seed, sealed with a gas-permeable tape, were stratified at 4 °C for 3–5 days in the dark. The plates were placed horizontally under constant dim light (35–50 $\mu Em^{-2}\ s^{-1}$) at 23 °C for 12 days. Images were obtained on the 12th day. Plants were also grown on soil in a long day chamber (200 $\mu Em^{-2}\ s^{-1}$, 16 h light/8 h dark) at 23 °C for 4–8 weeks as indicated in the experimental description.

Protocol

1. Photograph seedlings arranged on the square plates and save as RAW files. Any digital camera that stores images as RAW files can be used.

   Note 4: For this study, an Olympus digital camera E-3 captured the RGB color images and the images stored as ORF file, a type of RAW format. Settings for the digital camera were the following: aperture = f/7, shutter speed = 1/100 s, ISO = 400, quality = F, file storage = RAW. Shutter speed should be adjusted according to the actual ambient light condition.

   Note 5: Acquire images of the X-rite ColorChecker classic chart both immediately before and after acquiring images of the samples (Fig. 1a). The Color-Checker chart is to make sure the final data comparable despite different light conditions and cameras but may not be necessary. Before and after images of the chart are acquired to test whether the light condition is consistent during the photographing. Check the RGB values of the X-rite Colorchecker chart boards as described below. Should you find that the starting and ending values are different, it will be necessary to stabilize the light environment and re-acquire sample images. If the light condition is stable in the lab, it is not necessary to acquire two images every time. It is important to use the same settings and the same light conditions for all the images to be compared.

   Note 6: The image size for the plate is the same as for the ColorChecker chart, approximately 210 × 300 mm. The square plate is placed in the center of this field for imaging.

   Note 7: The light should be uniformly distributed. Tests for position effect in this field were determined to be a maximum of 4.8% (n = 36) (Additional file 5: S5) of the chlorophyll values. This value was calcu-

lated by estimating the chlorophyll content of 36 individual seedlings on a plate placed in the center and 4 corners of the field.

2. Convert the RAW files to DNG format using the DNG converter. There are many formats of RAW file; one is a DNG. If your camera stores the files in DNG format, nothing more is needed at this step.

3. Generate a DNG profile of the X-rite ColorChecker chart. This format can be used by Adobe PhotoShop and Lightroom programs. Launch the DNG profile editor, and then click on the 'Chart' tab. Then load the DNG file checkerchart. Use the mouse to position the four colored circles in the image at the centers of the four corner color panels of the chart. The colors of the circles should correspond to the colors of the patches. Leave the popup menu set at 'Both Color Tables' and click 'Create Color table' button. Then export the profile by selecting Export Profile in the File menu and name the profile as ColorCheckerChart.dcp. Make sure the file is saved in the CameraProfiles directory (default).

4. Using the Checker.dcp files generated in step 3, calibrate all seedling images using the DNG profile through the software Lightroom or Photoshop. These programs provide a step-by-step instructions for calibration. After the adjustment, export the images in jpg format labeled accordingly: IMAGE_ID.jpg.

5. Open the IMAGE_ID.jpg files in Adobe Photoshop. Clean or erase the background of the images with the eraser or magic wand/delete tools (Fig. 1b). Label files IMAGE_ID_cleaned.JPG The plugin will automatically remove slightly gray or otherwise imperfect backgrounds, but use of this capability should be carefully validated against test images with manually cleaned backgrounds.

   Note 8: Cleaning the background is an important step. Assure that shadows are completely eliminated and only leaves remain in the image. An example of a perfectly cleaned image is provided as Additional file 6: S6.

6. Open ImageJ and install the plugin Chloropyll_Imager provided in the Additional file 2: S2. Additional file 1: S1 is the .java code which could also be used if preferred.

7. Open ColorChecker.jpg in ImageJ and record the RGB values designated r, g, b of the white background. The r, g, and b values are obtained using the plugin labeled 'RGB_measure provided in Additional file 4: S4. Additional file 3: S3 is the .java code which could also be used if preferred. Select the white panel of the ColorChecker chart and run the plug-in. The corresponding r, g, b values of the white panel will appear in a table.

8. Open IMAGE_ID_cleaned.jpg files in ImageJ and run the plugin "Chorophyll_Image" which in step 6 you

a

b

c

Chlorophyll Imager inputs

-- Input image grid layout specification --

Number of data rows: 6

Number of data columns: 6

-- RGB value of the ColorChecker white panel --

r = 243.000

g = 243.000

b = 242.000

☐ Make normalized grayscale image

-- Coefficient values for chlorophyll estimation --

a1 = -0.0280

a2 = 0.0190

a3 = -0.0030

a4 = 5.780

Chlorophyll Imager v1.0

see Ying L et al (submitted) for details

OK   Cancel

d

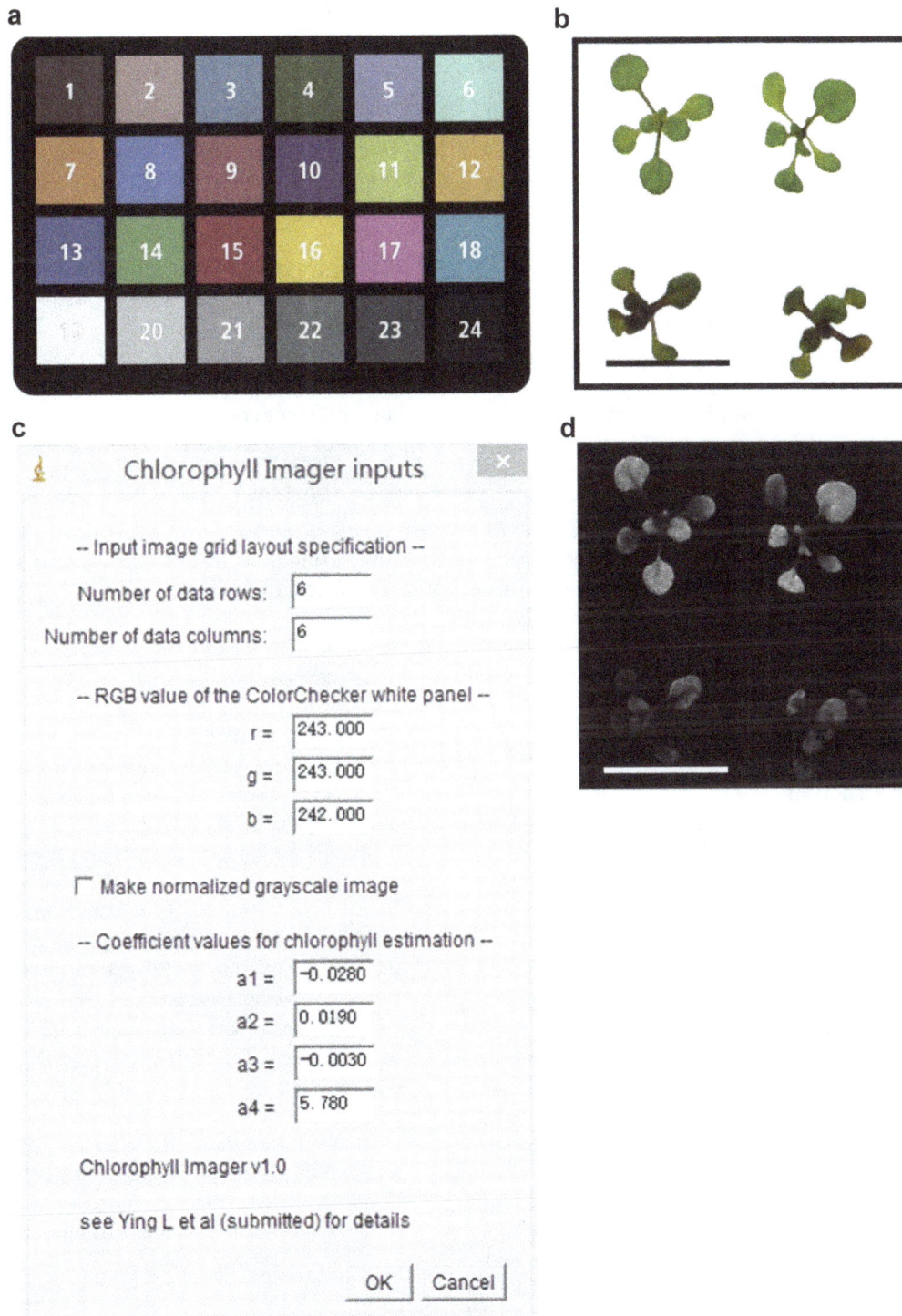

Fig. 1 Examples of images needed in converting RGB value to chlorophyll content. **a** The *picture* of the X-rite ColorChecker classic chart. **b** The original picture of the *Arabidopsis* seedlings grown on the plates under 4% glucose and 2 mM nitrogen for 12 days. **c** The dialogue box of the plugin in ImageJ. **d** The *grey scale pictures* represent the chlorophyll content, which is calculated by the equation estimate the chlorophyll content

had saved in the ImageJ 'plugins folder. A dialog box will appear asking for the number of rows and columns (Fig. 1c). Enter these values or use the default value of 6 rows × 6 columns. This step divides the images into 36 parts with 6 rows × 6 columns. The dialog box also will request the RGB values of the white panel recorded in step 7. Enter these values. By clicking "OK", you will generate a table of the chlorophyll content as ng/mm^2. The dialogue also asks if you want to make a normalized grayscale image (Fig. 1d). If so, click the box and a grey scale image will also be presented as output (see Additional file 7: S7 for an example). This greyscale image offer a spatial map of chlorophyll on a leaf; i.e. 2-dimensional information is provided.

Note 9: The default coefficient values for chlorophyll estimation are shown as a1, a2, a3, and a4 in the boxes of the plug-in menu. These values can be changed to fit other types of samples such as leaf pieces. Other coefficients can be determined as described in the "Do it yourself" section below.

9. Validation using a test sample. A cleaned image with 6 rows × 6 columns seedlings is provided as Additional file 6: S6. The RGB values of the white background for this figure are 171.666, 171.297, and 171.256, respectively. After running the plugin, you will obtain the chlorophyll content of the 36 seedlings provided in the test file. In order to confirm correct operation of the program, compare your calculated results with the values shown in Additional file 8: S8.

Chlorophyll extraction

For validation purposes, we compared our method to extracted chlorophyll. Chlorophyll content was estimated by spectrophotometry of samples prepared by 80% acetone extraction. The leaves were incubated at room temperature in a 1.5-mL tube with 1 mL 80% acetone solution for at least 24 h then clarified by centrifugation for 5 min at 15,000g. In this study, absorbance of the supernatant was measured at wavelengths 645, 646, and 663 nm (A_{645}, A_{646}, and A_{663}) with a Shimadzu UV-3000^{TM} dual-wavelength, double-beam spectrophotometer, although any spectrophotometer is suitable. Complete spectra were taken during development of this protocol in order to assure that the predominant absorbance was from chlorophyll; this is not routinely necessary. Samples having absorbance greater than 1 were diluted by half with 80% acetone and re-evaluated. Chlorophyll concentration was estimated following the Lichtenthaler's equations (A) [23] and the Arnon's equations (B) [24] as follows:

A. Chlorophyll a (μg/mL) $= -1.93\,A_{646} + 11.93\,A_{663}$
Chlorophyll b (μg/mL) $= 20.36\,A_{646} - 5.50\,A_{663}$
Total chlorophyll (μg/mL) $= 6.43\,A_{663} + 18.43\,A_{646}$

(1)

B. Chlorophyll a (μg/mL) $= 12.7\,(A_{663}) - 2.69\,(A_{645})$
Chlorophyll b (μg/mL) $= 22.9\,(A_{645}) - 4.68\,(A_{663})$
Total chlorophyll (μg/mL) $= 20.2\,(A_{645}) + 8.02\,(A_{663})$

(2)

The area of the leaves were measured by software Image J and chlorophyll content and the total chlorophyll content per leaf area was expressed as ng/mm^2.

Fitting parameters for the function and "Do it yourself" validation for other types of chlorophyll containing samples

The default coefficient values used above are described here in the event that the user needs to modify this tool to obtain different coefficients for other types of chlorophyll samples such as leaf pieces. A least squares method was used to search for the coefficients for the exponential function equation to estimate chlorophyll contents from RGB values. If the user desires to validate this method on their own using their own images and chlorophyll samples, then follow these steps:

1. Read the RGB values of each sample. A plugin labeled 'RGB_measure.java' is provided in Additional file 4: S4.
2. Measure the chlorophyll content using chemical extraction (see "Chlorophyll extract" section). The chlorophyll content should be expressed as total chlorophyll per leaf area, ng/mm^2.
3. Combine the datasets of R, G and B values obtained from step 1 with the chlorophyll content estimated by chemical extraction from step 2 to search for the coefficient with a least-squares method. An excel file, Additional file 9: S9, labeled EXAMPLE.xlsx, provides a sample dataset of RGB values with chlorophyll content data. Replace your datasets with the original ones. This spreadsheet provides the input chlorophyll and color data used to fit the best coefficients.
4. Open Matlab and upload the appropriate 'Leastsquareequation' script provided in Matlab code (Additional file 10: S10). The provided code finds the best fit of the coefficients using the following equation

$$Chl = EXP(a1*R*r/243 + a2*G*g/243 + a3*B*b/242 + a4)$$

(3)

where R, G and B refer to the color of the plants read from the plug-in described above, and r, g and b refer to the corresponding RGB values of the white background. Before running the scripts, change the input of r, g and b to these new values.

Having the new coefficients enables you to estimate chlorophyll in other samples nondestructively.

Note 10: For Matlab, paste the provided MatLab script 'Leastsquareequation.m' (Additional file 10: S10) and the EXAMPLE.xlsx (Additional file 9: S9) into the same folder.

Note 11: Other color charts may be used but the values of white may be different than for the X-rite ColorChart. In this case, replace the values 243, 243, 242, respectively in Eq. (3) with the corresponding white background values.

Results and discussion

From RGB value to chlorophyll content

We adapted the method of Riccardi et al. [5] to *Arabidopsis* seedlings grown on culture plates and compared this method to biochemical extraction of chlorophyll. We extracted chlorophylls with acetone, measured absorbance spectra, then calculated chlorophyll content in the solvent extract using both the Lichtenthaler's [23] and Arnon's equations [24] (Eqs. 1, 2).

Samples from 12-day-old *Arabidopsis* shoots from seedlings grown under different C/N treatments on square plates as described under plant growth were used to search for the parameters for the exponential function model (see Additional file 11: S11). To assure that the images taken by different cameras are comparable, we incorporated a standard to enable comparison of published data. The X-rite ColorChecker chart (Fig. 1a) was used to make a DNG profile and to adjust the white balance. Another critical variable to account for is the light intensity and color in the room, chamber, greenhouse or field. The X-rite ColorChecker chart solves these problems and makes the assay applicable to artificial and natural light. The software Image J was used to measure average R, G and B values of individual leaves. We used the exponential function model

$$Chl_i = e^{a_1 r_i + a_2 g_i + a_3 b_i + a_4} \tag{4}$$

to estimate the chlorophyll content where the (r_i, g_i, b_i) represents the R, G or B value for each sample where i accounts for the sample index [5]. We used a biochemical extraction with Lichtenthaler's (Eq. 1) equations to measure the chlorophyll content and fitted coefficients for the equation. We took samples from 4 independent experiments (n = 234 samples; Additional file 11: S11) and determined the coefficients: a_1 through a_4 in the equation (Eq. 4) using the least squares method in the MATLAB environment (Fig. 2a, b).

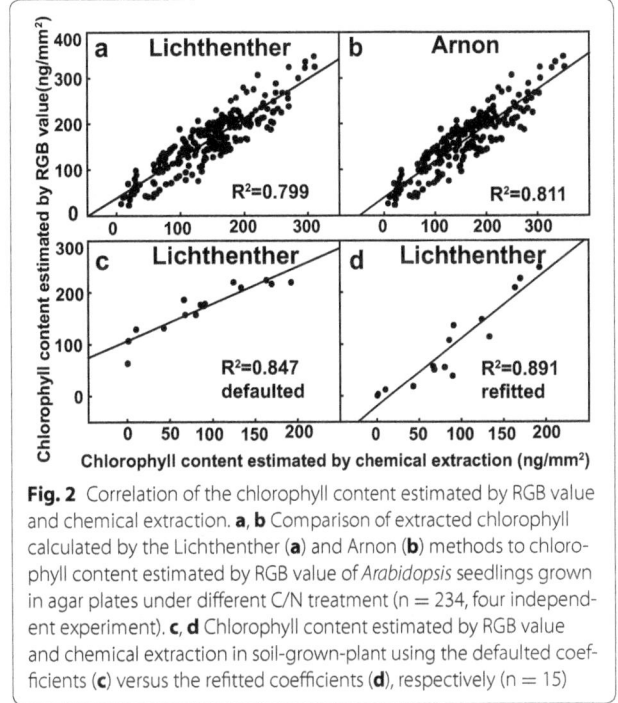

Fig. 2 Correlation of the chlorophyll content estimated by RGB value and chemical extraction. **a, b** Comparison of extracted chlorophyll calculated by the Lichthenther (**a**) and Arnon (**b**) methods to chlorophyll content estimated by RGB value of *Arabidopsis* seedlings grown in agar plates under different C/N treatment (n = 234, four independent experiment). **c, d** Chlorophyll content estimated by RGB value and chemical extraction in soil-grown-plant using the defaulted coefficients (**c**) versus the refitted coefficients (**d**), respectively (n = 15)

Robustness of the default parameters

Our method is optimized for *Arabidopsis* seedlings grown on agar plates, a common format for *Arabidopsis* researchers. Since the default parameter values were generated using chlorophyll extracted seedlings grown under one light condition, a concern is how well these values apply to other growth conditions. To test this, we compared chlorophyll estimation in two extreme growth conditions. In our lab, the thickness of 4–7 week-old leaves grown on soil under a long-day condition is almost twice as seedling leaves grown on plates under constant low light (176 vs. 100 μm, respectively). We compared chlorophyll estimation of these thicker leaf pieces using the default parameters to refitted parameters. The thicker leaf pieces were imaged as described above and the extracted chlorophyll used to refit the data to generate new parameters:

$$a_1 = -0.2032, \quad a_2 = 0.115409,$$
$$a_3 = 0.044964, \quad a_4 = 8.048844.$$

The optimized parameters increased the R^2 correlation coefficient only from 0.85 to 0.89 (cf. Fig. 2c, d). This indicates that the default values are robust with regard to different growth conditions that may affect leaf thickness. Nonetheless, when extreme accuracy is required, we recommend calculating the parameters by the "Do it yourself" fitting method described above.

Proof of utility: chlorophyll content of *Arabidopsis* seedlings grown under different C/N ratios

In order to determine how plants respond to different C/N ratios, we tested six different glucose concentrations (0, 1, 2, 4, 5 and 6% D-glucose) and five nitrogen concentrations (0.1, 0.3, 0.5, 2 and 6 mM KNO_3) in a matrix format. Figure 3 shows plant growth under these different C/N ratios and Table 1 shows the plants area in response to different C/N ratios. Plant area did not change in response to nitrogen under 0% glucose, although the average leaf area changed. As quantitated in Additional file 12: S12A, plant area at 0% glucose varied greatly but was statistically unchanged. When the medium contains glucose, plant area increased with increasing nitrogen. The optimal glucose concentration at the highest nitrogen concentration at 3 mM was 1–2% (Additional file 12: S12 panel A).

The chlorophyll content correlated with the nitrogen concentration in the presence of carbon supply. We estimated the chlorophyll content at different C/N ratio. As

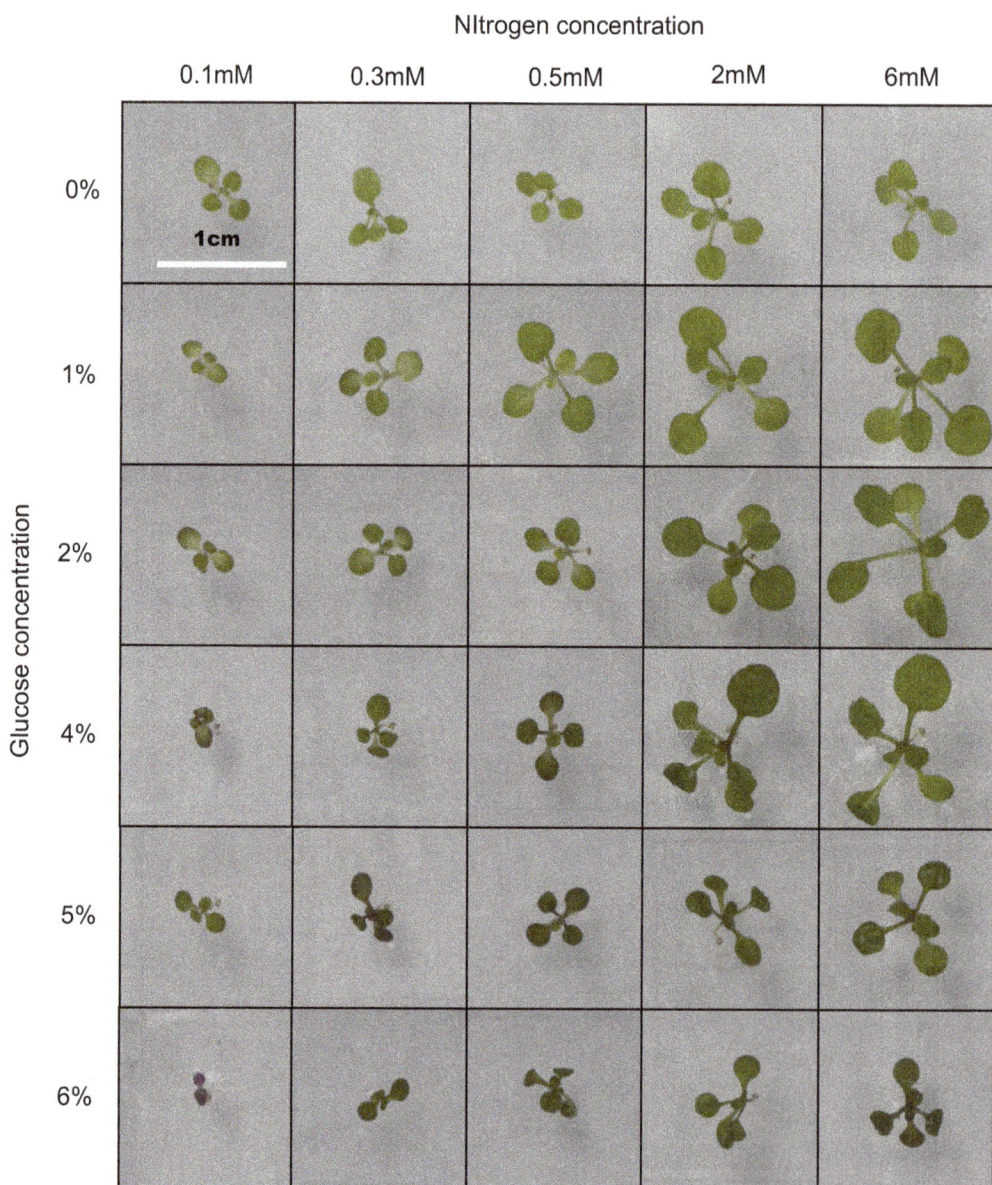

Fig. 3 Growth of the wild type *Arabidopsis* in response to C/N ratios. The image of the plate-grown plants under different C/N ratios, including six different glucose concentration (0, 1, 2, 4, 5, and 6%) and five nitrogen (0.1, 0.3, 0.5, 2 and 6 mM KNO_3) concentrations

Table 1 Plant size (mm^2) in response to glucose and nitrogen treatment

Glucose (w/v) (%)	Nitrogen concentration (mM)				
	0.1	0.3	0.5	2	6
0	8.94 a A	7.00 a A	13.59 a B	16.70 a DE	8.62 a D
1	5.93 d C	16.29 c AB	23.32 bc A	30.58 b BC	59.27 a A
2	6.74 d B	13.28 cd AB	21.20 c A	45.80 b A	55.73 a A
4	6.03 b B	9.02 b B	13.13 b B	31.85 a B	42.36 a B
5	6.75 b B	9.70 b AB	9.26 b B	22.32 a CD	23.60 a C
6	2.51 d D	6.20 cd C	8.15 bc B	11.59 b E	15.87 a CD

Growth condition and treatments are as described for Fig. 3. Data analysis is performed by software SAS8.0. Single factor analysis (n = 12). Capital letters represent similarity groups ($p > 0.05$) among glucose treatment and lowercase letters represent similarity groups among nitrogen treatments ($p > 0.05$)

Table 2 Chlorophyll content (ng/mm^2) in response to glucose and nitrogen treatment

Glucose (w/v) (%)	Nitrogen concentration (mM)				
	0.1	0.3	0.5	2	6
0	49.10 a B	38.27 a C	49.93 a C	53.85 a C	42.71 a D
1	31.82 d C	81.49 c AB	112.26 ab A	120.08 a B	98.58 b C
2	37.31 d BC	70.00 c B	72.02 c BC	125.64 a AB	105.37 b BC
4	49.36 c B	84.27 b AB	89.73 b AB	134.22 a AB	140.03 a A
5	71.66 c A	96.63 b A	104.63 b A	138.28 a A	133.65 a AB
6	44.74 b BC	99.69 a A	111.85 a A	131.07 a AB	111.33 a ABC

Data analysis is performed by software SAS8.0. Single factor analysis (n = 12). Different capital letters indicated significant differences among six glucose treatment ($p < 0.05$) and different lowercase letters indicated significant differences among five nitrogen treatment ($p < 0.05$)

shown in Table 2, the chlorophyll content of the Col-0 seedlings grown on 0% glucose did not change with increasing nitrogen in the medium. All the plates were grown under continuous dim light (35–50 µEm^{-2} s^{-1}) at 23 °C, which decreases the photosynthesis. However, even a slight amount of glucose dramatically changed this relationship. For example, in the presence of 1% glucose, the chlorophyll content increased in response to the nitrogen concentration, and slightly decreased when the nitrogen concentration was raised further. This indicates that the chlorophyll content of the leaves is slightly influenced by nitrogen concentration but highly influenced by the carbon availability. Also, the chlorophyll content increased as the glucose concentration increased; the seedlings grown with 4% or 5% D-glucose had dark green leaves. Based on plant area, optimal growth was at 1 and 2% glucose with 6 mM nitrogen. This condition did not produce the highest chlorophyll content because the seedlings were stressed. The highest concentration of 6% glucose, with 0.1 mM nitrogen, stressed seedlings further, as was apparent by the red color of leaves caused by excessive anthocyanin pigments.

Function of G protein signaling pathway in C/N sensing

In order to analyze whether G protein signaling pathway is important for C/N sensing/responsiveness, the null mutants of the G protein under various C/N condition were assayed. The leaf area differences between the mutants of *rgs1-2*, *gpa1-3* and *agb1-2* is not uniform under different C/N conditions (Additional file 13: S13). For example, *agb1-2* mutants grown under limited

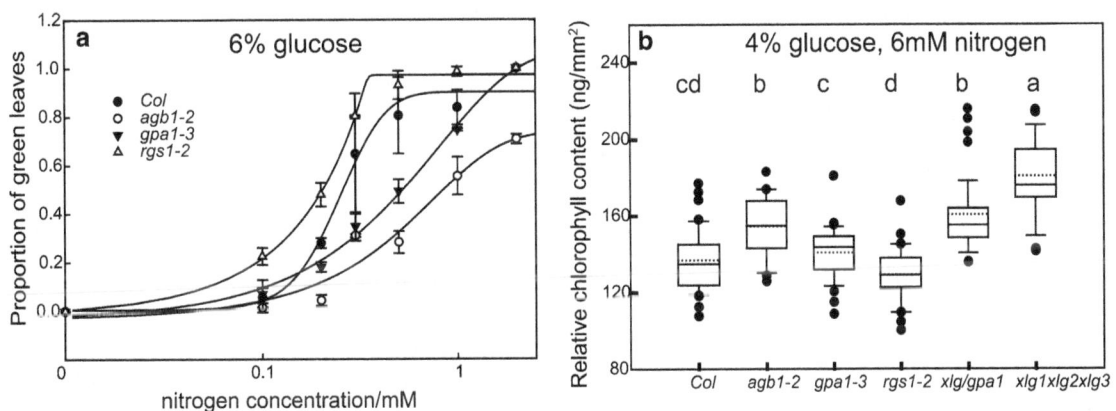

Fig. 4 The regulatory pathway of G protein signaling in the C/N sensing evaluated by RGB value. **a** Proportion of green leaves of the G protein mutants in response to nitrogen under 6% glucose. To distinguish from green and not green leaves, a threshold of was established (>15 ng/mm^2 = *green*) Experiments were repeated twice with 48 seedlings each. The curves were created through global curve fitting (sigmoid equation with four parameters) with SigmaPlot 12.5. **b** The chlorophyll content estimated by RGB value under moderate C/N stress (4% glucose and 6 mM nitrate, n = 40). For the box plot, *solid line* indicates the median and the *dotted line* indicates the mean value. *Different lowercase letters* indicate the significant differences among six genotypes ($p < 0.05$)

nitrogen (0.1 mM) are slightly larger than the wild type under 1 and 4% glucose but no difference under 2% glucose; when the glucose increased to 5 and 6%, the plant size of *agb1-2* is smaller than wild type.

We estimated the chlorophyll content in the G protein mutants. The optimal concentration for the growth of *Arabidopsis* grown on agar is 1–2% glucose (Additional file 12: S12B). The G protein complex is involve in glucose sensing [25] and AtRGS1 is a component of a glucose sensor [11, 12, 18]. Previous studies showed that the *rgs1-2* null mutant is tolerant to high glucose levels [7], however that report used a semi-quantitative method, the so-called "green-seedling" assay. In order to compare the present quantitative results to the published semi-quantitative results, we established a threshold value of 15 ng/mm^2 to distinguish yellow seedlings from green seedlings (Additional file 12: S12C; the threshold value was from inner fence of *agb1-2* mutants' box plot). Taking *rgs1-2* mutants as an example (Fig. 4a), the proportion of green leaves was fourfold greater than wild type when the nitrogen concentration was 0.1 mM (0.22 and 0.05 for *rgs1-2* and Col, respectively). When the nitrogen concentration was increased to 1 mM, the proportion of green leaves increased to 0.98 while the wild type was 0.83. Consistent with the results of Chen and Jones [7], the *rgs1-2* mutants showed more than 90% green seedlings versus wild type seedlings which scored less than 40%. In addition, as previously observed, the *agb1-2* mutants contain less chlorophyll than wild type at high glucose and low nitrogen growth conditions.

We also examined the chlorophyll content of the G protein mutants under moderate glucose and high nitrogen (4% glucose, 6 mM nitrogen) stress. Interestingly, as shown in Fig. 4b, the chlorophyll content of the xlg and agb1 mutants were significantly higher compared to the wild type under slight stress, whereas the *rgs1-2* mutant was not significantly different from wild type.

Conclusion

This method offer a non-destructive, sensitive, and quantitative way to estimate the chlorophyll content of *Arabidopsis* small seedlings grown on agar plates, and it is an effective way to evaluate the growth condition of the plants. This method provides a quantitative alternative to the qualitative 'green' versus not 'green' seedling assay [26, 27]. The use of the X-rite Color-Checker chart eliminates differences between cameras and environment, however, caution should be applied in comparing images from highly different environments. Although this method is dependent on the area of the seedlings, rather than the volume, it still

provides a good estimate of the relative chlorophyll content for the *Arabidopsis*.

Additional files

Additional file 1: S1. Chlorophyll_imager.java. This is the .java compiler for creating the .class file plug-in for imageJ to be used for the chlorophyll reads.

Additional file 2: S2. Chlorophyll_imager.class. This is the plug-in for imageJ to be used for the chlorophyll reads.

Additional file 3: S3. RGB_Measure.java This is the .java compiler for creating the .class file plug-in for imageJ to be used for the the RGB reads.

Additional file 4: S4. RGB_Measure.class. This is the plug-in for imageJ used for the RGB reads.

Additional file 5: S5. The chlorophyll content variation due to the position of the subject.

Additional file 6: S6. Cleanbackground.jpg. An example of the cleaned background JPG image.

Additional file 7: S7. Greyscaleimage.jpg. The output greyscale images created by imageJ indicated the chlorophyll content.

Additional file 8: S8. The chlorophyll content estimated by RGB value for Additional file 5: S5.

Additional file 9: S9. EXAMPLE.xlsx. An example of the data set format for the least square equation.

Additional file 10: S10. Leastsquareequation.m Script written for Matlab to apply the least square equation.

Additional file 11: S11. Raw data for fitting. The original data used for the chlorophyll content calculation with chemical extraction and RGB value, including 4 independent experiment data of the plates grown seedlings and 1 experiment for the plants grown in the soil.

Additional file 12: S12. The effect of different C/N ratios on *Arabidopsis* seedlings growth and chlorophyll content. **a** Box plot shows plant sizes in response to C/N ratio. Solid line indicates the median and the dotted line indicates the mean value. **b** The plant leave area of 1 mM nitrogen under different glucose concentrations. **c** The chlorophyll content of G protein mutants under 0.2 mM nitrogen and 6% glucose.

Additional file 13: S13. The effect of different C/N ratios on G protein mutants. Growth condition and treatments are as described for Fig. 3. Data analysis was performed by software SAS8.0. Single factor analysis (n = 48). Different capital letters indicate the similarity groups among four genotypes ($p < 0.05$) and lowercase letters indicated the similarity groups among five nitrogen treatments ($p < 0.05$).

Abbreviations
RGB: red, green and blue channel of the color; r, g, b: the values of the red, green and blue channels; r', g', b': the correction value of the red, green and blue channels; C/N: carbon to nitrogen ratio; MS medium: Murashige and Skoog medium; MES: 2-(N-morpholino) ethanesulfonic acid.

Authors' contributions
YL designed experiments, collected the data, prepared the figures, and wrote the manuscript. DU designed the experiments and edited the manuscript. KLL designed and wrote the Matlab code and fitted the model. TH designed and wrote the java code for the plugin. AMJ and YG edited the manuscript. All authors read and approved the final manuscript.

Author details
[1] Department of Biology, The University of North Carolina at Chapel Hill, Coker Hall, CB#3280, Chapel Hill, NC 27599-3280, USA. [2] College of Natural Resources and Environment, Northwest A&F University, Yangling 712100, Shaanxi, China. [3] Department of Pharmacology, University of North Carolina at Chapel Hill, Chapel Hill, NC 27599-3280, USA.

Acknowledgements

YL thanks the Chinese Scholarship Council (CSC) to offer the scholarship to do research in University of North Carolina for 2 years. We thank Wenhu Li and Xiaoyu Zhao for beta testing and Amanda Lohmann for Java programming assistance.

Competing interests

The authors declare that they have no competing interests.

Funding

This work was supported by a Grant DE-FG02-05er15671 to A.M.J. from the Division of Chemical Sciences, Geosciences, and Biosciences, Office of Basic Energy Sciences of the US Department of Energy.

References

1. Steele MR, Gitelson AA, Rundquist DC. A comparison of two techniques for nondestructive measurement of chlorophyll content in grapevine leaves. Agron J. 2008;100:779.
2. Li L, Zhang Q, Huang D. A review of imaging techniques for plant phenotyping. Sensors (Basel). 2014;14:20078–111.
3. Sass L, Majer P, Hideg É. Leaf hue measurements: a high-throughput screening of chlorophyll content. Methods Mol Biol. 2010;918:61–71.
4. Wang Y, et al. Estimating rice chlorophyll content and leaf nitrogen concentration with a digital still color camera under natural light. Plant Methods. 2014;10:1–11.
5. Riccardi M, et al. Non-destructive evaluation of chlorophyll content in quinoa and amaranth leaves by simple and multiple regression analysis of RGB image components. Photosynth Res. 2014;120:263–72.
6. Lu Y, et al. ABI1 regulates carbon/nitrogen-nutrient signal transduction independent of ABA biosynthesis and canonical ABA signalling pathways in *Arabidopsis*. J Exp Bot. 2015;66:2763–71.
7. Chen J-G, Jones AM. AtRGS1 Function in *Arabidopsis thaliana*. Methods Enzymol. 2004;389:338–50.
8. Kang J, Turano FJ. The putative glutamate receptor 1.1 (AtGLR1.1) functions as a regulator of carbon and nitrogen metabolism in *Arabidopsis thaliana*. Proc Natl Acad Sci USA. 2003;100:6872–7.
9. Moore B, et al. Role of the *Arabidopsis* glucose sensor HXK1 in nutrient, light, and hormonal signaling. Science. 2003;300:332–6.
10. Sun H, et al. Heterotrimeric G proteins regulate nitrogen-use efficiency in rice. Nat Genet. 2014;46:652–6.
11. Fu Y, et al. Reciprocal encoding of signal intensity and duration in a glucose-sensing circuit. Cell. 2014;156:1084–95.
12. Grigston JC, et al. D-Glucose sensing by a plasma membrane regulator of G signaling protein, AtRGS1. FEBS Lett. 2008;582:3577–84.
13. Ding L, Pandey S, Assmann SM. *Arabidopsis* extra-large G proteins (XLGs) regulate root morphogenesis. Plant J. 2008;53:248–63.
14. Pandey S, et al. G-protein complex mutants are hypersensitive to abscisic acid regulation of germination and postgermination development. Plant Physiol. 2006;141:243–56.
15. Chakravorty D, et al. An atypical heterotrimeric G-protein gamma-subunit is involved in guard cell K(+)-channel regulation and morphological development in *Arabidopsis thaliana*. Plant J. 2011;67:840–51.
16. Booker KS, et al. Glucose attenuation of auxin-mediated bimodality in lateral root formation is partly coupled by the heterotrimeric G protein complex. PLoS One. 2010;5:e12833.
17. Johnston CA, et al. GTPase acceleration as the rate-limiting step in *Arabidopsis* G protein-coupled sugar signaling. Proc Natl Acad Sci USA. 2007;104:17317–22.
18. Urano D, et al. Endocytosis of the seven-transmembrane RGS1 protein activates G-protein-coupled signalling in *Arabidopsis*. Nat Cell Biol. 2012;14:1079–88.
19. Ullah H. The beta-subunit of the *Arabidopsis* G protein negatively regulates auxin-induced cell division and affects multiple developmental processes. Plant Cell Online. 2003;15:393–409.
20. Chen J-G, et al. A seven-transmembrane RGS protein that modulates plant cell proliferation. Science. 2003;301:1728–31.
21. Jones AM, Ecker JR, Chen JG. A reevaluation of the role of the heterotrimeric G protein in coupling light responses in *Arabidopsis*. Plant Physiol. 2003;131:1623–7.
22. Urano D, et al. Saltational evolution of the heterotrimeric G protein signaling mechanisms in the plant kingdom. Sci Signal. 2016;9:ra93.
23. Lichtenthaler HK, Wellburn AR. Determination of total carotenoids and chlorophylls a and b of leaf extracts in different solvents. Biochem Soc Trans. 1983;11:591–2.
24. Arnon DI. Copper enzymes in isolated chloroplasts, polyphenoloxidase in beta vulgaris. Plant Physiol. 1949;24:1–15.
25. Chakravorty D, et al. Extra-large G proteins expand the repertoire of subunits in *Arabidopsis* heterotrimeric G protein signaling. Plant Physiol. 2015;169:512–29.
26. Lia P, et al. Fructose sensitivity is suppressed in *Arabidopsis* by the transcription factor ANAC089 lacking the membrane-bound domain. PNAS. 2011;108:3436–41.
27. Huang JP, et al. Cooperative control between AtRGS1 and AtHXK1 in a WD40-repeat protein pathway in *Arabidopsis thaliana*. Front Plant Sci. 2015;6:851.

4

A method to estimate plant density and plant spacing heterogeneity: application to wheat crops

Shouyang Liu[1*], Fred Baret[1], Denis Allard[2], Xiuliang Jin[1], Bruno Andrieu[3], Philippe Burger[4], Matthieu Hemmerlé[5] and Alexis Comar[5]

Abstract

Background: Plant density and its non-uniformity drive the competition among plants as well as with weeds. They need thus to be estimated with small uncertainties accuracy. An optimal sampling method is proposed to estimate the plant density in wheat crops from plant counting and reach a given precision.

Results: Three experiments were conducted in 2014 resulting in 14 plots across varied sowing density, cultivars and environmental conditions. The coordinates of the plants along the row were measured over RGB high resolution images taken from the ground level. Results show that the spacing between consecutive plants along the row direction are independent and follow a gamma distribution under the varied conditions experienced. A gamma count model was then derived to define the optimal sample size required to estimate plant density for a given precision. Results suggest that measuring the length of segments containing 90 plants will achieve a precision better than 10%, independently from the plant density. This approach appears more efficient than the usual method based on fixed length segments where the number of plants are counted: the optimal length for a given precision on the density estimation will depend on the actual plant density. The gamma count model parameters may also be used to quantify the heterogeneity of plant spacing along the row by exploiting the variability between replicated samples. Results show that to achieve a 10% precision on the estimates of the 2 parameters of the gamma model, 200 elementary samples corresponding to the spacing between 2 consecutive plants should be measured.

Conclusions: This method provides an optimal sampling strategy to estimate the plant density and quantify the plant spacing heterogeneity along the row.

Keywords: Wheat, Gamma-count model, Density, RGB imagery, Sampling strategy, Plant spacing heterogeneity

Background

Plant density at emergence is governed by the sowing density and the emergence rate. For a given plant density, the uniformity of plant distribution at emergence may significantly impact the competition among plants as well as with weeds [1, 2]. Plant density and uniformity is therefore a key factor explaining production, although a number of species are able to compensate for low plant densities by a comparatively significant development of individual plants during the growth cycle. For wheat crops which are largely cultivated over the globe, tillering is one of the main mechanisms used by the plant to adapt its development to the available resources that are partly controlled by the number of tillers per unit area. The tillering coefficient therefore appears as an important trait to be measured. It is usually computed as the ratio of the number of tillers per unit area divided by the plant density [3]. Plant density is therefore one of the first variables measured commonly in most agronomical trials.

Crops are generally sown in rows approximately evenly spaced by seedling devices. Precision seedling systems mostly used for crops with plants spaced on the row by

*Correspondence: Shouyang.Liu@inra.fr
[1] INRA, UMR-EMMAH, UMT-CAPTE, UAPV, 228 Route de l'aérodrome CS 40509, 84914 Avignon, France
Full list of author information is available at the end of the article

more than few centimeters (e.g. maize, sunflower or soybean) distribute seeds relatively evenly along the row. Conversely, for most crops with short distances among plants on the row, e.g. wheat, barley or canola, seeds are distributed non-evenly along the row. This can be attributed both to the mechanisms that free, at a variable frequency, the seed from the seed tank, and the trajectory of the seed that may also vary in the pipe that drives it from the seed tank to the soil. Further, once reaching the soil, the seed may also move with the soil displaced by the sowing elements penetrating the soil surface. Finally, some seeds may abort or some young plants may die because of pests or too extreme local environmental conditions (excess or deficit of moisture, low temperature etc.). The population density and its non-uniformity are therefore recognized as key traits of interests to characterize the canopy at the emergence stage. However, very little work documents the plant distribution pattern along the row, which is partly explained by the lack of dedicated device for accurate plant position measurement [4]. Electromagnetic digitizers are very low throughput and not well adapted to such field measurements [5]. Alternatively, algorithms have been developed to measure the inter-plant spacing along the row for maize crops from top-view RGB (Red Green Blue) images [6, 7]. Improvements were then proposed by using three dimensional sensors [8–10]. However, these algorithms were only validated on maize crops that show relatively simple plant architecture with generally fixed inter-plant spacing along the row.

Manual field counting in wheat crops is still extensively employed as the reference method. Measurements of plant population density should be completed when the majority of plants have just emerged and before the beginning of tillering when individual plants start to be difficult to be identified. Plants are counted over elementary samples corresponding either to a quadrat or to a segment [11]. The elementary samples need to be replicated in the plot to provide a more representative value [12]. For wheat crops, [3] suggested that at least a total of 3 m of rows (0.5 m segment length repeated 6 times) should be counted, while [13] proposed to sample a total of 6 m (segments made of 2 consecutive rows by one meter repeated 3 times in the plot). [14] proposed to repeat at least 4 times the counting in 0.25 m^2

quadrats corresponding roughly to a total of 6.7 m length of rows (assuming the rows are spaced by 0.15 m). In this case, quadrats may be considered as a set of consecutive row segments with the same length when the quadrat is oriented parallel to the row direction or with variable lengths when the quadrat is oriented differently. Although these recommendations are simple and easy to apply, they may not correspond to an optimal sampling designed to target a given precision level. They may either provide low precision if under sampled or correspond to a waste of human resources in the opposite case.

The sample size required to reach a given precision of the plant density will depend on the population density and the heterogeneity of plant positions along the row that may be described by the distribution of the distances between consecutive plants. This distribution is more likely to be skewed, which could be described by an exponential distribution or a more general one such as the Weibull or the gamma distributions. Fitting such random distribution functions provides not only access to the plant density at the canopy level, but also to its local variation that may impact the development of neighboring plants as discussed earlier.

The objective of this study is to propose an optimal sampling method for plant density estimation and to quantify the heterogeneity of plant spacing along the row. For this purpose, a model is first developed to describe the distribution of the plants along the row. The model is then calibrated over a number of ground experiments. Further, the model is used to compare several plant counting strategies and to evaluate the optimal sampling size to reach a given precision. Finally, the model was also exploited to design a method for quantifying the non-uniformity of plant distribution.

Methods

Field experiment

Three sites in France were selected in 2014 (Table 1): Avignon, Toulouse and Grignon. A mechanical seed drill was used in the three sites, which represents the standard practice for wheat crops. In Grignon, five plots were sampled, corresponding to different cultivars with a single sowing density. In Toulouse, five sowing densities were sampled with the same "Apache" cultivar. In Avignon, four sowing densities were sampled also with the same

Table 1 The experimental design in 2014 over the three sites

Sites	Latitude	Longitude	Cultivar	Density (seeds m^{-2})
Toulouse	43.5°N	1.5°E	Apache	100, 200, 300, 400, 600
Grignon	48.8°N	1.9°E	Premio; Attlass; Flamenko; Midas; Koréli	150
Avignon	43.9°N	4.8°E	Apache	100, 200, 300, 400

"Apache" cultivar. All measurements were taken at around 1.5 Haun stage [15], when most plants already emerged and were easy to identify visually. This stage is reached approximately 10–14 days after the germination for wheat in France [3]. A total of 14 plots are thus available over the 3 sites showing contrasted conditions in terms of soil, climate, cultivars, sowing density and sowing machine, with however a fixed row spacing of 17.5 cm. All the plots were at least 10 m length and 2 m width.

Image processing

A Sigma SD14 RGB camera with a resolution of 4608 by 3072 pixels was installed on a light moving platform (Fig. 1). The camera was oriented at 45° inclination perpendicular to the row direction and was focused on the central row from a distance of about 1.5 m (Fig. 1). The 50 mm focal length allowed to sample about 0.9 m of the row with a resolution at the ground level close to 0.2 mm. Images were acquired along the row with at least 30% overlap to allow stitching. A series of 20 pictures was collected that correspond to three to five rows over about 5 m length. The images were stitched using AutoStitch

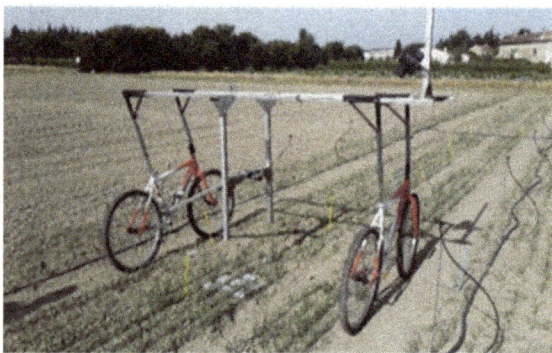

Fig. 1 The moving platform used to take the images in the field in 2014

(http://matthewalunbrown.com/autostitch/autostitch.html) [16]. For each site, one picture was taken over a reference chessboard put on the soil surface to calibrate the image: the transformation matrix derived from the chessboard image was applied to all the images acquired within the same site. It enables to remove perspective effects and to scale the pixels projected on the soil surface. The image correction and processing afterwards was conducted using MATLAB R2016a (code available on request). Coordinates of the plants correspond to the intersection between the bottom of the plant and the soil surface (Fig. 2). They were interactively extracted from the photos displayed on the screen. For each of the 14 plots, the coordinates of at least 150 successive plants from the same row were measured along (X axis) and across (Y axis) of the row. It took between 15 to 30 min to extract the plant coordinates, depending on the density. The precision on the coordinates values along the row is around 1.5 mm as estimated by independent replicates of the process over the same images. Some slightly larger deviations are observed marginally in case of occlusions by stones or straw in the field.

The coordinates x_n of plant n (noted Plant$_n$) along the row axis allow to compute the spacing $\Delta x_n = (x_n - x_{n-1})$ between Plant$_n$ and Plant$_{n-1}$. The actual plant density expressed in plants per square meter horizontal ground (plants m^{-2}) was computed simply as the number of plants counted on the segments, divided by the product of the length of the segments and the row spacing.

Development and calibration of the plant distribution model

Distribution of plant spacing

The autocorrelation technique was used to explore the spatial dependency of spacing between successive plants: the linear correlation between Δx_{n-m} and Δx_n where m is the lag is evaluated. Results illustrated in Fig. 3 over the

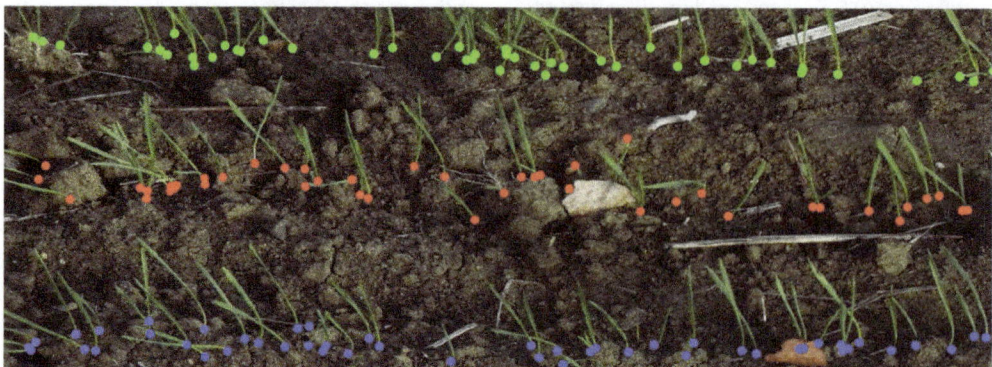

Fig. 2 Extraction of plants' coordinates from the image

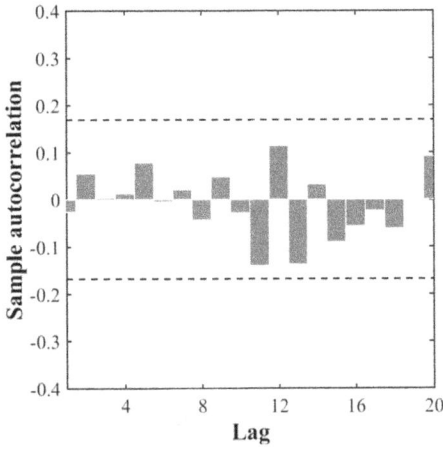

Fig. 3 The autocorrelation of the spacing among plants along the row direction illustrated with sowing density of 300 seeds m^{-2} observed over the Toulouse experiment. The lag is expressed as the number of plant spacing between 2 plants along the row direction (X axis). Lags 1–20 are presented. The *upper* and *lower horizontal line* represent the 95% confidence interval around 0

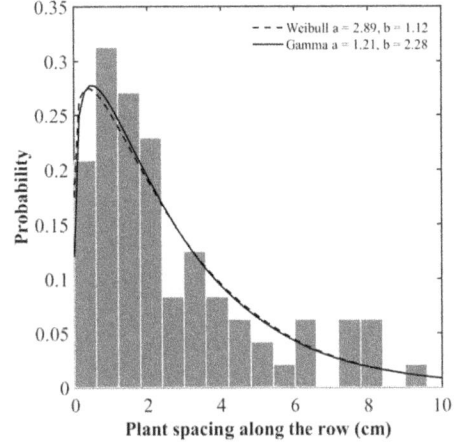

Fig. 4 Empirical histogram of the spacing along the row (*gray bars*). The *solid* (respectively *dashed*) line represents the fitted gamma (resp. Weibull) distribution. Case of the sowing density 300 seeds m^{-2} observed over the Toulouse experiment. *a* and *b* represent the shape and scale parameters respectively

Toulouse site show that the autocorrelation coefficient of inter-plant distance is not significant at 95% confidence interval. The same is observed over the other 13 plots acquired. It is therefore concluded that the positions among plants along the row direction are independent: each observation Δx could be considered as one independent realization of the random variable ΔX.

The distribution of the plant spacing is positively right-skewed (Fig. 4). A simple exponential distribution with only one scale parameter was first tentatively fitted to the data using a maximum likelihood method. However, the Chi square test at the 5% significance level showed that the majority of the 14 plots do not follow this simple exponential distribution law. Weibull and gamma distributions are both a generalization of the exponential distribution requiring an extra shape parameter. Results show that Weibull and gamma distributions describe well (Chi square test at the 5% significance level successful) the empirical distributions over the 14 plots (Fig. 4; Table 2). However, the gamma distribution will be preferred since it provides generally higher p value of Chi square test (Table 2) [17]. Besides, the tail of the Weibull distribution tends toward zero less rapidly than that of the gamma distribution: Weibull may show few samples with very large values [18], increasing the risk of overestimation for the larger plant spacing. The gamma distribution was therefore used in the following and writes [19]:

$$f(\Delta x | a, b) = \frac{1}{b^a \Gamma(a)} \Delta x^{a-1} e^{\frac{-\Delta x}{b}} \quad \Delta x, a, b \in R^+ \quad (1)$$

where a and b represent the shape and scale parameters respectively. The expectancy $E(\Delta X)$ and variance $Var(\Delta X)$ are simple expressions of the two parameters:

$$E(\Delta X) = a \cdot b \quad (2)$$

$$Var(\Delta X) = a \cdot b^2 \quad (3)$$

As a consequence, the coefficient of variation $CV(\Delta X) = \frac{\sqrt{Var(\Delta X)}}{E(\Delta X)}$ is a simple function of the shape parameter:

$$CV(\Delta X) = 1/\sqrt{a} \quad (4)$$

Modeling the distribution of the number of plants per row segment

The plant density evaluated over row segments needs to account for the uncertainties in row spacing. The variability of the row spacing is of the order of 10 mm as reported by [20] which corresponds to CV = 6% using a typical row spacing of 175 mm. For the sake of simplicity, the variability of row spacing will be neglected since it is likely to be small. Further, it is relatively easy to get precise row spacing measurements for each segment and to actually account for the actual row spacing values. Considering a given row spacing, the plant density depends only on the number of plants per unit linear row length. Estimating the number of plants within a row segment is a count data problem analogous to the estimation of the number of events during a specific time interval [19, 21]. Counts are common random variables that are assumed to be non-negative integer or continuous values

Table 2 Parameters of the fitted distributions

Sites	Sowing density (seeds m^{-2})	Cultivar	Gamma			Weibull		
			a	b	p value of Chi square test	a	b	p value of Chi square test
Avignon	100	Apache	1.14	6.38	0.27	7.44	1.07	0.29
	200	Apache	1.25	4.04	0.62	5.29	1.13	0.05
	300	Apache	0.99	2.53	0.38	2.51	1.00	0.56
	400	Apache	0.96	1.50	0.22	1.39	0.94	0.57
Toulouse	100	Apache	1.07	5.01	0.12	5.32	0.99	0.10
	200	Apache	1.39	1.95	0.17	2.86	1.15	0.12
	300	Apache	1.21	2.28	0.94	2.89	1.12	0.94
	400	Apache	1.24	1.37	0.51	1.76	1.10	0.40
	600	Apache	1.16	0.96	0.37	1.14	1.09	0.21
Grignon	150	Premio	1.12	3.37	0.70	3.85	1.06	0.68
	150	Attlass	1.13	2.48	0.69	2.87	1.05	0.67
	150	Flamenko	1.11	3.3	0.92	3.75	1.05	0.92
	150	Midas	1.24	3.03	0.21	3.92	1.12	0.24
	150	Koréli	1.15	2.89	0.24	3.48	1.15	0.18

representing the number of times an event occurs within a given spatial or temporal domain [22]. The gamma-count model suits well our problem with intervals independently following a gamma distribution as in our case. The probability, $P\{N_l = n\}$, to get n plants over a segment of length l, writes (Eqs. 5–8 were cited from [19, 21]):

$$P\{N_l = n\}$$

$$= \begin{cases} 1 - IG\left(a, \frac{l}{b}\right) & \text{for } n = 0 \\ IG\left(a \cdot n, \frac{l}{b}\right) - IG\left(a \cdot n + a, \frac{l}{b}\right) & \text{for } n = 1, 2, \ldots \end{cases}$$

(5)

where N_l is the number of plants over the segment of length l, and $IG\left(a \cdot n, \frac{l}{b}\right)$ is the incomplete gamma function:

$$IG\left(a \cdot n, \frac{l}{b}\right) = \frac{1}{\Gamma(a \cdot n)} \int_0^{l/b} t^{a \cdot n - 1} e^{-t} dt$$

(6)

where Γ is the gamma Euler function. The expectation and variance of the number of plants over a segment of length l is given by:

$$E(N_l) = \sum_{n=1}^{\infty} IG\left(a \cdot n, \frac{l}{b}\right)$$

(7)

$$Var(N_l) = \sum_{n=1}^{\infty} (2n - 1)IG\left(a \cdot n, \frac{l}{b}\right) - \left[\sum_{n=1}^{\infty} IG\left(a \cdot n, \frac{l}{b}\right)\right]^2$$

(8)

Finally, the expectation and variance of the plant density, D_l, estimated over a segment of length l can be expressed by introducing the row spacing distance, r, assumed to be known:

$$E(D_l) = \frac{E(N_l)}{l \cdot r}$$

(9)

$$Var(D_l) = \frac{Var(N_l)}{(l \cdot r)^2}$$

(10)

The expectation, $E(D_l)$, converges toward the actual density of the population when $l \to \infty$.

The transformed gamma-count model allows evaluating the uncertainty of plant density estimation as a function of the sampling size. The uncertainty can be characterized by the coefficient of variation (CV) as follows:

$$CV(D_l) = \frac{\sqrt{Var(D_l)}}{E(D_l)} = \frac{\sqrt{Var(N_l)}}{E(N_l)}$$

(11)

Several combinations of values of a and b may lead to the same plant density, but with variations in their distribution along the row (Fig. 5). The fitting of parameters a and b over the 14 plots using the transformed gamma-count model (Eq. 9) shows that the shape parameter, a, varies from 0.96 to 1.39 and is quite stable. Conversely, the scale parameter, b, appears to vary widely from 0.96 to 6.38, mainly controlling the plant density (Fig. 5). Since the CV depends only on the shape parameter a (Eq. 4), it should not vary much across the 14 plots considered. This was confirmed by applying a one-way analysis of variance on the CV values of the 14 plots available (F = 1.09, P = 0.3685): no significant differences are

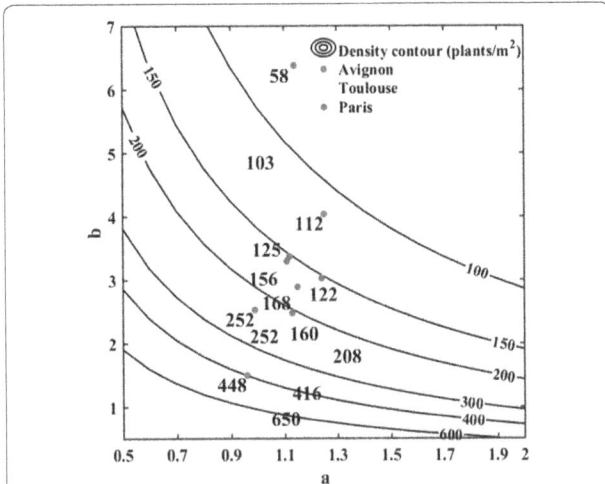

Fig. 5 Relationship between parameters a and b of the gamma-count model for a range of plant density (from Eqs. 6, 7, 9). The *lines* correspond to, 100, 150, 200, 300, 400 and 600 plants m^{-2}. The *dots'* color corresponds to the experimental sites

on the precision of the density estimates. The precision will be quantified here using the coefficient of variation (CV). The sample size can be expressed either as a given length of the segments where the (variable) number of plants should be counted, or as a (variable) length of the segment to be measured corresponding to a given number of consecutive plants. The two alternative sampling approaches will be termed FLS (Fixed Length of Segments) for the first one, and FNP (Fixed Number of Plants) for the second one.

When considering the FLS approach, the sample size is defined by the length of segment, L, where plants need to be counted. The optimal L value for a given target precision quantified by the CV will mainly depend on the current density as demonstrated in Fig. 6a: longer segments are required for the low densities. Conversely, shorter segments are needed for high values of the plant density to reach the same precision. The scale parameter, b, that controls the plant density drives therefore the optimal segment length L (Fig. 6a). Counting plants over $L = 5$ m (500 cm) provides a precision better than 10% for densities larger than 150 plants·m^{-2} for the most common conditions characterized by a shape coefficient $a > 0.9$. These figures agree well with the usual practice for plant counting as reviewed in the introduction [3, 13, 14]. Increasing the precision quantified by the CV will require longer segments L to be sampled (Fig. 7a).

When considering the FNP approach, the sample size is driven by the number, N, of consecutive plants that defines to a row segment whose length need to be

observed. This result may be partly explained by the fact that the same type of seed drill was used for all the three sites.

Results

Optimal sample size to reach a given precision for plant density estimation

The transformed gamma-count model provides a convenient way to investigate the effect of the sampling size

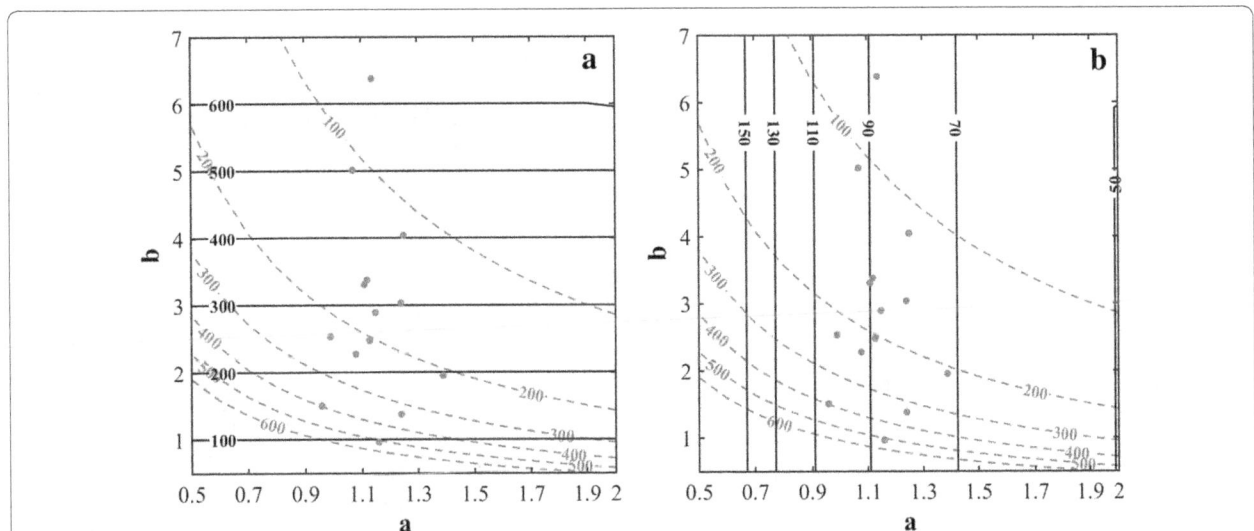

Fig. 6 a The optimal sampling size length (the *horizontal solid lines*, the length being indicated in cm) used in the FLS approach as a function of parameters a and b to get CV = 10% for the density estimation. **b** Idem as on the left but the sample is defined by the number of plants to be counted (the *vertical solid lines* with number of plants indicated) for the FNP approach. The *gray dashed lines* correspond to the actual plant density depending also on parameters a and b. The row spacing is assumed perfectly known and equal to 17.5 cm. The *gray points* represent the 14 plots measured

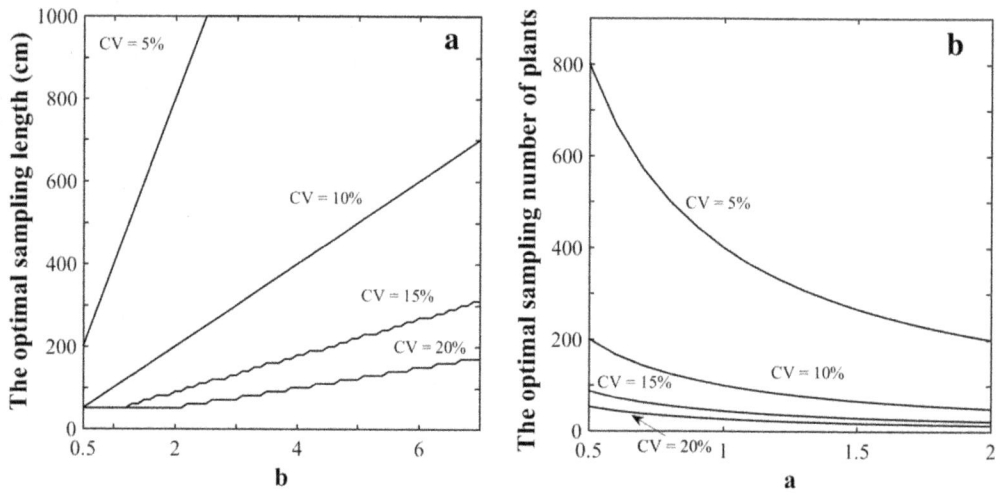

Fig. 7 The optimal sampling length for the FLS approach (**a**) and the number of plants for the FNP approach (**b**). The dominant parameter is used (the scale parameter for FLS and the shape coefficient for FNP). The precision is evaluated with the CV = 5, 10, 15 and 20%

measured. The simulations of the model (Fig. 6b) show that N is mainly independent from the plant density. For the 14 plots considered in this study, segments with $70 < N < 110$ plants should be measured to reach a precision of CV = 10%. The shape parameter a influences dominantly the sample size: more heterogeneous distribution of plants characterized by small values of the shape parameter will require more plants to be counted (Fig. 6b). To increase the precision (lower CV), more plants will also need to be counted (Fig. 7b).

The sampling approach FLS (Fixed Length of segments) is extensively used to estimate the plant density. The 600 cm segment length recommended by [13, 14] agrees well with our results (Figs. 6a, 7a) demonstrating that a precision better than 10% is ensured over large range of densities and non-uniformities. The optimal sampling length (FSL) and optimal number of plants sampled (FNP) was computed for other precision levels for a range of plant densities (Table 3). Results show that the FNP

method provides very stable values of the sampling size: it is easy to propose an optimal number of consecutive plants to count to reach a given precision. Conversely, the optimal length of the segment used in the FSL approach varies strongly with the plant density (Table 3): the FLS approach when applied with a segment length chosen a priori without knowing the plant density will result in a variable precision level.

Sampling strategy to quantify plant spacing variability on the row

The previous sections demonstrated that the scale and shape parameters could be estimated from the observed distribution of the plant spacing. However, the measurement of individual plant spacing is tedious and prone to errors as outlined earlier. The estimation of these parameters from the variability observed between small row segments containing a fixed number of plants will therefore be investigated here. This FNP approach is preferred

Table 3 Optimal sampling size for FSL and FNP over different densities (100, 150, 200, 300, 400 and 600 seeds m⁻²) and precisions (5, 10 and 15%)

Sowing density (seeds m⁻²)	Parameters		CV = 5%		CV = 10%		CV = 15%	
	a	b	FSL (cm)	FNP (Nb. Plt)	FSL (cm)	FNP (Nb. Plt)	FSL (cm)	FNP (Nb. Plt)
100	1.11	5.70	2478	363	620	90	250	39
150	1.15	3.01	1406	348	351	88	130	37
200	1.32	3.00	1398	308	350	78	130	33
300	1.10	2.41	1162	363	291	90	110	39
400	1.10	1.44	774	363	194	90	60	39
600	1.16	0.96	584	348	146	85	50	37

This was calculated using the average values of the parameters a and b of the gamma distribution derived for each density over the 14 plots available

here to the FSL one because there will be no additional uncertainties introduced by the position of the first and last plants of the segment with the corresponding start and end of the segment. These uncertainties may be significant in case of small segments in the FLS approach.

The probability distribution of a gamma distribution can be expressed as the sum of an arbitrary number of independent individual gamma distributions [23]. This property allows to compute the distribution of a segment of length L_n corresponding to n plant spacing between $(n + 1)$ consecutive plants with $L_n = \sum_{i=1}^{n} \Delta x_i$, as a gamma distribution with $n \cdot a$ as shape parameter and the same scale parameter b as the one describing the distribution of ΔX.

$$L_n \sim \text{Gamma}(n \cdot a, b) \qquad (12)$$

The parameters a and b will therefore be estimated by adjusting the gamma model described in Eq. 12 for the given value of $n + 1$ consecutive plants.

The effect of the sampling size on the precision of a and b parameters estimation was further investigated. A numerical experiment based on a Monte-Carlo approach was conducted considering a standard case corresponding to the average of the 14 plots sampled in 2014 with $a = 1.10$ and $b = 2.27$. The sampling size is defined by the number of consecutive plants for the FNP approach considered here and by the number of replicates. For each sampling size 300 samples were generated by randomly drawing in the gamma distribution (Eq. 12) and parameters a and b were estimated. The standard deviation between the 300 estimates of a and b parameters was finally used to compute the corresponding CV. This process was applied to a number of replicates varying between 20 to 300 by steps of 10 and a number of plants per segment varying between 2 (i.e. spacing between two consecutive plants) to 250 within 12 steps. This allows describing the variation of the coefficient of estimated values of parameters a and b as a function of the number of replicates and the number of plants (Fig. 8).

Results show that the sensitivity of the CV of estimates of parameters a and b are very similar (Fig. 8). The sensitivity of parameters a and b is dominated by the number of replicates: very little variation of CV is observed when the number of plants per segment varies (Fig. 8). Parameters a and b require about 200 replicates independently from the number of plants per segment. It seems therefore more interesting to make very small segments to decrease the total number of plants to count.

Additional investigations not shown here for the sake of brevity, confirmed the independency of the number of replicates to the number of plants per segment when parameters a and b are varying. Further, the number of replicates need to be increased as expected when the

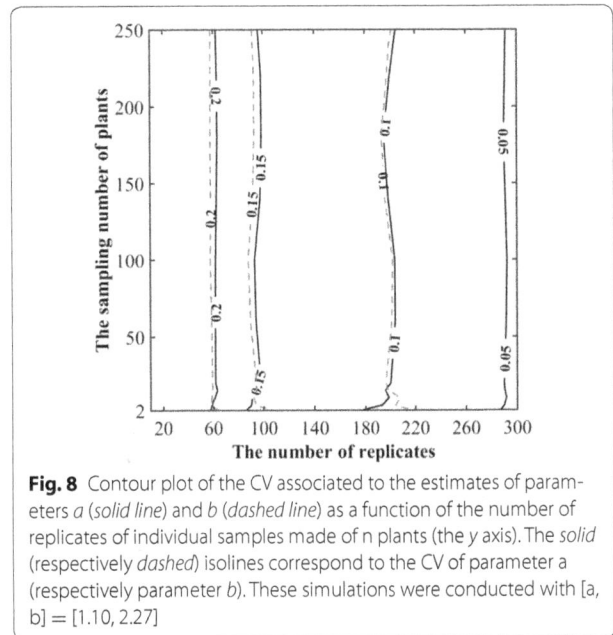

Fig. 8 Contour plot of the CV associated to the estimates of parameters a (*solid line*) and b (*dashed line*) as a function of the number of replicates of individual samples made of n plants (the y axis). The *solid* (respectively *dashed*) isolines correspond to the CV of parameter a (respectively parameter b). These simulations were conducted with [a, b] = [1.10, 2.27]

shape parameter a decreases (i.e. when the plant spacing is more variable) to keep the same precision on estimates of a and b parameters.

Discussion and conclusions

A method was proposed to estimate plant density and sowing pattern from high resolution RGB images taken from the ground. The method appears to be much more comfortable as compared with the standard outdoor methods based on plant counting in the field. Images should ideally be taken around Haun stage 1.5 for wheat crops when most plants have already emerged and tillering has not yet started. Great attention should be paid to the geometric correction in order to get accurate orthoimages where distances can be measured accurately. The processing of images here was automatic except the last step corresponding to the interactive visual extraction of the plants' coordinates in the image. However, recent work [24, 25] suggests that it will be possible to automatize this last step to get a fully high-throughput method.

The method proposed is based on the modeling of the plant distribution along the row. It was first demonstrated that the plant spacing between consecutive plants are independent which corresponds to a very useful simplifying assumption. The distribution of plant spacing was then proved to follow a gamma distribution. Although the Weibull distribution showed similar good performance, it was not selected because of the comparatively heavier tails of the distribution that may create artefacts. Further the Weibull model does not allow to simply derive the distribution law of the length of segments containing several consecutive plants [26]. The gamma model needs a scale parameter that drives mostly the intensity of the process, i.e. the plant density, and a shape parameter that governs the heterogeneity of plant spacing. This model was transformed into a count data model to investigate the optimal sampling required to get an estimate of plant density for a given precision level.

The adjustment of the gamma-count model on the measured plant spacing using a maximum likelihood method provides an estimate of the plant density (Eq. 9). The comparison to the actual plant density (Fig. 9) simply computed as the number of plants per segment divided by the area of the segments (segment length by row spacing), shows a good agreement, with RMSE ≈ 50 plants m^{-2} over the 14 plots available. The model performs better for the low density with a RMSE of 21 plants m^{-2} for density lower than 400 plants m^{-2}. These discrepancies may be mainly explained by the accuracy in the measurement of the position of individual plants (around 1–2 mm). Uncertainties on individual plant spacing will be high in relative values as compared to that associated with the measurement of the length of the segment used in the simple method to get the 'reference' plant density. Hence it is obviously even more difficult to get a good accuracy in plant spacing measurements for high density, i.e. with a small distance among plants. In addition, small deviations from the gamma-count model are still possible, although the previous results were showing very good performance.

The model proposed here concerns mainly relatively nominal sowing, i.e. when the sowing was successful on average on the row segments considered: portions of rows with no plants due to sowing problems or local damaging conditions (pests, temperature and moisture). The sowing was considered as nominal on most of the plots investigated in this study, with no obvious 'accidents'. However, it is possible to automatically identify from the images the unusual row segments with missing plants or excessive concentration of plants [25]. Rather than describing blindly the bulk plant density, it would be then preferred to get a nested sampling strategy: the unusual segments could be mapped extensively, and the plant density of nominal and unusual segments could be described separately using the optimal sampling proposed here.

This study investigated the sampling strategy to estimate the plant density with emphasis on the variability of plant spacing along the row, corresponding to the sampling error. However additional sources of error should be accounted for including measurement biases, uncertainties in row spacing or non-randomness in the sample selection [27–29]. Unlike sampling error, it could not be minimized by increasing sampling size. The non-sampling error may be reduced by combining a random sampling selection procedure with a measurement method ensuring high accuracy including accounting for the actual values of the row spacing measured over each segment [30].

Optimal sampling requires a tradeoff between minimum sampling error obtained with maximum sampling size and minimum cost obtained with minimum sampling size [31]. The optimal sampling strategy should first be designed according to the precision targeted here quantified by the coefficient of variation (CV) characterizing the relative variability of the estimated plant density between several replicates of the sampling procedure. The term 'optimal' should therefore be understood as the minimum sampling effort to be spent to achieve the targeted precision. Two approaches were proposed: the first one considers a fixed segment length (FSL) over which the plants have to be counted; the second one considers a fixed number of successive plants (FNP) defining a row segment, the length of which needs to be measured. The first method (FLS) is the one generally applied within most field experiments. However, we demonstrated that it is generally sub-optimal: since the segment length required to achieve a given CV depends mainly on the actual plant density: the sampling will be either too large for the targeted precision, or conversely too small, leading to possible degradation of the precision of plant density estimates. Nevertheless, for the plant density (>100 plants m^{-2}) and shape parameter ($a > 0.9$) usually experienced, a segment length of 6 m will ensure a precision better than 10%. The second approach (FNP) appears generally more optimal: it aims at measuring the length of the segment corresponding to a number of consecutive plants that will depend mainly on the targeted precision. Results demonstrate that in

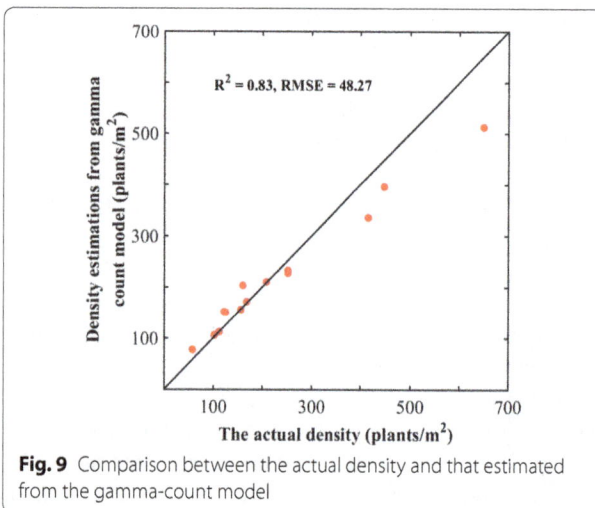

Fig. 9 Comparison between the actual density and that estimated from the gamma-count model

our conditions, the density should be evaluated over segments containing 90 plants to achieve a 10% precision. The sampling size will always be close to optimal as compared to the first approach where optimality requires the knowledge of the plant density that is to be estimated. Further, the FNP approach is probably more easy to implement with higher reliability: as a matter of facts, measuring the length of a segment defined by plants at its two extremities is easier than counting the number of plants in a fixed length segment, where the extremities could be in the vicinity of a plant and its inclusion or not in the counting could be prone to interpretation biases by the operator. The total number of plants required in a segment could be split into subsamples containing smaller number of plants that will be replicated to get the total number of plants targeted. This will improve the spatial representativeness. Overall, the method proposed meets the requirements defined by [32, 33] for the next genearation of phenotyping tools: increase the accuracy, the precision and the throughput while reducing the labor and budgetary costs.

The gamma-count model proved to be well suited to describe the plant spacing distribution along the row over our contrasted experimental situations. It can thus be used to describe the heterogeneity of plant spacing as suggested by [20]. This may be applied for detailed canopy architecture studies or to quantify the impact of the sowing pattern heterogeneity on inter-plant competition [1, 2]. The heterogeneity of plant spacing may be described by the scale and shape parameters of the gamma model. Quantification of the heterogeneity of plant spacing requires repeated measurements over segments defined by a fixed number of plants. Our results clearly show that the precision on estimates of the gamma count parameters depends only marginally on the number of plants in each segment. Conversely, it depends mainly on the number of segments (replicates) to be measured. For the standard conditions experienced in this study, the optimal sampling strategy to get a CV lower than 10% on the two parameters of the gamma distribution would be to repeat 200 times the measurement of plant spacing between 2 consecutive plants.

Authors' contributions

SL and FB designed the experiment and XJ, BA, PB, MH and AC contributed to the field measurement in different experimental sites. DA significantly contributed to the method development. The manuscript was written by SL and significantly improved by FB. All authors read and approved the final manuscript.

Author details

[1] INRA, UMR-EMMAH, UMT-CAPTE, UAPV, 228 Route de l'aérodrome CS 40509, 84914 Avignon, France. [2] UMR BioSP, INRA, UAPV, 84914 Avignon, France. [3] UMR ECOSYS, INRA, AgroParisTech, Université Paris-Saclay, 78850 Thiverval-Grignon, France. [4] UMR AGIR, INRA, INPT, 31326 Toulouse, France. [5] Hi-Phen, 84914 Avignon, France.

Acknowledgements

We thank the people from Grignon, Toulouse and Avignon who participated to the experiments. Great thanks to Paul Bataillon and Jean-Michel Berceron from UE802 Toulouse INRA for their help in field experiment. The work was completed within the UMT-CAPTE funded by the French Ministry of Agriculture.

Competing interests

The authors declare that they have no competing interests.

Funding

This study was supported by "Programme d'investissement d'Avenir" PHENOME (ANR-11-INBS-012) and Breedwheat (ANR-10-BTR-03) with participation of France Agrimer and "Fonds de Soutien à l'Obtention Végétale". The grant of the principal author was funded by the Chinese Scholarship Council.

References

1. Olsen J, Kristensen L, Weiner J. Influence of sowing density and spatial pattern of spring wheat (*Triticum aestivum*) on the suppression of different weed species. Weed Biol Manag. 2006;6:165–73.
2. Olsen J, Weiner J. The influence of Triticum aestivum density, sowing pattern and nitrogen fertilization on leaf area index and its spatial variation. Basic Appl Ecol. 2007;8:252–7.
3. Reynolds MP, Ortiz-Monasterio JI, McNab A. CIMMYT: application of physiology in wheat breeding. Mexico: CIMMYT; 2001.
4. Van der Heijden G, De Visser PHB, Heuvelink E. Measurements for functional-structural plant models. Frontis. 2007;22:13–25.
5. Ma Y, Wen M, Guo Y, Li B, Cournede P-H, de Reffye P. Parameter optimization and field validation of the functional-structural model GREENLAB for maize at different population densities. Ann Bot. 2007;101:1185–94.
6. Tang L, Tian L. Plant identification in mosaicked crop row images for automatic emerged corn plant spacing measurement. Trans ASABE. 2008;6:2181–91.
7. Tang L, Tian LF. Real-time crop row image reconstruction for automatic emerged corn plant spacing measurement. Trans ASABE. 1079;2008:51.
8. Jin J, Tang L. Corn plant sensing using real-time stereo vision. J Field Robot. 2009;26:591–608.
9. Nakarmi AD, Tang L. Automatic inter-plant spacing sensing at early growth stages using a 3D vision sensor. Comput Electron Agric. 2012;82:23–31.
10. Nakarmi AD, Tang L. Within-row spacing sensing of maize plants using 3D computer vision. Biosys Eng. 2014;125:54–64.
11. Norman DW: *The farming systems approach to development and appropriate technology generation.* Food & Agriculture Org.; 1995.
12. Marshall MN. Sampling for qualitative research. Fam Pract. 1996;13:522–6.
13. Gate P. Ecophysiologie du blé. Paris: Tec & Doc Lavoisier; 1995.
14. Whaley JM, Sparkes DL, Foulkes MJ, Spink JH, Semere T, Scott RK. The physiological response of winter wheat to reductions in plant density. Ann Appl Biol. 2000;137:165–77.
15. Haun J. Visual quantification of wheat development. Agron J. 1973;65:116–9.
16. Brown M, Lowe DG. Automatic panoramic image stitching using invariant features. Int J Comput Vision. 2006;74:59–73.
17. Plackett RL. Karl pearson and the Chi squared test. Int Stat Rev/Revue Internationale de Statistique. 1983;51:59–72.
18. Papalexiou SM, Koutsoyiannis D, Makropoulos C. How extreme is extreme? An assessment of daily rainfall distribution tails. Hydrol Earth Syst Sci. 2013;17:851–62.
19. Winkelmann R. A count data model for gamma waiting times. Stat Pap. 1996;37:177–87.
20. Liu S, Baret F, Andrieu B, Abichou M, Allard D, Solan Bd, Burger P. Modeling the distribution of plants on the row for wheat crops: consequences on the green fraction at the canopy level. Comput Electron Agric. 2016;136:147–56.
21. Winkelmann R. Duration dependence and dispersion in count-data models. J Bus Econ Stat. 1995;13:467–74.

22. Zeviani WM, Ribeiro PJ Jr, Bonat WH, Shimakura SE, Muniz JA. The Gamma-count distribution in the analysis of experimental underdispersed data. J Appl Stat. 2014;41:2616–26.

23. Steutel FW, Van Harn K. Infinite divisibility of probability distributions on the real line. Boca Raton: CRC Press; 2003.

24. Jin X, Liu S, Baret F, Hemerlé M, Comar A: Estimates of plant density from images acquired from UAV over wheat crops at emergence. Remote Sens Environ. 2016 (**accepted**).

25. Liu S, Baret F, Andrieu B, Burger P, Hemerlé M: Estimation of plant density from high resolution RGB imagery over wheat crops. Front Plant Sci. 2016 (**accepted**).

26. Rinne H. The Weibull distribution: a handbook. Boca Raton: CRC Press; 2008.

27. Lohr S. Sampling: design and analysis. Boston: Cengage Learning; 2009.

28. Israel GD: *Determining sample size*. University of Florida Cooperative Extension Service, Institute of Food and Agriculture Sciences, EDIS; 1992.

29. Särndal C-E, Swensson B, Wretman J. Model assisted survey sampling. New York: Springer Science & Business Media; 2003.

30. Scheuren F, Association AS: What is a survey? In. American Statistical Association; 2004.

31. Biemer PP. Total survey error: design, implementation, and evaluation. Public Opin Q. 2010;74:817–48.

32. Cobb JN, DeClerck G, Greenberg A, Clark R, McCouch S. Next-generation phenotyping: requirements and strategies for enhancing our understanding of genotype–phenotype relationships and its relevance to crop improvement. Theor Appl Genet. 2013;126:867–87.

33. Fiorani F, Schurr U. Future scenarios for plant phenotyping. Annu Rev Plant Biol. 2013;64:267–91.

5

Mid-infrared spectroscopy combined with chemometrics to detect Sclerotinia stem rot on oilseed rape (*Brassica napus* L.) leaves

Chu Zhang[1], Xuping Feng[1], Jian Wang[2], Fei Liu[1*], Yong He[1*] and Weijun Zhou[2]

Abstract

Background: Detection of plant diseases in a fast and simple way is crucial for timely disease control. Conventionally, plant diseases are accurately identified by DNA, RNA or serology based methods which are time consuming, complex and expensive. Mid-infrared spectroscopy is a promising technique that simplifies the detection procedure for the disease. Mid-infrared spectroscopy was used to identify the spectral differences between healthy and infected oilseed rape leaves. Two different sample sets from two experiments were used to explore and validate the feasibility of using mid-infrared spectroscopy in detecting Sclerotinia stem rot (SSR) on oilseed rape leaves.

Results: The average mid-infrared spectra showed differences between healthy and infected leaves, and the differences varied among different sample sets. Optimal wavenumbers for the 2 sample sets selected by the second derivative spectra were similar, indicating the efficacy of selecting optimal wavenumbers. Chemometric methods were further used to quantitatively detect the oilseed rape leaves infected by SSR, including the partial least squares-discriminant analysis, support vector machine and extreme learning machine. The discriminant models using the full spectra and the optimal wavenumbers of the 2 sample sets were effective for classification accuracies over 80%. The discriminant results for the 2 sample sets varied due to variations in the samples.

Conclusion: The use of two sample sets proved and validated the feasibility of using mid-infrared spectroscopy and chemometric methods for detecting SSR on oilseed rape leaves. The similarities among the selected optimal wavenumbers in different sample sets made it feasible to simplify the models and build practical models. Mid-infrared spectroscopy is a reliable and promising technique for SSR control. This study helps in developing practical application of using mid-infrared spectroscopy combined with chemometrics to detect plant disease.

Keywords: Mid-infrared spectroscopy, Second derivative spectra, Sclerotinia stem rot, Oilseed rape, Sample set validation

Background

Oilseed rape (*Brassica napus* L.) is one of the most important sources of edible oil and biodiesel. The growth of oilseed rape, a widely planted oil-bearing crop, is affected by many factors, including seed, soil, water supply, nutritional elements, weather conditions and diseases. Diseases are major threats to oilseed rape, resulting in yield and quality loss.

Sclerotinia stem rot (SSR) is a major disease affecting the oilseed rape growth and causing severe yield loss. The ascospores of SSR are produced by the apothecia in the soil, or the seeds are discharged into the air. Some of the ascospores are dispersed more widely from other fields into the surrounding crops. The spread of ascospores makes it difficult to control the disease completely before its onset. The detection of SSRs at an early stage provides an alternative for disease control.

The early detection of SSR on oilseed rapes is a priority for SSR control on oilseed rape plants. Traditional methods, such as polymerase chain reaction (PCR)

*Correspondence: fliu@zju.edu.cn; yhe@zju.edu.cn
[1] College of Biosystems Engineering and Food Science, Zhejiang University, 866 Yuhangtang Road, Xihu District, Hangzhou 310058, China
Full list of author information is available at the end of the article

[1], enzyme-linked immunosorbent assay (ELISA) [2], nucleic acid hybridization [3] and serological techniques [4], rely on the identification of spores by microscopy or culture-based techniques to detect plant diseases. These traditional methods applied in the detection of plant diseases are accurate and standard. However, these methods also have some limitations such as being time-consuming, requiring special operation skills, generating reagent waste and requiring complex sample preparation, which makes these methods unsuitable for large-scale field detection. Thus, new techniques for cheap, fast and accurate identification of plant diseases should be developed.

Spectroscopy techniques, such as visible/near-infrared spectroscopy [5, 6], mid-infrared spectroscopy [7–9], Raman spectroscopy [10] and fluorescence spectroscopy [11] have been studied to detect plant diseases. Mid-infrared spectroscopy provides the information about the fundamental vibrational bands of the functional groups in the samples. The plants affected by diseases experience internal physiological changes, which in turn results in changes in their mid-infrared spectra. Mid-infrared spectroscopy has been used as an effective technique for detecting plant diseases. Sankaran et al. [7] used mid-infrared spectroscopy to detect nitrogen deficiency and Huanglongbing of citrus leaves. Hawkins et al. [8] used Fourier transform infrared-attenuated total reflection spectroscopy for the detection of Huanglongbing in citrus leaves. Hawkins et al. [9] also used the Fourier transform infrared-attenuated total reflection spectroscopy to detect Huanglongbing, citrus leaf rugose virus, citrus tristeza virus, citrus psorosis virus, *Xanthomonas axonopodis* and nutritional deficiency.

Moreover, the use of mid-infrared spectroscopy in plant disease detection mainly focuses on the spectral differences or the discriminant results, and the feasibility of using mid-infrared spectroscopy for plant disease detection has been proven. However, there is a wide gap between the feasibility of this technique and its practical application is great. The rapid acquisition of spectra and simple sample preparation makes it possible to develop mid-infrared spectroscopy as a practical method for the rapid detection of plant diseases. The primary purpose of developing a practical application of mid-infrared spectroscopy depends on the calibration models. Robust and accurate models using informative wavenumbers with minimum colinearity and redundancy are required.

The objective of this study was to explore and validate the use and capacity of mid-infrared spectroscopy for detecting SSRs on oilseed rape leaves. The specific objectives were: (1) to evaluate the influence of different samples sets on mid-infrared spectroscopy, (2) to select and compare optimal wavenumbers in different sample sets, and (3) to develop and compare the optimal classification models in different sample sets.

Methods
Sample preparation
The seeds of the oilseed rape (*Brassica napus* L., cv. ZS758) were used in our study. The seeds were sown into the seedbed, and 200 oilseed rape plants were transplanted into the experimental pots after 30 days. Forty days after transplant, the oilseed rape leaves were suitable for *Sclerotinia sclerotiorum* infection. *Sclerotinia sclerotiorum* was cultured on a potato dextrose agar. The oilseed rape plants were kept in a controlled environment at a temperature of 20 °C and 80% humidity. Two experiments were conducted. For the first experiment, the oilseed rape leaves were inoculated with *Sclerotinia sclerotiorum*. Seventy-two hours later, when the disease symptoms on the leaves became visible, 60 infected leaves and 60 healthy leaves were collected and placed in an icebox to keep the leaves fresh. After the measurement of physiological parameters, the remaining leaves were dried in an oven at a temperature of 75 °C for 48 h. The dried leaves were then ground into a powder, sieved through a 100-mesh sieve, and stored in plastic bags. Seven days later, the second experiment (similar to the first) was conducted.

Mid-infrared spectra acquisition
The mid-infrared spectra of samples were acquired by a Jasco FT/IR-4100 spectrometer (Japan) in the spectral range of 400–4000 cm^{-1}. Before spectra collection, the potassium bromide (KBr) powders were dried in an oven at a temperature of 105 °C for 4 h. Then, 10 mg of each sample was mixed with 490 mg KBr powders, and the mixture was ground and mixed thoroughly. The mixture was then placed into a tablet machine for tabletting, and the sample tablets were used for transmittance mid-infrared spectral data collection. For each sample, 32 scans were applied with a resolution of 8 cm^{-1}, and the average of the 32 spectra was used as the transmittance spectrum of the sample.

Multivariate data analysis
Spectra preprocessing
The acquired mid-infrared spectra contained noises. An effective reduction in noises is significant for further analysis. Wavelet transform (WT) is an efficient denoising method in the spectral analysis [12]. WT with mother wavelet Daubechies was applied in this study to reduce the noises.

Principal component analysis

Principal component analysis (PCA) is a generally used method for feature extraction and qualitative analysis of the samples. PCA linearly transforms the original data into new orthogonal variables (called principal components, PCs). The first few PCs contains the maximum feature information, which could be used to observe the distribution of samples and identify their differences [13].

Classification models

To evaluate the performance of using mid-infrared spectroscopy for identifying the infected and healthy leaves of oilseed rape, we used the partial least square-discriminant analysis (PLS-DA) [14], support vector machine (SVM) [15] and extreme learning machine (ELM) [16] to establish the classification models.

PLS-DA is a widely used supervised pattern recognition method in spectral data analysis. PLS-DA is conducted in the manner of PLS regression (PLSR), with the integral category value as Y variables. PLSR linearly transforms the original data into new variables (called latent variables, LV), and the first few LVs carry the most useful information. The outputs of PLSR and PLS-DA are real numbers with decimals. Thus, the threshold value is needed to determine the category of the samples. Herein, the threshold value was set as 0.5.

SVM is also a widely used supervised pattern recognition method in spectral data analysis. The general concept of SVM is to transform the original data from the low dimension space to the high dimension space, and constructs a hyperplane to maximize the separation of the different sample classes. SVM could address linear and non-linear issues efficiently. The selection of the kernel function is important in SVM. In this study, radial basis function (RBF) was selected as the kernel function.

ELM is a feedforward neural network with a single hidden layer. ELM has shown advantages such as fast learning speed and good generalization ability. In ELM, only the number of neurons in the hidden layer should be set. The determination of the number of neurons in the hidden layer is critical in ELM. In this study, the number of neurons in the hidden layer was determined by a step by step search within a predefined range. The number of neurons corresponding to the best performance was selected.

Optimal wavenumber selection

The acquired mid-infrared spectra contained a large number of wavenumber variables, which may suffer from the risk of non-informative variables and variable collinearity. With a large number of wavenumber variables, the calibration models may become unstable, computation consuming, complex and difficult to interpret.

Wavenumber (wavelength) selection in spectral analysis for multivariate analysis is an important step in selecting the informative and noncollinear wavenumber variables. The wavenumber (wavelength) selection may improve the model performance while significantly reducing the number of variables, resulting in stable, simple and accurate models.

Second derivative spectrum (2nd spectrum) is a manual selection method based on the spectral profile of the samples [17]. The second derivative is generally used as an efficient preprocessing method in spectral analysis. Compared with the raw spectra, the 2nd spectra could improve spectral resolution, identify overlapping peaks and reduce the background information. Thus, the variables related to the chemical compositions were enhanced and highlighted as peaks and valleys within the 2nd spectra. Therefore, the peaks and valleys with differences between different sample classes were selected as the optimal wavenumbers.

Software and model evaluation

In this study, the second derivative preprocessing, PCA and PLS-DA were conducted on the Unscrambler® 10.1 (CAMO AS, Oslo, Norway). The WT preprocessing, SVM and ELM models were conducted on MATLAB (R2014b) software (The Math Works, Inc., Natik, MA, USA). The model performances were evaluated by the classification accuracy in the calibration set and the prediction set.

Results and discussion

Mid-infrared spectra

Due to the instrument and experiment conditions, the head and tail of the collected mid-infrared spectra contained obvious noises. Thus, only the spectra in the range of 900–3800 cm^{-1} were studied. Figure 1a, b show the raw spectra of the sample set 1 and 2, and noise could be observed in the two sets. WT was applied on raw spectra to reduce the noise. For sample set 1, WT using Daubechies 6 with a decomposition level of 5 was applied. For sample set 2, WT using Daubechies 5 with a decomposition level of 5 was applied. Figure 1c, d show the preprocessed spectra of the sample set 1 and sample set 2. Obvious denoising could be found in Fig. 1. The general spectral features of sample sets 1 and 2 were similar.

Figure 2a, b show the average spectra of healthy and infected leaves of sample sets 1 and 2. As detailed in Fig. 2a, the average transmittance spectra of healthy and infected leaves of sample set 1 showed differences in their transmittance value, and larger differences could be observed in the ranges of 900–1500 and 1800–2750 cm^{-1}. As shown in Fig. 2b, the average transmittance spectra of healthy and infected leaves of sample set 2 showed

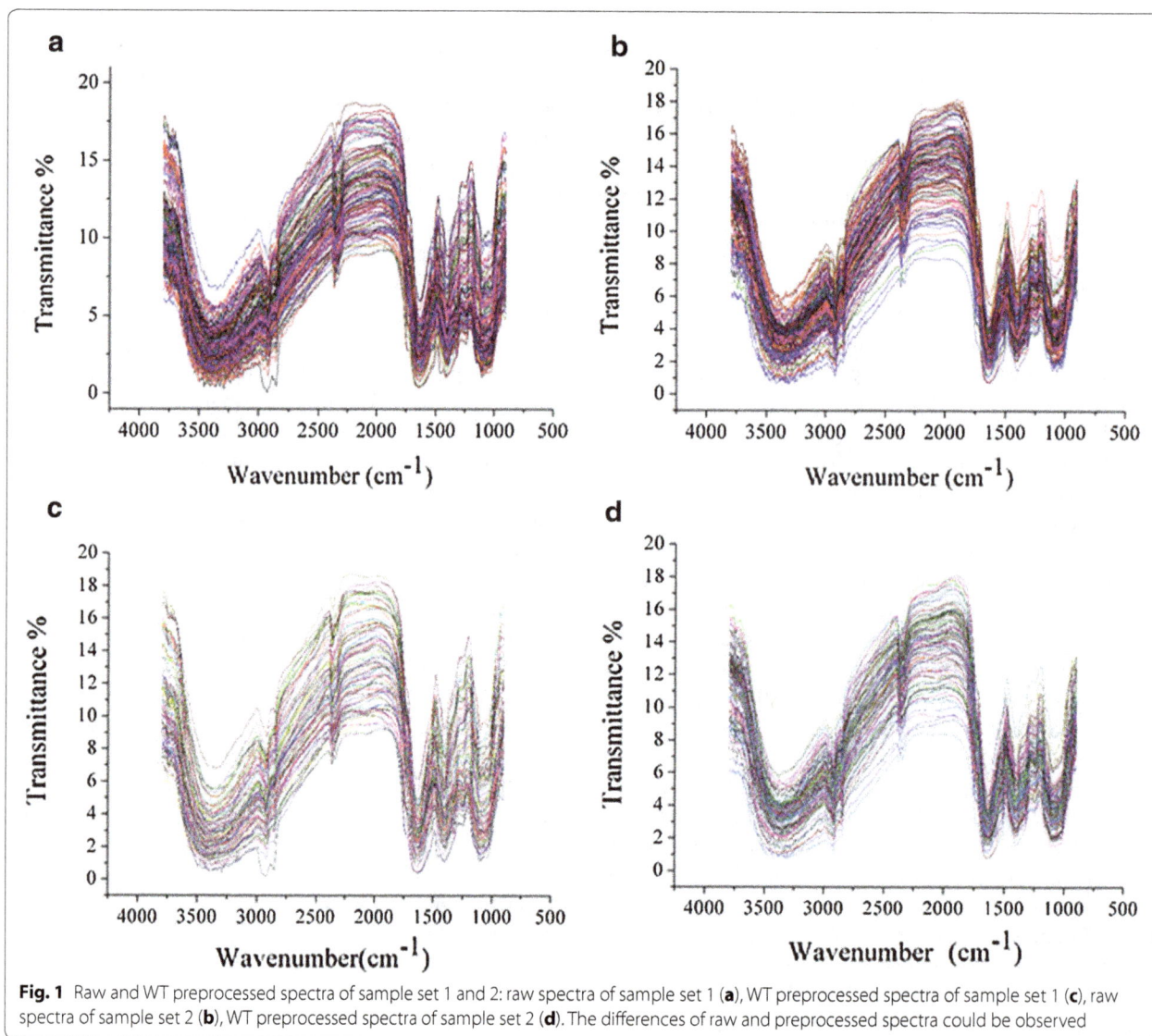

Fig. 1 Raw and WT preprocessed spectra of sample set 1 and 2: raw spectra of sample set 1 (**a**), WT preprocessed spectra of sample set 1 (**c**), raw spectra of sample set 2 (**b**), WT preprocessed spectra of sample set 2 (**d**). The differences of raw and preprocessed spectra could be observed

differences in the transmittance value, and larger differences could be observed in the range of 900–1500 cm^{-1}. The larger differences between the healthy and infected leaves of the two types of samples were observed in the ranges of 900–1500 and 1800–2750 cm^{-1}, the same as sample set 1.

PCA analysis

The samples of the sample sets 1 and 2 were randomly divided into the calibration and prediction sets at a ratio of 2:1. The healthy leaves were assigned the category value 1, and the infected leaves were assigned the category value 2.

PCA was performed on the preprocessed spectra of the calibration set of the sample sets 1 and 2 to visualize the distribution of healthy and infected samples. For sample set 1, PC1, PC2 and PC3 explained 71.021, 21.269 and 3.642% of the total variance, respectively. The first 3 PCs explained 95.931% of the total variance. The score scatter plots of PC1 and PC2, PC1 and PC3, and PC2 and PC3 are shown in Fig. 3a, c, e. Figure 3a, e demonstrated that the healthy samples could be easily differentiated from the infected samples.

For the sample set 2, PC1, PC2 and PC3 explained 81.619, 10.533 and 3.523% of the total variance, and the first 3 PCs explained 95.675% of the total variance. The scores scatter plots of PC1 and PC2, PC1 and PC3, and PC2 and PC3 are shown in Fig. 3b, d, f. Figure 3b, f indicate that the healthy samples could be differentiated from the infected samples with a few overlaps.

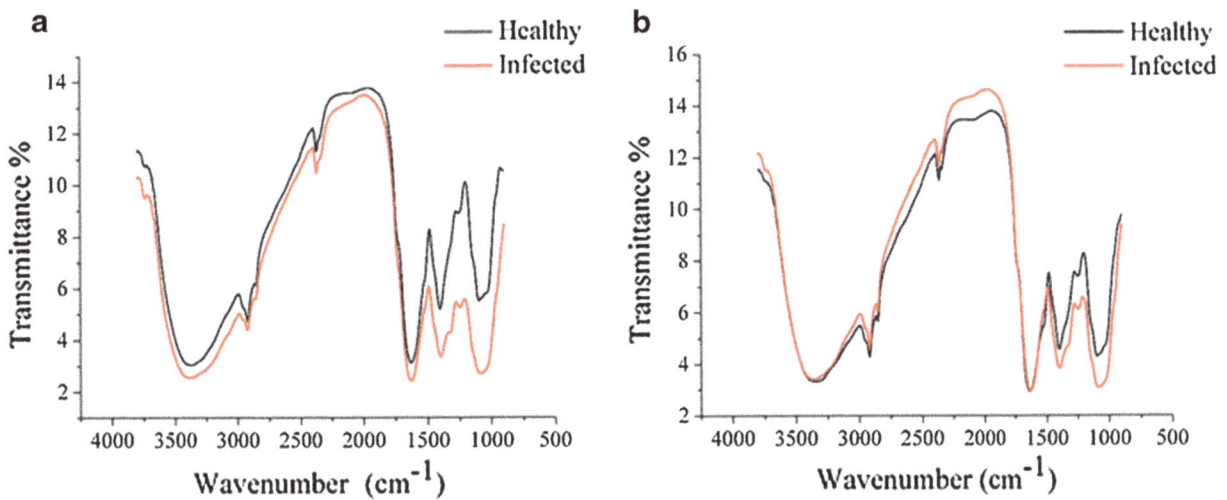

Fig. 2 Average spectra of healthy and infected leaves of sample set 1 (**a**), and average spectra of healthy and infected leaves of sample set 2 (**b**). The differences of healthy and infected leaves of two sample sets could be observed

PCA of sample set 1 and 2 indicated that healthy and infected leaves could be classified. The distribution of healthy and infected samples in the corresponding score scatter plot of the two sample sets were similar, and due to different sample sets, the separation differed. The distribution of healthy samples in the score scatter plot was observed to disperse more widely than the infected samples.

Discriminant models using full mid-infrared spectra

PCA provided visual distribution trends of samples, and the discriminant models were further needed for quantitative classification.

PLS-DA, SVM and ELM models were built by using the full mid-infrared spectra of the 2 sample sets to classify the healthy and infected leaves. A PLS-DA model was built using leave-one-out cross validation, and the number of optimal LVs was determined. SVM used RBF as the kernel function, and the optimal penalty coefficient (C) and the kernel function parameter gamma (g) were obtained by a grid-search procedure in the range of 2^{-8} to 2^8. The number of neurons in the hidden layer of ELM models were determined by comparing the performances of the ELM models by using different numbers of neurons from 1 to 80 with a step of 1. The ELM models with optimal performances were selected. The results of the discriminant models are shown in Table 1.

For the sample set 1, all discriminant models demonstrated good performances, with classification accuracies of 100% in the calibration set and over 80% in the prediction set. ELM showed the best results, with a classification accuracy of 92.5%. For the sample set 2, all

discriminant models demonstrated good performances, with classification accuracies over 90% in both the calibration and the prediction sets. PLS-DA models showed best results, with classification accuracies of 100% in the calibration and prediction sets.

The performances of the discriminant models in a sample set were different, and the discriminant results of a discriminant model between the 2 sample sets were also different. All discriminant models showed good performances, the sample sets affected the classification performances, and the selection of suitable discriminant models for practical application was imperative.

The discriminant results of the calibration sets of the two sample sets matched with the PCA analysis, and the general discriminant results of the calibration set of the sample set 1 performed slightly better than the calibration set of the sample set 2. Contrarily, the general prediction results of the sample set 2 were slightly better than those of sample set 1. The results were obtained due to the random division of the samples into the calibration and the prediction sets. The overall results indicated that it was feasible to detect SSR on oilseed rape leaves by using mid-infrared spectroscopy, and its practical application in detecting plant diseases was promising.

Optimal wavenumber selection

In this study, the mid-infrared spectra were acquired with a spectral resolution of 8 cm^{-1}. In the spectral range of 900–3800 cm^{-1}, there were 1504 wavenumber variables of the spectra. The selection of the informative wavenumber variables was important for better models.

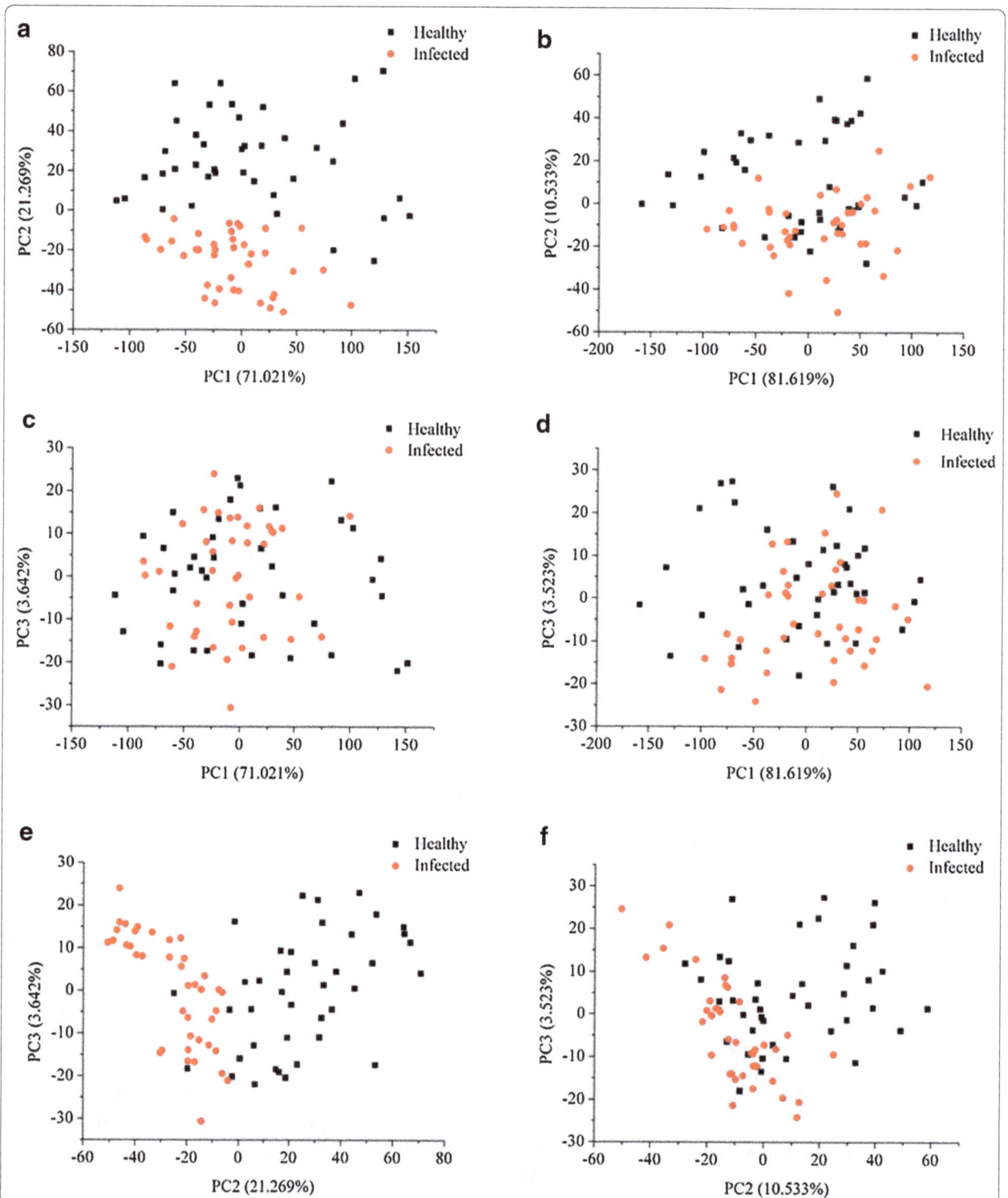

Fig. 3 Scores scatter plot of PC1 versus PC2 (**a**), PC1 versus PC3 (**c**) and PC2 versus PC3 (**e**) of sample set 1, and scores *scatter plot* of PC1 versus PC2 (**b**), PC1 versus PC3 (**d**) and PC2 versus PC3 (**f**) of sample set 2. The plots were used to explore the separability between healthy and infected samples qualitatively

Table 1 Results of discriminant models using full mid-infrared transmittance spectra of sample sets 1 and 2

Models	Sample set 1			Sample set 2		
	Par[a]	Cal[b] (%)	Pre[c] (%)	Par	Cal (%)	Pre (%)
PLS-DA	4	100	85	10	100	100
SVM	(1.7411, 0.0118)	100	80	(84.4485, 0.0039)	100	92.5
ELM	22	100	92.5	60	92.5	90

[a] Par means the parameters of the models, the number of LVs for PLS-DA, (C, g) for SVM and number of neurons for ELM

[b] Cal means the calibration set

[c] Pre means the prediction set

The second derivative with 7 smoothing points by the Savitzky–Golay algorithm was applied to the average spectra of the healthy and infected leaves of sample sets 1 and 2. The 2nd spectra were used to select optimal wavenumbers. Figure 4 show the 2nd spectra and the corresponding selected optimal wavenumbers of sample sets 1 and 2. The selected wavenumbers are also shown in Table 2.

Figure 4 and Table 3 demonstrate that the 2nd spectra of the sample sets 1 and 2 were quite similar, and the maximum optimal wavenumbers selected by the 2nd spectra of the 2 sample sets were similar or the same.

Some differences were also observed due to the variations among the different sample sets and the instrument condition. The selected optimal wavenumbers matched the spectra regions with differences of the average spectra shown in Fig. 2.

The optimal wavenumbers selected by the 2nd spectra of the 2 sample sets showed repeatability, indicating the efficiency for the optimal wavenumber selection by 2nd spectra. However, the number of samples used for optimal wavenumber selection were small, which was common in spectral analysis. More samples were needed to obtain the optimal wavenumbers for practical

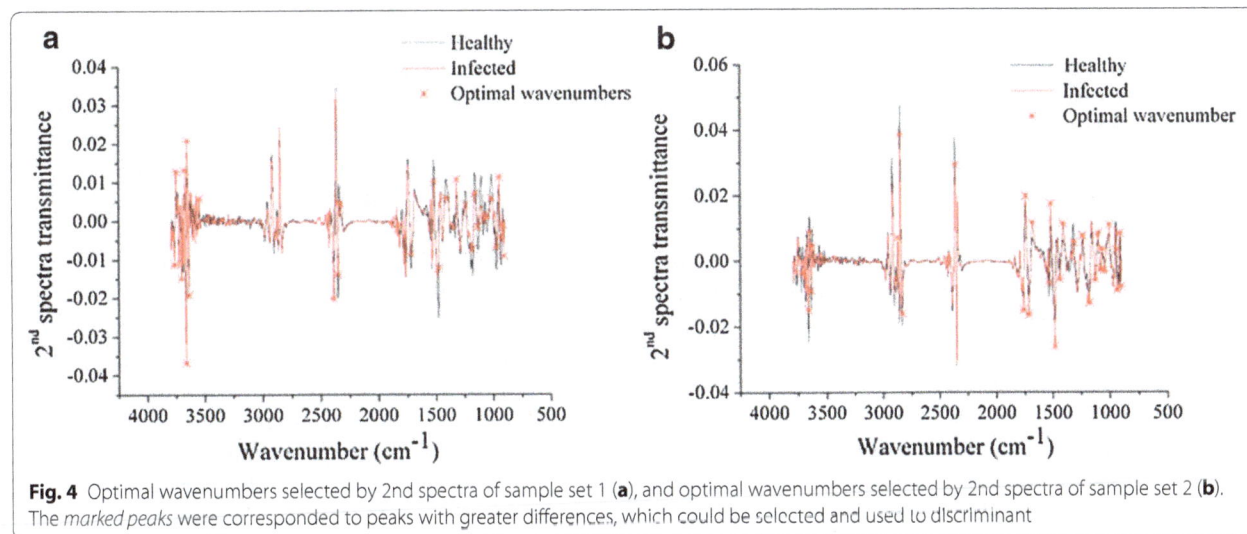

Fig. 4 Optimal wavenumbers selected by 2nd spectra of sample set 1 (**a**), and optimal wavenumbers selected by 2nd spectra of sample set 2 (**b**). The *marked peaks* were corresponded to peaks with greater differences, which could be selected and used to discriminant

Table 2 Optimal wavenumbers selected by the 2nd spectra of sample sets 1 and 2

	Number	Wavenumber (cm⁻¹)
Sample set 1	28	906.3795, 916.0218, 935.3065, 946.8773, 973.8758, 1018.2305, 1070.2992, 1083.7985, 1133.9386, 1159.0087, 1180.2218, 240.0043, 1317.1429, 1409.7094, 1479.1342, 1517.7035, 1716.3356, 2341.1589, 2350.8013, 2389.3706, 3546.4507, 3639.0171, 3650.5879, 3662.1587, 3671.801, 3700.728, 3747.0112, 3762.439
Sample set 2	31	906.3795, 916.0218, 937.2349, 948.8057, 1008.5882, 1049.0861, 1070.2992, 1085.7269, 1105.0116, 1132.0101, 1180.2218, 240.0043, 1317.1429, 1409.7094, 1440.5648, 1481.0626, 1517.7035, 1533.1312, 1685.4801, 1716.3356, 745.2626, 1764.5472, 2362.3721, 2834.8464, 2852.2026, 2875.3442, 3629.3748, 3639.0171, 3652.5164, 3662.1587, 3721.9412

Table 3 The results of the discriminant models using optimal wavenumbers from sample sets 1 and 2

Models	Sample set 1			Sample set 2		
	Par	Cal (%)	Pre (%)	Par	Cal (%)	Pre (%)
PLS-DA	6	100	82.5	9	100	100
SVM	(5.2780, 1)	100	82.5	(48.5029, 0.0359)	95	95
ELM	62	100	95	76	100	95

applications. The selection of optimal wavenumbers in the 2 sample sets indicated the possibility of selecting the widely accepted optimal wavenumbers by the 2nd spectra for practical application.

The selected peaks in the 900–1200 cm^{-1} region were attributed to the C–O stretching bands mainly from the carbohydrates [18], whereas those in the 1500–1700 cm^{-1} region were attributed to the amide bands of proteins [18]. The selected peaks in the 1200–1500 cm^{-1} region were assigned to the C–H bending modes [19]; the peak at 1716.336 cm^{-1} was assigned to the amide I of protein [20]; the peak at 1745.263 cm^{-1} was assigned to the COOR bond [17]; the peak at 1764.547 cm^{-1} was attributed to the symmetric C=O stretching of the ester group [21]; the peak at 2350 cm^{-1} was assigned to the asymmetric C=O bonds [22]. Moreover, peaks in the 2800–3000 cm^{-1} region were attributed to the lipid [23], and those in the 3000–3800 cm^{-1} region were attributed to the O–H stretching vibrations [24].

Discriminant models using optimal wavenumbers

To evaluate the performance of the selected optimal wavelengths in SSR detection, PLS-DA, SVM and ELM models were built. The modelling procedure was the same with the full spectra models. The results of the discriminant models are illustrated in Table 3.

For the sample set 1, all discriminant models showed good performances, with classification accuracies of 100% in the calibration set and over 80% in the prediction set. ELM models performed the best with a classification accuracy of 95% in the prediction set.

For the sample set 2, all discriminant models demonstrated satisfactory performances, with classification accuracies over 95% in both the calibration and the prediction sets. PLS-DA models performed best with a classification accuracy of 100% in both the calibration and the prediction sets.

Notably, the performances of the discriminant models using the optimal wavenumbers were different. Nevertheless, all discriminant models showed good performances.

The results of optimal wavenumber selection and the calibration models using the selected optimal wavenumbers of the 2 sample sets suggested the efficiency and the

reliability of the optimal wavenumber selection, indicating a great potential for practical application.

Comparison of the full spectra models and optimal wavenumber models

As presented in Tables 1 and 3, the discriminant models using the full spectra and the selected optimal wavenumbers all showed good performances. For the sample set 1, the discriminant models using the optimal wavenumbers showed similar results as the discriminant models using full spectra. However, the number of wavenumber variables was reduced from 1504 to 28, resulting in a reduction of 98.138%. These results indicated that optimal wavenumbers selected by the 2nd spectra could significantly reduce the number of wavenumber variables in the mid-infrared spectra, and the selected optimal wavenumbers were capable of keeping the model performances for sample set 1. For sample set 2, the discriminant models using the optimal wavenumbers showed similar results as the discriminant models using full spectra. Nonetheless, the wavenumber variables were reduced from 1504 to 31, resulting in a reduction of 99.939%. The results indicated that the optimal wavenumbers selected by the 2nd spectra could significantly reduce the number of wavenumber variables in the mid-infrared spectra, and the selected optimal wavenumbers were capable of keeping the model performances for sample set 2.

Considering that the optimal wavenumbers selected by the 2nd spectra for the 2 sample sets were similar, and the models using the optimal wavenumbers of the 2 sample sets showed good performances, mid-infrared spectroscopy combined with optimal wavenumber selection by the 2nd spectra was proven to be an efficient and promising technique for SSR detection of oilseed rapes. However, beyond the exploration and validation of using the mid-infrared spectroscopy combined with chemometrics for detecting plant diseases, the results of this study also indicated that mid-infrared spectroscopy was an efficient, reliable and promising technique with practical applications, and not just the feasibility of exploration.

Conclusion

Two sample sets of SSR infected oilseed rape leaves and their corresponding mid-infrared transmittance spectral information were studied to detect SSR on oilseed rape leaves. The differences in the mid-infrared spectra of the healthy and infected leaves indicated the differences in their physiological constituents in the corresponding samples. The discriminant results by different models indicated the feasibility of using mid-infrared spectra for detecting SSR on oilseed rape leaves. The results of discriminant models (including PLS-DA, SVM and ELM) and the optimal wavenumber selection method (2nd spectra), showed the effectiveness of mid-infrared spectroscopy combined with chemometrics in detecting SSR on oilseed rape leaves. The quite similar optimal wavenumbers selected by the 2nd spectra demonstrated the effectiveness of wavenumbers selection. The results of the 2 sample sets proved and validated that mid-infrared spectroscopy was a promising and reliable technique for SSR detection. Mid-infrared spectroscopy could be an efficient method for disease detection for real-world disease control, with a reliable and accurate selection of optimal calibration models and optimal wavenumbers.

Author details

[1] College of Biosystems Engineering and Food Science, Zhejiang University, 866 Yuhangtang Road, Xihu District, Hangzhou 310058, China. [2] College of Agriculture and Biotechnology, Zhejiang University, Hangzhou 310058, China.

Authors' contributions

CZ, FL, YH and WJZ designed the experiments; CZ, JW, and FL performed the experiments; CZ, XPF, FL and YH analysed and interpreted the data; CZ, XPF, FL and YH wrote the paper. All authors read and approved the final manuscript.

Acknowledgements

We thank Hongyan Zhu (College of Biosystems Engineering and Food Science, Zhejiang University) for critical reading of this manuscript.

Competing interests

The authors declare that they have no competing interests.

Funding

This study was funded by The National Natural Science Foundation of China (31471417), China Postdoctoral Science Foundation (2016M600466), Specialized Research Fund for the Doctoral Program of Higher Education (20130101110104863), Zhejiang Provincial Natural Science Foundation of China (LY15C130003).

References

1. Rogers SL, Atkins SD, West JS. Detection and quantification of airborne inoculum of Sclerotinia sclerotiorum using quantitative PCR (vol 58, pg 324, 2009). Plant Pathol. 2011;60(4):800.
2. Bom M, Boland GJ. Evaluation of polyclonal-antibody-based immunoassays for detection of Sclerotinia sclerotiorum on canola petals, and prediction of stem rot. Can J Microbiol. 2000;46(8):723–9.
3. Owens RA, Diener TO. Sensitive and rapid diagnosis of potato spindle tuber viroid disease by nucleic-acid hybridization. Science. 1981;213(4508):670–2.
4. Rochapena MA, Lee RF. Serological techniques for detection of citrus tristeza virus. J Virol Methods. 1991;34(3):311–31.
5. Jinendra B, Tamaki K, Kuroki S, Vassileva M, Yoshida S, Tsenkova R. Near infrared spectroscopy and aquaphotomics: novel approach for rapid in vivo diagnosis of virus infected soybean. Biochem Biophys Res Commun. 2010;397(4):685–90.
6. Sirisomboon P, Hashimoto Y, Tanaka M. Study on non-destructive evaluation methods for defect pods for green soybean processing by near-infrared spectroscopy. J Food Eng. 2009;93(4):502–12.
7. Sankaran S, Ehsani R, Etxeberria E. Mid-infrared spectroscopy for detection of Huanglongbing (greening) in citrus leaves. Talanta. 2010;83(2):574–81.
8. Hawkins SA, Park B, Poole GH, Gottwald T, Windham WR, Lawrence KC. Detection of citrus huanglongbing by Fourier transform infrared-attenuated total reflection spectroscopy. Appl Spectrosc. 2010;64(1):100–3.
9. Hawkins SA, Park B, Poole GH, Gottwald TR, Windham WR, Albano J, et al. Comparison of FTIR spectra between Huanglongbing (Citrus Greening) and other citrus maladies. J Agric Food Chem. 2010;58(10):6007–10.
10. Kim S, Lee S, Chi HY, Kim MK, Kim JS, Lee SH, et al. Feasibility study for detection of turnip yellow mosaic virus (TYMV) Infection of Chinese Cabbage Plants Using Raman Spectroscopy. Plant Pathol J. 2013;29(1):105–9.
11. Romer C, Burling K, Hunsche M, Rumpf T, Noga G, Plumer L. Robust fitting of fluorescence spectra for pre-symptomatic wheat leaf rust detection with Support Vector Machines. Comput Electron Agric. 2011;79(2):180–8.
12. Li XL, He Y. Discriminating varieties of tea plant based on Vis/NIR spectral characteristics and using artificial neural networks. Biosyst Eng. 2008;99(3):313–21.
13. Dong WJ, Ni YN, Kokot S. A near-infrared reflectance spectroscopy method for direct analysis of several chemical components and properties of fruit, for example, Chinese hawthorn. J Agric Food Chem. 2013;61(3):540–6.
14. Barker M, Rayens W. Partial least squares for discrimination. J Chemom. 2003;17(3):166–73.
15. Burges CJC. A tutorial on support vector machines for pattern recognition. Data Min Knowl Discov. 1998;2(2):121–67.
16. Ding S, Zhao H, Zhang Y, Xu X, Nie R. Extreme learning machine: algorithm, theory and applications. Artif Intell Rev. 2015;44(1):103–15.
17. Kamruzzaman M, Barbin D, ElMasry G, Sun DW, Allen P. Potential of hyperspectral imaging and pattern recognition for categorization and authentication of red meat. Innov Food Sci Emerg Technol. 2012;16:316–25.
18. Heraud P, Caine S, Sanson G, Gleadow R, Wood BR, McNaughton D. Focal plane array infrared imaging: a new way to analyse leaf tissue. New Phytol. 2007;173(1):216–25.
19. Yang J, Yen HE. Early salt stress effects on the changes in chemical composition in leaves of ice plant and Arabidopsis. A Fourier transform infrared spectroscopy study. Plant Physiol. 2002;130(2):1032–42.
20. Movasaghi Z, Rehman S, Rehman IU. Fourier transform infrared (FTIR) spectroscopy of biological tissues. Appl Spectrosc Rev. 2008;43(2):134–79.
21. D'Souza L, Devi P, DivyaShridhar MP, Naik CG. Use of fourier transform infrared (FTIR) spectroscopy to study cadmium-induced changes in Padina Tetrastromatica (Hauck). Anal Chem Insights. 2008;3:135–43.
22. Lu YZ, Du CW, Yu CB, Zhou JM. Determination of the contents of magnesium and potassium in rapeseeds using FTIR-PAS combined with least squares support vector machines and uninformative variable elimination. Anal Methods. 2014;6(8):2586–91.
23. Shoaib A, Akhtar N, Aqsa Aftab N. Fourier transform-infrared spectroscopy to monitor modifications in canola biochemistry caused by alternaria destruens. Pak J Phytopathol. 2013;25(2):105–9.
24. Lee CM, Kubicki JD, Fan BX, Zhong LH, Jarvis MC, Kim SH. Hydrogen-bonding network and oh stretch vibration of cellulose: comparison of computational modeling with polarized ir and sfg spectra. J Phys Chem B. 2015;119(49):15138–49.

6

The Tree Drought Emission MONitor (Tree DEMON), an innovative system for assessing biogenic volatile organic compounds emission from plants

Marvin Lüpke[1*] [ID], Rainer Steinbrecher[3], Michael Leuchner[1,4] and Annette Menzel[1,2]

Abstract

Background: Biogenic volatile organic compounds (BVOC) emitted by plants play an important role for ecological and physiological processes, for example as response to stressors. These emitted compounds are involved in chemical processes within the atmosphere and contribute to the formation of aerosols and ozone. Direct measurement of BVOC emissions requires a specialized sample system in order to obtain repeatable and comparable results. These systems need to be constructed carefully since BVOC measurements may be disturbed by several side effects, e.g., due to wrong material selection and lacking system stability.

Results: In order to assess BVOC emission rates, a four plant chamber system was constructed, implemented and throughout evaluated by synthetic tests and in two case studies on 3-year-old sweet chestnut seedlings. Synthetic system test showed a stable sampling with good repeatability and low memory effects. The first case study demonstrated the capability of the system to screen multiple trees within a few days and revealed three different emission patterns of sweet chestnut trees. The second case study comprised an application of drought stress on two seedlings compared to two in parallel assessed seedlings of a control. Here, a clear reduction of BVOC emissions during drought stress was observed.

Conclusion: The developed system allows assessing BVOC as well as CO_2 and water vapor gas exchange of four tree specimens automatically and in parallel with repeatable results. A canopy volume of 30 l can be investigated, which constitutes in case of tree seedlings the whole canopy. Longer lasting experiments of e.g., 1–3 weeks can be performed easily without any significant plant interference.

Keywords: Dynamic chambers, BVOC, Drought, Monoterpene, *Castanea sativa* Mill., Sweet chestnut

Background

Biogenic volatile organic compounds (BVOC) are emitted by the biosphere. The annual global flux of BVOC of 1.091 Gt a^{-1} for the year 2000 is estimated to consist of 49% isoprene, 14% monoterpene and 35% of various other volatile organic compounds (VOC) [1]. One major source of BVOC is the biochemical synthesis within plants; BVOC are then either stored or emitted directly [2]. Depending on the latter pathways BVOC emissions are strongly driven by light and/or temperature [3].

The production and emission of BVOC by plants is linked to a wide range of ecological functions, such as response to herbivore feeding by attracting potential predators or acting as repellent [4–7]; communication processes among plants or between plants and insects [8], e.g., BVOC related to herbivory induce the production of defense substances in non-attacked specimens [7, 9]; and attraction of pollinators to open flowers [5]. For the plant itself BVOC seem to reduce oxidative stress in case of heat waves or high ozone concentrations [10] and

*Correspondence: luepke@wzw.tum.de

[1] Ecoclimatology, Technische Universität München, Hans-Carl-von-Carlowitz-Platz 2, 85354 Freising, Germany
Full list of author information is available at the end of the article

other stress induced by the complex abiotic urban environment [11].

Beside their ecological functions, BVOC play a significant role in atmospheric chemistry [12], such as in formation of biogenic secondary organic aerosols (bSOA) [13, 14]; in O_3 formation in the presence of NO_x [15] a well as in O_3 destruction and OH reduction and production [16]. These processes can contribute to environmental pollution [17], thus influencing the global climate [18]. Oxidation of BVOC in the atmosphere may result in positive or negative feedbacks on the plants themselves and their BVOC production [19].

In order to model BVOC fluxes for different ecosystems [20–22] experimental data on the ecosystem-, tree- and leaf-level for parameterization and validation as well as a deeper process understanding are needed. BVOC fluxes at ecosystem-level are typically derived by micro-meteorological measurement techniques [23–29], whereas at plant- and leaf-level chamber/enclosure measurements [30–36] are used. Several excellent review articles [37–40] describe the relevant specifications and requirements for reproducible and accurate chamber experiments as well as potential sources of error. Ortega and Helmig [38] also gives a comprehensive overview on previously performed enclosure measurements. In general a dynamic chamber design with constant air exchange (mass flow controlled) is preferred, since this design may reach steady state conditions fast and consequently the built up of water vapor and extreme chamber heat is reduced [37–40]. Both factors are disadvantageous: water condensation in the chamber system would lead to compound losses and extreme heat would introduce stress for the plant [39], e.g., indicated by reduced transpiration and photosynthesis. Depending on the experiment location and design, regulation of temperature, CO_2 concentration and water vapor at inlets as well as illumination control should be considered. Thus, an effective and fast control of the environmental conditions for plants studied is desirable for achieving faster steady state conditions and thus stable gas exchange (see e.g., [41, 42]). In order to reduce wall losses or on-wall-reactions, inert materials should be used for constructing such a gas exchange study system, e.g., fluorinated plastics or stainless steel. In addition, a careful, fast, and accurate monitoring of the chamber environment and the plant status is needed for an exact quantification of leaf to air gas exchange.

The reported technical solutions range from simple branch bags [35, 43] to environmentally controlled inert chambers [36, 44–48]. Most studies use either commercial leaf chamber systems [32, 49, 50] or self-build chambers [44], yet multiple parallel (N > 2) chamber designs are rarely presented [51, 52]. Intensive BVOC screening studies or treatment-effect studies (e.g., stress vs.

control), however, would benefit from a greater number of simultaneously operated measurement chambers allowing larger sample sizes at a time or direct comparisons, respectively, and thus minimizing the number of (distracting) co-variables (e.g., growth or phenological development).

In order to investigate gas exchange of small trees under different environmental conditions and for different physiological states, the dynamic enclosure system Tree Drought Emission MONitor (Tree DEMON) was developed and evaluated. Using Tree DEMON BVOC emissions of four potted trees with a crown volume of up to 30 l were measured in parallel. Additionally, CO_2 and water vapor gas exchange as well as environmental parameters, such as air temperature, light, soil moisture, are monitored and controlled with an integrated data acquisition and control system. The focus of this study lies on the Tree DEMON development, its rigid performance testing and two case studies on BVOC emissions of sweet chestnut (*Castanea sativa* Mill.) demonstrating the power of the whole system for plant ecophysiological experiments.

Materials and methods
Tree DEMON system
System layout

The Tree Drought Emission MONitor (Tree DEMON) can be split into four functional units: (1) purge air supply and conditioning (Fig. 1a), (2) four dynamic plant chambers with environmental sensors (Fig. 1b), (3) BVOC sampler unit with four sample strings holding each four sample ports with adsorption tubes (Fig. 1c), and (4) CO_2 and water vapor gas exchange unit (Fig. 1d). This set-up allows measuring BVOC and CO_2 and water vapor gas exchange rates including chamber environment and other plant key parameters four times in parallel.

All system parts were selected for high material inertness, low gas permeability and, if possible, as industrial standard parts in order to ensure long-term maintenance. For further improving inertness almost all metal parts in contact with the chamber outlet air were chosen in grade 316 stainless steel (exceptions will be mentioned separately in the text). Since some polymers and rubber sealing can adsorb/desorb compounds [53, 54] PFA (perfluoralkoxy polymer) was used as tubing with 8 mm outer diameter. O-rings in valves and chambers consisted of FKM (fluoroelastomere) or PTFE (polytetrafluorethylene), respectively.

Air supply and conditioning

Chamber inlet air was conditioned in multiple steps from pressurized supply air in order to perform repeatable and reproducible gas exchange measurements (see Fig. 1a).

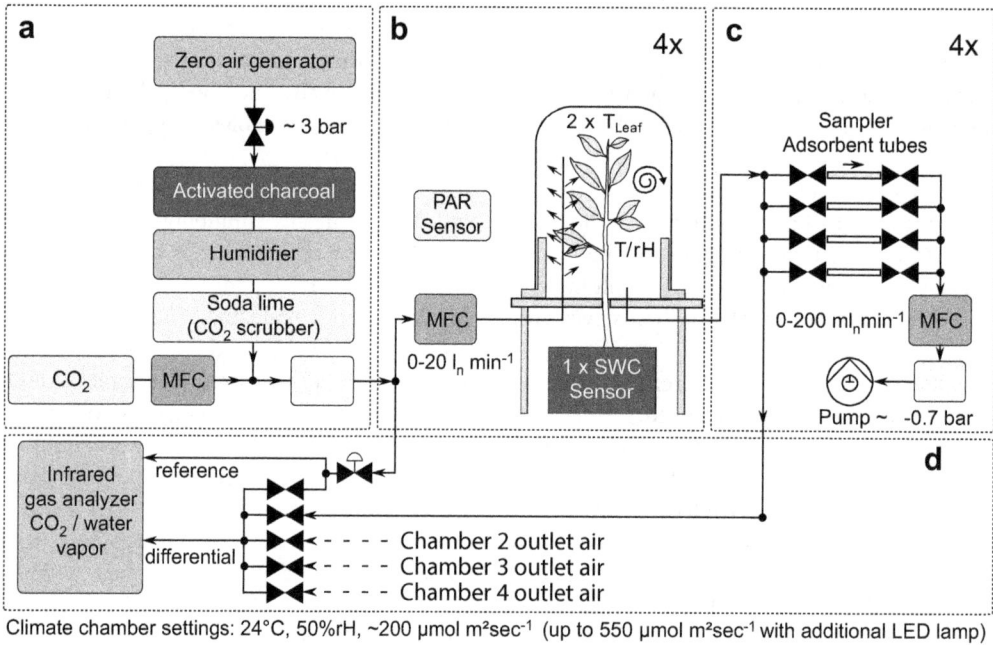

Fig. 1 System schematic. The Tree DEMON gas exchange study system design. **a** Purge air supply and conditioning; **b** chambers with sensors; **c** BVOC sampler unit containing four sampler strings with four ports each; **d** CO_2 and water vapor gas exchange unit. *Arrows* indicate the direction of air flow. *MFC* mass flow controller, *T/rH* temperature and relative humidity sensor, T_{leaf} leaf temperature sensor, *PAR sensor* photosynthetically active radiation sensor, *SWC sensor* soil water content sensor. A PID (proportional–integral–derivative) controlled valve was used to stabilize pressure to a constant level

More specifically, VOC-free air was produced by scrubbing the pressurized air with a zero air generator (AER-O40LS-80, PEUS INSTRUMENTS GmbH, Gaggenau, Germany, with up to 80 l min^{-1}) followed by an additional adsorption system of activated charcoal (VWR International GmbH, Darmstadt, Germany) in a 10 l stainless steel tank (RP216-1II-10-20-D, THIELMANN UCON GmbH, Hausach, Germany). Since supplied air pressure fluctuated due to the compressor intervals, pressure was stabilized with a proportional valve (PV22-20S, Aircom, Ratingen, Germany) which was controlled by a software based PID (proportional–integral–derivative) controller coupled with a pressure transducer (DRTR-ED-10V-R6B, B+B Thermo-Technik GmbH, Donaueschingen, Germany). This improved the pressure stability in the air supply for the chambers to 3 bars with a standard deviation of ±0.01 bar.

After VOC cleaning, the air was humidified by purging it through a 10 l tank filled with ultrapure water. The humidified air was further filtered through a soda lime filled 10 l tank to adsorb all ambient CO_2. Under continuous operation the filter needed replacement after approximately two months. After air cleaning and

humidification pure CO_2 from a gas cylinder (99.995% purity, Westfalen Gas, Münster, Germany) was added to the air stream over a nozzle using a mass flow controller (SMART6 GSC, Vögtlin Instruments AG, Aesch, Switzerland, 0–200 ml_n min^{-1}) to set the desired CO_2 concentrations for the chamber air. Since micro fluctuation of the mass flow may lead to unsteady CO_2 levels, a downstream 2 l stainless steel tank (Festo AG, Esslingen, Germany) was used for mixing and stabilizing the CO_2 level. Finally, the preconditioned air was fed into the four plant chambers using four mass flow controllers (MFC) (SMART4S GSC, Vögtlin Instruments AG, Aesch, Switzerland, 0–20 l_n min^{-1}).

Plant chambers and environmental sensors

The plant chamber system separated the above ground parts of the plants from the surrounding environment of the climate chamber and the root space in the pot. This set-up ensured controlled and repeatable conditions during the course of the experiments (Fig. 2). In each plant chamber air was typically exchanged with 15 l_n min^{-1} to reach fast steady-state conditions for the studied plants.

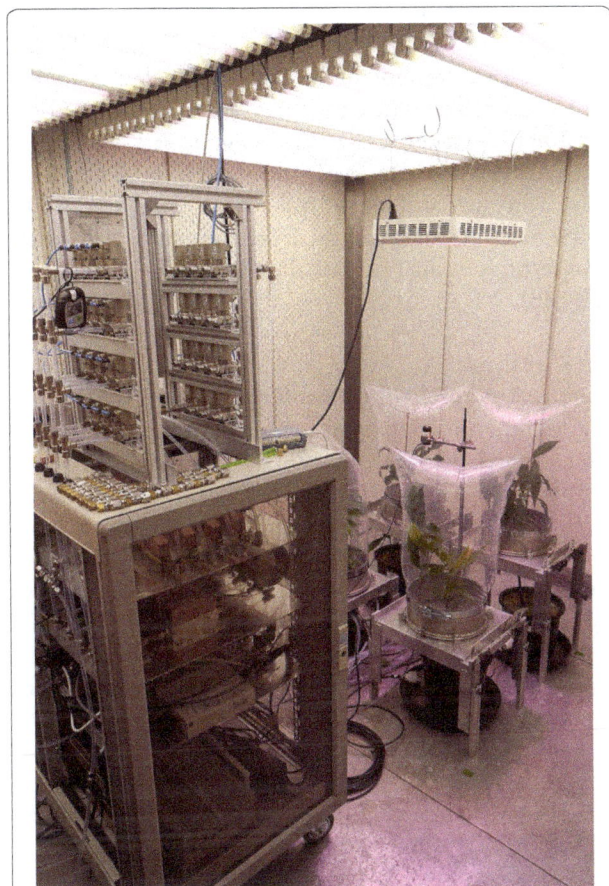

Fig. 2 Tree DEMON set-up in a climate controlled growth chamber. *Left side* BVOC sampler unit on top of a 19 in. rack with mass flow controllers, data acquisition unit, control pc and infrared gas analyzers (not visible, located behind the rack). *Right side* four plant chamber systems with built-in trees. *Top right side* LED panel for increasing light intensity received by the plants

The top part (hood) of the chambers was built of a 25 l polyvinylidenfluorid plastic air sampling bag (Supel™ Inert Film, Sigma-Aldrich Co. LLC, St. Louis, Missouri, USA). The hood then was mounted onto a stainless steel flange (BEVAB GmbH, Bergisch Gladbach, Germany) with a diameter of 272 mm and a height of 100 mm; fixed with a metal tension band and sealed by a FKM rubber band. The resulting plant chamber volume was approximately 30 l. The lower part of the chamber consisted of two ground plates made of polished duralumin metal to separate the tree top (crown, stem) from the pot (stem, roots). Both plates were placed on a height-adjustable table. This construction was extremely solid and stable and thus allowed a fast exchange of the plants.

Air inlet and outlet to the chambers as well as the combined air temperature and relative humidity sensor (FF-IND-10V-TE1, B+B Thermo-Technik GmbH,

Donaueschingen, Germany) were mounted on one of the ground plates. Leaf temperature was measured with two type K precision fine wire thermocouples (L-0044K-IEC, OMEGA ENGINEERING LTD, Northbank, Irlam, Manchester, UK) at the leaf reverse side at each tree. One soil moisture probe (SM-300, Delta-T Devices, Burwell, Cambridge, UK) was installed in each pot to measure volumetric soil water content. Photosynthetically active radiation (PAR) was measured by one PAR sensor (HOPL SKL 2620, Skye Instruments Ltd, Llandrindod Wells, Powys, UK) centered between the four chambers at mid-chamber height. Since the lamp in the climate chamber, in which the Tree DEMON system was placed for the experiments, only provided PAR around 250 µmol m^{-2} s^{-1}, an additional high power multi-spectral LED lamp (Bloom Power white 360, SPLED GmbH, Flensburg, Germany) was used to increase PAR up to 550 µmol m^{-2} s^{-1}.

The chamber air inlet was made of a stainless steel tube (1/4 in., closed at the end) with multiple micro nozzles (0.25 mm diameter) to generate turbulent mixing conditions within the chamber. This reduced the built-up of local concentration fields and ensured faster steady-state conditions. Both ground plates and the tree stem were sealed with a PTFE string/tape and the chamber flange were sealed by a flat FKM ring. Additionally, the chambers were operated at a slight overpressure (about 25 mbar) to prevent outside air leaking into the system. Finally, the outlet air was fed to the BVOC sample unit and CO_2 and water vapor gas exchange unit.

BVOC sample unit

The BVOC enriched outlet air was sampled via a bypass with the sampler unit (Fig. 2) with four separate sampler strings. Each sampler string had four ports holding the adsorption tubes (AT). Within a port the AT was separated from the bypass by two normally closed pneumatic stainless steel valves (VXA2120M-01F-1-B, SMC Pneumatik GmbH, Gröbenzell, Germany). ATs were connected via PTFE ferrules to a VCO® connector system (FITOK GmbH, Offenbach am Main, Germany) which allowed fast installation and reduced potential contamination.

At each port BVOC were sampled by drawing mass-flow-controlled outlet air (SMART4S GSC, Vögtlin Instruments AG, Aesch, Switzerland, 0–200 ml$_n$ min^{-1}) through the AT with a vacuum pump (GD-Thomas, Memmingen, Germany). The sample flow (standard: 150 ml$_n$ min^{-1}) was only active if both valves were open, so damage on the adsorption material due to a rapid pressured drop at sampling start was avoided. Furthermore, sample duration (standard: 60 min) and timing was completely customizable.

CO_2 and water vapor gas exchange unit

After the sampling unit, the net photosynthesis and transpiration rate of the enclosed trees was calculated using differences of CO_2 and water vapor in chamber inlet and outlet air monitored by a CIRAS-2 DC (PP-System, Amesbury, Massachusetts, USA) differential infrared gas analyzer system (IRGA). The IRGA measured conditioned chamber inlet air continuously at the reference channel (see Fig. 1d). Air downstream the chamber was fed via a magnetic valve manifold (E111AAV20/A/301 Fluid Concept GmbH, Karlsdorf-Neuthard, Germany) to the difference channel. This configuration allowed a subsequent monitoring of all four chambers. IRGA stability and off-set was checked by measuring reference air at both channels every 5 min. Net photosynthesis rate and transpiration rate were calculated according to von Caemmerer and Farquhar [55].

Software and measurement hardware

All functional units of the Tree DEMON were integrated into a LABVIEW (National Instruments, Austin, Texas, USA) based control and data measurement software (Fig. 3a, b). All sensors and relays for valve and lamp control were connected to a NI PCIe-6323 measurement card (National Instruments, Austin, Texas, USA); thermocouples were connected to a USB-TEMP measurement box (Measurement Computing, Norton, Massachusetts, USA) and MFCs were controlled over a digital MODBUS® system. The IRGA was controlled and logged via a serial connection. The software controlled and recorded automatically timing and settings and signals of all MFCs, valves and lamps (Fig. 3b).

BVOC analysis

Compound adsorption and thermal desorption

Inert silica coated stainless steel AT (CAMSCO, Houston, Texas, USA) with a two bed configuration of 40 mg Carbograph® 5TD and 70 mg Tenax® TA and a mesh size of 60/80 were used to sample and pre-concentrate emitted compounds of interest. No breakthrough of target isoprenoids compounds was detected. This was checked by placing two AT in a row and calculating the recovery rate on the first AT using sample calibration gas under standard sample settings and checking the second tube for compounds.

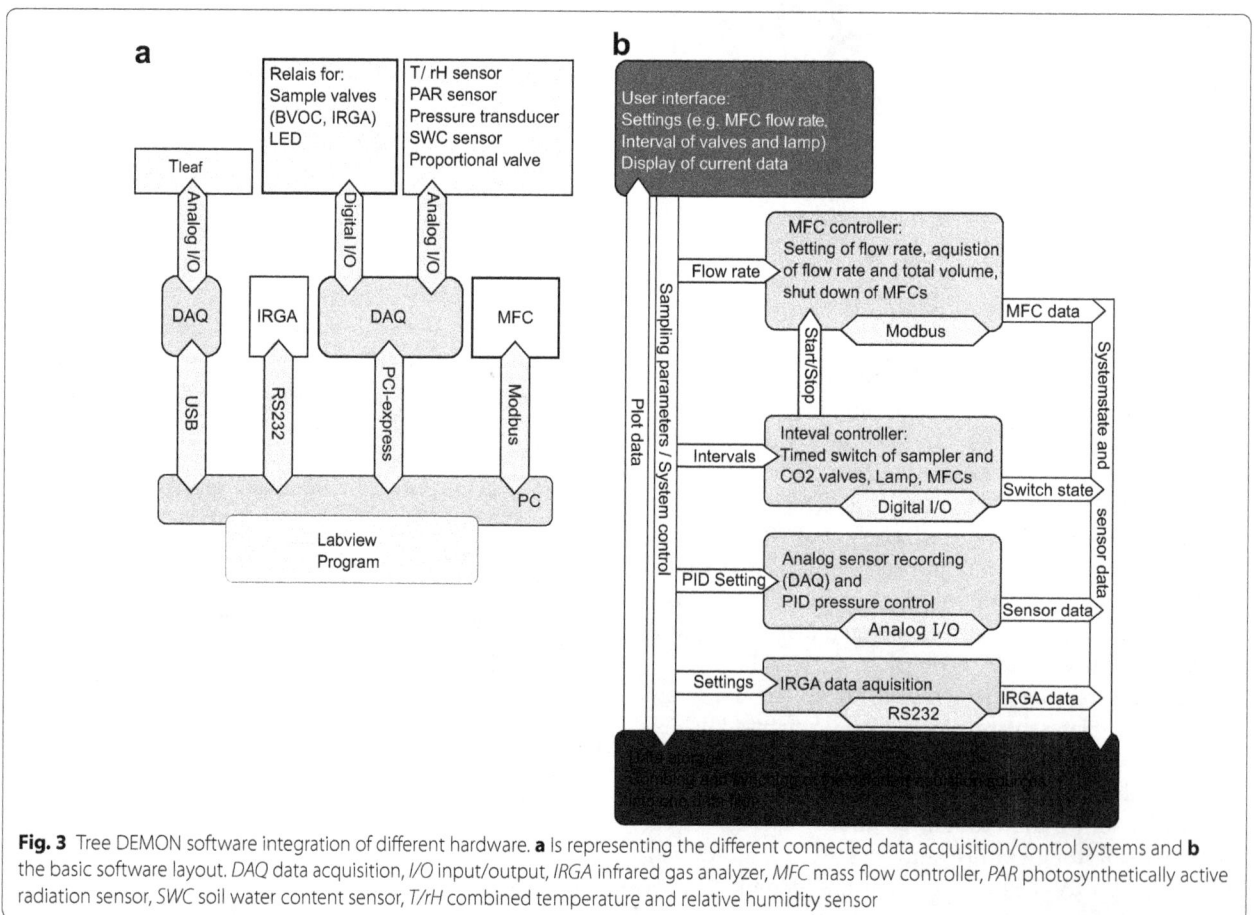

Fig. 3 Tree DEMON software integration of different hardware. **a** Is representing the different connected data acquisition/control systems and **b** the basic software layout. *DAQ* data acquisition, *I/O* input/output, *IRGA* infrared gas analyzer, *MFC* mass flow controller, *PAR* photosynthetically active radiation sensor, *SWC* soil water content sensor, *T/rH* combined temperature and relative humidity sensor

Compounds from the AT were transferred to the gas chromatograph (GC, see "Compound analysis" section) using an automatic thermal desorber (ATD 650, Perkin Elmer, Waltham, Massachusetts, USA). The AT was first dry purged with helium for 3 min and then the compounds were desorbed with 25 ml min^{-1} for 10 min at 280 °C with a split flow of 2 ml min^{-1}. A Peltier-cooled cold trap (TurboMatrix Air Monitoring Trap™, Perkin Elmer, Waltham, Massachusetts, USA) pre-focused the desorbed compounds at −30 °C. The pre-focused sample was then injected via a transfer line (silica capillary maintained at 250 °C) with a helium carrier gas flow of 1.5 ml min^{-1} onto the separation column of the GC during ballistic heating (40 °C s^{-1}) of the cold trap up to 300 °C. During the analysis of the compounds the respective AT was reconditioned with helium at 300 °C in the ATD.

Compound analysis

For analyzing the sample matrix, a GC/MS-FID (CLARUS® SQ8, Perkin Elmer, Waltham, Massachusetts, USA) system with an Elite 5 MS column (33 m, 250 µm, 95% Methylpolysiloxane, 5% Phenyl, Perkin Elmer, Waltham, Massachusetts, USA) was used. The gas stream was split after leaving the separation column and compounds were detected using a flame ionization detector (FID) and a quadrupole electron ionization mass spectrometer in parallel. Compounds were separated by the following GC temperature program: initial temperature of 40 °C for 4 min, first ramp with 15 °C min^{-1} up to 100 °C, second ramp with 5 °C min^{-1} up to 230 °C and a 4 min hold at the end. The mass spectrometer was set to full scan mode from 33 to 330 DA and 70 eV to detect and quantify unknown compounds. The FID was set to 300 °C with a flow of 40 ml min^{-1} H$_2$ and 400 ml min^{-1} zero air. Unknown substances were identified according to the fragmentation patterns using the NIST database 08 [56] and if available by respective pure standards. Calibration was done by sampling a 16-component BVOC (C5–C12) gas standard (1.81–2.22 ± 0.09–0.30 nmol mol^{-1}, expanded uncertainty, NPL, Teddington, Middlesex, UK) onto the AT over an extra sampler system which was identical to the one used in the Tree DEMON. Additionally, 50 ml of internal standard of Δ^2-carene (87 ± 10.4 nmol mol^{-1}, expanded uncertainty, Siad Austria GmbH, St. Pantaleon, Austria) were added onto each AT prior to air sampling. (Chromatograms of the standards can be found in Additional file 1: Fig. S1). For quantification the relative response factor (RRF) between each compound of the calibration gas and the internal standard were calculated from the FID signal of the respective compound. For compounds not present in the calibration gas, a structural equivalent RRF was used. A Level of Quantification

(LOQ) of 0.004 nmol mol^{-1} was achieved and values below LOQ limit were handled as zero.

Emission rate calculation

Emission rate EM (nmol m^{-2} s^{-1}) is determined by following equation Eq. 1 [39]:

$$EM = (\chi_{out} - \chi_{in}) F_M A_{Leaf}^{-1} \tag{1}$$

here EM it is derived by the difference between out- and inlet BVOC mixing ratios χ_{out} and χ_{in} (nmol mol^{-1}) which are multiplied by the inlet volume passing through the chamber within one second F_M (mol s^{-1}) (determined by the mass flow controller) and divided by is the leaf area A_{Leaf} (m^2) (see leaf area determination in the next chapter). However, it was assumed that incoming BVOC mixing ratio χ_{in} was zero due to air cleaning. Furthermore, increased water vapor induced by plant transpiration E (mol m^{-2} s^{-1}) within the chamber was corrected. This is necessary since induced water vapor dilutes the BVOC concentration in the chamber and the inlet volume does not include the added water vapor, thus the mass balance has to be corrected. Inlet air cleaning and water vapor correction resulted in Eq. 2:

$$EM = \chi_{out} F_M A_{Leaf}^{-1} + \chi_{out} E \tag{2}$$

Finally emissions were standardized to 30 °C and 1000 µmol m^{-2} s^{-1} for better comparability with Eq. 3 [57]:

$$EM_{std} = \frac{EM}{f_{Tl} f_Q} \tag{3}$$

with the correction algorithm f_{Tl} and f_Q adjusting for the effects of leaf temperature and PAR, respectively (see Additional file 1: Equation S1, S2, and S3 for detailed algorithm). For comparability to other studies the emission rates were also converted into mass based emission rates (mass of emitted compound per dry mass leaf and hour in µg g$_{dw}^{-1}$ h^{-1}).

System evaluation and characterization tests

Table 1 provides an overview of conducted BVOC sampling performance tests and tests on potential chamber side effects.

Sampler unit repeatability and reproducibility

The repeatability and reproducibility between each sampler string was evaluated by sampling Δ^2-carene enriched air on each sample unit simultaneously. For this test, gas from the Δ^2-carene standard instead of CO$_2$ was added over a MFC (~50 ml$_n$ min^{-1}) into the Tree DEMON supply air. The Δ^2-carene concentration in the enriched air

Table 1 Performance tests of the Tree DEMON for BVOC emission studies

Test	Method	N	Settings
BVOC sampler unit repeatability and reproducibility	Sampling of Δ^2-carene enriched air direct at chamber inlet	48 (4 sampler strings, 4 ports, three repetitions)	Sample rate 150 ml$_n$ min^{-1}, Sample duration: 10 min, Standard gas addition: 50 ml$_n$ min^{-1} of Δ^2-carene Chamber flow rate: 5 l$_n$ min^{-1} (~0.22 nmol mol^{-1} Δ^2-carene mixing ratio in the inlet air)
Chamber wall effects	Sampling Δ^2-carene enriched air direct at chamber inlet and outlet simultaneously at 2 chambers	20 (2 chambers with each 2 sampler strings at inlet and outlet, 5 measurements)	See above
Residence time of compounds	Sampling system air after 1 h stop of Δ^2-carene enrichment	12 (2 chambers with each 2 sampler strings at inlet and outlet, 3 measurements)	See above

N number of measurements

was estimated to be at 0.22 nmol mol^{-1} and was sampled directly after the four inlet MFCs (5 l$_n$ min^{-1}) with the sampler unit. In each sampler four samples with three repetitions (N = 12 per sampler) were taken. The duration for each sampling was 10 min and a flowrate was set to 150 ml$_n$ min^{-1}. Afterwards the FID area count for the Δ^2-carene was determined for each AT and was used to perform the comparison statistics. Repeatability was determined for one repetition with four samples for each sampler string by mean FID area count and its standard deviation. Repeatability for the whole sample unit was determined with the mean and its standard deviation from the mean FID area count of each sample string. Reproducibility of each sampler string was described by the mean and the standard deviation of the mean FID area count of each sampler string calculated for each of the three replications. Reproducibility for the whole sample unit was determined with the mean and its standard deviation from the mean FID area count of each sample string of all three repetitions.

Chamber wall effects and compound residence time

Potential chamber-wall effects were tested with the same the Δ^2-carene addition used in the sampler repeatability and reproducibility test. Analysis of the AT was performed as described above. However, for this test the Tree DEMON sample layout was reconfigured, so that always two sample units were used for sampling inlet and outlet air of two empty chambers in parallel. The inlet and outlet chamber air was sampled five times for 10 min with a flowrate of 150 ml$_n$ min^{-1} and flow rate the chamber was 5 l$_n$ min^{-1}.

Background air screening and residence time of the Δ^2-carene standard was checked by taking three samples three times at in- and outlet for each chamber at 1 h before and at 1 h after feeding 50 ml$_n$ min^{-1} Δ^2-carene standard gas with an inlet air flow rate of 5 l$_n$ min^{-1} into the chambers (see settings at sampler repeatability). Both

tests were performed at one third of the normal chamber flow rates and at shorter sample. These settings were necessary, since with lower flow rates the Tree DEMON pressure stabilizing system regulated faster and standard gas enrichment was more stable. Additionally since no trees were installed a much higher tightness was achieved, so lower chamber flow rates could be used. Also here, the FID area counts for the Δ^2-carene were determined for each AT and were used to perform the comparison statistics.

Air mixing, exchange rates and overpressure in the chambers

For reproducible measurements of gas exchange and BVOC emissions a well-mixed chamber was required in order to achieve rapidly steady-state exchange conditions and reduce local concentration fields. Air mixing of chamber was visually checked and recorded with a camera (Nex 6, Sony, Tokyo, Japan) by injecting white smoke into one chamber. The smoke was formed by passing humidified air via sulfuric acid in a smoke tube (Dräger Safety AG & Co. KGaA, Lübeck, Germany) into the middle of the chamber. Inlet air flow rate was first set to zero for a static state and switched to 10 l$_n$ min^{-1} to show the dynamic mixed state.

Chambers were also tested in terms of exchange time of CO_2 by changing CO_2 inlet mixing ratios from 400 to 0 µmol mol^{-1} and measuring the duration until the difference channel was zero again. The chamber flow rate was set to 5 l$_n$ min^{-1}, which was considered as minimum chamber flow rate.

Overpressure was measured by placing a pressure sensor (LPS25H barometer sensor, Geneva, Switzerland, ST Microelectronics) into an empty chamber and measuring air pressure at 0 and 15 l$_n$ min^{-1} air flow, respectively.

Lamp characterization and chamber PAR transmissivity

Two different types of light sources were installed and light distribution and spectral characteristics were tested, since these factors directly affects emission rates for light-dependent emitted compounds as well as photosynthesis rates. In the test set-up PAR was measured at three height levels (0/30/60 cm) above the plant chambers in a 20 cm × 20 cm grid for an 6400 cm^2 area where all chambers fitted in. Distance from the chamber top to each light source was 65 cm for the LED and 105 cm for the climate chamber neon tubes. The spectral characteristics of the neon tubes, a 2:1 mixture of Lumilux Cool White (OSRAM, Munich, Germany) and plant lights Fluora (OSRAM, Munich, Germany), and additionally the neon tubes together with the LED were measured with a spectral radiometer (LI-1800, LI-COR, Lincoln, Nebraska, USA). Due to a slight opaqueness of the chamber material used, PAR transmissivity tests were performed. A piece of the chamber material was placed onto the spectral radiometer as well as and the extinction was measured by determining the differences of the sensors response with and without chamber material at two light levels (PAR sensor level) of 250 μmol m^{-2} s^{-1} for the neon tubes and 550 μmol m^{-2} s^{-1} for neon tube and LED, respectively. Due to the lower position of radiometer reported, which was at around 10 cm above the chamber bottom, the reported light levels are lowered to 149 and 295 μmol m^{-2} s^{-1}, respectively. PAR transmissivity of the chamber was measured by placing a piece of chamber foil on the PAR sensor at light level of 187 μmol m^{-2} s^{-1}.

Case studies

The value of Tree DEMON for plant gas exchange studies was demonstrated by two case studies using sweet chestnut (*Castanea sativa* L.) trees: (1) a BVOC screening to investigate the emission composition per specimen and (2) a drought experiment to demonstrate likely impacts on CO$_2$ and water vapor exchange and possibly also on BVOC emission of sweet chestnut. For both studies a total of 40 one-year-old seedlings were planted into 5 l pots with a substrate mixture of 70% sand and 30% humus already in November 2013. All pots were arranged within a greenhouse and were irrigated during wintertime by hand and in summertime by a dripping water system (Netafim Ltd, Tel Aviv, Israel) with 0.5 l per pot between two to four day intervals depending on meteorological conditions. Further, the plants were fertilized with a 0.5‰ solution (FERTY® 2 and 3, Planta Düngemittel GmbH, Regenstauf, Germany) six times from May to July. For the case studies, 20 of the 40 trees were randomly selected and at a time four of those were studied in parallel.

BVOC screening

The BVOC screening study on the trees was performed in June 2014 after complete leaf development. Plants were checked for insect or fungal infestations and carefully cleaned from dust before installation into the Tree DEMON to ensure that no optical visible biological stressors impacted the results. 20 well-watered 2-year-old trees were sampled at 5 days from 17.06 to 25.06.2014 each twice at two light levels (250 and 550 μmol m^{-2} s^{-1}) and subsequently, emission rates and compound composition were determined (detailed experiment description below).

Drought application

In July 2014, the effect of decreasing water availability on BVOC emission rates and CO$_2$ and water vapor gas exchange was investigated on four sweet chestnut trees from the screening experiment. Trees were installed at 11.07.2014 in the afternoon and were allowed to acclimate to the climate chamber for 17 h until the first sampling. During the first two days of the experiment starting at 12.07.2014 emission of all trees were considered as non-stressed. Next, watering of two of the four specimens was stopped for six days until reaching a soil water content (SWC) of 0.04 m^3 m^{-3}. The watering of the other two trees took place between 13:00 and 15:00 with each with 250 ml of tap water. The gas exchange was assessed for each tree four times a day using the settings described below.

Sampling procedure and environmental settings

Plants were installed at least 12 h before the first BVOC sample to ensure acclimation to the climate chamber environment. Environmental parameters of the climate chamber (dimension h × l × w: 2.25 m × 4.45 m × 2.75 m) were set to a constant temperature of 24 °C, 50% relative humidity and a simulated 14 h day and 10 h night pattern. Light intensity of the neon tubes was controlled by a ramped program to simulate a diurnal distribution. The initial light intensity (PAR) started with 75 μmol m^{-2} s^{-1} for 1 h, followed by an increase to 150 μmol m^{-2} s^{-1} for 1 h, and a 2 h light intensity of 250 μmol m^{-2} s^{-1}. From early noon to mid-afternoon an additional LED light source was used to raise the light intensity up to 550 μmol m^{-2} s^{-1} for 6 h. Light intensity in the evening was reduced with the same but reversed steps as in the morning.

BVOC sampling was conducted at two light intensities at around 250 μmol s^{-1} m^{-2} at 9:10 and 10:15 and at 550 μmol s^{-1} m^{-2} at 11:15 and 12:30 for 60 min each with a sample flow rate of 150 ml$_n$ min^{-1} and a chamber air flow rate of 15 l$_n$ min^{-1}. Determined emission rates were standardized to 30 °C and 1000 μmol m^{-2} s^{-1} by Eq. 3.

Biomass assessment

Leaf area of the 20 specimens in the BVOC screening study was estimated non-invasively by measuring length and width of each leaf (screening study) since the trees were (partly) further used in the drought experiment. These parameters were converted to leaf area with a regression function fitted by the method of [58]. After the drought application, leaf area was determined with ImageJ [59] by first harvesting and then scanning the fresh leaves. Dry leaf weight was determined after drying leaves for 48 h at 60 °C. The difference between both methods was not significant tested with paired Student t test.

Statistical analysis

Data processing and statistical analysis was performed with R 3.1 [60].

For assessing the repeatability and reproducibility in the system evaluation, the relative standard deviation within and between sample units was checked and an ANOVA with a post hoc Tukey's test was performed to check for differences between the samplers. The difference between chamber in- and outlet air was checked with the paired Student's t test.

For the screening study the average standardized relative emission of each tree was clustered by partitioning the relative compound information around medoids (PAM) with R package cluster [61]. The optimal cluster number was selected by the highest average silhouette size from one to ten calculated clusters.

Results
System evaluation
Tests of sampler unit

The air sampler test experiment revealed a relative standard deviation (RSD, n = 4) of the repeatability ranging between 0.74 and 2.26% using Δ^2-carene as tracer (Table 2). Reproducibility of each sample unit showed a RSD between 0.63 and 2.15%. On average, the RSD for all sample units was 0.56% for the reproducibility and 0.62% for the single repeatability test, respectively. Further, an additional Tukey's post hoc test on the ANOVA results for the area counts of each sampler string revealed no significant differences between each other.

Chamber wall effects and compound residence time

Samples before Δ^2-carene addition did not show any Δ^2-carene or coeluting contaminations. Further no chamber wall effects were observed since there were no significant differences in the FID area counts between sampled Δ^2-carene enriched air at the in- and outlet of the chambers (paired t test, $p = 0.16$, df = 9).

The residence time of Δ^2-carene was less than 1 h, since samples taken 1 h after the end of internal standard addition did not show any residual Δ^2-carene.

Additional performed blank test with preconditioned air for each chamber showed only little contamination with some compounds. These trace compound contamination occurred, however, in chromatogram windows outside the retention times for the target compounds (see also Additional file 1: Fig. S2). Possibly, the applied air filtering system was not effective enough and processes to adjust the humidity and CO_2 concentration in the chamber air may have caused additional contamination.

Chamber mixing

The first test with zero chamber flow confirmed that there was almost no mixing of injected sulfuric acid particles (see Additional file 1: Fig. S3, left). After setting the air flow rate in the chamber to 10 l_n min^{-1} (the standard flow was 15 l_n min^{-1}) the chamber air was already well mixed (see Additional file 1: Fig. S3, right). The mixing was completed within seconds after the onset of the chamber air flow (see Additional file 2: Video).

Chamber air exchange rates and tightness

The complete air exchange of one chamber took around 45 min with an inlet air flow rate of 5 l_n min^{-1} (one third of the standard flow rate of 15 l_n min^{-1}), which was shown by the CO_2 removal from a mixing ratio of 400 to 0 μmol mol^{-1} (see also Additional file 1: Fig. S4). In case of installed trees the outlet flow amounted to 50–75% of the inlet flow due to leaks in the chamber system, such as at the stem sealing. Slight overpressure in the system reduced the risk of contamination of the chamber air with outside air. The overpressure was 25 mbar

Table 2 Repeatability and reproducibility tests

Sampler string	Repeatability (mean FID counts)	RSD (%)	N	Reproducibility (mean FID counts)	RSD (%)	N
1	3690.25	2.26	4	3706.17	0.63	3
2	3664.75	0.74	4	3664.92	0.89	3
3	3642.00	1.58	4	3674.17	0.83	3
4	3688.75	1.90	4	3661.33	2.15	3
Average	3671.44	0.62	4	3676.65	0.56	4

Tests of the sampler strings with FID counts of repeated samples of 1.5 l air with 0.22 nmol mol^{-1} Δ^2-carene

RSD relative standard deviation, *N* number of repetitions

in a pressure test with a chamber inlet flow rate of 15 l_n min^{-1}. The corresponding outlet flow rate was 9.5 l_n min^{-1}. In tests with empty chambers no target compounds were detected. Also in-chamber air CO_2 mixing ratio (chamber: 404 µmol mol^{-1}, ambient mixing ratio 400–1000 µmol mol^{-1}) remained constant indicating no inflow from outside air. Even at 0 µmol mol^{-1} CO_2 in chamber air no CO_2 diffusion from outside was observed.

Lamp characterization and chamber film PAR transmissivity

The neon tube mixture showed local peaks in 436, 546, 612 and 812 nm; additionally a local increase at 490 and 650 nm was visible in Fig. 4a. The LED lamp showed its local maximal intensities at 455 and 666 nm wavelength as seen in Fig. 4a.

The upper part of the chambers was built from transparent PVDF plastic with a PAR transmissivity of 97%; therefore outside measured PAR had to be reduced by 3% for inner chamber estimates. As shown in Fig. 4a, the foil showed same reductions of PAR at all wavelengths from 400 to 700 nm. The light intensity and distribution is shown in Fig. 4b. At height levels of 0 and 30 cm we see a uniform distribution under LED + climate chamber light, whereas at 60 cm the intensity was less uniform and increased towards the center spot of the LED light.

Case studies

Results of screening study

In total 20 trees were screened in June 2014 with focus on monoterpenes and in total 15 different compounds were identified by the NIST library or by gas standards (see Fig. 5).

The standardized total monoterpene emission rate was on average 0.14 ± 0.16 nmol m^{-2} s^{-1} (0.45 ± 0.93 µg g_{dw}^{-1}) and ranged from almost below the detection limit [0.01 nmol m^{-2} s^{-1} (0.07 µg g_{dw}^{-1} h^{-1})] up to 0.68 nmol m^{-2} s^{-1} (3.93 µg g_{dw}^{-1} h^{-1}; see Fig. 5a).

Analysis of the relative compound emissions by PAM clustering and silhouette width resulted in three clusters (see Fig. 5b). These clusters could be separated into a trans-β-ocimene (>50%) dominated cluster (cluster 1), an intermediate cluster with higher shares of α-/β- and γ-terpinens and α-thujene (cluster 2), and α- and β-pinene dominated (>25%) cluster (cluster 3).

Results of drought experiments

In Fig. 6 daytime averaged soil water content, transpiration rate, net photosynthesis rate and BVOC emission rate for all measurements is shown for the drought application experiment. SWC served as proxy of the drought stress experienced by two trees (SWC < 0.09 m^3 m^{-3}). Non-watering of the selected trees led to a fast

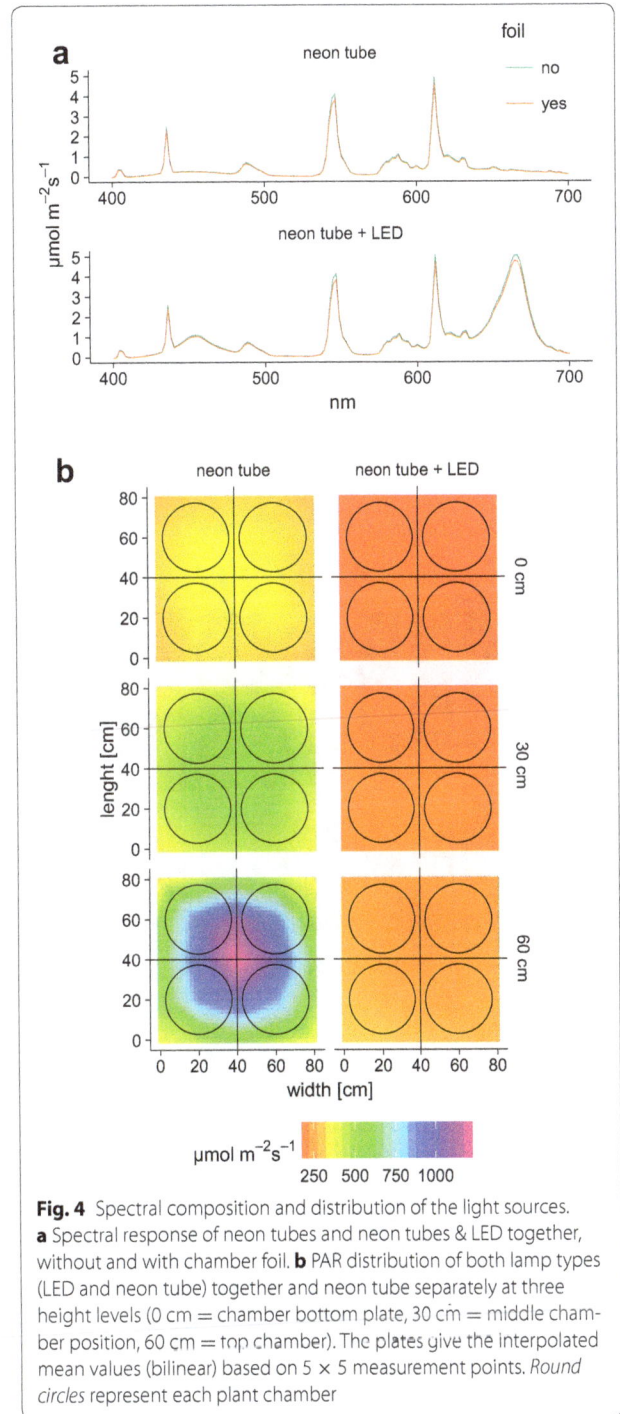

Fig. 4 Spectral composition and distribution of the light sources. **a** Spectral response of neon tubes and neon tubes & LED together, without and with chamber foil. **b** PAR distribution of both lamp types (LED and neon tube) together and neon tube separately at three height levels (0 cm = chamber bottom plate, 30 cm = middle chamber position, 60 cm = top chamber). The plates give the interpolated mean values (bilinear) based on 5 × 5 measurement points. *Round circles* represent each plant chamber

decline of water availability and the permanent wilting point at a SWC of 0.06 m^3 m^{-3} was reached between day five and seven. In average an inlet concentration was measured of 403.85 ± 1.40 µmol mol^{-1} for CO_2 and 8.47 ± 0.17 mmol mol^{-1} for water vapor at all chambers during the experiment. With respect to gas exchange, one of the control trees (#3) showed a slight decrease of

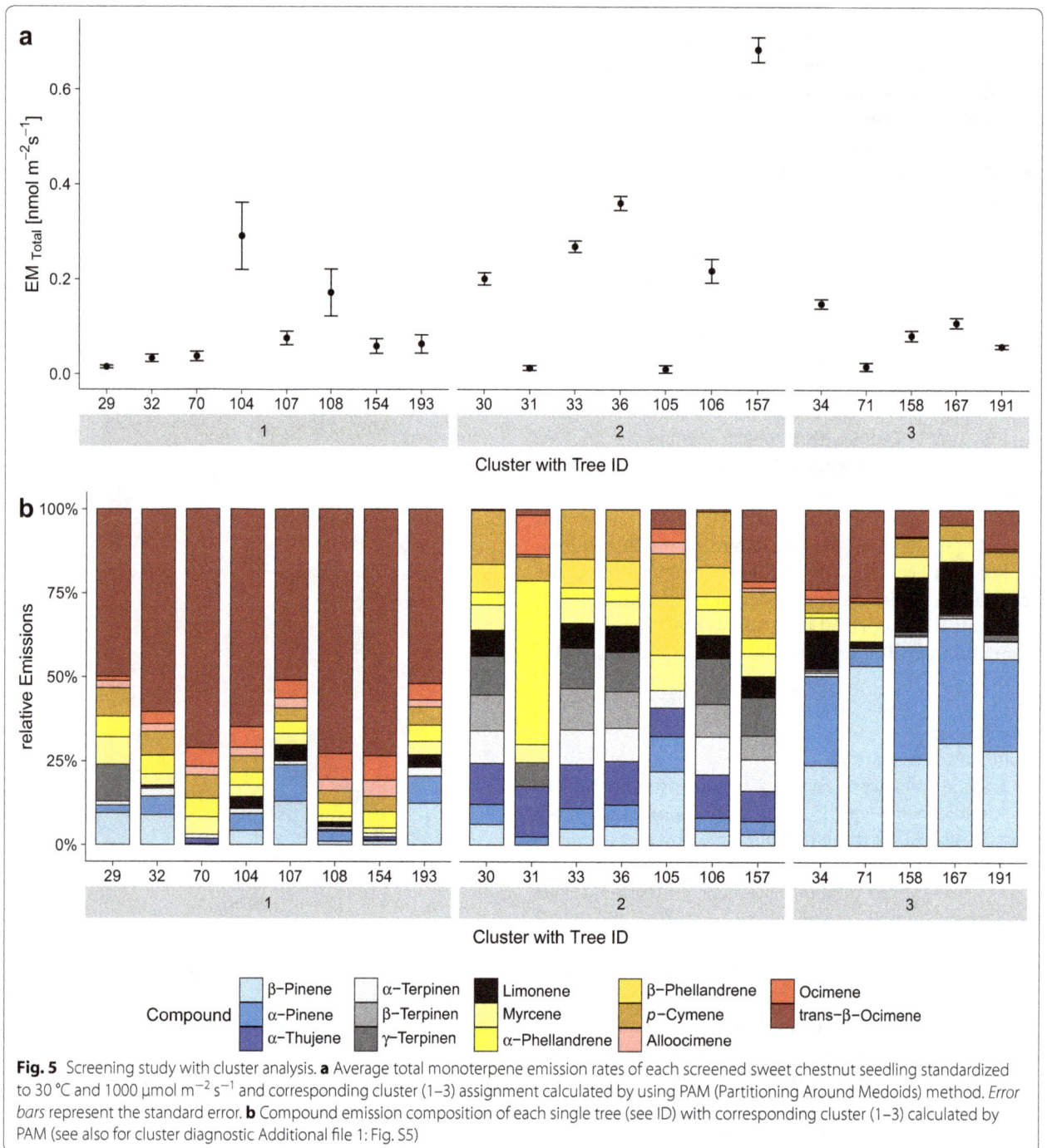

Fig. 5 Screening study with cluster analysis. **a** Average total monoterpene emission rates of each screened sweet chestnut seedling standardized to 30 °C and 1000 µmol m^{-2} s^{-1} and corresponding cluster (1–3) assignment calculated by using PAM (Partitioning Around Medoids) method. *Error bars* represent the standard error. **b** Compound emission composition of each single tree (see ID) with corresponding cluster (1–3) calculated by PAM (see also for cluster diagnostic Additional file 1: Fig. S5)

the transpiration rates from 1.90 to 1.72 mmol m^{-2} s^{-1} and of photosynthesis rates from 7.10 to 5.35 µmol m^{-2} s^{-1} during the 9 days of the experiment, whereas tree #4 showed transpiration rates ranging between 1.48 and 1.60 mmol m^{-2} s^{-1} and photosynthesis rates between 6.53 and 5.47 µmol m^{-2} s^{-1}. Concerning the stressed trees, #1 showed stable transpiration rates of around 1.40 mmol m^{-2} s^{-1} and photosynthesis rates of 5.95 µmol m^{-2} s^{-1} for the first three days and then during drought application a rapid decrease until the end of the experiment. Tree #2 showed a similar but later decrease of gas exchange rates during the drought application. There is no obvious explanation for the short drop of 50% in gas exchange observed on day 2 (#2).

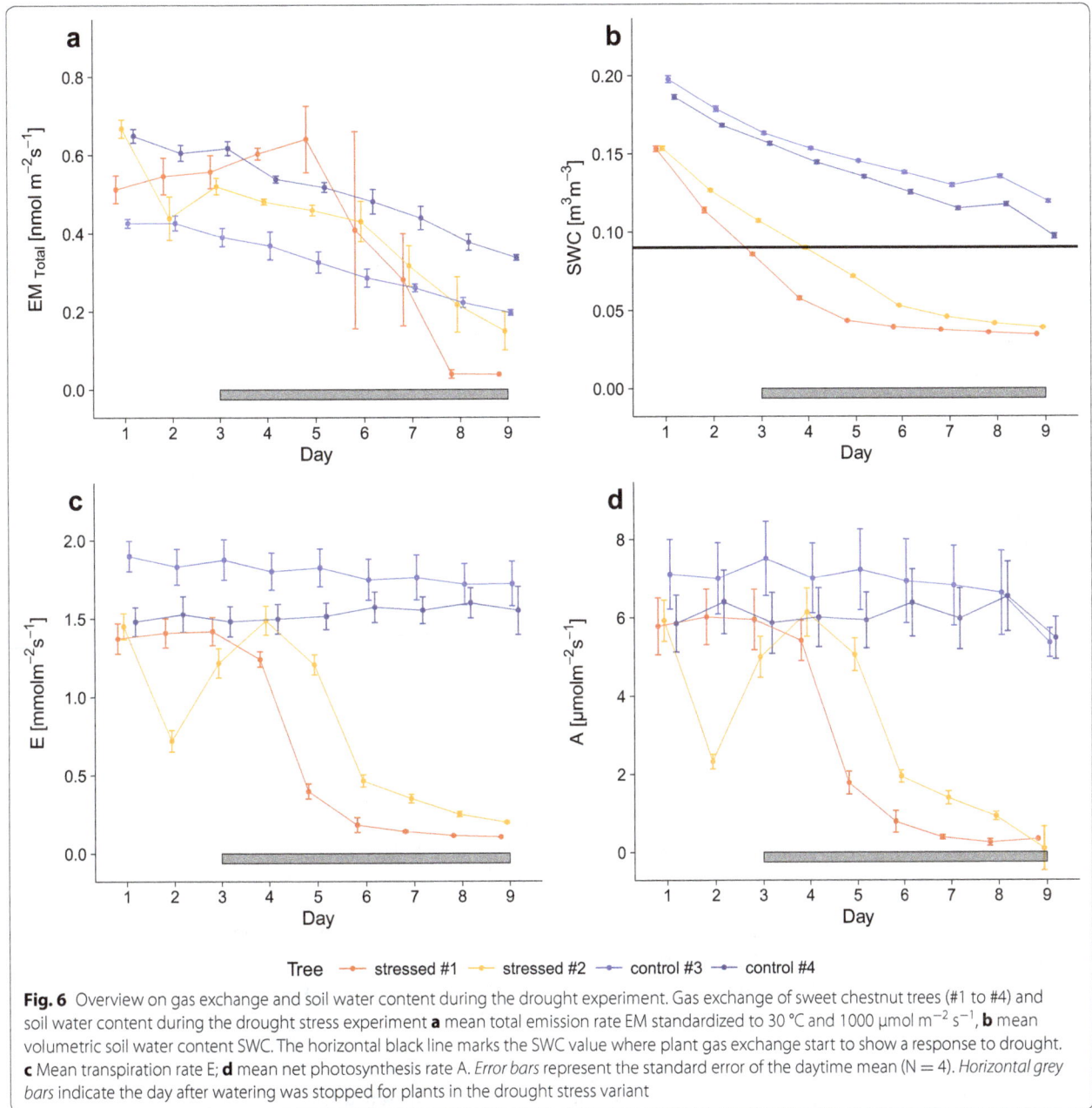

Fig. 6 Overview on gas exchange and soil water content during the drought experiment. Gas exchange of sweet chestnut trees (#1 to #4) and soil water content during the drought stress experiment **a** mean total emission rate EM standardized to 30 °C and 1000 μmol m^{-2} s^{-1}, **b** mean volumetric soil water content SWC. The horizontal black line marks the SWC value where plant gas exchange start to show a response to drought. **c** Mean transpiration rate E; **d** mean net photosynthesis rate A. *Error bars* represent the standard error of the daytime mean (N = 4). *Horizontal grey bars* indicate the day after watering was stopped for plants in the drought stress variant

Monoterpene emissions rate of all trees decreased during the drought experiment, despite watering of the control trees. However, drought-stressed trees showed a much stronger response to decreasing SWC. Within the first three days of the experiment, when all trees could be considered as non-stressed, emission rates EM ranged between 0.43 and 0.68 nmol m^{-2} s^{-1}. At the end of the experiment the emission decreased for non-stressed trees by 50% from 0.43 to 0.20 nmol m^{-2} s^{-1} for #3 and from 0.65 to 0.34 nmol m^{-2} s^{-1} for #4, respectively. The emission rates of the stressed trees decreased from 0.52

to 0.038 nmol m^{-2} s^{-1} for #1 and 0.67–0.14 nmol m^{-2} s^{-1} for #2, respectively. However for #1, first an increase of emission was observed followed by a sharp decrease to 0.038 nmol m^{-2} s^{-1} at day 8 and 9.

Discussion

Tree DEMON BVOC sampling performance

Compared to other existing chamber systems (for instance two chamber system [36, 44]; three chamber system [45]) the Tree DEMON is able to investigate up to four young trees in parallel similarly to a much smaller system used for investigating leaves described by Ghirardo et al. [62]. The four chambers enable an increased sample size and allow experimental designs with two treated and two control trees investigated in parallel. The Tree DEMON takes four samples in sequence per chamber automatically, resulting in a total number of 16 samples per day, which was the optimal workload for the following chemical analysis that allowed a continuous operation over several weeks. The sample number was in the range of other automatic sorbent tube samplers, where up to 10 automatic samples [63], up to 20 automatic samples [64] and up to 24 automatic samples [37] could be taken. Performance of the sampler strings was shown with a repeatability test, in which the relative standard deviations of 0.7–2.3% were in the same range of other sample systems such as proposed by [64]. The calibration with gas standards in an identically constructed external sampler run in the laboratory compensated for likely remaining not accountable uncertainties through the sampling system, such as by dead volume and potential wall losses of valves, connectors and tubing.

The chamber supply air filtering techniques, also used by other studies [37, 44, 47, 65], allowed to maintain background VOC concentration low and free from target compounds, so an additional sampling of inlet air concentration was not necessary, thus halving the number of samples per measuring point. Additionally, the chamber air test with addition of Δ^2-carene, serving as a proxy for monoterpenes, did not show any significant wall and memory effects of the used construction materials, reducing the number of needed blank chamber samplings. For studying other BVOC emission target compounds not measured in this study such as sesquiterpenes [63] or other aromatic species [66], it is recommended to evaluate specifically the wall and chamber effects for the selected target compounds.

Tree DEMON system stability

For long-term investigations of treatment effects on gas exchange of plants highly reproducible settings of the assessment conditions are required. Therefore, the system was placed into a climate-controlled chamber environment and air supply for the gas exchange system was conditioned and automatically supervised to generate stable and reproducible CO_2 and water vapor inlet air concentrations. The humidification system used in the Tree DEMON is a rather simple technique with a bubbling tank used for generating a humidification level ranging from 27 to 29% relative humidity in the inlet air due to constant pressure, temperature, and flow rates achieved with the proposed set-up. Preconditioned dry VOC free air is humidified prior to CO_2 removal since the CO_2 scrubber requires water to adsorb CO_2 and to

set the water content in chamber air to the desired levels. For providing the required CO_2 amounts, pure CO_2 was fed via a mass flow-controlled manifold to the CO_2 depleted air stream. The highest CO_2 scrubber efficiency was observed after 24 h conditioning time with humidified air. Over time, the water in the bubbling tank depleted and the humidity in the air downstream the humidifier may decrease slowly. Real-time supervision and control of the inlet air water content as well as CO_2 amount ensured stable gas concentrations during the course of the experiments.

In case a more variable temperature regime with constant relative air humidity for each temperature step is needed, the proposed humidifying procedure is not ideal due to a long response time of the system. Here, other methods may be better suited e.g., a cold trap in the water saturated air stream [37] or mixing a humid air stream with a dry gas stream [67]. Another method of humidifying air has been proposed by Sun et al. [68] with refillable headspace humidifiers, where heated water generates a constant reservoir of water vapor.

If experiments in the chambers last for several days, plant physiology may change due to the artificial chamber conditions and may bias the results [69]. In case that chamber walls get in touch with parts of the plant material, even VOC emission may be induced by mechanical stress [39]. The chamber construction of the Tree DEMON ensured no or only low contact with the plant material, primarily with the PTFE sealed stem, the inlet air distribution tube inside the chamber and leaf temperature sensors. Here, only very light pressure marks were visible on the leaves, but no wounding was observed. The constructed chamber covered the whole tree top, thus investigations were more representative compared to single leaf measurements since the whole environment around the canopy is evenly controlled and not only the leaf in the chamber.

The regulation of environmental conditions in the chambers described is not as fast as for a small leaf cuvette (for instance [32, 50, 70, 71]), since air exchange rates of the chambers are smaller. Thus, the larger volume chambers are not very well suited to conduct experiments requiring fast (in the order of seconds) environmental changes. For the proposed set-up the intended CO_2 amount in the chamber air was achieved in 5–45 min, e.g., it took 45 min with a flowrate of 5 l min^{-1}) to increase the CO_2 concentration change from 0 to 400 µmol mol^{-1}. Additionally, the light regime sensed by the trees is more variable than in a single leaf chamber, since leaves have different angles and distance to the light source with self-shading effects as PAR measurements in and at different chamber positions have shown. Therefore, only a mean PAR, corrected by the chamber foil effect, was continuously recorded by placing the PAR sensor in the middle between the chambers at half of the chamber height where most of the leaf biomass was located.

Case studies

The case studies on the sweet chestnut showed two potential applications of the Tree DEMON: (1) a BVOC emission screening study for 20 sweet chestnut trees and (2) a soil desiccation drought experiment to investigate the impact of SWC on the gas exchange. For sweet chestnut only very few emissions studies have been conducted in the past [36, 72]. Both studies showed a significant amount of monoterpene emissions, which was confirmed by this study. Yet, the total emission amount was much lower with 0.45 µg g_{dw}^{-1} in our study compared to the literature values of 14.2 µg g_{dw}^{-1} h^{-1} [36] and 8.41 µg g_{dw}^{-1} h^{-1} from [72]. A reason for this difference could be the age of the trees (2 years) compared to adult trees used by the other studies [36, 72]. Furthermore, Pio et al. [72] used cut-off branches or leaves possibly inducing some additional emission by mechanical stress [73]. Also seasonality may affect the emission patterns and amounts as already shown for other species such as *Quercus ilex* L. [74, 75] or *Fagus sylvatica* L. [76]. Main emitted compounds from sweet chestnut trees were dominated in the study of Pio et al. [72] by β-pinene and in the study from Aydin et al. [36] by sabinene and ocimene. In our study, the emission pattern was also variable with the third identified emission cluster similar to the composition shown by Pio et al. [72]. Cluster 1, however, was *trans*-β-ocimene dominated. Paré and Tumlinson [6] reported that aphid feeding on plant induce *trans*-β-ocimene emissions. We cannot completely exclude this as a factor for the observed *trans*-β-ocimene emissions in our study since some aphid infestation was observed on a few other specimens in the greenhouse a week before the measurements started, however not on the individuals selected for our case studies.

In the soil drought experiment monoterpene emissions as well as CO_2 and water vapor gas exchange rates declined as expected for the drought stressed plants (e.g., [70, 77]), however a slight decrease was observed for the non-stressed trees too, indicating a high sensitivity of monoterpene emissions of sweet chestnut trees to SWC. The higher standard errors for the CO_2 and water vapor gas exchange shown in Fig. 6 are due to the two light levels within each daily measurement. The small decrease in gas exchange over time was due to the decreasing SWC, which was regulated manually and plants may have transpired more water than added.

Conclusion

The Tree DEMON was developed and evaluated as a versatile instrument for assessing gas exchange of whole plants including BVOC emission. It allowed a high number of replicates in a short time period. Two case studies demonstrated the satisfying excellent performance of the Tree DEMON. The reliable, robust sorbent air sampling system in combination with simultaneously measured CO_2 and water vapor gas exchange of the plants operated in a controlled environment can be used to perform mid- to long-term studies in which e.g., SWC is manipulated and confounding side effects are excluded. Furthermore, the system is easy expandable through a modular hardware and software design, so e.g., additional sample ports or more sophisticated humidity control of the chamber air can easily be implemented. The Labview software was designed to control, measure, and monitor a complete experiment to improve reproducibility and reduce potential user errors. All in one, the Tree DEMON offers an integrated solution to assess BVOC emission of plants.

Additional files

> **Additional file 1: Figure S1.** FID chromatograms of empty trap, tubes, and gas standards. **Figure S2.** FID chromatograms of internal gas standard and empty chambers. **Figure S3.** Gas mixing test. **Figure S4.** Chamber exchange rates and chamber/system tightness test with CO_2. **Figure S5.** Cluster analysis diagnostics. **Equation S1.** General standardization algorithm. **Equation S2.** Leaf temperature correction term. **Equation S3.** Light correction term.
>
> **Additional file 2.** Video of air mixing test in one chamber.

Authors' contributions

MLP developed and tested the Tree DEMON system and performed the data analysis. MLP was the main author of this manuscript. RS, ML and AM conceived the study, participated in its design and coordination and helped to draft the manuscript. All authors read and approved the final manuscript.

Author details

[1] Ecoclimatology, Technische Universität München, Hans-Carl-von-Carlowitz-Platz 2, 85354 Freising, Germany. [2] TUM Institute for Advanced Study, Lichtenbergstraße 2 a, 85748 Garching, Germany. [3] Department of Atmospheric Environmental Research (IMK-IFU), Institute of Meteorology and Climate Research, Karlsruhe Institute of Technology (KIT), Kreuzeckbahnstraße 19, 82467 Garmisch-Partenkirchen, Germany. [4] Present Address: Springer Science+Business Media B.V., Van Godewijckstraat 30, 3311 GX Dordrecht, The Netherlands.

Acknowledgements

The study was performed with the support of the Technische Universität München—Institute for Advanced Study, funded by the German Excellence Initiative. Special thanks to Florian Soutschek helping with the experimental work and analysis during his master thesis. Further thanks for help with statistical questions to Michael Matiu and for helping with the setup and care of the plants within greenhouse to Hannes Seidel. A very special thank you goes out to Nikolaus Hofmann, who greatly supported the mechanical and electrical construction of the Tree DEMON.

Competing interests

The authors declare that they have no competing interests.

Source of funding

This study was financed by the European Research Council under the European Union Seventh Framework Program (FP7/2007–2013/ERC Grant Agreement No. 282250). Funding for travel und conference costs by MICMoR (Mechanisms and Interactions of Climate Change in Mountain Regions Helmholtz Research School) is appreciated. This work was supported by the German Research Foundation (DFG) and the Technical University of Munich (TUM) in the framework of the Open Access Publishing Program.

References

1. Guenther AB, Jiang X, Heald CL, Sakulyanontvittaya T, Duhl T, Emmons LK, Wang X. The Model of Emissions of Gases and Aerosols from Nature version 2.1 (MEGAN2.1): an extended and updated framework for modeling biogenic emissions. Geosci Model Dev. 2012;5:1471–92. doi:10.5194/gmd-5-1471-2012.

2. Kreuzwieser J, Schnitzler JP, Steinbrecher R. Biosynthesis of organic compounds emitted by plants. Pant Biol. 1999;1:149–59. doi:10.1111/j.1438-8677.1999.tb00238.x.

3. Niinemets Ü, Monson RK, Arneth A, Ciccioli P, Kesselmeier J, Kuhn U, et al. The leaf-level emission factor of volatile isoprenoids: caveats, model algorithms, response shapes and scaling. Biogeosciences. 2010;7:1809–32. doi:10.5194/bg-7-1809-2010.

4. Holopainen JK. Multiple functions of inducible plant volatiles. Trends Plant Sci. 2004;9:529–33. doi:10.1016/j.tplants.2004.09.006.

5. Schiestl FP. The evolution of floral scent and insect chemical communication. Ecol Lett. 2010;13:643–56. doi:10.1111/j.1461-0248.2010.01451.x.

6. Paré PW, Tumlinson JH. Plant volatiles as a defense against insect herbivores. Plant Physiol. 1999;121:325–32. doi:10.1104/pp.121.2.325.

7. Dicke M, Baldwin IT. The evolutionary context for herbivore-induced plant volatiles: beyond the 'cry for help'. Trends Plant Sci. 2010;15:167–75. doi:10.1016/j.tplants.2009.12.002.

8. Arneth A, Niinemets Ü. Induced BVOCs: how to bug our models? Trends Plant Sci. 2010;15:118–25. doi:10.1016/j.tplants.2009.12.004.

9. Baldwin IT, Kessler A, Halitschke R. Volatile signaling in plant–plant–herbivore interactions: what is real? Curr Opin Plant Biol. 2002;5:351–4. doi:10.1016/S1369-5266(02)00263-7.

10. Loreto F, Schnitzler J-P. Abiotic stresses and induced BVOCs. Trends Plant Sci. 2010;15:154–66. doi:10.1016/j.tplants.2009.12.006.

11. Ghirardo A, Xie J, Zheng X, Wang Y, Grote R, Block K, et al. Urban stress-induced biogenic VOC emissions and SOA-forming potentials in Beijing. Atmos Chem Phys. 2016;16:2901–20. doi:10.5194/acp-16-2901-2016.

12. Atkinson R, Arey J. Gas-phase tropospheric chemistry of biogenic volatile organic compounds: a review. Atmos Environ. 2003;37:197–219. doi:10.1016/S1352-2310(03)00391-1.

13. Claeys M, Graham B, Vas G, Wang W, Vermeylen R, Pashynska V, et al. Formation of secondary organic aerosols through photooxidation of isoprene. Science. 2004;303:1173–6. doi:10.1126/science.1092805.

14. Emanuelsson EU, Hallquist M, Kristensen K, Glasius M, Bohn B, Fuchs H, et al. Formation of anthropogenic secondary organic aerosol (SOA) and its influence on biogenic SOA properties. Atmos Chem Phys. 2013;13:2837–55. doi:10.5194/acp-13-2837-2013.

15. Calfapietra C, Fares S, Manes F, Morani A, Sgrigna G, Loreto F. Role of Biogenic Volatile Organic Compounds (BVOC) emitted by urban trees on ozone concentration in cities: a review. Environ Pollut. 2013;183:71–80. doi:10.1016/j.envpol.2013.03.012.

16. Goldstein AH. Forest thinning experiment confirms ozone deposition to forest canopy is dominated by reaction with biogenic VOCs. Geophys Res Lett. 2004. doi:10.1029/2004GL021259.

17. Leuchner M, Rappenglück B. VOC source–receptor relationships in Houston during TexAQS-II. Atmos Environ. 2010;44:4056–67. doi:10.1016/j.atmosenv.2009.02.029.

18. D'Andrea SD, Acosta Navarro JC, Farina SC, Scott CE, Rap A, Farmer DK, et al. Aerosol size distribution and radiative forcing response to anthropogenically driven historical changes in biogenic secondary organic aerosol formation. Atmos Chem Phys. 2015;15:2247–68. doi:10.5194/acp-15-2247-2015.

19. Blande JD, Holopainen JK, Niinemets U. Plant volatiles in polluted atmospheres: stress responses and signal degradation. Plant Cell Environ. 2014;37:1892–904. doi:10.1111/pce.12352.

20. Guenther A, Hewitt CN, Erickson D, Fall R, Geron C, Graedel T, et al. A global model of natural volatile organic compound emissions. J Geophys Res. 1995;100:8873. doi:10.1029/94JD02950.

21. Arneth A, Miller PA, Scholze M, Hickler T, Schurgers G, Smith B, Prentice IC. CO_2 inhibition of global terrestrial isoprene emissions: potential implications for atmospheric chemistry. Geophys Res Lett. 2007. doi:10.1029/2007GL030615.

22. Steinbrecher R, Smiatek G, Köble R, Seufert G, Theloke J, Hauff K, et al. Intra- and inter-annual variability of VOC emissions from natural and semi-natural vegetation in Europe and neighbouring countries. Atmos Environ. 2009;43:1380–91. doi:10.1016/j.atmosenv.2008.09.072.

23. Steinbrecher R, Klauer M, Hauff K, Stockwell R, Jaeschke W, Dietrich T, Herbert F. Biogenic and anthropogenic fluxes of non-methane hydrocarbons over an urban-impacted forest, Frankfurter Stadtwald, Germany. Atmos Environ. 2000;34:3779–88. doi:10.1016/S1352-2310(99)00518-X.

24. Karl TG, Spirig C, Rinne J, Stroud C, Prevost P, Greenberg J, et al. Virtual disjunct eddy covariance measurements of organic compound fluxes from a subalpine forest using proton transfer reaction mass spectrometry. Atmos Chem Phys. 2002;2:279–91. doi:10.5194/acp-2-279-2002.

25. Valentini R, Greco S, Seufert G, Bertin N, Ciccioli P, Cecinato A, et al. Fluxes of biogenic VOC from Mediterranean vegetation by trap enrichment relaxed eddy accumulation. Atmos Environ. 1997;31:229–38. doi:10.1016/S1352-2310(97)00085-X.

26. Rinne H, Guenther AB, Greenberg JP, Harley PC. Isoprene and monoterpene fluxes measured above Amazonian rainforest and their dependence on light and temperature. Atmos Environ. 2002;36:2421–6. doi:10.1016/S1352-2310(01)00523-4.

27. Laffineur Q, Aubinet M, Schoon N, Amelynck C, Müller J-F, Dewulf J, et al. Isoprene and monoterpene emissions from a mixed temperate forest. Atmos Environ. 2011;45:3157–68. doi:10.1016/j.atmosenv.2011.02.054.

28. Kuhn U, Andreae MO, Ammann C, Araújo AC, Brancaleoni E, Ciccioli P, et al. Isoprene and monoterpene fluxes from Central Amazonian rainforest inferred from tower-based and airborne measurements, and implications on the atmospheric chemistry and the local carbon budget. Atmos Chem Phys. 2007;7:2855–79. doi:10.5194/acp-7-2855-2007.

29. Greenberg JP, Peñuelas J, Guenther A, Seco R, Turnipseed A, Jiang X, et al. A tethered-balloon PTRMS sampling approach for surveying of landscape-scale biogenic VOC fluxes. Atmos Meas Tech. 2014;7:2263–71. doi:10.5194/amt-7-2263-2014.

30. Steinbrecher R, Hauff K, Rabong R, Steinbrecher J. Isoprenoid emission of oak species typical for the Mediterranean area: source strength and controlling variables. Atmos Environ. 1997;31:79–88. doi:10.1016/S1352-2310(97)00076-9.

31. Bäck J, Hari P, Hakola H, Juurola E, Kulmala M. Dynamics of monoterpene emissions in Pinus sylvestris during early spring. Boreal Environ Res. 2005;10:409–24.

32. Blanch J-S, Peñuelas J, Llusià J. Sensitivity of terpene emissions to drought and fertilization in terpene-storing Pinus halepensis and non-storing Quercus ilex. Physiol Plant. 2007;131:211–25. doi:10.1111/j.1399-3054.2007.00944.x.

33. Yassaa N, Custer T, Song W, Pech F, Kesselmeier J, Williams J. Quantitative and enantioselective analysis of monoterpenes from plant chambers and in ambient air using SPME. Atmos Meas Tech. 2010;3:1615–27. doi:10.5194/amt-3-1615-2010.

34. Šimpraga M, Verbeeck H, Demarcke M, Joó É, Pokorska O, Amelynck C, et al. Clear link between drought stress, photosynthesis and biogenic volatile organic compounds in Fagus sylvatica L. Atmos Environ. 2011;45:5254–9. doi:10.1016/j.atmosenv.2011.06.075.

35. Baghi R, Helmig D, Guenther A, Duhl T, Daly R. Contribution of flowering trees to urban atmospheric biogenic volatile organic compound emissions. Biogeosciences. 2012;9:3777–85. doi:10.5194/bg-9-3777-2012.

36. Aydin YM, Yaman B, Koca H, Dasdemir O, Kara M, Altiok H, et al. Biogenic volatile organic compound (BVOC) emissions from forested areas in Turkey: determination of specific emission rates for thirty-one tree species. Sci Total Environ. 2014;490:239–53. doi:10.1016/j.scitotenv.2014.04.132.

37. Komenda M. Measurements of biogenic VOC emissions: sampling, analysis and calibration. Atmos Environ. 2001;35:2069–80. doi:10.1016/S1352-2310(00)00502-1.

38. Ortega J, Helmig D. Approaches for quantifying reactive and low-volatility biogenic organic compound emissions by vegetation enclosure

techniques—part A. Chemosphere. 2008;72:343–64. doi:10.1016/j.chemosphere.2007.11.020.

39. Niinemets Ü, Kuhn U, Harley PC, Staudt M, Arneth A, Cescatti A, et al. Estimations of isoprenoid emission capacity from enclosure studies: measurements, data processing, quality and standardized measurement protocols. Biogeosciences. 2011;8:2209–46. doi:10.5194/bg-8-2209-2011.

40. Materic D, Bruhn D, Turner C, Morgan G, Mason N, Gauci V. Methods in plant foliar volatile organic compounds research. Appl Plant Sci. 2015. doi:10.3732/apps.1500044.

41. Sharkey TD, Loreto F, Delwiche CF. High carbon dioxide and sun/shade effects on isoprene emission from oak and aspen tree leaves. Plant Cell Environ. 1991;14:333–8. doi:10.1111/j.1365-3040.1991.tb01509.x.

42. Wilkinson MJ, Monson RK, Trahan N, Lee S, Brown E, Jackson RB, et al. Leaf isoprene emission rate as a function of atmospheric CO_2 concentration. Global Change Biol. 2009;15:1189–200. doi:10.1111/j.1365-2486.2008.01803.x.

43. Genard-Zielinski A-C, Boissard C, Fernandez C, Kalogridis C, Lathière J, Gros V, et al. Variability of BVOC emissions from a Mediterranean mixed forest in southern France with a focus on *Quercus pubescens*. Atmos Chem Phys. 2015;15:431–46. doi:10.5194/acp-15-431-2015.

44. Kreuzwieser J, Scheerer U, Rennenberg H. Metabolic origin of acetaldehyde emitted by poplar (*Populus tremula* × *P. alba*) trees. J Exp Bot. 1999;50:757–65. doi:10.1093/jxb/50.335.757.

45. Joó É, van Langenhove H, Šimpraga M, Steppe K, Amelynck C, Schoon N, et al. Variation in biogenic volatile organic compound emission pattern of *Fagus sylvatica* L. due to aphid infection. Atmos Environ. 2010;44:227–34. doi:10.1016/j.atmosenv.2009.10.007.

46. Mentel TF, Wildt J, Kiendler-Scharr A, Kleist E, Tillmann R, Dal Maso M, et al. Photochemical production of aerosols from real plant emissions. Atmos Chem Phys. 2009;9:4387–406. doi:10.5194/acp-9-4387-2009.

47. Dal Maso M, Liao L, Wildt J, Kiendler-Scharr A, Kleist E, Tillmann R, et al. A chamber study of the influence of boreal BVOC emissions and sulfuric acid on nanoparticle formation rates at ambient concentrations. Atmos Chem Phys. 2016;16:1955–70. doi:10.5194/acp-16-1955-2016.

48. Hohaus T, Kuhn U, Andres S, Kaminski M, Rohrer F, Tillmann R, et al. A new plant chamber facility, PLUS, coupled to the atmosphere simulation chamber SAPHIR. Atmos Meas Tech. 2016;9:1247–59. doi:10.5194/amt-9-1247-2016.

49. Sharkey TD, Singsaas EL, Vanderveer PJ, Geron C. Field measurements of isoprene emission from trees in response to temperature and light. Tree Physiol. 1996;16:649–54. doi:10.1093/treephys/16.7.649.

50. Filella I, Peñuelas J, Seco R. Short-chained oxygenated VOC emissions in *Pinus halepensis* in response to changes in water availability. Acta Physiol Plant. 2009;31:311–8. doi:10.1007/s11738-008-0235-6.

51. Pape L, Ammann C, Nyfeler-Brunner A, Spirig C, Hens K, Meixner FX. An automated dynamic chamber system for surface exchange measurement of non-reactive and reactive trace gases of grassland ecosystems. Biogeosciences. 2009;6:405–29. doi:10.5194/bg-6-405-2009.

52. Morfopoulos C, Sperlich D, Penuelas J, Filella I, Llusia J, Medlyn BE, et al. A model of plant isoprene emission based on available reducing power captures responses to atmospheric CO_2. New Phytol. 2014;203:125–39. doi:10.1111/nph.12770.

53. Allaire SE, Yates SR, Ernst F, Papiernik SK. Gas-phase sorption-desorption of propargyl bromide and 1,3-dichloropropene on plastic materials. J Environ Qual. 2003;32:1915. doi:10.2134/jeq2003.1915.

54. Harogoppad SB, Aminabhavi TM. Diffusion and sorption of organic liquids through polymer membranes. II. Neoprene, SBR, EPDM, NBR, and natural rubber versus n-alkanes. J Appl Polym Sci. 1991;42:2329–36. doi:10.1002/app.1991.070420824.

55. von Caemmerer S, Farquhar GD. Some relationships between the biochemistry of photosynthesis and the gas exchange of leaves. Planta. 1981;153:376–87. doi:10.1007/BF00384257.

56. Stein S. NIST Standard Reference Database 1A. Gaithersburg; 2008.

57. Niinemets Ü, Arneth A, Kuhn U, Monson RK, Peñuelas J, Staudt M. The emission factor of volatile isoprenoids: stress, acclimation, and developmental responses. Biogeosciences. 2010;7:2203–23. doi:10.5194/bg-7-2203-2010.

58. Serdar Ümit, Demirsoy Hüsnü. Non-destructive leaf area estimation in chestnut. Sci Hortic. 2006;108:227–30. doi:10.1016/j.scienta.2006.01.025.

59. Schneider CA, Rasband WS, Eliceiri KW. NIH Image to ImageJ: 25 years of image analysis. Nat Methods. 2012;9:671–5. doi:10.1038/nmeth.2089.

60. R Core Team. R: a language and environment for statistical computing 2014. Vienna, Austria.

61. Maechler M, Rousseeuw P, Struyf A, Hubert M, Hornik K. Cluster: cluster analysis basics and extensions 2016.

62. Ghirardo A, Gutknecht J, Zimmer I, Bruggemann N, Schnitzler J-P. Biogenic volatile organic compound and respiratory CO_2 emissions after 13C-labeling: online tracing of C translocation dynamics in poplar plants. PLoS ONE. 2011;6:e17393. doi:10.1371/journal.pone.0017393.

63. Helmig D, Bocquet F, Pollmann J, Revermann T. Analytical techniques for sesquiterpene emission rate studies in vegetation enclosure experiments. Atmos Environ. 2004;38:557–72. doi:10.1016/j.atmosenv.2003.10.012.

64. Kuhn U, Dindorf T, Ammann C, Rottenberger S, Guyon P, Holzinger R, et al. Design and field application of an automated cartridge sampler for VOC concentration and flux measurements. J Environ Monit. 2005;7:568–76. doi:10.1039/B500057B.

65. Crespo E, Graus M, Gilman JB, Lerner BM, Fall R, Harren F, Warneke C. Volatile organic compound emissions from elephant grass and bamboo cultivars used as potential bioethanol crop. Atmos Environ. 2013;65:61–8. doi:10.1016/j.atmosenv.2012.10.009.

66. Ortega J, Helmig D, Daly RW, Tanner DM, Guenther AB, Herrick JD. Approaches for quantifying reactive and low-volatility biogenic organic compound emissions by vegetation enclosure techniques—part B: applications. Chemosphere. 2008;72:365–80. doi:10.1016/j.chemosphere.2008.02.054.

67. Behrendt T, Veres PR, Ashuri F, Song G, Flanz M, Mamtimin B, et al. Characterisation of NO production and consumption: new insights by an improved laboratory dynamic chamber technique. Biogeosciences. 2014;11:5463–92. doi:10.5194/bg-11-5463-2014.

68. Sun S, Moravek A, von der Heyden L, Held A, Sörgel M, Kesselmeier J. Twin-cuvette measurement technique for investigation of dry deposition of O_3 and PAN to plant leaves under controlled humidity conditions. Atmos Meas Tech. 2016;9:599–617. doi:10.5194/amt-9-599-2016.

69. Niederbacher B, Winkler JB, Schnitzler JP. Volatile organic compounds as non-invasive markers for plant phenotyping. J Exp Bot. 2015;66:5403–16. doi:10.1093/jxb/erv219.

70. Llusià J, Peñuelas J, Alessio GA, Ogaya R. Species-specific, seasonal, inter-annual, and historically-accumulated changes in foliar terpene emission rates in *Phillyrea latifolia* and *Quercus ilex* submitted to rain exclusion in the Prades Mountains (Catalonia). Russ J Plant Physiol. 2011;58:126–32. doi:10.1134/S1021443710061020.

71. Song W, Staudt M, Bourgeois I, Williams J. Laboratory and field measurements of enantiomeric monoterpene emissions as a function of chemotype, light and temperature. Biogeosciences. 2014;11:1435–47. doi:10.5194/bg-11-1435-2014.

72. Pio CA, Nunes TV, Brito S. Volatile hydrocarbon emissions from common and native species of vegetation in Portugal. In: Slanina J, Angeletti G, editors. General assessment of biogenic emissions and deposition of nitrogen compounds, sulphur compounds and oxidants in Europe: Joint Workshop CEC/BIATEX, May 4–7, 1993, Aveiro, Portugal. Brussels: Guyot; 1993. pp. 291–8.

73. Fall R, Karl T, Hansel A, Jordan A, Lindinger W. Volatile organic compounds emitted after leaf wounding: on-line analysis by proton-transfer-reaction mass spectrometry. J Geophys Res. 1999;104:15963–74. doi:10.1029/1999JD900144.

74. Penuelas J, Llusia J. Seasonal emission of monoterpenes by the Mediterranean tree *Quercus ilex* in field conditions: relations with photosynthetic rates, temperature and volatility. Physiol Plant. 1999;105:641–7. doi:10.1034/j.1399-3054.1999.105407.x.

75. Llusia J, Peñuelas J, Seco R, Filella I. Seasonal changes in the daily emission rates of terpenes by *Quercus ilex* and the atmospheric concentrations of terpenes in the natural park of Montseny, NE Spain. J Atmos Chem. 2012;69:215–30. doi:10.1007/s10874-012-9238-1.

76. Holzke C, Dindorf T, Kesselmeier J, Kuhn U, Koppmann R. Terpene emissions from European beech (shape *Fagus sylvatica*~L.): pattern and Emission Behaviour Over two Vegetation Periods. J Atmos Chem. 2006;55:81–102. doi:10.1007/s10874-006-9027-9.

77. Llusià J, Peñuelas J. Changes in terpene content and emission in potted Mediterranean woody plants under severe drought. Can J Bot. 1998;76:1366–73. doi:10.1139/b98-141.

7

Non-invasive absolute measurement of leaf water content using terahertz quantum cascade lasers

Lorenzo Baldacci[1†], Mario Pagano[2*†] 🆔, Luca Masini[1], Alessandra Toncelli[4], Giorgio Carelli[3], Paolo Storchi[2] and Alessandro Tredicucci[4]

Abstract

Background: Plant water resource management is one of the main future challenges to fight recent climatic changes. The knowledge of the plant water content could be indispensable for water saving strategies. Terahertz spectroscopic techniques are particularly promising as a non-invasive tool for measuring leaf water content, thanks to the high predominance of the water contribution to the total leaf absorption. Terahertz quantum cascade lasers (THz QCL) are one of the most successful sources of THz radiation.

Results: Here we present a new method which improves the precision of THz techniques by combining a transmission measurement performed using a THz QCL source, with simple pictures of leaves taken by an optical camera. As a proof of principle, we performed transmission measurements on six plants of *Vitis vinifera* L. (cv "Colorino"). We found a linear law which relates the leaf water mass to the product between the leaf optical depth in the THz and the projected area. Results are in optimal agreement with the proposed law, which reproduces the experimental data with 95% accuracy.

Conclusions: This method may overcome the issues related to intra-variety heterogeneities and retrieve the leaf water mass in a fast, simple, and non-invasive way. In the future this technique could highlight different behaviours in preserving the water status during drought stress.

Keywords: Terahertz quantum cascade laser, Water content, Draught stress, *Vitis vinifera* L.

Background

There is an increasing need to improve the knowledge of the water resources of plant varieties, owing to the variable rainfall after climate change, in order to assess the need for irrigation [1]. In this perspective, leaves are essential organs for the water balance [2]. Leaf morphology comes from a long evolutionary process of a polyhedral anatomical structure in which the veins are definitely at the heart of this organic evolution [3]. This structure is an essential tool for the mechanical support of the anatomical organization but it also plays a crucial role in the photosynthesis efficiency [4, 5] and in the consumption of water. Furthermore, the capillary branching of the veins in the leaf allows better cooling [6] with potential benefits for the photosynthesis performance [7]. Recent studies show that the vein density per unit area [8] and the thickness of the mesophyll [9] may be involved in the efficiency of hydraulic performance. A decreasing hydraulic functionality, due to water stress or to a limited vein network density, can involve a loss of production. The leaf water potential measurement, defined as the measure of the free energy per unit volume of water (J m^{-3}) [10], explains the level of plant drought stress. The first water potential measurement came in 1960 with two basic instruments: the thermocouple psychrometer and the Scholander pressure chamber. Leaf water potential

---

*Correspondence: mario.pagano@crea.gov.it

[†]Lorenzo Baldacci and Mario Pagano contributed equally to this work

[2] Consiglio per la ricerca in agricoltura e l'analisi dell'economia agraria, Centro di ricerca per la Viticoltura e l'Enologia, Viale Santa Margherita 80, 52100 Arezzo, Italy

Full list of author information is available at the end of the article

measured with the Scholander pressure bomb [11] is considered the reference method by plant physiologists. However, this device needs a destructive sampling of the leaf and presents different operating limits: it requires a cylinder with a propellant of nitrogen (or compressed air), cumbersome instrumentation, long operating times and the measurements cannot be automated. The thermocouple psychrometer, instead, is based on the principle that the relative vapour pressure of a solution or piece of plant material is related to its water potential. The requirement of tight temperature control restricted the use of uncompensated thermocouple psychrometers in field studies to situations where good laboratory facilities were at hand. The water content of plants is another basic parameter which describes the plant water deficit and is commonly determined by weighing the material immediately after sampling, drying at 105 °C, and reweighing 24 h later; a process which is destructive, time-consuming and hard to be automated.

A growing attention in the scientific community has been directed to terahertz spectroscopic techniques, mostly time domain spectroscopy or confocal microscopy [12], as a non-invasive tool for measuring leaf water content [13, 14] and related quantities such as drought stress [15, 16] and dehydration kinetics [17]. The potential of THz absorption measurements was also demonstrated in the study of the hydration of biomolecules [18–20] and ions [21–24]. The terahertz spectral region is particularly promising in this branch of studies: thanks to the large absorption coefficient of water [25], opposite to the relatively small absorption coefficient of the leaf dry matter [13] terahertz techniques are more sensitive to changes in water content than near infrared and microwave techniques, because they suffer less disturbances from changes in concentration of soluble substances, such as inorganic salts [26]. Terahertz techniques have to face the major challenge of the translation from lab tables to the field. The measurement method must be thought of for non-ideal working conditions, hence robust and reliable, in addition to non-invasive [27].

In this work, we report on a method for measuring the leaf wet mass using a terahertz quantum cascade laser [28] (THz QCL) and a camera. This method is based on a linear law which relates the leaf wet mass to the product between its optical depth and projected area. We compared this new method with the one based on the sole terahertz transmission, by studying six plants of Vitis vinifera L. (cv "Colorino"). A linear regression model was employed to fit the data obtained from the two methods. In the first case we had the sole optical depth as function of the leaf wet mass, and the least square fit of the data set produced an adjusted coefficient of determination of 0.31. With the new method instead, the linear regression model was adopted

to fit the product between the leaf optical depth and the leaf projective area, as function of the leaf wet mass. In this case, the least square fit of the dataset has a coefficient of determination of 0.95, which means that the new method is reliable also when leaves come from different plants.

Methods
Plant preparation
All the experiments were performed in September 2015 on 2 year old vines. Six plants were pruned as a single-cordon and grafted onto SO4 (V. berlandieri × V. riparia) rootstock. Vitis vinifera L. (cv "Colorino") were used for the trials. The plants were arranged to 2 L pots filled with a peat:sand mix (2:1). The vines were acclimated to the same environmental condition (24 °C; 45.5% of relative humidity; approximately $660 \, \mu mol \, m^{-2} \, s^{-1}$ P.A.R.) inside the laboratory of the Department of physics at the University of Pisa, Italy. The measurements were performed under the same environmental conditions. Photon Flux measurements were conducted using em50 data logger (Decagon Devices, Inc.) equipped with a calibrated QSO-S PAR Photon Flux sensor. The environmental conditions, temperature and humidity, were measured with a calibrated usb temperature and humidity data logger (model IMD 100, Imagintronix Inc.). During the experiments, the photoperiod was 13 h 15 min day and 10 h 45 min night. For each of six plants all leaves fully extended and developed were sampled from the main shoot and, according to Kapos [29], the water content of leaves was determined by weighing the leaves immediately after sampling, drying at 105 °C, and reweighing 24 h later.

Measurement setup
Leaf optical depths were measured by means of a simple transmission setup, like the one sketched in Fig. 1a. THz laser radiation produced by a THz QCL is shined onto the leaf. The fraction of light passing through the leaf tissues is then collected by a THz detector. The ratio between the total incident power and the transmitted one gives the experimental optical transmission data. The optical source was a THz QCL emitting at frequencies around 2.5 THz, cryo-cooled down to 30 K and driven by a current pulser. For each leaf, we measured its transmittance and thickness; then, right after detachment, leaf fresh mass and projective area were measured. In order to sample most of the leaf structure, the transmittance data were sampled over four different regions with a diameter of 8 mm: one from the petiolar sinus to the first bifurcation; one on the first order vase in the distal part; two on the lateral lobes, left and right. The regions are loosely defined in order to assess the robustness of the measurement procedure when different leaves are measured. The readout signal is measured by a Stanford

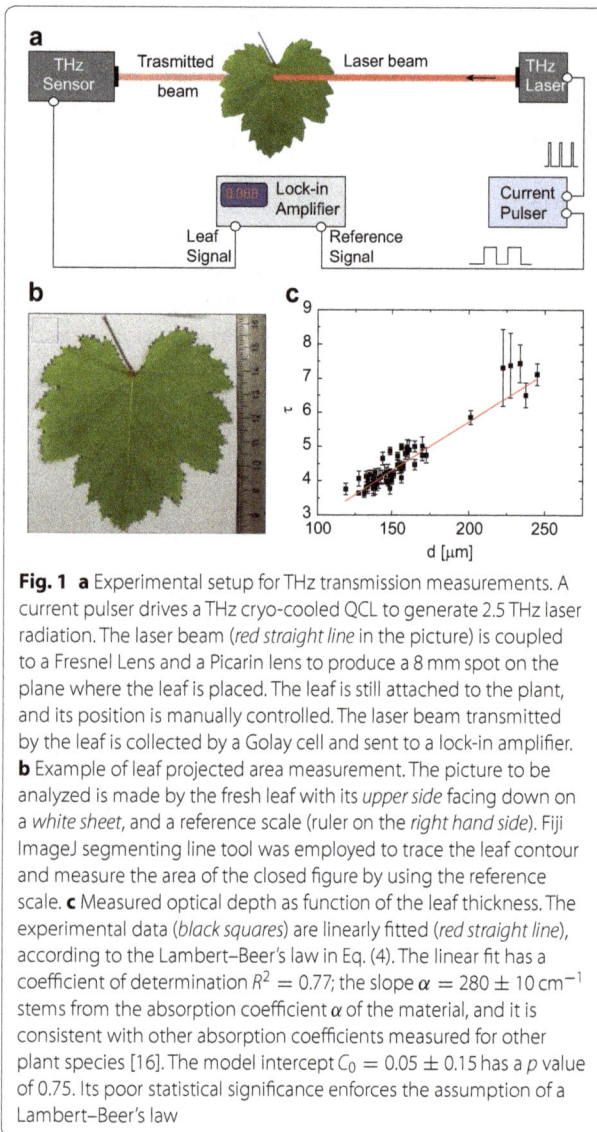

Fig. 1 **a** Experimental setup for THz transmission measurements. A current pulser drives a THz cryo-cooled QCL to generate 2.5 THz laser radiation. The laser beam (*red straight line* in the picture) is coupled to a Fresnel Lens and a Picarin lens to produce a 8 mm spot on the plane where the leaf is placed. The leaf is still attached to the plant, and its position is manually controlled. The laser beam transmitted by the leaf is collected by a Golay cell and sent to a lock-in amplifier. **b** Example of leaf projected area measurement. The picture to be analyzed is made by the fresh leaf with its *upper side* facing down on a *white sheet*, and a reference scale (ruler on the *right hand side*). Fiji ImageJ segmenting line tool was employed to trace the leaf contour and measure the area of the closed figure by using the reference scale. **c** Measured optical depth as function of the leaf thickness. The experimental data (*black squares*) are linearly fitted (*red straight line*), according to the Lambert–Beer's law in Eq. (4). The linear fit has a coefficient of determination $R^2 = 0.77$; the slope $\alpha = 280 \pm 10\,\text{cm}^{-1}$ stems from the absorption coefficient α of the material, and it is consistent with other absorption coefficients measured for other plant species [16]. The model intercept $C_0 = 0.05 \pm 0.15$ has a p value of 0.75. Its poor statistical significance enforces the assumption of a Lambert–Beer's law

SR830 lock-in amplifier synchronized with the QCL current supply. To further suppress unwanted background and residual scattered radiation on the detector, a black screen with a 8 mm hole was placed behind the leaf sampling region. The black screen was made of an aluminum plate with a rough surface. In order to choose the hole width, a lever iris was placed along the optical path (the aluminum plate was removed) and the signal recorded by the detector was measured with increasing iris apertures, until the recorded signal did not change its least significant digit with respect to the case without the iris. During the transmission measurement the sample was held by hand. At the cost of some precision, the hand offers many translational and rotational degrees of freedom, enabling hardly reachable leaves to be placed in the optical path. For each leaf the average optical depth was calculated starting from the transmittance data, by averaging over the four samples. The leaf thickness d_j was measured with a contact caliper by averaging over four 6 mm diameter regions, selected with the same criteria as the regions for the transmission measurement. Note that the thickness was measured over a region without major veins; in this way the thickness is underestimated by a factor 5–10% according to the literature [30]. The plant material was then detached and weighted to acquire the leaf fresh mass. Each projective area was measured immediately after weighting, with a procedure explained in Fig. 1b. The leaf was placed with its lower side facing up onto a white paper sheet with a scale reference on a side, and photographed. Its projective area was measured using Fiji ImageJ segmented line tool on the image collected [31]. Finally the leaf was dried for 24 h at a temperature of 105 °C and weighted to acquire its dry mass.

Results

The main scope of our work is to find a relation between the leaf optical properties at THz frequencies and the leaf water content, which is reproducible for different plants of the same variety. We examined six plants of *Vitis vinifera* L. (cv "Colorino") by measuring 58 leaves, chosen according to the Plant Preparation Section. The optical transmittance was sampled over four regions for each leaf, according to the Measurement setup Section. Given the j-th leaf, the optical transmittance of the k-th sample region is defined as the ratio between the laser beam intensity transmitted by the leaf to the detector, $S_{j,k}$, and the incident one, I_j:

$$T_{j,k} = \frac{S_{j,k}}{I_j}. \tag{1}$$

In order to perform statistical data analysis, for each leaf we calculate its average experimental optical depth $\tau_{j,k}$, defined by the Lambert–Beer's law [32]

$$T_{j,k} = \exp[-\tau_{j,k}], \tag{2}$$

therefore $\tau_{j,k}$ accounts for material losses. The average optical depth τ_j can be calculated as

$$\tau_j = \frac{1}{4} \sum_{k=1}^{4} \tau_{j,k}. \tag{3}$$

Within an effective medium approximation [33, 34], τ_j is proportional to the effective leaf absorption coefficient α_j and the leaf average thickness d_j:

$$\tau_j = \alpha_j d_j. \tag{4}$$

If all the leaves had the same distribution of water, air, and dry materials, their α_j should be the same. In Fig. 1c all the measured τ_j are plotted against d_j, each leaf contributing with one point. The red straight line reports the linear best fit, resulting from a two parameter linear regression in which data are statistically weighted with the inverse of their variances, to account for heteroscedasticity. We found that our linear regression has a coefficient of determination of $R^2 = 0.77$, which means it is able to explain 77% of the overall τ_j fluctuations around their mean value; the resulting effective absorption coefficient $\alpha = 280 \pm 10 \, \mathrm{cm}^{-1}$ is similar to the ones found for other species (for example, measured *Coffea Arabica* effective absorption coefficient is $\alpha \simeq 250 \, \mathrm{cm}^{-1}$ at a frequency of 1.8 THz [16]). Also the high p value related to the intercept parameter (C_0 in Fig. 1c) confirms the consistency of the assumed Lambert–Beer's law. It is worth to point out that we chose not to report the p value of those parameters having p value smaller than 10^{-10}.

Leaf water content and optical depth

Trying to extract the leaf water mass from global optical depth is not trivial, because air, water and dry materials might be combined in different fractions and give approximately the same absorption coefficient. This concept is graphically explained in Fig. 2a, where the experimental τ has been plotted as function of the experimental leaf water mass M_W, which was measured for each leaf according to the Mesurement Setup section. In this case the linear slope is statistically different from 0 with p value of $3.8e - 6$, but the linear regression

explains only 31% of the total data fluctuations. Better effective medium models could improve the predictions on leaf water mass. Recently, Landau–Lifshitz–Looyenga model has been successfully employed on *Coffea Arabica* and *Hordeum Vulgare* plants [13, 16], at the expense of new variables to be measured, such as the leaf dry material refractive index, which requires the leaf to be dehydrated and pelletized. In this work we found a simpler model to enhance the description of the experimental data, by combining M_W with the leaf projective area A. If the optical depth is dominated by water absorption, Eq. (4) may be rewritten as:

$$\tau = \alpha_{\mathrm{eff}} d_W, \tag{5}$$

where α_{eff} is the effective absorption coefficient and $d_W = V_W/A$ is the effective water thickness inside the leaf, expressed as the ratio between the volume of water inside the leaf and the leaf projective area. Water volume may be safely expressed as function of the water mass M_W, because water density is known and can be approximated to $\rho_W = 1000 \, \mathrm{mg} \, \mathrm{cm}^{-3}$; Eq. (5) then becomes

$$\tau = \mathbb{K} \frac{M_W}{A}, \tag{6}$$

where $\mathbb{K} = \alpha_{\mathrm{eff}}/\rho_W$. Looking at Fig. 2b, we see the linear regression of τ as function of $M_W \cdot A^{-1}$. In this case the best fit of linear regression (red line) returns $\mathbb{K} = 0.33 \pm 0.03 \, \mathrm{cm}^2 \, \mathrm{mg}^{-1}$, which means an absorption coefficient $\alpha_{\mathrm{eff}} = 330 \pm 30 \, \mathrm{cm}^{-1}$, but also an intercept $C_1 = 1.0 \pm 0.1$, wich is statistically different from 0 with very small p value; this result entails a non-negligible contribution to losses from dry mass and/or vapor absorption, and from scattering due to the heterogeneity of the leaf tissues. The coefficient of determination is greatly improved with respect to the sole use of M_W, with $R^2 = 0.74$, which means that optical depth data describe almost 75% of the $M_W \cdot A^{-1}$ data fluctuations. If the intercept is set to 0, from the linear fit (green line in Fig. 2b) we find $\alpha_{\mathrm{eff}} = 444 \pm 2 \, \mathrm{cm}^{-1}$; in this case the accuracy of the model is slightly reduced, because the residual sum of squares $\sum_{j=1}^{58}(\tau_j - \tau_j(\mathrm{model}))$ increases by $\sim$29%. By adding a term which is inversely proportional to A, Eq. (6) is transformed into

$$\tau = \frac{C_1}{A} + C_2 \frac{M_W}{A}, \tag{7}$$

where C_1 and C_2 are constant coefficients. By moving A on the left hand side of the equation we obtain a linear relation

$$\tau A = C_1 + C_2 M_W. \tag{8}$$

In Fig. 3c we report the linear regression best fit of the experimental data $\tau_j A_j$ as function of the leaf

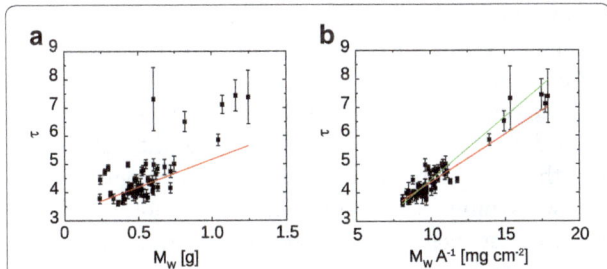

Fig. 2 a Measured optical depth τ as function of the leaf water mass M_W. The experimental data (*black squares*) are linearly fitted (*red straight line*) as $\tau = C_1 + C_2 \cdot M_W$. The linear fit has a coefficient of determination $R^2 = 0.31$; the coefficients are $C_1 = 3.2 \pm 0.1$ and $C_2 = 2.0 \pm 0.1 \, \mathrm{g}^{-1}$. **b** Measured optical depth versus $M_W A^{-1}$. The coefficient of determination is greatly improved with respect to the previous model, by the simple measurement of the leaf projective area A. In this case the linear model produces a best fit with $R^2 = 0.74$. There are two statistically significant parameters: the intercept $C_1 = 1.0 \pm 0.1$ may be ascribed to residual scattering and absorption from the leaf dry mass and vapor, whereas the slope $K = 0.33 \pm 0.03 \, \mathrm{cm}^2 \, \mathrm{mg}^{-1}$ is linked to the effective absorption coefficient of water; according to Eqs. (5) and (6) $\alpha_{\mathrm{eff}} = 330 \pm 30 \, \mathrm{cm}^{-1}$. If the intercept is set to 0 as proposed in Eq. (6), the absorption coefficient changes to $\alpha_{\mathrm{eff}} = 444 \pm 2 \, \mathrm{cm}^{-1}$, and the model fit accuracy is reduced (see the *green line in the graph*)

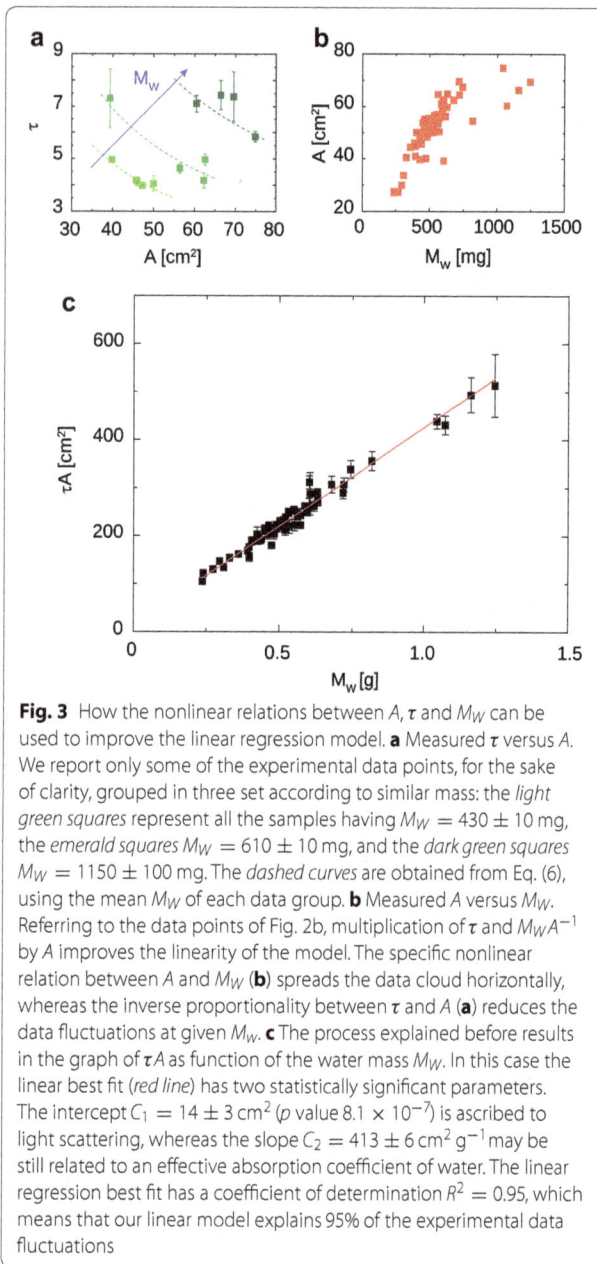

Fig. 3 How the nonlinear relations between A, τ and M_W can be used to improve the linear regression model. **a** Measured τ versus A. We report only some of the experimental data points, for the sake of clarity, grouped in three set according to similar mass: the *light green squares* represent all the samples having $M_W = 430 \pm 10$ mg, the *emerald squares* $M_W = 610 \pm 10$ mg, and the *dark green squares* $M_W = 1150 \pm 100$ mg. The *dashed curves* are obtained from Eq. (6), using the mean M_W of each data group. **b** Measured A versus M_W. Referring to the data points of Fig. 2b, multiplication of τ and $M_W A^{-1}$ by A improves the linearity of the model. The specific nonlinear relation between A and M_W (**b**) spreads the data cloud horizontally, whereas the inverse proportionality between τ and A (**a**) reduces the data fluctuations at given M_W. **c** The process explained before results in the graph of τA as function of the water mass M_W. In this case the linear best fit (*red line*) has two statistically significant parameters. The intercept $C_1 = 14 \pm 3$ cm^2 (p value 8.1×10^{-7}) is ascribed to light scattering, whereas the slope $C_2 = 413 \pm 6$ cm^2 g^{-1} may be still related to an effective absorption coefficient of water. The linear regression best fit has a coefficient of determination $R^2 = 0.95$, which means that our linear model explains 95% of the experimental data fluctuations

water mass data M_j. We found that the variable τA describes the water mass experimental data with 95% of accuracy, and both coefficients have statistically relevant values, $C_1 = 14 \pm 3$ cm^2 (p value 8.1×10^{-7}) and $C_2 = 413 \pm 6$ cm^2 g^{-1}. The inverse proportionality with A of the term $C_1 A^{-1}$ in Eq. (7) could be ascribed to the nonlinear correlation between the leaf projective area and the water mass, as shown in Fig. 3a (lower graph): the larger is M_W, the slower is the increase of A, and the contribution of bulk absorption becomes more important.

Discussions

The reason why the use of Eq. (8) improves the linear regression, as shown in Fig. 3c, can be found inside the nonlinear relation between A, M_W, and τ, and it is graphically explained in Fig. 3, panels a and b. Here we report the two graphs of τ versus A (panel a) and A versus M_W (panel b). Multiplication by A on both sides of Eq. (7) produces two synergical actions. It has a horizontal action, because lower values of A grow more steeply with M_W than higher values, as shown in Fig. 3b; this effect tends to horizontally elongate the data cloud in Fig. 2b. There is also a vertical action, which can be explained looking at Fig. 3a. Here we reported three groups of data points, based on similarity of M_W. The first group (light green squares) has $M_W = 430 \pm 10$ mg, the second (emerald squares) $M_W = 610 \pm 10$ mg, the third (dark green squares) $M_W = 1150 \pm 100$ mg. The dashed lines indicate the curves found using Eq. (6) and the mean M_W of each group. Since the curves reasonably fit the data points, we may consider A and τ to be inversely proportional; consequently, multiplication of τ by A reduces the data fluctuations at given M_W. The good agreement between the model Eq. (8) and the experimental data tells us that the model is robust enough to describe different plants of the same variety. During our experiment, leaves were detached from the plants after measuring the THz transmission, in order to perform the gravimetric measurements. Our treatment provided the leaf projective area by simply placing the leaf onto a white page with a scale bar on it. This method can of course be improved by placing a scalebar directly in the THz transmission setup. With this simple adaptation, the measurement of both the optical depth and the leaf projective area would become telemetric, thus exploitable for non-invasive measurements. Two other important points to take care of are the cryogenic temperature of the THz source and the leaf temperature. The cryo-cooler employed in this work mounts a Gifford-McMahon motor, which is free from cryogenic liquids such as liquid helium or liquid nitrogen; despite that, its size and weight would be impractical for in field measurements, requiring relatively large water-cooled compressors. This issue can be overcome by using low-current, high-efficiency THz QCLs [35]: they can be operated on board of compact, self-contained Stirling cryo-coolers, that necessitate only of limited electrical power ($\sim$10 W) and air cooling. Temperature is also an important factor in determining the water absorption coefficient, whose variation is of the order of 10% ranging from 25 to 35 °C. In this case a model describing how the dielectric permittivity of water changes with temperature is discussed in [36]: after the absorption coefficient is defined as function of the temperature, a conversion table can be arranged in order to normalize the in-field data with respect to the leaf temperature conditions, which could be probed for example by using an infrared thermometer [37]. Finally, our results unlock an interesting question about the measurements. The confidence of our linear fit depends on the specific relation between leaf water mass and leaf projective area. This relation could change with the plant species, and it will be the subject of future theoretical and experimental studies. The main result of this work can now be summarized: using Eq. (8) we are able to measure the water mass inside a grapevine leaf by means of two potentially telemetric measurements, which are fast, non-invasive, and reliable.

Conclusions

In this work, we developed a new method for measuring the leaf water mass using a THz QCL, based on a leaf statistical sampling and a linear equation which relates the leaf water mass to the product between the optical depth and the leaf projected area. The results are in very good agreement with the model proposed here, proving the robustness and reliability of our new technique, which is under patent filing (application no. 102016000106179). Our method can be employed when plant averaged or cumulative data are required. For example, the canopy water mass can be combined with other plant parameters such as the biomass fresh and dry weight or the grain yield, and used as diagnostic indicators for cultivars under drought stress [38]. In future experiments, further calibration curves will be performed on other plant genotypes in order to prove our law or find other ones, whereas new relations between the shape of the scaling law and different plant behaviours could be highlighted, e.g. when preserving the water status during drought stress. By combining our technique with a gas exchange analyser, experiments can be carried out monitoring plant physiological aspects such as stomatal conductance and water potential directly in the laboratory or in the field. Furthermore, the different vein architecture among polyhedral leaves may be related to different management of water demand within the hydraulic network.

Authors' contributions

LB prepared the terahertz transmission setup, the experimental protocol, and performed the data analysis; MP prepared the plant material and the experimental protocol; LM prepared the terahertz transmission setup and the experimental protocol. MP, LB and LM performed the measurements. AT performed the data analysis and supervised the work. GC prepared the experimental setup. PS and AT supervised the work. All authors contributed to writing the article. All authors read and approved the final manuscript.

Author details
[1] NEST, CNR Istituto Nanoscienze and Scuola Normale Superiore, Piazza San Silvestro 12, 56127 Pisa, Italy. [2] Consiglio per la ricerca in agricoltura e l'analisi dell'economia agraria, Centro di ricerca per la Viticoltura e l'Enologia, Viale Santa Margherita 80, 52100 Arezzo, Italy. [3] Dipartimento di Fisica, Università di Pisa, Largo Pontecorvo 3, 56127 Pisa, Italy. [4] NEST, CNR Istituto Nanoscienze and Dipartimento di Fisica, Università di Pisa, Largo Pontecorvo 3, 56127 Pisa, Italy.

Acknowledgements
Not applicable.

Competing interests
The authors declare that they have no competing interests.

Funding
Funding information is not applicable.

Use of plant material
We did not use any genetically modified material nor opiaceous. Our plant treatments were not subject to any particular guideline or legislation.

References
1. Fischer G, Tubiello FN, Van Velthuizen H, Wiberg DA. Climate change impacts on irrigation water requirements: effects of mitigation, 1990–2080. Technol Forecast Soc Change. 2007;74(7):1083–107.
2. Schulze E-D, Robichaux R, Grace J, Rundel P, Ehleringer J. Plant water balance. Bioscience. 1987;37:30–7.
3. Brodribb TJ, Feild TS, Jordan GJ. Leaf maximum photosynthetic rate and venation are linked by hydraulics. Plant Physiol. 2007;144(4):1890–8.
4. Price CA, Symonova O, Mileyko Y, Hilley T, Weitz JS. Leaf extraction and analysis framework graphical user interface: segmenting and analyzing the structure of leaf veins and areoles. Plant Physiol. 2011;155(1):236–45.
5. Pagano M, Corona P, Storchi P. Image analysis of the leaf vascular network: physiological considerations. Photosynthetica. 2016;54:567–71. doi:10.1007/s11099-016-0238-2.
6. Pagano M, Storchi P. Leaf vein density: a possible role as cooling system. J Life Sci. 2015;9:299–303.
7. Chaves MM, Pereira JS, Maroco J, Rodrigues ML, Ricardo CPP, Osório ML, Carvalho I, Faria T, Pinheiro C. How plants cope with water stress in the field. Photosynthesis and growth. Ann Bot. 2002;89(7):907–16.
8. Sack L, Frole K. Leaf structural diversity is related to hydraulic capacity in tropical rain forest trees. Ecology. 2006;87(2):483–91.
9. Aasamaa K, Sõber A, Rahi M. Leaf anatomical characteristics associated with shoot hydraulic conductance, stomatal conductance and stomatal sensitivity to changes of leaf water status in temperate deciduous trees. Funct Plant Biol. 2001;28(8):765–74.
10. Taiz L, Zeiger E. Plant physiology. 5th ed. Sunderland: Sinauer Assoc; 2010.
11. Scholander PF, Bradstreet ED, Hemmingsen E, Hammel H. Sap pressure in vascular plants negative hydrostatic pressure can be measured in plants. Science. 1965;148(3668):339–46.
12. Jepsen PU, Cooke DG, Koch M. Terahertz spectroscopy and imaging-modern techniques and applications. Laser Photon Rev. 2011;5(1):124–66.
13. Gente R, Koch M. Monitoring leaf water content with thz and sub-thz waves. Plant Methods. 2015;11(1):15.
14. de Cumis US, Xu J-H, Masini L, Degl'Innocenti R, Pingue P, Beltram F, Tredicucci A, Vitiello MS, Benedetti PA, Beere HE, et al. Terahertz confocal microscopy with a quantum cascade laser source. Opt. Express. 2012;20(20):21924–31.
15. Born N, Behringer D, Liepelt S, Beyer S, Schwerdtfeger M, Ziegenhagen B, Koch M. Monitoring plant drought stress response using terahertz time-domain spectroscopy. Plant Physiol. 2014;164(4):1571–7.
16. Jördens C, Scheller M, Breitenstein B, Selmar D, Koch M. Evaluation of leaf water status by means of permittivity at terahertz frequencies. J Biol Phys. 2009;35(3):255–64.

17. Castro-Camus E, Palomar M, Covarrubias A. Leaf water dynamics of arabidopsis thaliana monitored in-vivo using terahertz time-domain spectroscopy. Sci Rep. 2013;3:2910.

18. Heugen U, Schwaab G, Bründermann E, Heyden M, Yu X, Leitner D, Havenith M. Solute-induced retardation of water dynamics probed directly by terahertz spectroscopy. Proc Nat Acad Sci. 2006;103(33):12301–6.

19. Arikawa T, Nagai M, Tanaka K. Characterizing hydration state in solution using terahertz time-domain attenuated total reflection spectroscopy. Chem Phys Lett. 2008;457(1):12–7.

20. Grossman M, Born B, Heyden M, Tworowski D, Fields GB, Sagi I, Havenith M. Correlated structural kinetics and retarded solvent dynamics at the metalloprotease active site. Nat Struct Mol Biol. 2011;18(10):1102–8.

21. Schmidt DA, Birer O, Funkner S, Born BP, Gnanasekaran R, Schwaab GW, Leitner DM, Havenith M. Rattling in the cage: ions as probes of sub-picosecond water network dynamics. J Am Chem Soc. 2009;131(51):18512–7.

22. Heisler IA, Meech SR. Low-frequency modes of aqueous alkali halide solutions: glimpsing the hydrogen bonding vibration. Science. 2010;327(5967):857–60.

23. Tielrooij K, Van Der Post S, Hunger J, Bonn M, Bakker H. Anisotropic water reorientation around ions. J Phys Chem B. 2011;115(43):12638–47.

24. Funkner S, Niehues G, Schmidt DA, Heyden M, Schwaab G, Callahan KM, Tobias DJ, Havenith M. Watching the low-frequency motions in aqueous salt solutions: the terahertz vibrational signatures of hydrated ions. J Am Chem Soc. 2011;134(2):1030–5.

25. Hale GM, Querry MR. Optical constants of water in the 200-nm to 200-μm wavelength region. Appl Opt. 1973;12(3):555–63.

26. Ulaby FT, Jedlicka R. Microwave dielectric properties of plant materials. IEEE Trans Geosci Remote Sens. 1984;4:406–15.

27. Fiorani F, Schurr U. Future scenarios for plant phenotyping. Annu Rev Plant Biol. 2013;64:267–91.

28. Köhler R, Tredicucci A, Beltram F, Beere HE, Linfield EH, Davies AG, Ritchie DA, Iotti RC, Rossi F. Terahertz semiconductor-heterostructure laser. Nature. 2002;417(6885):156–9.

29. Kapos V. Effects of isolation on the water status of forest patches in the brazilian amazon. J Trop Ecol. 1989;5(02):173–85.

30. Sack L, Scoffoni C, McKown AD, Frole K, Rawls M, Havran JC, Tran H, Tran T. Developmentally based scaling of leaf venation architecture explains global ecological patterns. Nat Commun. 2012;3:837.

31. Schindelin J, Arganda-Carreras I, Frise E, Kaynig V, Longair M, Pietzsch T, Preibisch S, Rueden C, Saalfeld S, Schmid B, et al. Fiji: an open-source platform for biological-image analysis. Nat Methods. 2012;9(7):676–82.

32. Bhatt M, Ayyalasomayajula KR, Yalavarthy PK. Generalized Beer–Lambert model for near-infrared light propagation in thick biological tissues. J Biomed Opt. 2016;21(7):076012.

33. Scheller M, Jansen C, Koch M, et al. Applications of effective medium theories in the terahertz regime. London: INTECH Open Access Publisher; 2010.

34. Kocsis L, Herman P, Eke A. The modified Beer–Lambert law revisited. Phys Med Biol. 2006;51(5):91.

35. Amanti MI, Scalari G, Castellano F, Beck M, Faist J. Low divergence terahertz photonic-wire laser. Opt Express. 2010;18(6):6390–5.

36. Ellison W. Permittivity of pure water, at standard atmospheric pressure, over the frequency range 0–25 thz and the temperature range 0–100 c. J Phys Chem Ref Data. 2007;36(1):1–18.

37. Winterhalter L, Mistele B, Jampatong S, Schmidhalter U. High throughput phenotyping of canopy water mass and canopy temperature in well-watered and drought stressed tropical maize hybrids in the vegetative stage. Eur J Agron. 2011;35(1):22–32.

38. Elsayed S, Darwish W. Hyperspectral remote sensing to assess the water status, biomass, and yield of maize cultivars under salinity and water stress. Bragantia. 2017;76(1):62–72.

8

Improved axenic hydroponic whole plant propagation for rapid production of roots as transformation target tissue

Kyle Benzle and Katrina Cornish*[ID]

Abstract

Background: Plant roots are used as an efficient target tissue for plant transformation assays. In root propagable species transformed roots are able to regenerate into whole plants without the addition of exogenous hormones, thus avoiding somaclonal variation associated with many plant transformation protocols. Plants grown in soil or soilless solid medium have roots that tend to be extremely delicate and are difficult to sterilize in advance of plant transformation experiments. Axenic tissue culture plants grown on semi-solid media are slow to produce large amounts of biomass compared to plants grown in solution-based media.

Methods: Seeds were germinated and grown for 14 days on half-strength semi-solid Murashige and Skoog medium containing 1% sucrose. Seedlings were then transferred to Magenta™ GA7 vessels containing either liquid or semi-solid ½ MS medium with 0.25, 0.5, 1, 2 or 3% sucrose. In the hydroponics (liquid medium) treatments, expanded clay balls were used to anchor seedlings. Hydroponic vessels were fitted with a sterile air aeration hose and filled ¾ full (100 mL) with liquid ½ MS media. Liquid media were replaced after 7 days. All plants were grown under fluorescent lights for 14 days.

Results: We have developed an improved axenic hydroponic propagation system for producing large quantities of plant roots for use in transformation assays using *Taraxacum kok-saghyz* as a model for root propagable species. Plants grew significantly faster in liquid media than on solid media. Addition of sucrose from 0.25 to 2% was correlated with an increase in biomass accumulation in plants grown in liquid media.

Conclusions: Our improved axenic hydroponic method yields sufficient quantities of roots for extensive plant transformation/molecular studies.

Background

Taraxacum kok-saghyz (TK) is a root propagable plant of interest because of the high quality natural rubber that is produced and stored in its roots [1, 2]. However, this species is recently wild-collected and modern molecular methods are needed to accelerate its domestication. Previously, a complimentary rapid plant transformation method was developed using *Agrobacterium rhizogenes* root transformation [3]. In both stable and transient root transformation methods, a common hurdle is the production of sufficient amounts of roots for

experimentation. Traditionally, sterile tissue cultured plants are grown on semi-solid medium in autoclavable vessels which allow for microbe-free gas exchange through a vented lid. Plants grown under such conditions tend to grow slowly and methods must be optimized for each individual species, or even cultivar, targeted because species vary in their ability to regenerate under selection [4]. Cells targeted for transformation should be actively dividing and ideally near the tissue surface to be accessible to the transformation vector [5] and, in most cases, a high degree of replication is necessary to produce successful stable transformants.

Although hydroponic culture has been used successfully in many plants, overall adoption is low because the

*Correspondence: cornish.19@osu.edu
Department of Horticulture and Crop Science, The Ohio State University, 1680 Madison Avenue, Wooster, OH, USA

nutrient rich growth medium used encourages growth of unwanted micro-organisms and thus is difficult to maintain. Axenic plant hydroponic culture has generally been limited to cell culture or callus grown in suspension. However, axenic culture of whole plants has the added benefit of avoiding negative effects of somaclonal variation which can occur in cell and callus culture [6]. Whole plant cultures must be oxygenated but if this is accomplished using a shaker table, as is standard for callus tissue, delicate root structures may be damaged: such damage does not occur when a filtered aeration system is used [7]. Previously described systems have used glass jars and stainless mesh to anchor plantlets in liquid media [8] but these are difficult to maintain and do not produce large amounts of harvestable tissue. In this paper, we describe an easily implementable axenic culturing system. This system is suited to small-scale, rapid root production from seedlings to fully developed plants.

Methods

Growth media

In a preliminary experiment, TK tissue culture plants were grown on semi-solid media containing various concentrations of MS nutrients and sucrose. MS medium at $1\times$ reduced adventitious root growth compared to $0.5 \times$ MS medium. Therefore, in this study, solid and liquid media were made with $0.5 \times$ Murashige and Skoog [9] medium (½ MS) (half strength MS micro- and macro-salts (Caisson Laboratories. North Logan, UT, USA) with $1 \times$ Gamborg's B5 vitamins and 10 g L^{-1} sucrose) [10] at pH 5.7 and 3.5 mM MES (25 mg L^{-1}). Semi-solid growth media included plant tissue culture agar (Sigma-Aldrich, St. Louis, MO, USA) at 1% (w/v). Seeds, for both semi-solid and hydroponic cultures, were germinated on semi-solid ½ MS (see next section). Because 1% sucrose is the concentration normally used to screen genetically transformed plants [11], this was used to germinate seeds on semi-solid media (½ MS plates). Plants were grown in ½ MS with or without 1% agar at sucrose levels of 0.25, 0.5, 1, 2 and 3%.

Plant material

All tissue manipulation was performed with ethanol-flamed forceps in a sterile laminar flow hood. *Taraxacum kok-saghz*, accession TK-17 [12] seeds were surface-sterilized in 25 mL 30% (v/v) domestic bleach with 0.03% Triton® X-100 for 8 min at room temperature. The seeds were rinsed three times with 25 mL sterile double distilled water. About 100 seeds were sown on semi-solid ½ MS medium in Petri dishes (50 mL volume, 15 mm × 150 mm, Thermo Fisher Scientific Inc., Waltham, MA, USA) with 5–10 seeds per plate. Dishes

were sealed with 3 M Micropore™ surgical tape (Thermo Fisher) and left for 48 h at 22 °C under ambient light. Seeded dishes then were placed under 16 h day fluorescent lighting conditions (PAR = 80–120 μmol m^{-2} s^{-1}) for 7 days at 22 °C. Seedlings then were moved to either ½ MS liquid or semi-solid media in Magenta vessels and grown for an additional 14 days at 22 °C.

Hydroponics apparatus

Modified Magenta G7 vessels were vertically connected using MK-5 10 mm Connector Lids (Caisson Laboratories) (Fig. 1a). The lower unit was filled with washed and autoclaved 10–30 mm expanded clay balls (Fig. 1b). A 5 mm hole was drilled in the top unit to insert a sterile air hose. Ten of the connected vessels were placed into a Nalgene™ Autoclavable Polypropylene Pan (Thermo Fisher) fitted with sufficient 5 mm Tygon tubing (Sigma-Aldrich) to reach the bottom of each vessel interior and have approximately 15 cm of tubing on the outside. Autoclavable Tygon splitters (Fig. 1c) were used to branch the single sterile air input into 10 separate lines, one for each vessel. The single air input line was fitted with an autoclavable 20 μM Acro® 50 Vent Filter (Pall Laboratory, Port Washington, NY, USA) or Millipore Lab Millistak Mini, MCOHC23HH3, (Millipore Corporation, Billerica, MA) to sterilize incoming air (Fig. 1d).

Once fitted with the tubing, the entire apparatus was autoclaved and allowed to cool in a laminar flow hood before adding five treatment media in duplicate. Using aseptic technique, one seedling was placed in each of the 10 liquid medium treatment vessels and the 10 semi-solid medium filled vessels. The hydroponic apparatus was then fitted with a 12-W air pump and constantly aerated. The hydroponic vessels and semi-solid vessels were placed under 16 h day fluorescent lighting for 14 days at 22 °C. A range of sucrose concentrations was evaluated to maximize root growth. Media were replaced after the first 7 days, taking care to maintain sterile conditions. Tissue was harvested aseptically by sterilizing the outside of the vessel with 70% ethanol and removing the plants from their vessels using 30 cm forceps, briefly blotting them on sterile paper towels, and separating the roots and shoots using forceps. Root/shoot fresh weights and lengths were quantified using a balance (to nearest 0.01 g) and ruler (to nearest 1 mm).

Results and discussion

Hydroponically-grown plants grew up to 10 times larger than those on semi-solid media (Fig. 2a–d). Among treatments, the highest root and shoot biomass was achieved in hydroponically-grown TK plants in 2% sucrose with $0.5 \times$ MS. One-way analysis of variance (ANOVA) was

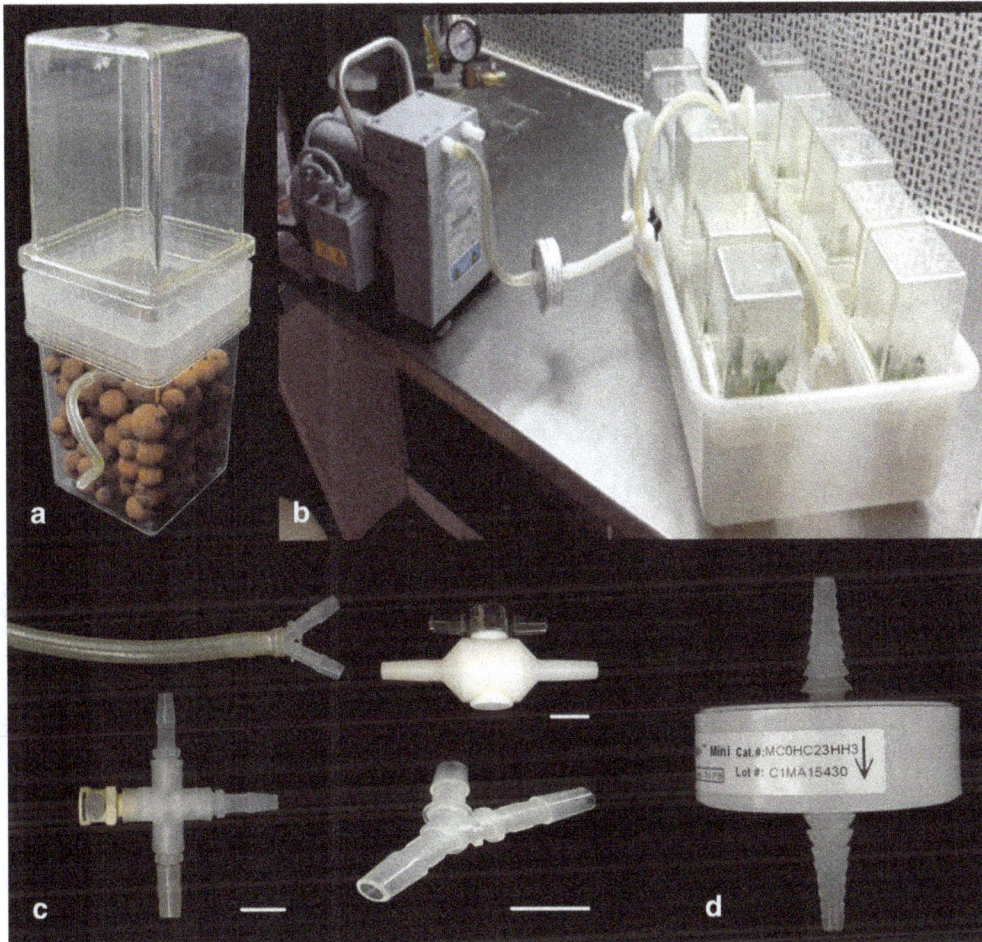

Fig. 1 Hydroponics apparatus. Hydroponic apparatus consisting of: **a** liquid medium in autoclavable Magenta vessel with expanded clay balls fitted with 20 cm aeration hose. **b** A set of 10 vessels connected with Tygon tubing and aerated with air passed through a 20 μm filter. **c** Tygon tubing, connectors, splitters and valve used in the apparatus. **d** Millipore autoclavable filter. *Line* is approximately 1 cm

used to compare methods, Tukey's Honestly Significant Difference (HSD) post hoc test was then used to compare groups and identify which samples differed significantly (with $p < 0.05$). Analysis was done with R v. 3.4.0 software (R Development Core Team, 2017). Statistically significant differences in plant root weight were found between sucrose levels of 0.25 and 3% when plants were grown in liquid culture that were not seen when a semi-solid medium was used.

All plants appeared to be morphologically normal, and primary and secondary roots were produced. In liquid medium, the largest plants were observed when plants were grown with 1–2% sucrose (Fig. 2a, b). However, the sucrose concentrations used in this experiment had very little effect on seedling growth on semi-solid media (Fig. 2c, d) although all supported more growth than no sucrose (data not shown). Growth of seedlings was inhibited by 3% sucrose in liquid media (Fig. 2b). Root biomass increased with the level of sucrose, but shoot biomass did not significantly increase (Fig. 3a–c). Visually, plants in liquid media (Fig. 4a–e) grew larger than those on semi-sold media (Fig. 4f–j), with biomass increasing more

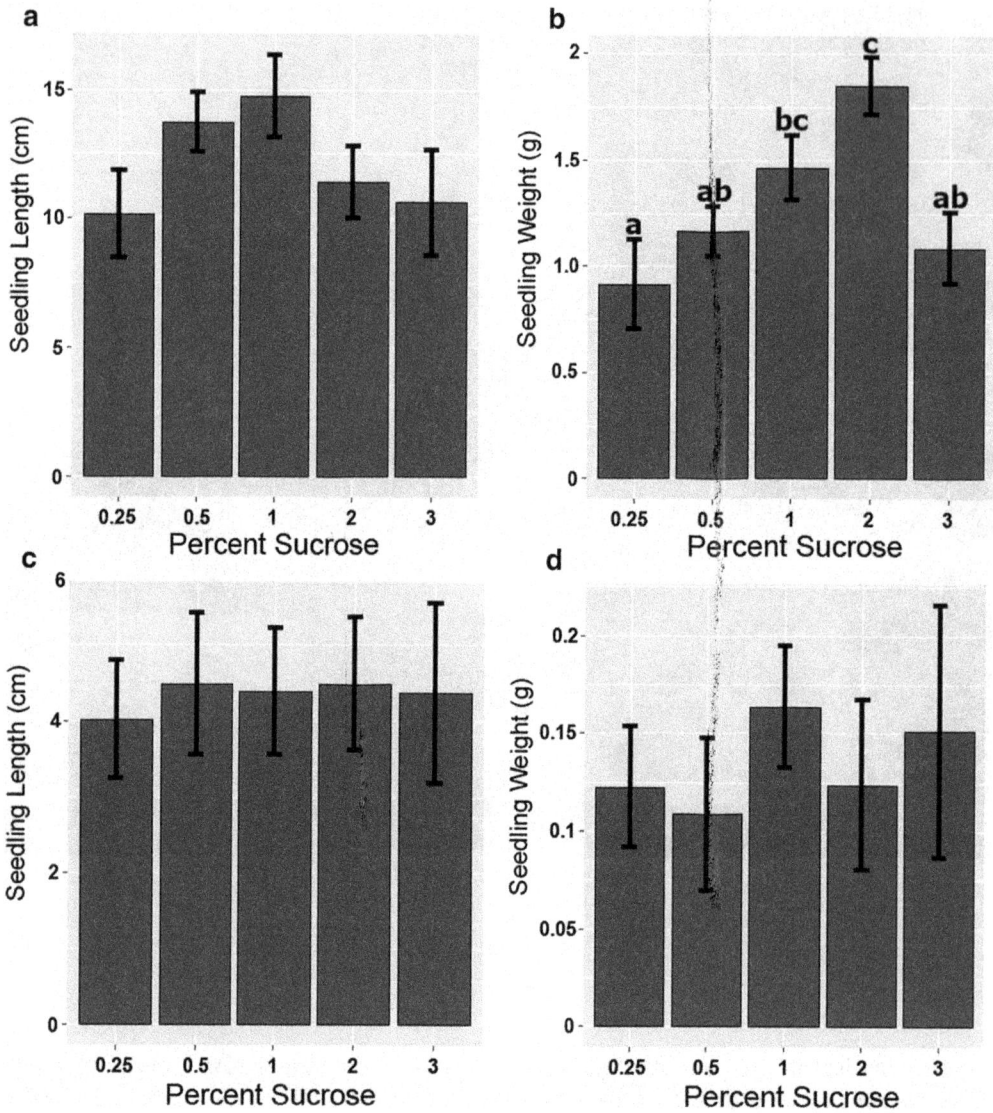

Fig. 2 Comparison of hydroponic and semi-solid growth under varying sucrose levels. Analysis of root and shoot tissue under liquid (**a**, **b**) or semi-solid (**c**, **d**) culture treatments. **a** Seedling total length per individual plant for selected treatments. **b** Fresh weight biomass. **c** Seedling total length. **d** Fresh weight biomass. ANOVAs were conducted for all comparisons and Tukey's HSD means comparison *letters* are shown where significant differences were found. *Letters* denote significant differences ($p < 0.05$), values are the mean $\pm$ SE (n = 8)

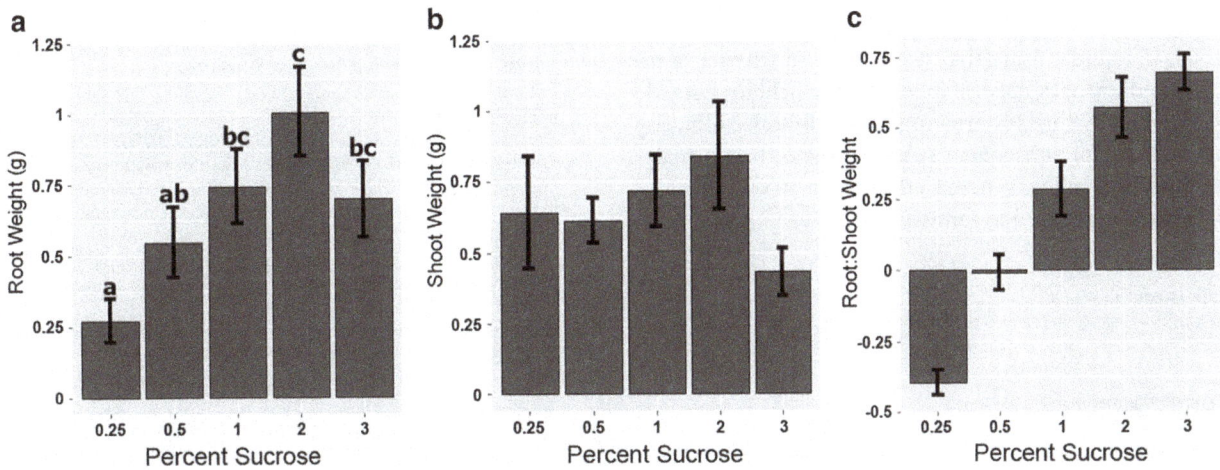

Fig. 3 Hydroponic root-to-shoot weight ratios. Root and shoot fresh weights by percent sucrose in liquid media **a** root weight, **b** shoot weight, **c** root:shoot ratio. ANOVAs were completed for both comparisons and Tukey's HSD means comparison *letters* are shown only where significant differences were found, values are the mean ± SE (n = 8)

Fig. 4 Hydroponic and semi-solid grown TK. TK seedlings, 21 days old, grown in liquid (**a–e**) and semi-solid (**f–j**) medium. *Letters* indicate sucrose levels of 0.25, 0.5, 1, 2 and 3% for **a/f**, **b/g**, **c/h**, **d/i** and **e/j** respectively. *Bar* 1 cm

rapidly than seedling length, producing stouter plants. Nutrient levels and lighting conditions may be further optimized in future experiments.

Conclusions

In conclusion, we have developed an improved and simple procedure for axenic plant culture. Statistically significant differences in plant root weight were found between sucrose levels of 0.25 and 3% when plants were grown in liquid culture that were not seen when a semi-solid medium was used. Expanded clay pebbles provide excellent anchorage for growing roots while also allowing easy harvest for subsequent subculturing or tissue analysis. Root biomass was increased by up to $10\times$ by this liquid system compared to semisolid culture.

Authors' contributions

Both authors read and approved the final manuscript.

Acknowledgements

I would like to thank The Ohio State University and The United States Department of Agriculture, National Institute of Food and Agriculture, Hatch Project 230837 for providing resources for this work.

Competing interests

The authors declare that they have no competing interests.

Funding

Funding for this work was provided by The Ohio State University and The United States Department of Agriculture, National Institute of Food and Agriculture, Hatch Project 230837.

References

1. Schmidt T, Lenders M, Hillebrand A, van Deenen N, Munt O, Reichelt R, et al. Characterization of rubber particles and rubber chain elongation in *Taraxacum koksaghyz*. BMC Biochem. 2010;11:1–11.
2. Cornish K. Biochemistry of natural rubber, a vital raw material, emphasizing biosynthetic rate, molecular weight and compartmentalization, in evolutionarily divergent plant species. Nat Prod Rep. 2001;18:182–9.
3. Zhang Y, Iaffaldano BJ, Xie W, Blakeslee JJ, Cornish K. Rapid and hormone-free *Agrobacterium rhizogenes*-mediated transformation in rubber producing dandelions *Taraxacum kok-saghyz* and *T. brevicorniculatum*. Ind Crop Prod. 2015;66:110–8.
4. Anami S, Njuguna E, Coussens G, Aesaert S, Van Lijsebettens M. Higher plant transformation: principles and molecular tools. Int J Dev Biol. 2013;57:483–94.
5. Birch RG. Plant transformation: problems and strategies for practical application. Annu Rev Plant Biol. 1997;48:297–326.
6. Karp A. Somaclonal variation as a tool for crop improvement. Euphytica. 1995;85:295–302.
7. Hetu M, Tremblay LJ, Lefebvre DD. High root biomass production in anchored Arabidopsis plants grown in axenic sucrose supplemented liquid culture. Biotechniques. 2005;39:345.
8. Alatorre-Cobos F, Calderón-Vázquez C, Ibarra-Laclette E, Yong-Villalobos L, Pérez-Torres C-A, Oropeza-Aburto A, et al. An improved, low-cost, hydroponic system for growing Arabidopsis and other plant species under aseptic conditions. BMC Plant Biol. 2014;14:1–13.
9. Murashige T, Skoog F. A revised medium for rapid growth and bio assays with tobacco tissue cultures. Physiol Plant. 1962;15:473–97.
10. Gamborg OL, Miller R, Ojima K. Nutrient requirements of suspension cultures of soybean root cells. Exp Cell Res. 1968;50:151–8.
11. Bhojwani SS, Dantu PK. Plant tissue culture: an introductory text. Berlin: Springer; 2013.
12. Hellier B. Collecting in Central Asia and the Caucasus: US national plant germplasm system plant explorations. Hort Sci. 2011;46(11):1438–9.

9

In vivo label-free mapping of the effect of a photosystem II inhibiting herbicide in plants using chlorophyll fluorescence lifetime

Elizabeth Noble[1,2,3]* ⓘ, Sunil Kumar[1], Frederik G. Görlitz[1], Chris Stain[4], Chris Dunsby[1,5]‡ and Paul M. W. French[1]‡

Abstract

Background: In order to better understand and improve the mode of action of agrochemicals, it is useful to be able to visualize their uptake and distribution in vivo, non-invasively and, ideally, in the field. Here we explore the potential of plant autofluorescence (specifically chlorophyll fluorescence) to provide a readout of herbicide action across the scales utilising multiphoton-excited fluorescence lifetime imaging, wide-field single-photon excited fluorescence lifetime imaging and single point fluorescence lifetime measurements via a fibre-optic probe.

Results: Our studies indicate that changes in chlorophyll fluorescence lifetime can be utilised as an indirect readout of a photosystem II inhibiting herbicide activity in living plant leaves at three different scales: cellular (~μm), single point (~1 mm²) and macroscopic (~8 × 6 mm² of a leaf). Multiphoton excited fluorescence lifetime imaging of *Triticum aestivum* leaves indicated that there is an increase in the spatially averaged chlorophyll fluorescence lifetime of leaves treated with Flagon EC—a photosystem II inhibiting herbicide. The untreated leaf exhibited an average lifetime of 560 ± 30 ps while the leaf imaged 2 h post treatment exhibited an increased lifetime of 2000 ± 440 ps in different fields of view. The results from in vivo wide-field single-photon excited fluorescence lifetime imaging excited at 440 nm indicated an increase in chlorophyll fluorescence lifetime from 521 ps in an untreated leaf to 1000 ps, just 3 min after treating the same leaf with Flagon EC, and to 2150 ps after 27 min. In vivo single point fluorescence lifetime measurements demonstrated a similar increase in chlorophyll fluorescence lifetime. Untreated leaf presented a fluorescence lifetime of 435 ps in the 440 nm excited chlorophyll channel, CH4 (620–710 nm). In the first 5 min after treatment, mean fluorescence lifetime is observed to have increased to 1 ns and then to 1.3 ns after 60 min. For all these in vivo plant autofluorescence lifetime measurements, the plants were not dark-adapted.

Conclusions: We demonstrate that the local impact of a photosystem II herbicide on living plant leaves can be conveniently mapped in space and time via changes in autofluorescence lifetime, which we attribute to changes in chlorophyll fluorescence. Using portable fibre-optic probe instrumentation originally designed for label-free biomedical applications, this capability could be deployed outside the laboratory for monitoring the distribution of herbicides in growing plants.

Keywords: Fluorescence spectroscopy, Plant, FLIM, Herbicide, Photosystem II, Chlorophyll fluorescence lifetime

Background

Understanding the specific fates of agrochemicals in plants is crucial to optimise their function. The absorption, distribution, transportation and storage of agrochemicals within and through the various cellular structures will greatly affect the extent and effectiveness of their interaction with plant metabolism [1] and so the ability to map the distribution or effect of agrochemicals in vivo could help optimise the formulation of new agrochemicals.

In vitro methods that are based on solvent extraction or cell culture analysis, such as biochemical assays or mass

*Correspondence: e.noble@imperial.ac.uk

†Chris Dunsby and Paul M. W. French contributed equally to this work

² Department of Chemistry, Imperial College London, London SW7 2AZ, UK

Full list of author information is available at the end of the article

spectrometry [2, 3], can provide highly specific chemical information but are destructive and time consuming. Furthermore, any physical or chemical perturbation to the natural state of a plant can result in stress, potentially triggering various reflex mechanisms that could lead to incorrect inferences [4] and such measurements can fail to report the true spatio-temporal distribution and interaction of agrochemicals. Analytical methods such as radiolabelling [5] can provide highly specific information on herbicide metabolism and over all absorption rates, but do not offer spatially resolved herbicide distribution data. Scanning electron microscopy (SEM) [6] can provide higher spatial resolution but can be prone to artefacts arising from sample fixation and mechanical sectioning and cannot follow dynamic events.

Fluorescence imaging techniques, such as widefield fluorescence imaging [7], confocal laser scanning microscopy (CLSM) [8, 9] and two-photon excitation microscopy (TPEM) [10, 11], provide the ability to follow dynamic events, including in live plants. Hence, it is desirable to explore non-invasive fluorescence imaging techniques, capable of providing qualitative and quantitative information concerning the in vivo distribution of agrochemicals, in the laboratory and in the field. In principle, this could be realised by fluorescence imaging of agrochemicals that are intrinsically fluorescent. However, it is generally challenging to distinguish such fluorescence from background autofluorescence arising from chlorophyll and/or other endogenous fluorophores. Direct fluorescent labelling of the agrochemical could improve the contrast but such labels are comparable to or larger in size than the agrochemicals themselves and may compromise their mode of action. An alternative approach is to map the local action of an agrochemical through its impact on the intrinsic (chlorophyll) fluorescence and thereby infer information concerning the agrochemical distribution. Here we demonstrate this approach using fluorescence lifetime imaging of chlorophyll to map the local action of a PS II inhibiting herbicide.

Herbicides represent a major fraction of all agrochemicals used, of which PS II inhibiting herbicides are an important class. Light energy absorbed by leaves can be expended driving the photochemistry of photosynthesis, emitted as chlorophyll fluorescence or dissipated as heat [12]. When a chlorophyll molecule absorbs light and is promoted to its first singlet excited state, the excited state energy can be transferred to other chlorophyll molecules and ultimately to the photosynthetic reaction centres (PS II and PS I) through Förster resonance energy transfer [13]. The presence of a photosynthetic inhibitor suppressing this pathway leads to an increase in the lifetime of singlet excited chlorophyll, which increases the probability of inter-system crossing to long lived chlorophyll

triplet excited states. This increases the probability of transfer of energy from the triplet excited excited states of chlorophyll to ground state molecular oxygen, which results in the formation of singlet state molecular oxygen that can subsequently lead to the production of radical species such as ^3Chl, $^1O_2$, H_2O_2 and O_2^-. These radicals are phytotoxic in nature and can hinder important biological processes or cause membrane damage. Treating a plant with a PS II inhibitor will trigger an increased production of such radical species and can result in phytotoxicity. This is the basis for the mode of action of most of PS II inhibiting herbicides [14].

Chlorophyll fluorescence can thus provide a readout of the action of PS II inhibitors. In general, the balance between photochemistry, chlorophyll fluorescence and heat dissipation following the absorption of light has been studied using a range of chlorophyll fluorometric techniques that are sensitive to the photosynthetic electron transfer rate and therefore provide information on the overall PS II efficiency [15]. Early studies were based on the assumption that the light energy absorbed in leaves was divided equally between the PS I and PS II light harvesting complexes [15] and therefore the factor, F_v/F_m, which is the ratio of fluorescence variation to maximal fluorescence, gives a measure of maximum PS II efficiency [16]. Note that $F_v = F_m - F_o$, where F_o is the fluorescence from the leaf in the presence of weak measuring light (~0.1 µmol photons m^{-2} s^{-1}) and F_m is the maximum fluorescence from a dark-adapted leaf when excited by a saturating light flash, i.e. one with sufficient intensity to drive a high proportion of the PSII centres into the "closed" state where they are not capable of photochemistry. For Arabidopsis, a saturating photon flux density of ~4000 µmol photons m^{-2} s^{-1} is required [17] but this value may vary between plant species. Subsequent studies, however, showed that this assumption was not valid due to PS I fluorescence varying independently of photosynthetically active photon flux density, unlike PS II [15]. Furthermore, the PSI and PSII fluorescence vary differently as a function of temperature [18]. Therefore estimates of the photosynthetic electron transfer rate based on measurements of photosynthetic yield, F_v/F_m [17] could be erroneous. Nevertheless, chlorophyll fluorescence-based approaches have been successfully applied at different scales to study the dynamics of basic photosynthetic reactions, including biotic/abiotic stress responses [19–21], and efforts are being made to translate this optical signal from laboratory to field phenotyping [22] in plants, linking microscopic observations to macroscopic and to leaf level dynamics of photosynthetic reactions. Chlorophyll fluorescence imaging has previously been used to obtain qualitative readouts of metabolic changes correlated to herbicide action, as reported in [17], although the instrumentation used could not localise herbicide distribution with high resolution. Higher resolution fluorescence imaging studies [15, 23, 24], have demonstrated that the uptake of the herbicide Diuron can be monitored in plant leaves using chlorophyll fluorescence imaging following dark adaption, although these utilised wide-field imaging studies and so did not provide depth-resolved imaging of changes to chlorophyll fluorescence. Optically sectioned imaging can be provided by confocal laser scanning fluorescence microscopy and used to image up to a depth of 100–150 μm but image quality deteriorates when imaging deeper than ~100 μm into the sample owing to background fluorescence and scattering effects caused by the leaf tissues [25]. Improved performance in terms of imaging depth and reduced background fluorescence can be provided by two photon excitation, due to the increased ability of near-infrared light to penetrate biological samples and the limiting of excitation to the focal plane, as has been applied in plant leaves e.g. [26]. However, optical scattering and aberrations still impact the quantification of intensity-based chlorophyll fluorescence readouts.

Fluorescence lifetime measurements [27] can provide quantitative readouts even when image information is degraded and absolute intensity measurements are compromised by optical scattering, sample absorption (inner filter effect) and/or variations in fluorophore concentration. Fluorescence lifetime provides a direct readout of the impact of the local fluorophore environment on relaxation pathways following excitation and so provides powerful sensing capabilities. In biomedicine, fluorescence lifetime measurements and fluorescence lifetime imaging (FLIM) are used to study changes in tissue autofluorescence [28]. However, fluorescence lifetime techniques have not been widely utilized for plant studies although FLIM has been used to study photosynthesis [26, 29, 30], and the uptake of minerals [31, 32] and was previously used to study the effect of a photosynthetic inhibitor DCMU (3-(3,4-dichlorophenyl)-1,1-dimethylurea) in *Chlamydomonas reinhardtii* [29]. We report here the application of fluorescence lifetime imaging of chlorophyll to provide a label-free in vivo means to non-invasively map the effect of a PS II inhibiting herbicide in *Triticum aestivum* (Winter Wheat) on different spatial scales through its local impact on chlorophyll fluorescence.

The action of inhibitors on the photosynthetic electron transport chain has previously been observed using non-imaging, time-resolved fluorescence spectroscopy techniques: Petrasek et al. [33] observed an overall increase in chlorophyll fluorescence lifetime under the stress of a PS II inhibitor, DCMU, and Hunsche et al. [34] reported a significant increase in the mean lifetime

of the fluorescence measured from plants treated with PS II inhibiting herbicides. FLIM experiments have also shown that the inhibition of PS II by a herbicide leads to an increase of the chlorophyll fluorescence intensity and lifetime—attributed to a reduction in photochemical quenching—but no spatially resolved lifetime data was presented [26].

Here we explore the potential to utilise the change in chlorophyll fluorescence lifetime as an indirect read out of herbicide activity and thereby map the time-dependent uptake of the herbicide by plants in vivo—and ultimately in the field—using FLIM or single-point probe lifetime measurements. We present in vivo readouts from non-dark adapted plants at three different scales: cellular (~μm), single point (~1 mm^2) and macroscopic (~8 × 6 mm^2 of a leaf) to study the distribution of the effect of a PS II inhibiting herbicide. We first utilised two-photon laser scanning microscopy (TPLSM) to perform in vivo FLIM in leaves excited at 900 nm, taking advantage of the deeper optical sectioning and reduced out-of-plane photobleaching and phototoxic effects compared to single photon laser scanning confocal microscopy [35]. While it is difficult to make a generalization of the typical imaging depth achievable, since plant leaves have a complex structural anatomy that varies from species to species, we note that TPLSM imaging depths of ~200 μm have been reported [10, 36]. We also explored the potential to repurpose instrumentation originally developed for autofluorescence lifetime studies of human tissue for medical applications, noting that this instrumentation is already relative portable and could be engineered for application in the field. Specifically, we applied a custom-built time-resolved spectrofluorometer, which was originally developed to make single point measurements of tissue via a fibre-optic probe [37] to make measurements of plant leaves treated and untreated with the PSII inhibiting herbicide, Flagon 400 EC. This approach can provide mapping of the action of herbicides with a spatial resolution on the order of millimetres. Finally, for macroscopic imaging, we applied wide-field time-gated FLIM technology [38], obtaining maps of herbicide action on a time scale of seconds. We note that spectrally resolved lifetime measurements can provide further contrast, which could address the interference from other plant pigments, e.g. [39, 40].

Methods

Plant material and growth conditions

For the studies conducted, plants of *Triticum aestivum* (Winter Wheat cv. Hereward), a cool climate monocot food crop (Growth room conditions: 20/16 °C (day/night temp) 16 h of daylight, approx. 65% relative humidity, lighting-150 μmol/m^2/s or 31.2 W/m^2) were used. Seeds were obtained from Syngenta Jealott's Hill International Research Centre Bracknell, Berkshire, UK.

Chemicals

PS II inhibiting herbicide Flagon 400 EC was provided in solution by Syngenta (Jealott's Hill International Research Centre Bracknell, Berkshire, UK) and was diluted in milliQ water.

Sample preparation and plant treatment

Before the plants are treated with active ingredient (AI) formulation, a non-fluorescent felt pen (Berol Toughpoint) was used to mark the desired treatment area with four spots on a square with side approximately 10 mm so that the treated area could be easily located again in subsequent fluorescence measurements. Flagon EC 400 was diluted in water to a concentration of ~49 ppm/1.22 M and two closely spaced 2 μl droplets (a close approximation to the the size of typical spray droplets used in the field) were applied using a Hamilton micropipette (Hamilton Bonaduz AG, Bonaduz, Switzerland) in the centre of the four marked spots.

For measuring the leaf autofluorescence excitation/emission matrix, a new leaf was used for each emission spectrum to avoid any photochemical changes to the leaf caused by the excitation light. For each set of experiments, all the leaves were from plants of the same age and grown under the same growth conditions. Unless otherwise specified, we used 18 days old plants for the measurements reported here.

Measurement of fluorescence spectra

The autofluorescence of *Triticum aestivum* plant leaves and also the fluorescence properties of the herbicide were characterised in order to optimise the excitation wavelengths and fluorescence detection spectral windows for these studies. The fluorescence excitation and emission spectra of Flagon EC 400 and the plant leaves were measured using a commercial UV/VIS Spectrofluorometer (RF-5301PC, Shimadzu, Japan).

Characterisation of Flagon 400 EC fluorescence

Time-resolved fluorescence measurements of Flagon EC 400 were undertaken using the custom-built multidimensional spectrofluorometer described in [41], of which a schematic diagram of the optical set-up of the system is provided in Additional file 1. A supercontinuum laser source (SPC-400, 20 MHz Fianium, UK) operating at a repetition rate of 20 MHz provides tunable picosecond pulses for excitation. For measurements of Flagon EC 400, a bandpass filter centered at 400 and 40 nm bandwidth was used to select the excitation radiation

that passes through a polarizer before reaching sample solution in a cuvette. The detection beam path is at right angles to the excitation beam path and the resultant fluorescence decay is recorded for 11 distinct emission wavelengths spaced 2 nm apart over the range 480–500 nm, using time-correlated single photon counting (TCSPC), with an average of ~6000 photon counts over an acquisition time of 180 s per decay. All measurements were performed with a polarizer in the emission path placed at the magic angle polarization to remove any fluorescence anisotropy effects. The whole system is controlled by a custom-written LabVIEW software (LabVIEW, National Instruments). The instrument response function (IRF) for the fluorometer is measured using a scatterer, LUDOX (a solution of colloidal silica, Sigma-Aldrich, UK). The fluorescence decay curves are analysed using the *FLIMfit* software tool developed at Imperial College London [42] and the maximum likelihood iterative method was used to fit the experimental data to model fluorescence decay profiles.

Multiphoton excited fluorescent lifetime imaging (MPE-FLIM)

MPE-FLIM in plant cells was undertaken both in vivo in a live plant leaf and in situ in a recently removed plant leaf in order to obtain high (subcellular) resolution maps of chlorophyll fluorescence lifetime. FLIM measurements were made using a Leica SP5 system (TCS SP5, Leica Microsystems GmbH, Germany) with a tunable (690–1020 nm) Ti:Sapphire laser (Spectra-Physics, Broadband Mai Tai) providing 100 fs pulses at 80 MHz for multiphoton excitation and FLIM implemented with TCSPC. The optimum excitation wavelength for mulitphoton imaging of chlorophyll in a plant leaf was determined using a detection band of 680–735 nm and by scanning the excitation wavelength over the range 850–990 nm. The highest signal was obtained for excitation at 900 nm and so this was chosen for all subsequent multiphoton imaging. Treated and untreated leaf samples were imaged in situ and in vivo using multiphoton excitation at 900 nm and detection in the band 600–730 nm, with typical acquisition times of 40 s being required to acquire images with 256 × 256 pixels. The leaves were treated with two closely spaced 2 μl droplets of Flagon EC 400 and were imaged after 2.5 h for the in situ experiment. For in vivo time course experiments, images were taken at 5 min intervals starting from 5 min after treatment to 45 min.

The instrument response function (IRF) of the system was recorded using gold nano-rods and a background image (with no excitation) with the same experimental parameters as FLIM measurements. The fluorescence decay data were analyzed using the *FLIMfit* software [42].

Multispectral point-probe fluorescence lifetime measurements in vivo

To explore the potential to monitor herbicide distribution in the field, we used a portable time-resolved spectrofluorometer incorporating a fibre-optic probe that was originally developed for clinical and preclinical studies [37], to make single point measurements of leaves on live plants. This instrument, which is depicted in Fig. 1, incorporates two picosecond pulsed excitation diode lasers: a laser diode (LDH-P−C-375B, PicoQuant GmbH, Germany) that provides 70 ps pulses at 372 nm with an average output power of 3.3 mW and a laser diode (LDH-P−C-440B, PicoQuant GmbH, Germany) providing 90 ps pulses at 440 nm with 3.5 mW average power. The repetition rates of both the lasers were set to 20 MHz. The laser beams are coupled into a custom-made optical fiber bundle (FiberTech Optica, Canada) consisting of three excitation fibers and fourteen detection fibers arranged in a hexagonal structure around the excitation fibers.

For these experiments the fibre-optic probe was held by a clamp to the leaf and fluorescence was collected by the detection fibers into three spectrally resolved detection channels implemented using a set of dichroic mirrors and band-pass filters. The leaves were treated with two drops (2 μl) of the herbicide Flagon EC 400 and measurements were made 5 min afterwards. At 372 nm excitation, all the three channels are used: the first channel collects light from 400 to 420 nm (CH1); the second channel collects fluorescence light from 430 to 480 nm (CH2); and the third channel collects light from 620 to 710 nm (CH3). For the 440 nm excitation light, only the third (620–710 nm) channel is active and this is referred as "channel 4" (CH4). Since 440 nm is an optimum excitation wavelength for chlorophyll fluorescence

in leaves, CH4 is the spectral channel of primary interest for this study. The integration time for each acquisition was about 30 s for both UV excitation as well as 440 nm excitation. 1.7 μW of 375 nm excitation and 29 μW of 440 nm excitation at the distal tip of the fibre probe were used for measurements of plant leaves. IRF measurements were made for each detection channels using a scattering sample under the same conditions. This fibre-optic probe interrogates an area of ~1 mm^2 on the leaf and so can provide mm spatial resolution.

In vivo wide-field macroscopic time-gated fluorescence imaging

For imaging herbicide distribution with higher spatial resolution and on faster timescales, we constructed a wide field FLIM macroscope for in vivo imaging of plants. A schematic diagram describing the setup is shown in Fig. 2. Our system design partially follows previously described wide field FLIM endoscope instrumentation [43, 44], but here free space illumination using a diverging lens was used for conveniently exciting the leaf samples. The excitation source was a gain-switched diode laser (PicoQuant, LDH-P−C-440B with driver PDL-800-B), which provided pulses of <500 ps at 40 MHz with average powers of ~4 mW. The emission from the diode laser was passed through a spectral clean-up filter (F1, Semrock, FF02-438/24) to suppress out of band radiation from the diode laser and the elliptical beam was expanded by a diverging lens to illuminate a FOV of ~30 × 6 mm on a leaf held in place by a clamp while attached to a live plant. A black anodized piece of metal was fixed to the clamp behind the leaf to block any unwanted signals or reflections. Neutral density filters were inserted inbetween the clean-up filter and diverging

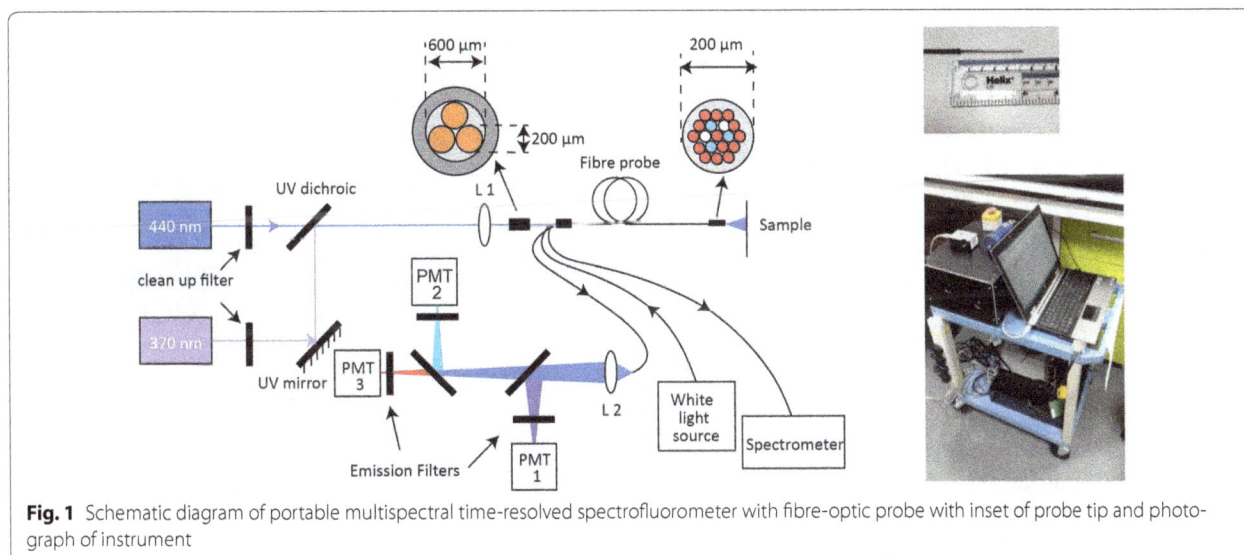

Fig. 1 Schematic diagram of portable multispectral time-resolved spectrofluorometer with fibre-optic probe with inset of probe tip and photograph of instrument

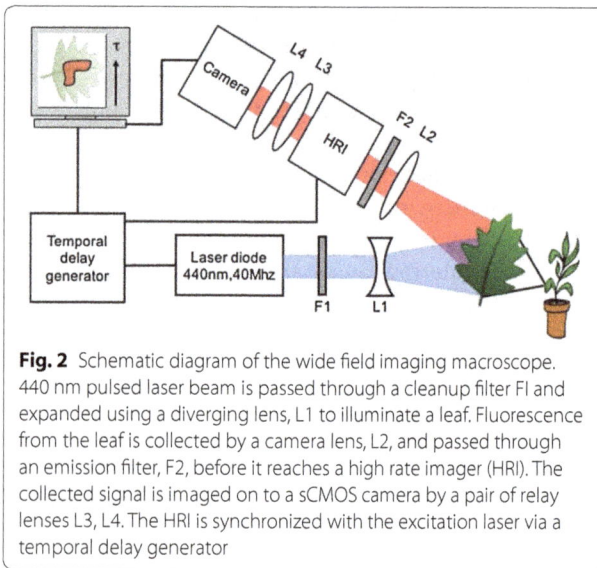

Fig. 2 Schematic diagram of the wide field imaging macroscope. 440 nm pulsed laser beam is passed through a cleanup filter Fl and expanded using a diverging lens, L1 to illuminate a leaf. Fluorescence from the leaf is collected by a camera lens, L2, and passed through an emission filter, F2, before it reaches a high rate imager (HRI). The collected signal is imaged on to a sCMOS camera by a pair of relay lenses L3, L4. The HRI is synchronized with the excitation laser via a temporal delay generator

lens to reduce the excitation beam power to 0.3 mW. The fluorescence from the leaf was collected by a camera lens (Pentax 12.5–75 mm) and passed through an emission filter (F2, Semrock, 641/75 nm bandpass filter) chosen to overlap the chlorophyll emission peak at 690 nm. The filtered fluorescence signal was focused onto a gated optical intensifier (Kentech Instruments, model HRI) read out by a sCMOS camera (Andor, Zyla 5.5) via two relay lenses (L3, L4). Wide-field time-gated FLIM, entails synchronizing the pulsed laser source with the HRI and adjusting the delay between excitation and time-gated detection to sample the fluorescence intensity decay profiles in each pixel. For this study we sampled the fluorescence signal with 9 time gates of 1 ns duration at increasing time delays after the excitation pulse. The read out camera was operated with an exposure time of 200 ms per frame. Overall, each FLIM acquisition required ~3 s.

The leaf sample was first imaged once without any treatment and then a time-series of FLIM images was acquired at 3 min intervals after application of two 2 μl drops of Flagon EC 400. Image acquisition was controlled by the openHCA-FLIM μManager plug-in developed in the Photonics group at Imperial College London [45]. An IRF based on the excitation pulses was measured under the same experimental conditions using a scattering sample.

Results

Plant autofluorescence characterization

The fluorescence excitation-emission matrix for *Triticum aestivum* leaves measured using the Shimadzu spectrofluorometer is shown in Fig. 3. There are two major spectral emission bands: a blue-green fluorescence (ex

~320 nm, em ~450 nm) that could be due to the presence of cinnamic acids [46] and a fluorescence band attributed to chlorophyll with two prominent peaks at 690 nm and 720 nm [47]. For the subsequent fibre optic probe and wide-field imaging fluorescence lifetime measurements discussed below, we chose to concentrate on the 690 nm emission peak attributed to fluorescence from chlorophyll in leaves at 440 nm excitation as the spectral region of interest. This matched the capabilities of our instrumentation and we detected no significant fluorescence from the Flagon EC 400 in this spectral detection band, see below.

Characterisation of Flagon EC 400 fluorescence

The emission spectrum of Flagon EC 400 in the range 400–800 nm was measured (49 ppm solution) using the Shimadzu spectrofluorometer with excitation at 440 nm, see Fig. 4a. The fluorescent decay profiles of Flagon EC 400 were measured (49 ppm solution) using the custom-built multidimensional spectrofluorometer as described in Manning et al. [41] with excitation at 405 nm (20 MHz repetition rate) and detection at 482 nm. The data were fitted to a double exponential decay model using *FLIMfit* which returned lifetime components of 6214 ps (18%) and 2239 ps (82%) with an average χ^2 value of 1.17. An exemplar decay profile is shown in Fig. 4b. We investigated how the fluorescence decay profile varied across the emission spectrum

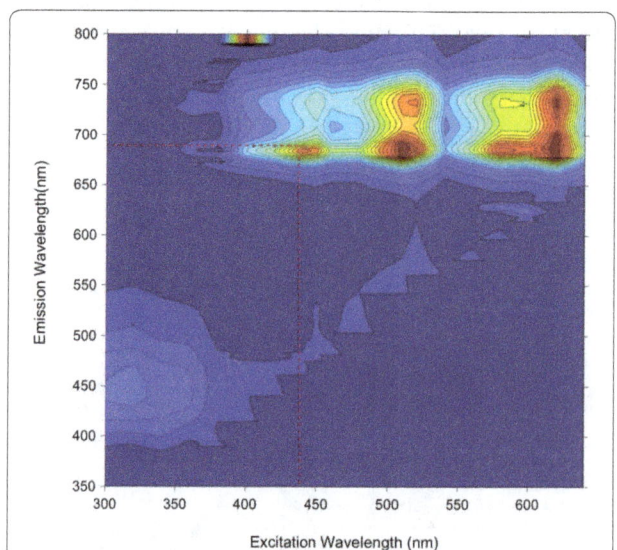

Fig. 3 Excitation emission matrix of fluorescence from *Triticum aestivum* leaves measured using a spectrofluorometer. Fluorescence excitation was scanned over a wavelength range 300–650 nm at 10 nm intervals and corresponding emission spectrum measured for each excitation wavelength. A different leaf was used for each emission scan

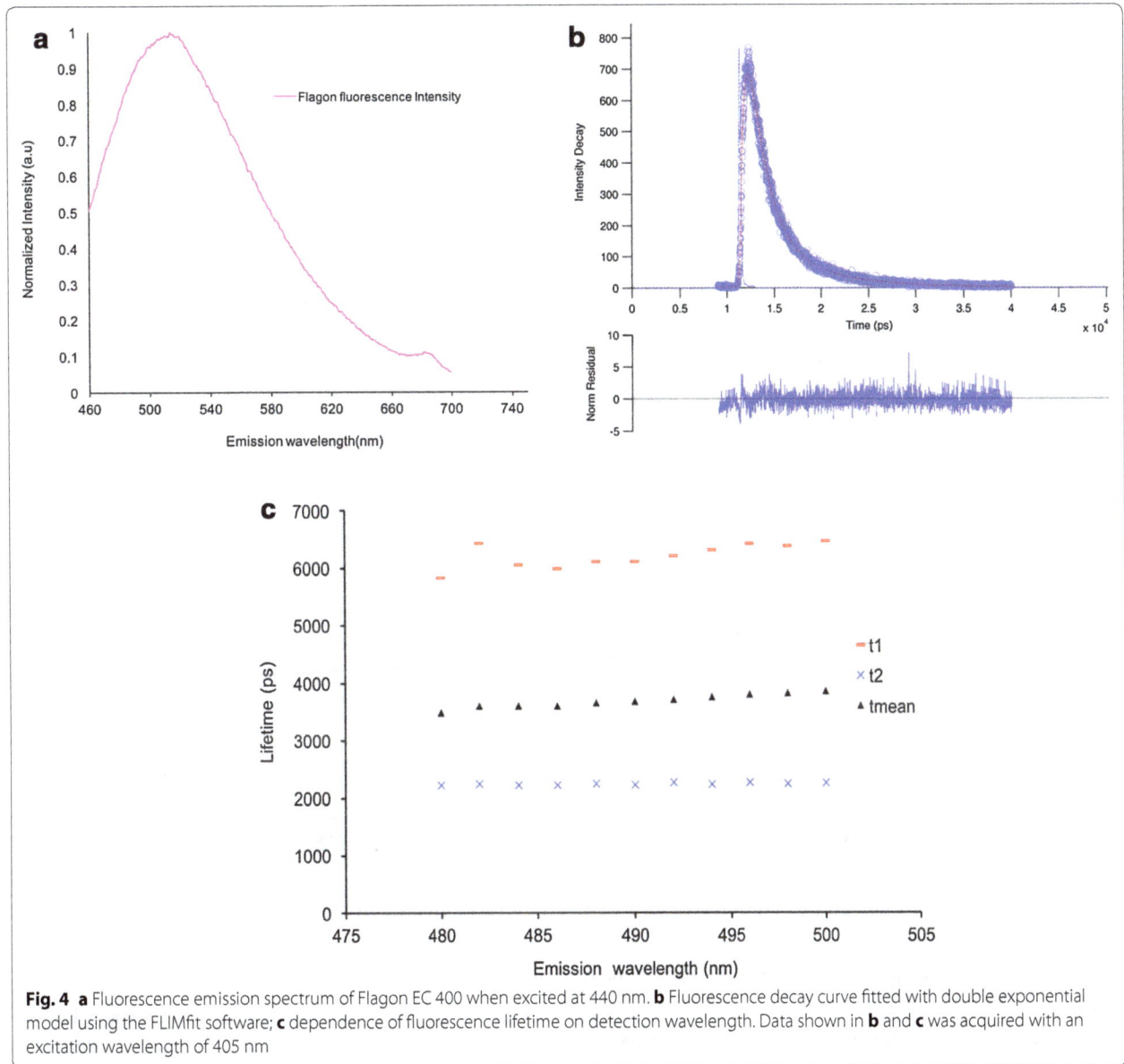

Fig. 4 **a** Fluorescence emission spectrum of Flagon EC 400 when excited at 440 nm. **b** Fluorescence decay curve fitted with double exponential model using the FLIMfit software; **c** dependence of fluorescence lifetime on detection wavelength. Data shown in **b** and **c** was acquired with an excitation wavelength of 405 nm

and the results are presented in Fig. 4c. Both lifetimes remain approximately constant over the detection range 480–500 nm.

In order to show that fluorescence from Flagon EC400 is small compared to that from the *Triticum aestivum* leaf when using fluorescence excitation at 440 nm, the two emission spectra were scaled according to the measurements of their relative emission intensities using the multispectral point-probe spectrofluorometer, see Fig. 5. Here the emission intensity of Flagon EC 400 was measured when it was dried on to a black anodised aluminium surface. From this data the relative fluorescence signal from Flagon EC 400 is only 0.58% of chlorophyll fluorescence from a leaf obtained in the same spectral channel.

In vivo multiphoton FLIM showing effect of herbicide at a cellular level

To explore the indirect read out of the presence of the Flagon EC 400 herbicide, we used two photon excitation at 900 nm to acquire high resolution FLIM images of chlorophyll fluorescence in *Triticum aestivum* leaves. Imaging of treated (Flagon EC 400) and untreated plant leaves was performed in situ using TCSPC on the Leica SP5 multiphoton system with a detection spectral band from 600 to 730 nm and implemented with leaf samples from the same plant fixed to a microscope slide. We imaged 4 fields of view (FOVs) in the sample of untreated leaf and 4 FOVs in the sample of treated leaf at 2.5 h post treatment. Figure 6a, c show typical autofluorescence

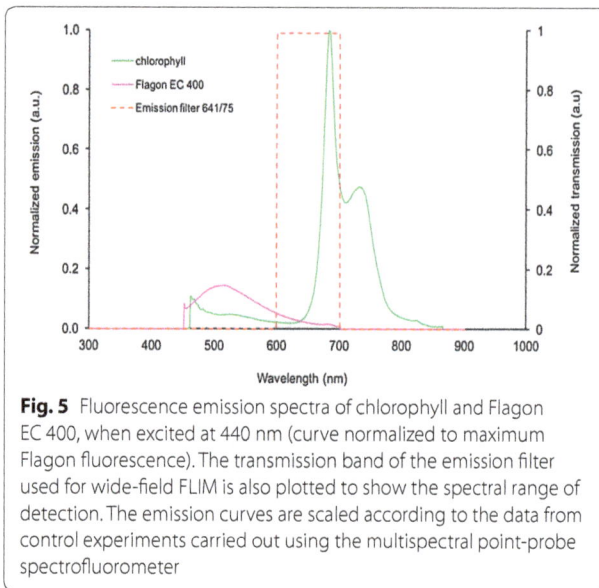

Fig. 5 Fluorescence emission spectra of chlorophyll and Flagon EC 400, when excited at 440 nm (curve normalized to maximum Flagon fluorescence). The transmission band of the emission filter used for wide-field FLIM is also plotted to show the spectral range of detection. The emission curves are scaled according to the data from control experiments carried out using the multispectral point-probe spectrofluorometer

intensity images and Fig. 6b, d show the corresponding intensity-merged fluorescence lifetime images of the untreated and treated leaves respectively. These images show mesophyll cells (larger ovoid shapes) and chloroplasts (smaller circular disc like structures) within them. It can be seen that the presence of the herbicide within the cells results in an increase in autofluorescence lifetime and an increase in the heterogeneity of the fluorescence lifetime. In the images shown in Fig. 6d, this heterogeneity is clearly visible across the different FOVs. These images were all acquired at a depth of 100 μm from the leaf surface. Fluorescence decay profiles were fitted to a double exponential decay model, as in [29], using the *FLIMfit* software and the intensity-weighted mean lifetimes (τ_m) averaged over the FOV were calculated. This analysis indicated that the untreated FOVs exhibits an average τ_m of 560 ± 30 ps while the treated leaf sample exhibited increased τ_m of 2000 ± 440 ps across different FOVs as shown in Fig. 7, which is consistent with the chlorophyll fluorescence lifetime increasing in the presence of a PS II inhibiting herbicide.

Following these in situ fixed endpoint measurements, we undertook a time course of two photon excited FLIM (TPE-FLIM) measurements of treated and untreated plants in vivo. Figure 8a shows intensity merged fluorescence lifetime images of (a) an untreated leaf and (b) a treated leaf at different time points. As expected, the treated leaf exhibits an increase in the mean chlorophyll fluorescence lifetime in contrast to that of an untreated leaf.

A plot of the weighted mean lifetimes obtained from the analysis is presented in Fig. 9. The untreated leaf

control time-course experiment yielded images with a τ_m of 351 (±1 SE) ps averaged over the FOV and the time course of a treated leaf yielded images with an increased mean lifetime of 1021 ps at 5 min to 1110 ps at 45 min post treatment.

In vivo multispectral point-probe autofluorescence lifetime measurements

The multispectral fibre-optic point-probe was applied to in vivo studies of untreated and treated (with Flagon EC 400) leaves of *Triticum aestivum*, for which the results are shown in Fig. 10. The measured autofluorescence decay profiles were fitted to a double exponential model with the untreated leaves presenting a mean τ_m of 480 (±16 SE) ps in the 440 nm excited chlorophyll channel, CH4 (620–710 nm). For leaves treated with Flagon EC 400, an increase in τ_m up to 1438 ps was observed in this channel.

In CH3, which detects fluorescence (620–710 nm) excited at 372 nm corresponding to the chlorophyll emission wavelengths, an average τ_m of 490(±10 S.E.) ps was obtained for untreated leaves and 1170 (±140 S.E.) ps for treated leaves.

The autofluorescence decay profiles measured in CH2 (430–480 nm, excited at 372 nm) presented a τ_m of ~750 ps for both treated and untreated leaves. This autofluorescence could be attributed to cinnamic acids or lignin [48].

Fluorescence collected from channel CH1 (400–420 nm, excited at 372 nm) is most strongly affected by background fluorescence from the optical fibres and does not follow a definite trend. The variation between measurements is attributed to the low signal level in this channel.

A time course experiment to study the changes in fluorescence lifetime after treating a leaf with Flagon EC 400 was undertaken and also analyzed using a biexponential fit model. The results are shown in Fig. 11. Treatment of the leaf with Flagon was carried out at time = 0 and the time of the "untreated" leaf measurement was 5 min before treatment. In the first 5 min after treatment, τ_m is observed to have increased to 1 ns and then to 1.3 ns after 60 min; thereafter it remained approximately constant up to 180 min. Control experiments where the treatment, i.e. 2 drops of Flagon EC 400, was applied to a black anodized metal sheet resulted in no significant signal detected in the chlorophyll channel, CH4. The resultant fluorescence collected in CH4 from Flagon was found to be insignificant compared to the chlorophyll fluorescence as shown in Fig. 5. We studied the variations in fluorescence lifetimes in untreated lifetime in plants of different age groups and the weighted mean fluorescence lifetimes calculated from CH4 are shown in Additional

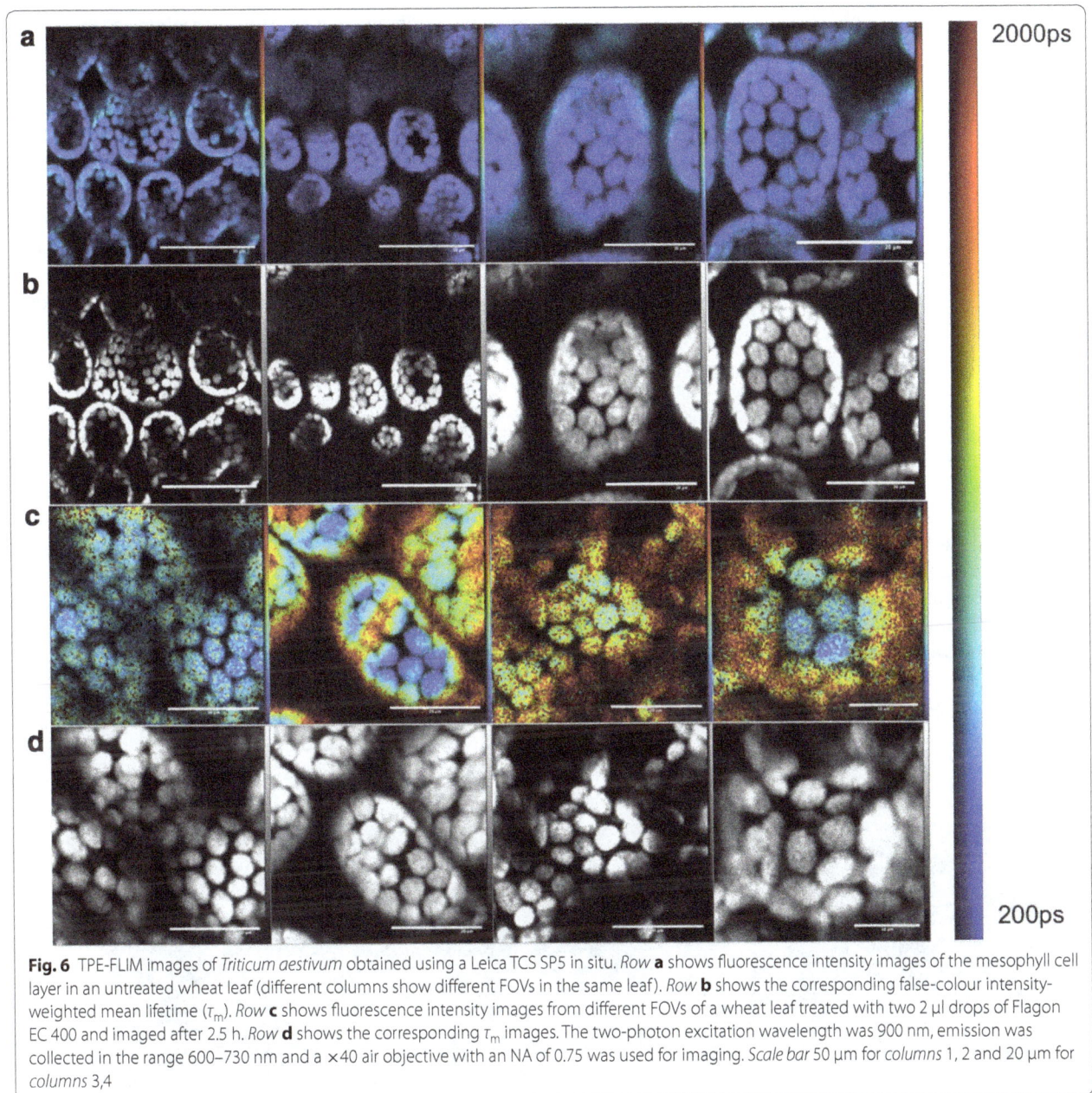

Fig. 6 TPE-FLIM images of *Triticum aestivum* obtained using a Leica TCS SP5 in situ. *Row* **a** shows fluorescence intensity images of the mesophyll cell layer in an untreated wheat leaf (different columns show different FOVs in the same leaf). *Row* **b** shows the corresponding false-colour intensity-weighted mean lifetime (τ_m). *Row* **c** shows fluorescence intensity images from different FOVs of a wheat leaf treated with two 2 µl drops of Flagon EC 400 and imaged after 2.5 h. *Row* **d** shows the corresponding τ_m images. The two-photon excitation wavelength was 900 nm, emission was collected in the range 600–730 nm and a ×40 air objective with an NA of 0.75 was used for imaging. *Scale bar* 50 µm for *columns* 1, 2 and 20 µm for *columns* 3,4

file 2. It can be seen that there are minor variations from plant to plant and in different age groups, but the general distribution could be seen to be lying in the 350–600 ps range.

Wide-field FLIM to map herbicide effect in *Triticum aestivum*

In vivo wide-field time-gated FLIM of *Triticum aestivum* plant leaves before and following treatment with Flagon EC 400 was undertaken. To orientate this study, Fig. 12a

shows a color photograph of the leaf with the treatment area (enclosed by the red-dotted circle) marked with four dots of a black felt pen. Figure 12b shows a fluorescence intensity image of the leaf before it is treated with Flagon EC 400. To confirm that there is no contribution to the fluorescent signal due to fluorescence of the herbicide, we performed a control experiment using the same treatment of Flagon EC 400 applied to a black anodized metal surface and did not detect any significant fluorescence.

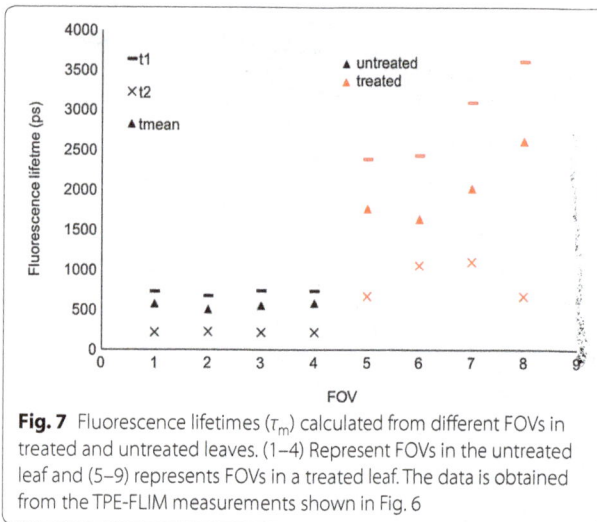

Fig. 7 Fluorescence lifetimes (τ_m) calculated from different FOVs in treated and untreated leaves. (1–4) Represent FOVs in the untreated leaf and (5–9) represents FOVs in a treated leaf. The data is obtained from the TPE-FLIM measurements shown in Fig. 6

histograms obtained from the experiment at each time point are presented in Fig. 13. The correlation plot shown in Fig. 14 shows 2D histograms of the autofluorescence intensity versus mean lifetime for the treated and untreated regions using the same data set as for Fig. 13. For the pixels with a short mean lifetime (associated with the untreated leaf) the plots show a vertical distribution of points indicating that there is no strong correlation between fluorescence intensity and fluorescence lifetime for this measurement.

Figure 15 presents the evolution of τ_m averaged over either the red-dashed region of interest (treated region) indicated in Fig. 12a or the segmented area of the leaf outside the red-dashed region of interest. The leaf presents an initial τ_m of 521 ps before treatment in the red-dashed region of interest. After the treatment with herbicide, the mean autofluorescence lifetime is observed to increase over time, as shown in Figs. 13 and 15. At 3 min after treatment, the mean lifetime was 1000 ps. At 27 min after treatment, τ_m averaged over the red-dashed region of interest was 2150 ps, whereas in the region outside the red dotted circle (untreated region), the fluorescence lifetime remained approximately constant with an average τ_m value of 530 ($\pm$10 S.E.) ps.

A separate time course experiment was conducted on a different untreated plant and a leaf was imaged under the same conditions and at intervals of 3 min and the

Wide-field autofluorescence intensity and lifetime images were acquired of a leaf 3 min before treatment and then at 3 min intervals after the treatment up to 27 min. The autofluorescence decay profiles were fitted pixel wise to a double exponential model using the *FLIMfit* software. Time-integrated autofluorescence intensity images and the corresponding intensity-merged autofluorescence lifetime images and lifetime

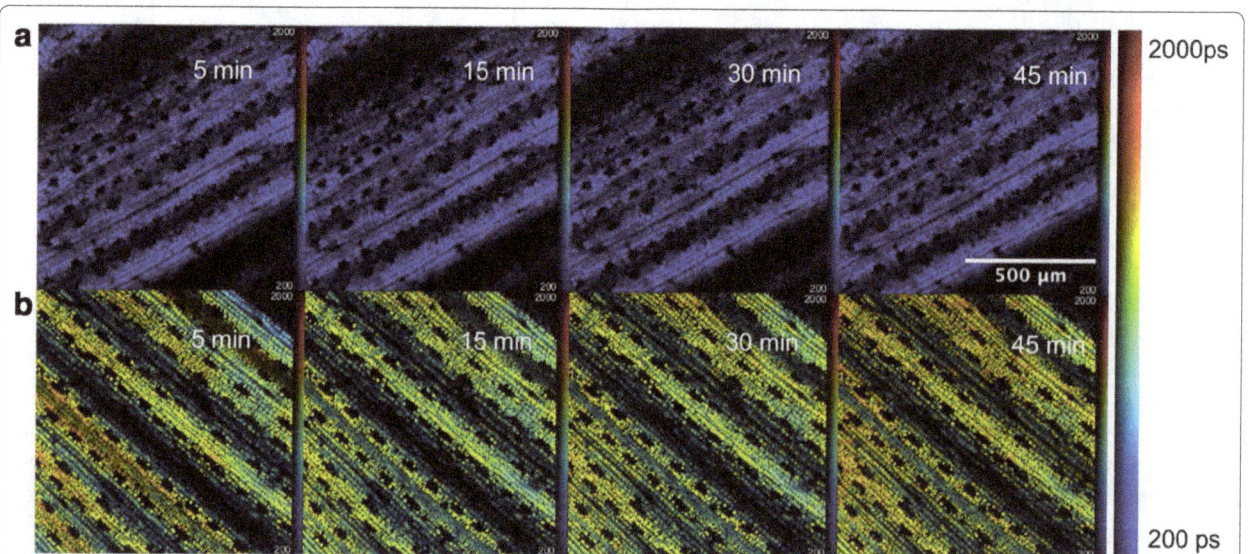

Fig. 8 TPE-FLIM images obtained from in vivo time course experiments in *Triticum aestivum* leaves. **a** Shows intensity merged fluorescence lifetime images of an untreated leaf at different time points. **b** Shows intensity merged lifetime images of a treated leaf at different time points showing an increase in lifetime compared to **a**. All the measurements were carried out using a Leica TCS SP5 system and analyzed using FLIMfit software. Imaging was performed using MPE excitation at 900 nm and a ×10 air objective with an NA of 0.30. Each FOV is 1 mm × 1 mm

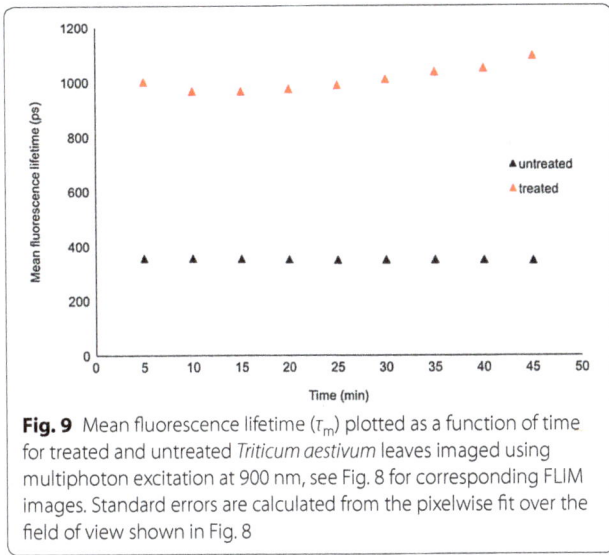

Fig. 9 Mean fluorescence lifetime (τ_m) plotted as a function of time for treated and untreated *Triticum aestivum* leaves imaged using multiphoton excitation at 900 nm, see Fig. 8 for corresponding FLIM images. Standard errors are calculated from the pixelwise fit over the field of view shown in Fig. 8

These lifetime values obtained are in agreement to the multiphoton imaging data. It can be seen from the lifetime frequency histograms that there is a significant decrease in the population fraction of the short lifetime (~580 ps) component and the lifetime distribution shifts towards longer lifetimes (1000–2500 ps) over time after treatment.

Figure 16a shows the evolution over time of maps of the autofluorescence intensity and fluorescence lifetime of an untreated leaf imaged using wide field FLIM. Figure 16b shows the evolution over time of the mean fluorescence lifetime averaged over the whole of the leaf for the treated leaf presented in Fig. 13 and for the untreated leaf presented in Fig. 16a, again demonstrating the observation of a longer autofluorescence lifetime of chlorophyll in herbicide treated leaves.

Discussion

Although plant autofluorescence is highly complex and is not yet fully understood, this study indicates that it is possible to utilise lifetime measurements of autofluorescence attributed to chlorophyll to read out and map the action of the PS II inhibiting herbicide, Flagon EC 400. Our empirical in vivo measurements across multiphoton microscopy, point-probe lifetime measurements

results are shown in Fig. 16. The fluorescence lifetimes of the untreated leaf remained fairly constant over a time period of 27 min and yielded an average τ_m of 580 ($\pm$10 S.E.) ps, in reasonable agreement with the τ_m values of the untreated regions of the treated leaf.

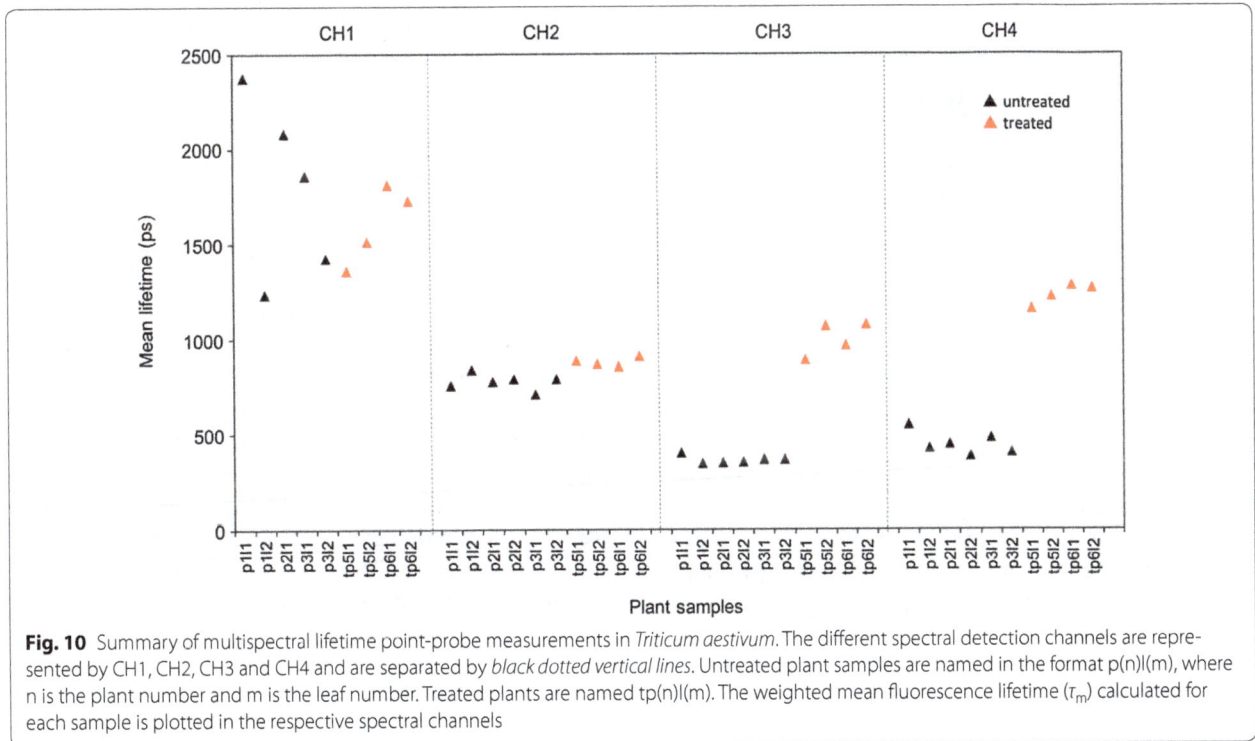

Fig. 10 Summary of multispectral lifetime point-probe measurements in *Triticum aestivum*. The different spectral detection channels are represented by CH1, CH2, CH3 and CH4 and are separated by *black dotted vertical lines*. Untreated plant samples are named in the format p(n)l(m), where n is the plant number and m is the leaf number. Treated plants are named tp(n)l(m). The weighted mean fluorescence lifetime (τ_m) calculated for each sample is plotted in the respective spectral channels

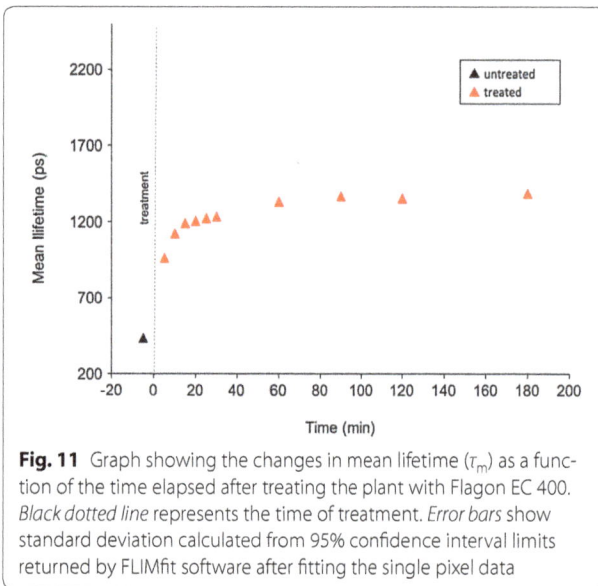

Fig. 11 Graph showing the changes in mean lifetime (τ_m) as a function of the time elapsed after treating the plant with Flagon EC 400. *Black dotted line* represents the time of treatment. *Error bars* show standard deviation calculated from 95% confidence interval limits returned by FLIMfit software after fitting the single pixel data

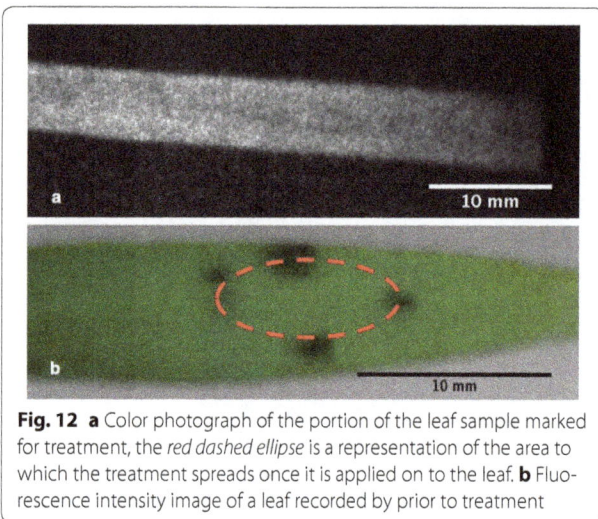

Fig. 12 a Color photograph of the portion of the leaf sample marked for treatment, the *red dashed ellipse* is a representation of the area to which the treatment spreads once it is applied on to the leaf. **b** Fluorescence intensity image of a leaf recorded by prior to treatment

and wide-field FLIM, all support the use of the change in lifetime of the autofluorescence in the chlorophyll emission spectral band to report the presence of the PS II herbicide through its localised impact on chlorophyll fluorescence. In the multiphoton microscopy images shown in Fig. 6d, a heterogeneity of lifetimes is clearly

visible across the different FOVs. The τ_m values along the edges of the mesophyll cells are higher (Fig. 6d, especially the second, third and fourth images) compared to the insides of the cells. This could be explained by the active ingredient penetrating the cell membranes and inhibiting the PS II centres, which leads to an increased fluorescence lifetime of chlorophyll molecules in the vicinity of the corresponding herbicide binding sites, i.e. PS II centres. The observed increase in chlorophyll fluorescence lifetime in the presence of a PSII inhibitor may be explained by the blocking of energy transfer to this photosynthetic pathway and the consequently decreased rate of loss of energy through only the radiative and non-radiative decay channels. In principle, it should be possible to relate these observations to previous studies undertaken to model PS II [30] and light harvesting antenna complexes-II (LHC-II) [49, 50] but this would require disentangling contributions from different compartments in the leaf that are integrated in the analysis of our in vivo measurements. Multispectral multiphoton FLIM microscopy may enable this challenge to be addressed by segmenting different compartments in the leaf and will be the subject of future studies.

Our fluorescence measurement spectral windows from ~600 to 710 nm overlap with emission peaks from chlorophyll *a* and chlorophyll *b*. However, since previous studies of energy transfer kinetics of photosynthetic pigments have indicated that chlorophyll *b* molecules transfer most of their excitation energy to chlorophyll *a* in less than 2 ps [51, 52], we believe that our measurements are dominated by chlorophyll *a* fluorescence.

We note that these chlorophyll fluorescence lifetime measurements were undertaken with the leaf in a light-adapted state, for which the different light doses applied using the various methodologies are provided in the Additional file 3: Table S1. This indicates the potential of fluorescence lifetime measurements to provide quantitative readouts of the herbicide presence without subjecting the plant to dark adaptation, which avoids practical difficulties associated with spectral ratiometric measurements that do require dark adaption. We therefore see chlorophyll fluorescence lifetime measurements as a promising methodology for in field measurements.

We note that the distribution of herbicides could, in principle, be mapped by directly imaging the fluorescence

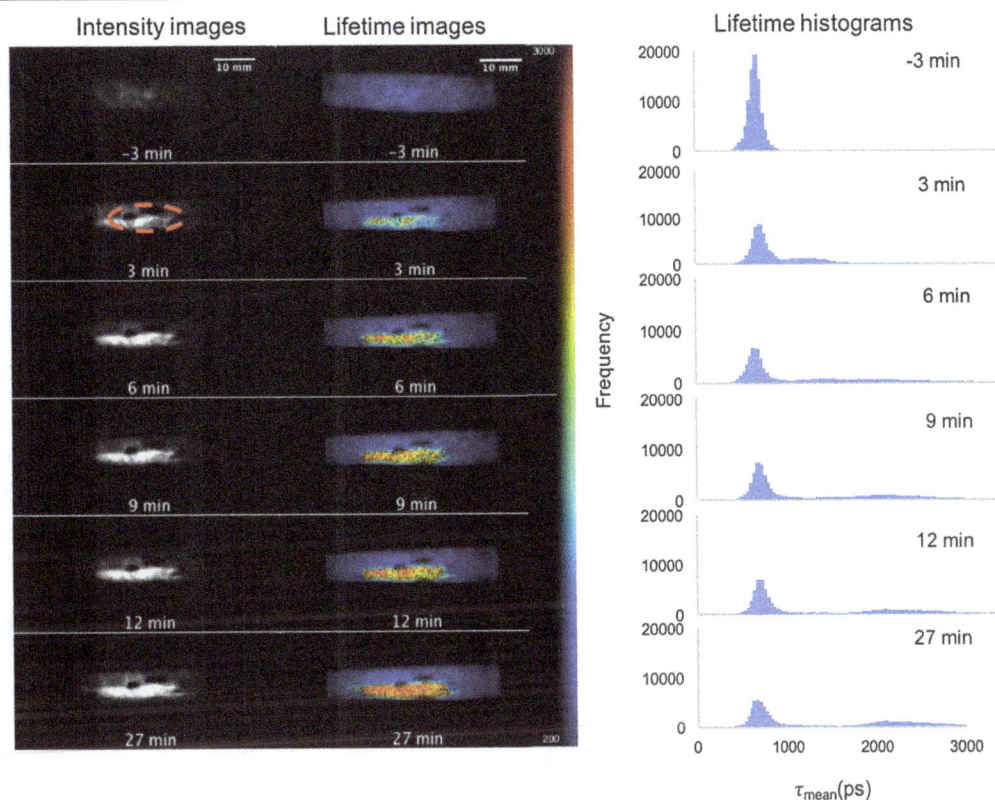

Fig. 13 Widefield macroscope images from time course experiments on a leaf. *Each row* shows the fluorescence intensity image, corresponding fluorescence lifetime image and corresponding fluorescence lifetime *histogram* over region of interest for one time point. Images were obtained using the time-gated wide-field macroscope. Images are acquired by exciting the sample at 440 nm (repetition rate 40 MHz) and collected using 9 gates of 1 ns width each. *Red dotted circle* is used to outline the treatment area marked by a black felt pen

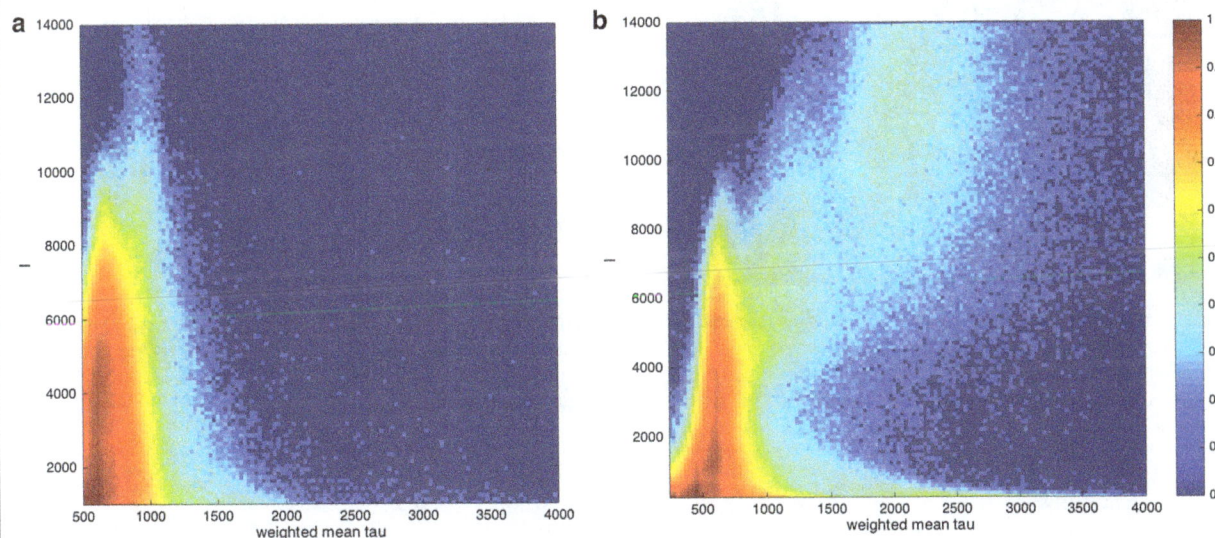

Fig. 14 Correlation plot of fluorescence intensity against weighted mean fluorescence lifetime for **a** the untreated region and **b** the treated region from the data acquired from all the time points after application of Flagon EC 400 using the wide field FLIM macroscope (same dataset as shown in Fig. 13)

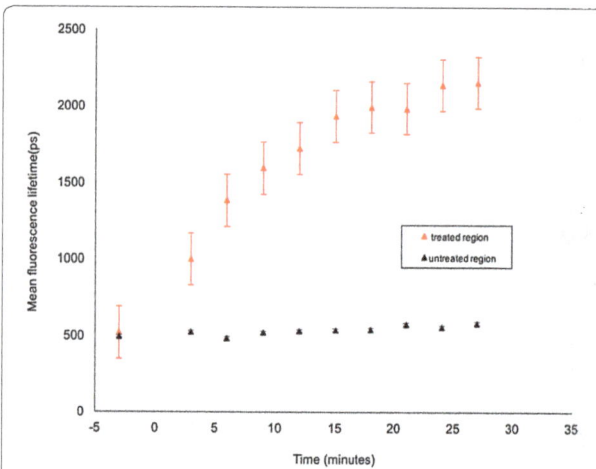

Fig. 15 Graph showing the changes in weighted mean lifetime, τ_m obtained from the wide field macroscope measurements as a function of the elapsed time after treating the plant with Flagon EC 400 for the region of interest marked with *red dotted circles* (treated region) and the region outside it (untreated region) separately. *Black dotted line* represents the time of treatment. Fluorescence lifetime calculations are made separately for the region of interest marked by the *red dotted line* and the region outside it. Standard errors are calculated from pixelwise fit over the region of interest

of the herbicide but for most herbicides, including Flagon EC 400, this emission is very weak relative to the plant autofluorescence. As indicated in Fig. 3, the fluorescence of Flagon EC 400 is very weak in the chlorophyll detection band and we verified that no significant fluorescence signals were detected from the herbicide for the multispectral point probe measurements and the wide-field FLIM experiments.

Conclusions

We have demonstrated the in vivo readout of the local action of a PS-II inhibiting herbicide using multiphoton FLIM microscopy, a fibre-optically delivered time-resolved spectrofluorometer for single point lifetime measurements and single photon excited wide-field FLIM applied to leaves of *Triticum aestivum*. We consistently observed an increase in fluorescence lifetime in the chlorophyll autofluorescence detection band following treatment with Flagon EC 400 that can serve as an indirect marker of the herbicide action. The multispectral point probe system already has a form factor suitable to be used as a portable diagnostic tool in greenhouses and in the field. The wide-field FLIM instrument, which

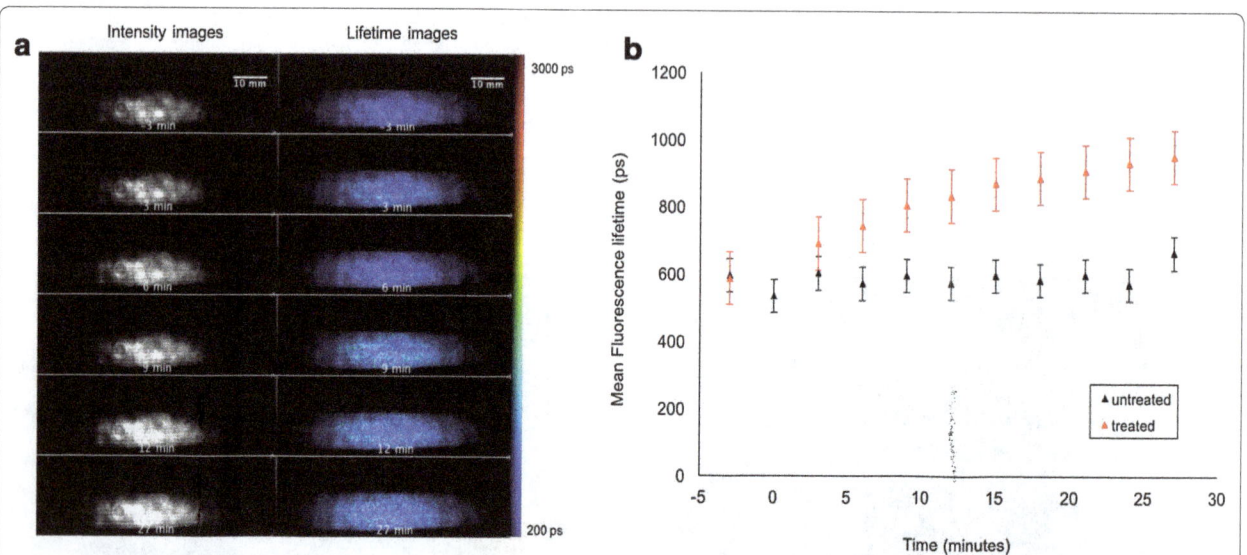

Fig. 16 **a** Widefield macroscope images from time course experiments on an untreated leaf. *Each row* shows the fluorescence intensity image and corresponding fluorescence lifetime images at a particular time point. **b** Changes in weighted mean lifetime, τ_m averaged over the whole leaf obtained from the wide field macroscope measurements as a function of the elapsed time from an untreated life and a separate leaf treated with Flagon EC 400. Fluorescence lifetime values reported are the average over the entire image. Standard errors are calculated from pixelwise fit over the entire field of view

enables spatiotemporal mapping of the distribution of this agrochemical over a large FOV comparable to an entire leaf, could also be engineered to be portable for in field studies.

Additional files

Additional file 1: Figure S1. Schematic representation of the optical set-up of multidimensional spectrofluorometer as described in Manning et al. [41].

Additional file 2: Figure S2. Distribution of fluorescence lifetimes in untreated *Triticum aestivum* plants of different age groups calculated from multispectral lifetime point-probe measurements in the spectral channel CH4 (excitation at 440 nm, detection wavelengths 620-710 nm). Data points from different age groups are represented by different colours. Plant samples are named in the format p(n)l(m), where n is the plant number and m is the leaf number. The weighted mean fluorescence lifetime (τ_m) calculated for each sample is plotted here.

Additional file 3: Table S1. Table of light doses applied using the various methodologies.

Abbreviations

PS II: Photosystem II; FLIM: Fluorescence lifetime imaging; SEM: Scanning electron microscopy; CLSM: Confocal laser scanning microscopy; TPEM: Two-photon excitation microscopy; PS I: Photosystem I; DCMU: 3-(3,4-Dichlorophenyl)-1,1-1-dimethylurea; AI: Active ingredient; TCSPC: Time-correlated single photon counting; IRF: Instrument response function; CH1: Channel 1; CH2: Channel 2; CH3: Channel 3; CH4: Channel 4; FOV: Field of view; TPE-FLIM: Two photon excited FLIM; LHC-II: Light harvesting antenna complexes-II.

Authors' contributions

EN, CS, CD and PF designed the experiments, which were performed by EN, SK and FG; EN analysed the data supervised by CD and PF; EN, CD and PF wrote the manuscript. All authors read and approved the final manuscript.

Author details

[1] Photonics Group, Department of Physics, Imperial College London, London SW7 2AZ, UK. [2] Department of Chemistry, Imperial College London, London SW7 2AZ, UK. [3] Institute of Chemical Biology, Imperial College London, London SW7 2AZ, UK. [4] Syngenta, Jealott's Hill International Research Centre, Bracknell, Berkshire RG42 6EY, UK. [5] Centre for Pathology, Imperial College London, London SW7 2AZ, UK.

Acknowledgements

We acknowledge research funding from the European Union Seventh Framework Programme (FP7/2007-2013) under Grant Agreement Number 607466, the UK Biotechnology and Biological Sciences Research Council (BBSRC, BB/M006786/1) and the Engineering and Physical Sciences Research Council (EPSRC) grant EP/I02770X/1. FG acknowledges a PhD studentship from the Engineering and Physical Sciences Research Council. Seeds of *Triticum aestivum* (Winter Wheat cv.Hereward) and Flagon EC 400 were obtained from Syngenta Jealott's Hill International Research Centre Bracknell, Berkshire, UK.

Competing interests

The authors declare that they have no competing interests.

Funding

The research leading to these results has received funding from the European Union Seventh Framework Programme (FP7/2007-2013) part of Marie (Skłodowska-)Curie actions research fellowships under grant agreement number 607466, the UK Biotechnology and Biological Sciences Research Council (BBSRC, BB/M006786/1) and the UK Engineering and Physical Sciences Research Council (EPSRC, EP/I02770X/1). FG acknowleddges a Ph.D. studentship from the EPSRC.

References

1. Wild E, Dent J, Thomas GO, Jones KC. Real-time visualization and quantification of PAH photodegradation on and within plant leaves. Environ Sci Technol. 2005;39:268–73.
2. Mullen AK, Clench MR, Crosland S, Sharples KR. Determination of agrochemical compounds in soya plants by imaging matrix-assisted laser desorption/ionisation mass spectrometry. Rapid Commun Mass Spectrom. 2005;19:2507–16.
3. Sarsby J, Towers MW, Stain C, Cramer R, Koroleva OA. Mass spectrometry imaging of glucosinolates in arabidopsis flowers and siliques. Phytochemistry. 2012;77:110–8.
4. Kim J, Jung J, Lee C. In vivo monitoring of the incorporation of chemicals into cucumber and rice leaves by chlorophyll fluorescence imaging. Plant Biotech. 2002;4:173–9.
5. Siminszky B, Corbin FT, Ward ER, Fleischmann TJ, Dewey RE. Expression of a soybean cytochrome P450 monooxygenase cDNA in yeast and tobacco enhances the metabolism of phenylurea herbicides. Proc Natl Acad Sci. 1999;96:1750–5.
6. Yilmaz G, Dane F. Phytotoxicity induced by herbicide and surfactant on stomata and epicuticular wax of wheat. Rom Biotechnol Lett. 2012;17:7757–65.
7. Lichtenthaler HK, Lang M, Sowinska M, Summ P, Heisel F, Miehe JA. Uptake of the herbicide diuron as visualised by the fluorescence imaging technique. Bot Acta. 1997;110:158–63.
8. Hepler PK, Gunning BES. Confocal fluorescence microscopy of plant cells. Protoplasma. 1998;201:121–57.
9. Hutzler P, Fischbach R, Heller W, Jungblut TP, Reuber S, Schmitz R, Veit M, Weissenbock G, Schnitzler J-P. Tissue localization of phenolic compounds in plants by confocal laser scanning microscopy. J Exp Bot. 1998;49:953–65.
10. Wild E, Dent J, Thomas GO, Jones KC. Use of two-photon excitation microscopy and autofluorescence for visualizing the fate and behavior of semivolatile organic chemicals within living vegetation. Environ Toxicol Chem. 2007;26:2486–93.
11. Feijó JA, Moreno N. Imaging plant cells by two-photon excitation. Protoplasma. 2004;223:1–32.
12. Maxwell K, Johnson GN. Chlorophyll fluorescence—a practical guide. J Exp Bot. 2000;51:659–68.
13. Holub O, Seufferheld MJ, Gohlke C, Govindjee, Heiss GJ, Clegg RM. Fluorescence lifetime imaging microscopy of *Chlamydomonas reinhardtii*: non-photochemical quenching mutants and the effect of photosynthetic inhibitors on the slow chlorophyll fluorescence transient. J Microsc. 2007;226(Pt 2):90–120.
14. Fuerst EP, Norman MA. Interactions of herbicides with photosynthetic electron transport. Weed Sci. 1991;39:458–64.
15. Perkins RG, Oxborough K, Hanlon ARM, Underwood GJC, Baker NR. Can chlorophyll fluorescence be used to estimate the rate of photosynthetic electron transport within microphytobenthic biofilms? Mar Ecol Prog Ser. 2002;228:47–56.
16. Baker NR. Chlorophyll fluorescence: a probe of photosynthesis in vivo. Annu Rev Plant Biol. 2008;59:89–113.
17. Barbagallo RP, Oxborough K, Pallett KE, Baker NR. Rapid, noninvasive screening for perturbations of metabolism and plant growth using chlorophyll fluorescence imaging. Plant Physiol. 2003;132:485–93.
18. Agati G, Cerovic ZG, Moya I. The effect of decreasing temperature up to chilling values on the in vivo F685/F735 chlorophyll fluorescence ratio in *Phaseolus vulgaris* and *Pisum sativum*: the role of the photosystem i contribution to the 735 nm fluorescence band &#xB6. Photochem Photobiol. 2000;72:75–84.
19. Baker NR, Rosenqvist E. Applications of chlorophyll fluorescence can improve crop production strategies: an examination of future possibilities. J Exp Bot. 2004;55:1607–21.
20. Baker NR, Oxborough K. Chlorophyll fluorescence as a probe of photosynthetic productivity. In: Papageorgiou GC, Govindjee, editors. Chlorophyll a fluorescence: a signature of photosynthesis. Dordrecht: Springer; 2004. p. 65–82.
21. Mishra KB, Mishra A, Novotná K, Rapantová B, Hodaňová P, Urban O, Klem K, Somerville C, Briscoe L, Lobell D, Schlenker W, Costa-Roberts J, AghaKouchak A, Feldman D, Hoerling M, Huxman T, Lund J, Thomson J, Valliyodan B, Nguyen H, Boyer J, Campbell B, Bray E, Gargallo-Garriga A, Sardans J, Pérez-Trujillo M, Oravec M, Urban O, Jentsch A, Kreyling J, et al.

Chlorophyll a fluorescence, under half of the adaptive growth-irradiance, for high-throughput sensing of leaf-water deficit in *Arabidopsis thaliana* accessions. Plant Methods. 2016;12:46.

22. Mishra KB, Mishra A, Klem K. Govindjee: plant phenotyping: a perspective. Indian J Plant Physiol. 2016;21:514–27.

23. Lichtenthaler HK, Miehe JA. Fluorescence imaging as a diagnostic tool for plant stress. Elsevier Sci. 1997;2:6–10.

24. Lichtenthaler HK, Langsdorf G, Buschmann C. Uptake of diuron and concomitant loss of photosynthetic activity in leaves as visualized by imaging the red chlorophyll fluorescence. Photosynth Res. 2013;116:355–61.

25. Pawley JB, editor. Handbook of biological confocal microscopy. 3rd ed. Boston: Springer; 2006.

26. Broess K, Borst JW, van Amerongen H. Applying two-photon excitation fluorescence lifetime imaging microscopy to study photosynthesis in plant leaves. Photosynth Res. 2009;100:89–96.

27. Lakowicz JR, editor. Principles of fluorescence spectroscopy. 3rd ed. Boston: Springer; 2006.

28. Marcu L, French PMW, Elson DS, editors. Fluorescence lifetime spectroscopy and imaging: principles and applications in biomedical diagnostics. Boca Raton: CRC Press; 2014.

29. Holub O, Seufferheld MJ, Gohlke C, Govindjee, Clegg RM. Holub Clegg fluorescnce lifetime imaging in real time- a new technique for photosynthesis research 2000.pdf. Photosynthetica. 2000;38:581–99.

30. Iwai M, Yokono M, Inada N, Minagawa J. Live-cell imaging of photosystem II antenna dissociation during state transitions. Proc Natl Acad Sci. 2010;107:2337–42.

31. Babourina O, Rengel Z. Uptake of aluminium into Arabidopsis root cells measured by fluorescent lifetime imaging. Ann Bot. 2009;104:189–95.

32. Hotzer B, Ivanov R, Bauer P, Jung G. Investigation of copper homeostasis in plant cells by fluorescence lifetime imaging microscopy. Plant Signal Behav. 2012;7(April):521–3.

33. Petrásek Z, Schmitt F-J, Theiss C, Huyer J, Chen M, Larkum A, Eichler HJ, Kemnitz K, Eckert H-J. Excitation energy transfer from phycobiliprotein to chlorophyll d in intact cells of *Acaryochloris marina* studied by time- and wavelength-resolved fluorescence spectroscopy. Photochem Photobiol Sci. 2005;4:1016–22.

34. Hunsche M, Bürling K, Noga G. Spectral and time-resolved fluorescence signature of four weed species as affected by selected herbicides. Pestic Biochem Physiol. 2011;101:39–47.

35. Xu C, Zipfel W, Shear JB, Williams RM, Webb WW. Multiphoton fluorescence excitation: new spectral windows for biological nonlinear microscopy. Proc Natl Acad Sci. 1996;93:10763–8.

36. Littlejohn GR, Mansfield JC, Christmas JT, Witterick E, Fricker MD, Grant MR, Smirnoff N, Everson RM, Moger J, Love J. An update: improvements in imaging perfluorocarbon-mounted plant leaves with implications for studies of plant pathology, physiology, development and cell biology. Front Plant Sci. 2014;5(April):140.

37. Lagarto J, Dyer BT, Talbot C, Sikkel MB, Peters NS, French PMW, Lyon AR, Dunsby C. Application of time-resolved autofluorescence to label-free in vivo optical mapping of changes in tissue matrix and metabolism associated with myocardial infarction and heart failure. Biomed Opt Express. 2015;6:324–46.

38. McGinty J, Galletly NP, Dunsby C, Munro I, Elson DS, Requejo-Isidro J, Cohen P, Ahmad R, Forsyth A, Thillainayagam AV, Neil MAA, French PMW, Stamp GW. Wide-field fluorescence lifetime imaging of cancer. Biomed Opt Express. 2010;1:627.

39. Cerovic ZG, Nicolas FM, Moya I. Ultraviolet-induced fluorescence for plant present state and prospects monitoring. Agronomie. 1999;19:543–78.

40. Meyer S. UV-induced blue-green and far-red fluorescence along wheat leaves: a potential signature of leaf ageing. J Exp Bot. 2003;54:757–69.

41. Manning HB, Kennedy GT, Owen DM, Grant DM, Magee AI, Neil MAA, Itoh Y, Dunsby C, French PMW. A compact, multidimensional spectrofluorometer exploiting supercontinuum generation. J Biophoton. 2008;1:494–505.

42. Warren SC, Margineanu A, Alibhai D, Kelly DJ, Talbot C, Alexandrov Y, Munro I, Katan M, Dunsby C, French PMW. Rapid global fitting of large fluorescence lifetime imaging microscopy datasets. PLoS One. 2013;8:e70687.

43. Sparks H, Warren S, Guedes J, Yoshida N, Charn TC, Guerra N, Tatla T, Dunsby C, French P. A flexible wide-field FLIM endoscope utilising blue excitation light for label-free contrast of tissue. J Biophoton. 2015;8:168–78.

44. Requejo-Isidro J, Mcginty J, Munro I, Elson DS, Galletly NP, Lever MJ, Neil MAA, Stamp GWH, French PMW, Kellett PA, Hares JD. High-speed widefield time-gated endoscopic fluorescence-lifetime imaging. Opt Lett. 2004;29:2249–51.

45. OpenFLIM-HCA Acquisition Software. http://www.imperial.ac.uk/photonics/research/biophotonics/instruments–software/openflim-hca/.

46. Harris PJ, Hartley RD. Detection of bound ferulic acid in cell walls of the Gramineae by ultraviolet fluorescence microscopy. Nature. 1976;259:508–10.

47. Rabinowitch E, Govindjee. Photosynthesis. New York: Wiley; 1969.

48. Coble PG, Lead J, Baker A, Reynolds DM, Spencer RG. Aquatic organic matter fluorescence. Cambridge: Cambridge University Press; 2014.

49. Pascal AA, Liu Z, Broess K, van Oort B, van Amerongen H, Wang C, Horton P, Robert B, Chang W, Ruban A. Molecular basis of photoprotection and control of photosynthetic light-harvesting. Nat Lett. 2005;436:134–7.

50. Leupold D, Teuchner K, Ehlert J, Irrgang K-D, Renger G, Lokstein H. Two-photon excited fluorescence from higher electronic states of chlorophylls in photosynthetic antenna complexes: a new approach to detect strong excitonic chlorophyll a/b coupling. Biophys J. 2002;82:1580–5.

51. Croce R, Müller MG, Bassi R, Holzwarth AR. Chlorophyll b to chlorophyll a energy transfer kinetics in the CP29 antenna complex: a comparative femtosecond absorption study between native and reconstituted proteins. Biophys J. 2003;84:2508–16.

52. Gradinaru CC, van Stokkum IHM, Pascal AA, van Grondelle R, van Amerongen H. Identifying the pathways of energy transfer between carotenoids and chlorophylls in LHCII and CP29. A multicolor, femtosecond pump–probe study. J Phys Chem B. 2000;104:9330–42.

10

Leucine-rich-repeat-containing variable lymphocyte receptors as modules to target plant-expressed proteins

André C. Velásquez[1], Kinya Nomura[1], Max D. Cooper[2], Brantley R. Herrin[2] and Sheng Yang He[1,3,4,5]*

Abstract

Background: The ability to target and manipulate protein-based cellular processes would accelerate plant research; yet, the technology to specifically and selectively target plant-expressed proteins is still in its infancy. Leucine-rich repeats (LRRs) are ubiquitously present protein domains involved in mediating protein–protein interactions. LRRs confer the binding specificity to the highly diverse variable lymphocyte receptor (VLR) antibodies (including VLRA, VLRB and VLRC types) that jawless vertebrates make as the functional equivalents of jawed vertebrate immunoglobulin-based antibodies.

Results: In this study, VLRBs targeting an effector protein from a plant pathogen, HopM1, were developed by immunizing lampreys and using yeast surface display to select for high-affinity VLRBs. HopM1-specific VLRBs (VLR$_{M1}$) were expressed *in planta* in the cytosol, the *trans*-Golgi network, and the apoplast. Expression of VLR$_{M1}$ was higher when the protein localized to an oxidizing environment that would favor disulfide bridge formation (when VLR$_{M1}$ was not localized to the cytoplasm), as disulfide bonds are necessary for proper VLR folding. VLR$_{M1}$ specifically interacted *in planta* with HopM1 but not with an unrelated bacterial effector protein while HopM1 failed to interact with a non-specific VLRB.

Conclusions: In the future, VLRs may be used as flexible modules to bind proteins or carbohydrates of interest *in planta*, with broad possibilities for their use by binding directly to their targets and inhibiting their action, or by creating chimeric proteins with new specificities in which endogenous LRR domains are replaced by those present in VLRs.

Keywords: Protein targeting, Leucine-rich repeat, Variable lymphocyte receptor, Modules, HopM1

Background

In order to relay signals and interact with other molecules, proteins have acquired certain commonly used repetitive domains. One domain that has been shown to be involved in mediating protein–protein interactions is the leucine-rich repeat (LRR) domain, which is present in a variety of proteins from all domains of life including bacteria, eukaryotes, and even viruses [1]. Each LRR domain contains a conserved segment with the consensus sequence LxxLxLxxN/CxL (where L, C, and N stand for leucine, cysteine, and asparagine, respectively, while x stands for any amino acid) and adopts the secondary structures of a β strand and an α helix connected by a loop [2]. Multiple LRR domains arranged in tandem form a crescent-shaped structure, in which a continuous β-sheet on the concave side forms the most common surface for protein–protein interactions [1]. The versatility of LRR domains in mediating protein–protein interactions is exemplified by the vast functions that proteins containing this domain may fulfill. For example, in plants, LRR domains are present in transmembrane receptor proteins of the LRR-receptor kinase (LRR-RK) and LRR-receptor protein (LRR-RP) classes [3–6], in intracellular nucleotide-binding site–LRR proteins (NBS–LRR) [7], in F-box proteins including hormone receptors of the E3 ubiquitin ligase class [8, 9], in a component of

*Correspondence: hes@msu.edu
[1] DOE Plant Research Laboratory, Michigan State University, East Lansing, MI 48824, USA
Full list of author information is available at the end of the article

nucleocytoplasmic transport (RanGAP) [10], in intracellular proteins involved in pollen development (PIRLs) [11], and in extracellular cell wall LRR-extensins [12] and polygalacturonase-inhibiting proteins (PGIP) [13]. The first three classes have radiated exponentially in plants and account for almost 3% of all genes in Arabidopsis [14, 15], and have been shown to be primarily involved in defense signaling and recognition, and plant development.

Variable lymphocyte receptors (VLRs) are non-self recognition receptors present in jawless vertebrates (Agnatha, which includes hagfishes and lampreys), involved in detecting invading microbes. They are the functional equivalent of immunoglobulin-based antigen receptors and antibodies in jawed vertebrates (Gnathostomata, which includes all other vertebrates from cartilaginous fish to mammals) [16–19]. Contrary to the immunoglobulin domains used by gnathostomes, VLR antibodies primarily bind to antigens using the concave surface formed by their LRR domains [20]. VLR proteins have the following domains: a signal peptide, an N-terminal LRR (LRRNT), multiple LRRs with variable sequence (up to 10 have been observed in a mature VLR [21]; the first LRR and the last one are referred as LRR1 and LRRVe, respectively), an incomplete LRR (connecting peptide, LRRCP), a C-terminal LRR (LRRCT), and a flexible, invariant stalk followed by a transmembrane or glycosylphosphatidylinositol (GPI)-anchor region [21]. Both the N- and C-terminal LRR domains have two characteristic disulfide bridges to stabilize the fold of the protein [20]. Agnathans possess T-like and B-like lymphocytes in which each differentiated lymphocyte carries a unique set of variable LRR sequences in their mature *VLR* gene [22]. The high variability in the LRR region of VLRs has been estimated to allow a potential repertoire of 10^{14}–10^{17} VLR variants, a feat that is achieved by somatic diversification through the step-wise incorporation of different LRR donor sequences into the incomplete germline gene until an in-frame functional mature VLR is formed [23].

Three different VLRs exist in lampreys and hagfishes; VLRA, VLRB, and VLRC; with individual lymphocyte lineages only expressing a single functional VLR type [22, 24]. *VLRA* and *VLRC* are expressed by lymphocytes that resemble jawed vertebrate T cells. After antigen stimulation, these T-like lymphocytes proliferate and increase expression of proinflammatory cytokines, while their antigen receptors always remain attached to the cell surface [22, 25]. In contrast, *VLRB*-expressing lymphocytes differentiate into plasmablasts that secrete their VLRB receptors as disulfide-linked multimers that serve as the functional equivalent of jawed vertebrate antibodies [26, 27].

In this study, the feasibility of using LRR-containing lamprey-derived VLRBs to target *in planta*-expressed proteins was investigated. The VLRBs were shown to accumulate in different cellular compartments, and VLRBs that were targeted through the plant secretory pathway were indeed able to interact *in planta* with their target, HopM1, a bacterial effector protein from a plant pathogen. These results provide a proof-of-concept demonstration for engineering VLR-based protein-targeting LRR modules *in planta*.

Results

The methodology developed for producing VLR-based LRR modules for targeting plant-expressed proteins starts by expressing and purifying the protein of interest, typically using at least two chromatographic purification steps to have a high-purity target protein, so that non-specific VLRs against contaminants are not also produced (Fig. 1). The purified protein is conjugated to an adjuvant (mammalian Jurkat T cells), since in lampreys, soluble proteins are weakly immunogenic on their own. The conjugated target protein is injected into lampreys for inducing the production of VLRB antibodies and, after confirming using ELISA that VLRB antibodies are present in the plasma of immunized lampreys, a yeast surface display (YSD) library is prepared by cloning the lymphocyte *VLRB* transcripts. The cloned *VLRBs* lack the N-terminal signal peptide and C-terminal anchor regions, and are fused to yeast protein Aga2p, so that they become attached to the cell wall of yeast cells after secretion [28]. VLRBs are also fused to a C-terminal c-Myc epitope for VLRB detection during yeast surface display. VLRB binding to the antigen of interest on the surface of yeast cells is detected by flow cytometry using a biotinylated antigen and fluorescently labeled streptavidin.

Typically, only 0.1–0.5% of VLRBs in the immune-stimulated YSD library have sufficient affinity for detection of antigen binding by flow cytometry. Therefore, the YSD library is enriched for antigen-specific high-affinity VLRBs using one or two rounds of magnetic-activated cell sorting (MACS) with streptavidin-conjugated magnetic beads before using fluorescence-activated cell sorting (FACS) to specifically isolate the highest affinity clones. The FACS-sorted yeast cells are plated and individual yeast clones are tested for antigen binding. The nucleotide sequence of the *VLRBs* from high-affinity antigen-binding clones is determined and the *VLRBs* are cloned into plant expression vectors. Transient expression or stable transformants are then generated through *Agrobacterium*-mediated transformation of plants, after which the *in planta* binding of the VLRB to the antigen of interest and any phenotypes of interest can be evaluated.

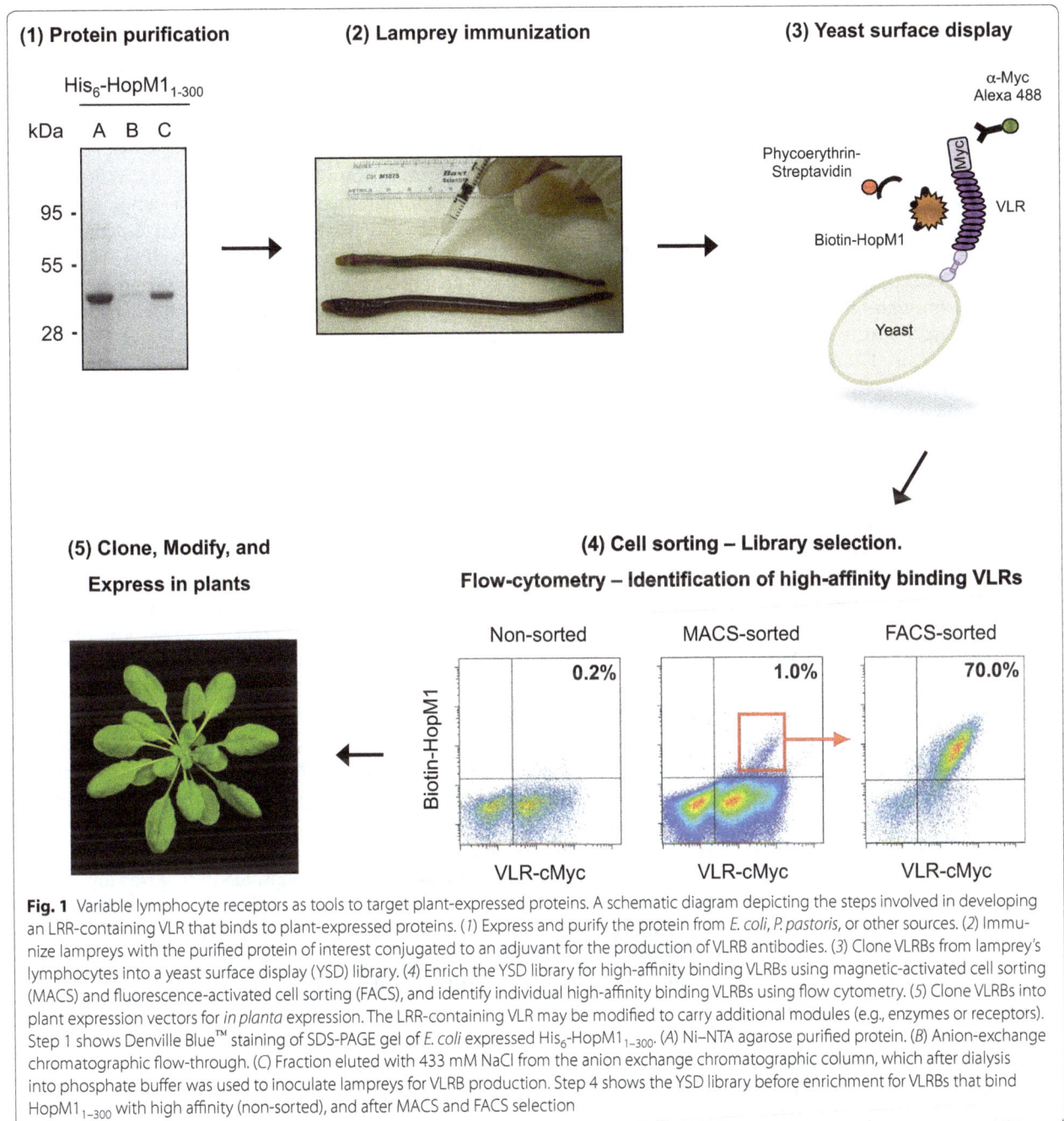

Fig. 1 Variable lymphocyte receptors as tools to target plant-expressed proteins. A schematic diagram depicting the steps involved in developing an LRR-containing VLR that binds to plant-expressed proteins. (1) Express and purify the protein from *E. coli*, *P. pastoris*, or other sources. (2) Immunize lampreys with the purified protein of interest conjugated to an adjuvant for the production of VLRB antibodies. (3) Clone VLRBs from lamprey's lymphocytes into a yeast surface display (YSD) library. (4) Enrich the YSD library for high-affinity binding VLRBs using magnetic-activated cell sorting (MACS) and fluorescence-activated cell sorting (FACS), and identify individual high-affinity binding VLRBs using flow cytometry. (5) Clone VLRBs into plant expression vectors for *in planta* expression. The LRR-containing VLR may be modified to carry additional modules (e.g., enzymes or receptors). Step 1 shows Denville Blue™ staining of SDS-PAGE gel of *E. coli* expressed His_6-HopM1$_{1-300}$. (*A*) Ni–NTA agarose purified protein. (*B*) Anion-exchange chromatographic flow-through. (*C*) Fraction eluted with 433 mM NaCl from the anion exchange chromatographic column, which after dialysis into phosphate buffer was used to inoculate lampreys for VLRB production. Step 4 shows the YSD library before enrichment for VLRBs that bind HopM1$_{1-300}$ with high affinity (non-sorted), and after MACS and FACS selection

Development of VLRBs against the bacterial effector HopM1

HopM1 is an effector from *Pseudomonas syringae* encoded in the conserved effector locus (*CEL*) [29]. HopM1 is not only one of the most conserved effectors in *P. syringae* strains [30], but also its *in planta* localization and target are known [31, 32]. We decided to test the feasibility of using LRR-containing VLRBs *in planta* to target HopM1. The N-terminus of HopM1 (amino acids 1–300; HopM1$_{1-300}$) fused to an N-terminal hexa-histidine tag was expressed and purified from *Escherichia coli* (Fig. 1). HopM1$_{1-300}$ was used instead of full-length HopM1 because of increased protein solubility and ease of purification. Purification was performed by using Ni–NTA agarose beads and ion-exchange chromatography. Purified N-terminal HopM1 was covalently conjugated to paraformaldehyde-fixed Jurkat T cells (as an adjuvant) and used to inject lamprey larvae to induce production of VLRB antibodies against HopM1 (VLR$_{M1}$). Three lampreys were immunized a total of three times at 2-week intervals. After the final immunization, blood plasma was collected from the lampreys and tested for binding to HopM1$_{1-300}$ by ELISA. Plasma from lamprey-1 had the highest binding to HopM1$_{1-300}$ (at almost a 1 in a 1000 dilution of the plasma; Additional file 1: Figure S1), and as such, the *VLRB* repertoire from this lamprey was PCR amplified from total lymphocyte cDNA and used to construct a YSD library (of approximately 1.1 × 10^6 clones) to select for VLR$_{M1}$ clones. The YSD library was enriched for clones with high-binding affinity for HopM1 by one round of MACS sorting using 100 nM of biotinylated HopM1$_{1-300}$, before FACS sorting for yeast cells expressing higher affinity VLR$_{M1}$ clones were selected (Fig. 1).

Forty randomly selected VLR$_{M1}$-expressing yeast colonies from the FACS-sorted library were individually tested for binding to HopM1. The strengths of binding varied among these clones (Fig. 2a, b). The *VLRB* gene from nine colonies with the highest binding affinity to HopM1 was sequenced. All nine *VLRB* clones carried a strikingly similar sequence in which less than 2% of nucleotides were polymorphic, which translated into only 4 amino acids (out of 168; 2.4%) being different (Fig. 2c). VLR$_{M1}$ carried 3 LRRs (LRR1; LRRV, for LRR variable; and LRRVe) flanked by N-terminal and C-terminal LRRs. This number of LRR domains is very close to the average number of LRRs, 3.81, observed in VLRBs [20]. We performed homology modeling of the structure of VLR$_{M1}$ (the uppermost VLR$_{M1}$ sequence from Fig. 2c was used for this analysis and for the remainder of the experiments, unless indicated otherwise) using a lysozyme-specific VLRB (VLR$_{HEL}$) [16]. This analysis revealed the characteristic structure for VLRBs, a solenoid forming an arc, in which the β-strands in the concave surface (with the sequence xxLxLxx, in which L stands for leucine and x for any amino acid) are predicted to be involved in the binding interaction with HopM1 (Fig. 2d).

In planta expression and visualization of VLR$_{M1}$

VLR$_{M1}$ was expressed in plants under the control of the cauliflower mosaic virus (CaMV) 35S promoter. No accumulation of cytoplasmic VLR$_{M1}$ was observed for any of the three HopM1 high-affinity sequences expressed (Fig. 3a). However, accumulation was detected when VLR$_{M1}$ was fused to YFP, albeit at a low level (Fig. 3b; compare to expression to an unrelated effector from *P. syringae*, HopK1). Since disulfide bond formation in plants occurs in the endoplasmic reticulum (ER) or at the cell wall (except for mitochondrial and chloroplast proteins) [33], and VLRBs have 4 intramolecular disulfide bonds necessary for proper protein folding [20], we decided to express *VLR$_{M1}$* fused to a signal peptide (from

AtPR1; At2g14610), so that the protein would be targeted to the plant secretory pathway. Contrary to cytoplasmic VLR$_{M1}$ accumulation, SP-VLR$_{M1}$ accumulation was readily detectable (Fig. 3a). We also evaluated if targeting VLR$_{M1}$ to a specific cell compartment without utilizing the secretory pathway would increase protein accumulation. Indeed, fusion of VLR$_{M1}$ to syntaxin SYP61, a tail-anchored protein involved in vesicle selection and fusion localized to the early endosome/*trans*-Golgi network (TGN) [34], increased VLR$_{M1}$ accumulation (Fig. 3c). As a tail-anchored protein, SYP61 is inserted post-translationally into the membrane through its hydrophobic C-terminus [35]. If required, VLR$_{M1}$ could be targeted towards the lumen of the TGN, by simply fusing SYP61 to the N-terminus of the protein [36], instead of the C-terminus as was done in this study.

Visualization of fluorescently labeled VLR$_{M1}$ (VLR$_{M1}$-YFP) revealed that the protein localized to the cytoplasm and to large aggregates that did not seem to correspond to the nucleus (Fig. 4a). These aggregates might reflect the accumulation of unfolded VLR$_{M1}$ proteins, as the cytoplasm is not conducive for disulfide bond formation [33]. When VLR$_{M1}$ was targeted to the TGN (VLR$_{M1}$-SYP61-YFP), a punctuate pattern was observed instead (Fig. 4a). The same pattern was observed for SYP61, as had been observed before in transgenic Arabidopsis plants [37] (Additional file 1: Figure S2). To visualize secreted VLR$_{M1}$ in plant cells, SP-VLR$_{M1}$ was fused to mRFP1, as mRFP1 is mostly insensitive to pH changes in the physiological range [38] while YFP fluorescence is quenched at acidic pH (i.e., in the apoplast), and as such, GFP and its derivatives may not be used as fluorescent tags for extracellular proteins. SP-VLR$_{M1}$ was found to be localized to the periphery of the cell with a similar localization to that observed for secreted mRFP1 (SP-mRFP1) (Fig. 4b), and different from that of cytoplasmic and nuclear localized mRFP1.

In planta interaction between VLR$_{M1}$ and HopM1

We next evaluated the feasibility of *in planta* interaction between VLR$_{M1}$ and HopM1 through co-immunoprecipitation experiments. We used apoplast-localized SP-VLR$_{M1}$ for these experiments, as its expression was higher than that observed for cytoplasmic VLR$_{M1}$ (Fig. 3a). As a negative control, we used a randomly selected VLRB sequence from a YSD library that was prepared from lampreys immunized against Toll-like receptor 5 (TLR5, which recognizes bacterial flagellin) [39] (Fig. 2c, note that this VLRB carries 4 LRR domains instead of the 3 observed for VLR$_{M1}$). For the co-immunoprecipitation experiments, a signal peptide for protein secretion was fused to the N-terminus of HopM1 (HopM1$_{1-300}$), so that both HopM1$_{1-300}$ and VLR$_{M1}$ would be localized to the

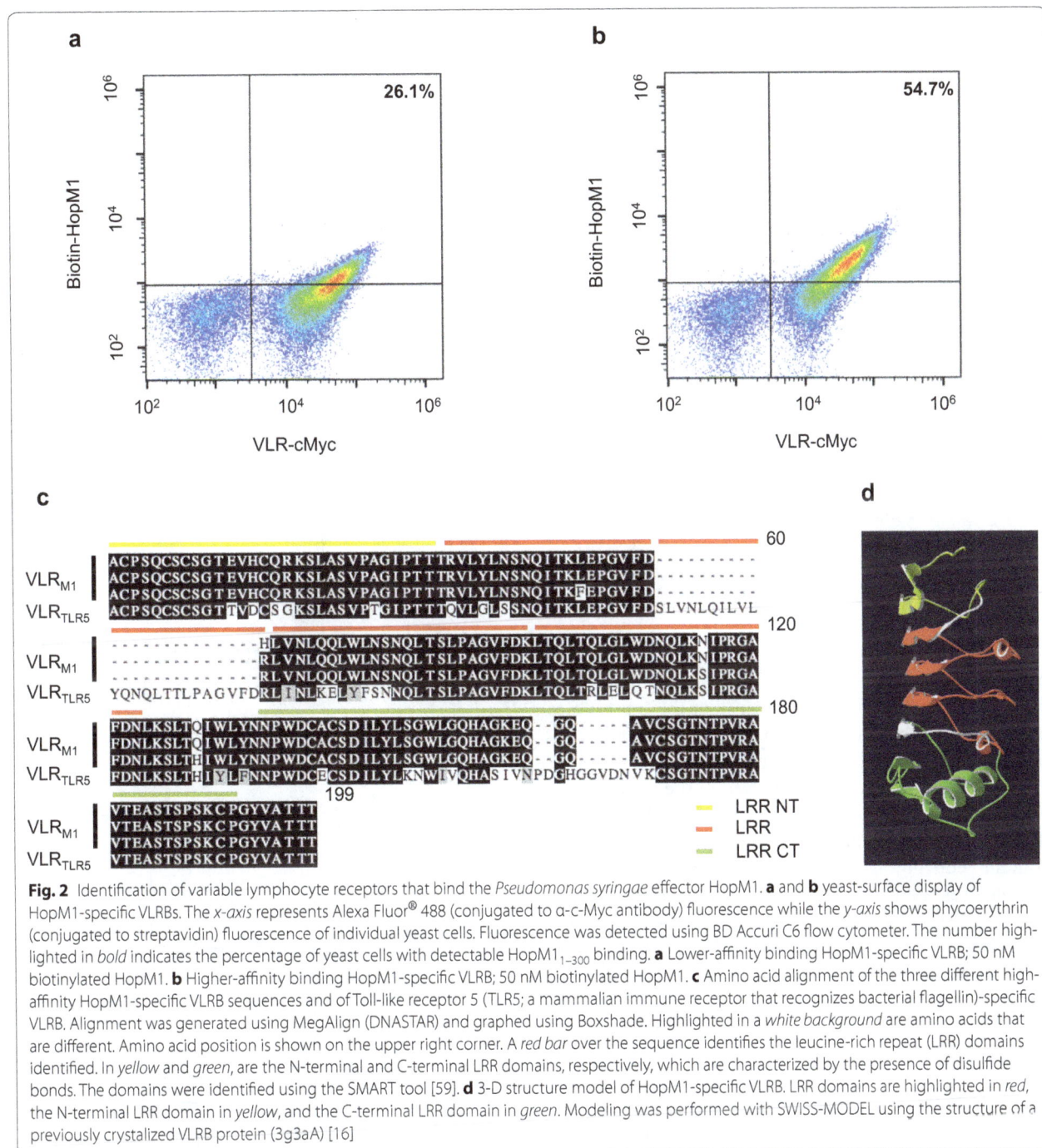

Fig. 2 Identification of variable lymphocyte receptors that bind the *Pseudomonas syringae* effector HopM1. **a** and **b** yeast-surface display of HopM1-specific VLRBs. The *x-axis* represents Alexa Fluor® 488 (conjugated to α-c-Myc antibody) fluorescence while the *y-axis* shows phycoerythrin (conjugated to streptavidin) fluorescence of individual yeast cells. Fluorescence was detected using BD Accuri C6 flow cytometer. The number highlighted in *bold* indicates the percentage of yeast cells with detectable HopM1$_{1-300}$ binding. **a** Lower-affinity binding HopM1-specific VLRB; 50 nM biotinylated HopM1. **b** Higher-affinity binding HopM1-specific VLRB; 50 nM biotinylated HopM1. **c** Amino acid alignment of the three different high-affinity HopM1-specific VLRB sequences and of Toll-like receptor 5 (TLR5; a mammalian immune receptor that recognizes bacterial flagellin)-specific VLRB. Alignment was generated using MegAlign (DNASTAR) and graphed using Boxshade. Highlighted in a *white background* are amino acids that are different. Amino acid position is shown on the upper right corner. A *red bar* over the sequence identifies the leucine-rich repeat (LRR) domains identified. In *yellow* and *green*, are the N-terminal and C-terminal LRR domains, respectively, which are characterized by the presence of disulfide bonds. The domains were identified using the SMART tool [59]. **d** 3-D structure model of HopM1-specific VLRB. LRR domains are highlighted in *red*, the N-terminal LRR domain in *yellow*, and the C-terminal LRR domain in *green*. Modeling was performed with SWISS-MODEL using the structure of a previously crystalized VLRB protein (3g3aA) [16]

same compartment. As negative controls, we used free YFP and HopK1, the latter of which is a bacterial effector that does not share sequence similarity with HopM1. All proteins (except for YFP and HopK1) were tagged with either YFP or four c-Myc tags so that reciprocal co-immunoprecipitations could be performed. Even though expression of each protein was variable (Fig. 5a, b), the amount of the YFP-tagged proteins immunoprecipitated

was equivalent between all samples (Fig. 5c). Specific interaction between HopM1-specifc VLRB and HopM1$_{1-300}$ was clearly observed (Fig. 5d; Additional file 1: Figure S3). No interaction was observed with the unrelated VLRB recognizing TLR5, nor against HopK1 or YFP.

It is important to note that the immunoprecipitations did not use any reducing agents, as attempts to

Fig. 3 *In planta* accumulation of VLR$_{M1}$ is higher if the protein goes through the secretory pathway. **a** Western blot of cytoplasmic and apoplastic VLRBs fused at their C-terminus with three HA tags detected with α-HA antibodies. Eight μg of total protein were loaded per well. Expression of three different high-affinity HopM1-specific VLRBs (done in duplicate) with slightly different amino acid sequences (see Fig. 2c) was performed for the cytoplasmic (VLR$_{M1}$-HA$_3$) and apoplastic (SP-VLR$_{M1}$-HA$_3$) versions of the HopM1-specific VLRB. A Ponceau S staining of the membrane is shown below the blot to confirm similar sample loading of the gel. Accumulation of only SP-VLR$_{M1}$ was observed. *SP* signal peptide. **b** Western blot of cytoplasmic VLRB fused at its C-terminus with YFP detected with α-GFP antibodies. Twenty-nine μg of total protein were loaded per sample. Expression of three different high-affinity HopM1-specific VLRBs with slightly different amino acid sequences was performed. YFP and HopK1-YFP (K) were used as positive controls, while an *A. tumefaciens* strain (At) devoid of any plant expression vectors was used as a negative control. The *asterisk* represents the position of a non-specific band. A Ponceau S staining of the membrane is shown below the blot to confirm similar sample loading of the gel. **c** Western blot of cytoplasmic VLRB fused at its C-terminus with syntaxin SYP61 (*At1g28490*) and three HA tags detected with α-HA antibodies. Twenty-three μg of total protein were loaded per well. Proteins were extracted from four T$_1$ transgenic *A. thaliana* Col-0 plants and an untransformed Col-0 plant. Asterisks represent the position of non-specific bands. A Ponceau S staining of the membrane is shown below the blot to confirm similar sample loading

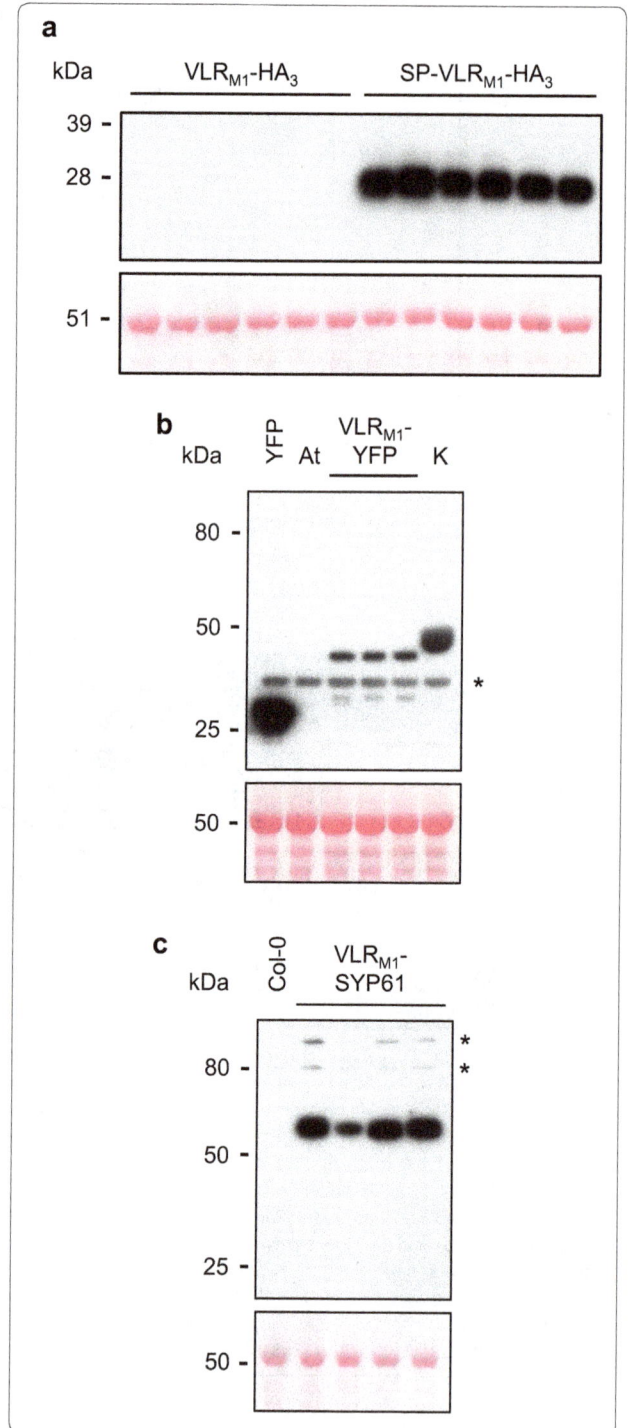

perform the immunoprecipitations with dithiothreitol in the buffer failed, probably because of the importance of the disulfide bonds for proper VLR folding. In the co-immunoprecipitation experiments, a protein band of approximately 100 kDa was observed in the western blot (Fig. 5d; the lithium dodecyl sulfate (LDS) buffer in which immunoprecipitated proteins were resuspended did not contain reducing agents either). The expected molecular weight of a dimer between VLR$_{M1}$ and HopM1$_{1-300}$ is about 90–93 kDa (depending on the tags used), which is very close to the molecular weight of this specific band (protein multimers have been observed before even in denaturing SDS-PAGE conditions [40]), providing further support for the *in planta* interaction between these two proteins.

Discussion

We have described an original method for targeting plant-expressed proteins using LRR-containing VLRBs (Fig. 1). After lamprey immunization with the protein of interest, yeast surface display is used to identify high-affinity VLRBs (Fig. 2b). Cloning the *VLRB* into a vector suitable for plant expression and *Agrobacterium*-mediated transformation of plants allows the targeting of specific proteins. In this study, we have successfully targeted the N-terminus of a bacterial effector from a plant pathogen, HopM1, by expressing an appropriate specific VLRB (VLR$_{M1}$) *in planta* (Fig. 5d). VLR$_{M1}$ interacted with

HopM1 and not with an unrelated effector, and HopM1 failed to interact with a non-specific VLRB (VLR$_{TLR5}$).

The different high-affinity sequences identified in this study for VLR$_{M1}$ clones had very few polymorphisms. This lack of variability is not surprising, as only 3% nucleotide differences had been observed for 50 *VLRA*

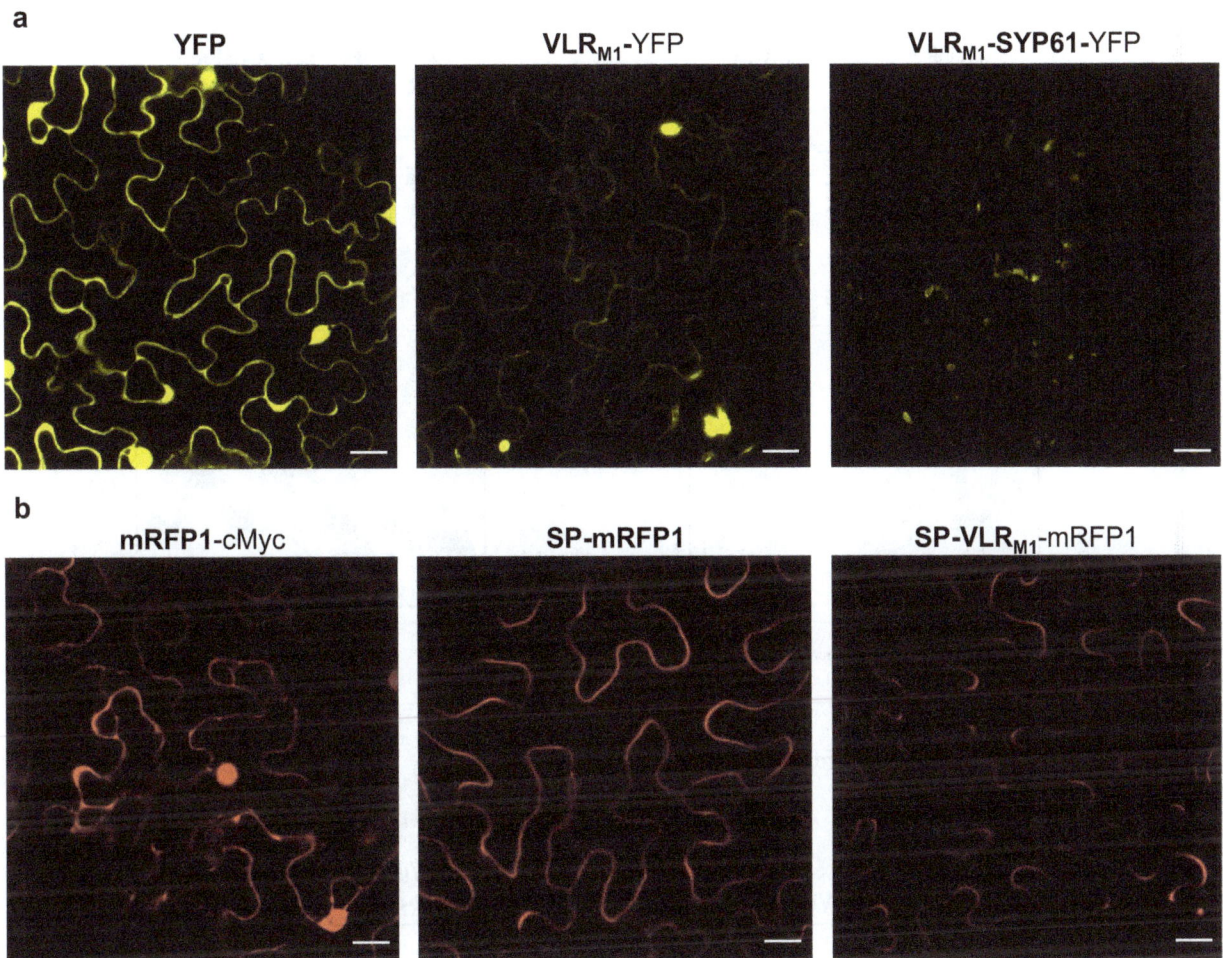

Fig. 4 Visualization of *in planta* VLRB protein expression. **a** Expression of intracellular YFP, HopM1-specific VLRB (VLR$_{M1}$), and VLR$_{M1}$ fused to *A. thaliana* syntaxin SYP61 (*At1g28490*) in *Nicotiana benthamiana*. Images were taken with the Olympus FluoView 1000 confocal microscope using a 515 nm laser for YFP excitation, while emission was collected between 530 and 569 nm. Fusing VLR$_{M1}$ to SYP61 targets the VLRB to intracellular compartments, most likely the *trans*-Golgi network. **b** Expression of intracellular mRFP1, and predicted extracellular SP-mRFP1 and SP-VLR$_{M1}$-mRFP1 in *N. benthamiana*. mRFP1 and HopM1-specific VLRB were targeted to the apoplast by fusing the VLRB to the signal peptide (SP) of *Arabidopsis thaliana* PR1 (*At2g14610*). Accumulation on the periphery of the cells was observed. Images were taken with the Olympus FluoView 1000 confocal microscope using a 559 nm laser for excitation and collecting the emission between 570 and 600 nm. *White bar length* represents 20 μm. Image brightness increased 40%

mRNAs (specific for hen egg lysozyme) [41], which are expressed by the lamprey T-like cells [25], but is in contrast with the finding that more than 30% of amino acids were different between seven VLRBs specific for the BclA *Bacillus anthracis* spore-coat protein [27]. The anti-BclA VLRB clones were screened as multivalent, secreted proteins that bound with high avidity, even though the monomeric subunits had low affinity. In contrast, the HopM1- and hen egg lysozyme-specific VLRs were isolated using yeast display to select for the highest affinity clones, which are uncommon in the repertoire and therefore, have more limited sequence diversity. The β-strands in the concave region of VLRB, which, except for the leucines of the LRR, are highly variable in sequence and

confer binding specificity [16], were clearly divergent in amino acid sequence when comparing VLR$_{M1}$, VLR$_{TLR5}$ and VLR$_{HEL}$. The overall amino acid sequence identity between VLR$_{M1}$ and VLR$_{TLR5}$ was 68%, and between VLR$_{M1}$ and VLR$_{HEL}$ was 77%, while the average amino acid identity between the variable amino acids (non-leucine) in the β-strands of LRR1, LRRRV, and LRRVe was only 20 and 19%, respectively.

VLRBs can be highly specific to the target being recognized, as VLRBs have been observed to differentiate between proteins that were 89% identical [27]. VLRBs have been shown to bind not only proteins but also carbohydrates [20, 40], and as such, VLRBs could be used to target specific carbohydrate moieties in plants. If desired,

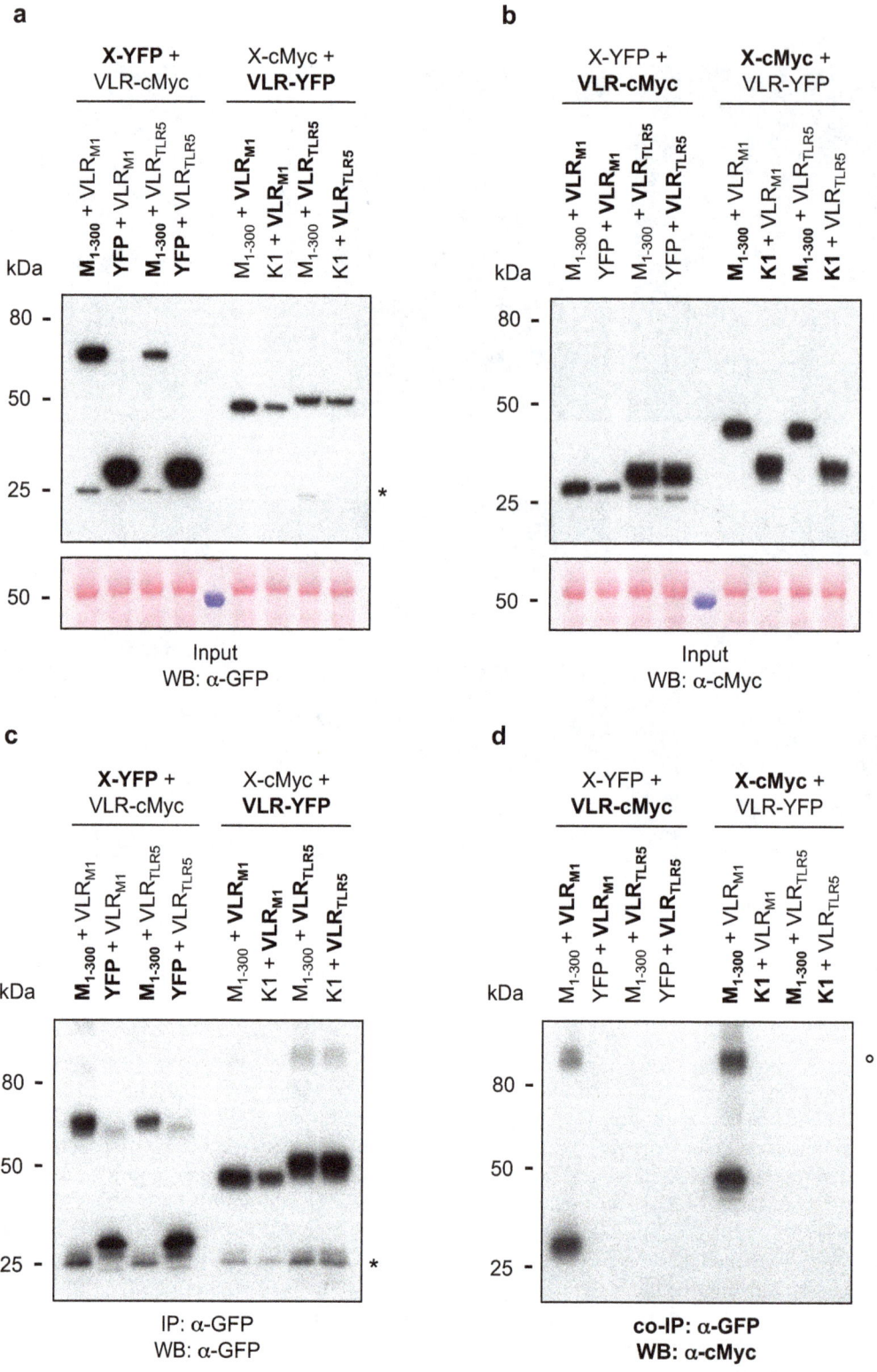

the binding specificity of the VLRs may be improved in vitro by random mutation of the amino acids responsible for the interaction in their corresponding LRRs. A more than a 1000-fold increase in binding has been observed using this method for identifying VLRs that recognize hen egg lysozyme [41]. In addition, VLRBs form high-avidity multimeric binding structures composed of 8–10 identical VLRBs as they are secreted [27]. The cysteines necessary for forming these higher order multimeric structures are not present in VLR_{M1}, as this truncated protein lacks the invariant stalk region containing the cysteines and as such, is unable to form these multimeric structures. Multimeric secreted VLRBs with potentially higher affinity could be produced in plants by adding the stalk region to the plant-expressed VLRs.

Cytoplasmic expression of VLR_{M1} was relatively low, especially when compared to secreted VLR_{M1} (Fig. 3a, b). Expression of cytoplasmic immunoglobulin antibodies in plants has also encountered the same problem, as even when the immunoglobulin gene is highly transcribed, the accumulation of cytoplasmic immunoglobulin proteins is barely detectable in plants (with a more than 300-fold difference in protein concentration being observed when comparing cytoplasmic and secreted antibodies) [42]. Nanobodies®, the recombinant variable binding domain of heavy-chain only antibodies (V_HH) from Camelids [43] offer another alternative to target plant proteins. However, high expression of nanobodies has only been observed in the apoplast [44] and chloroplasts [45]. The low antibody accumulation in the cytoplasm probably reflects the inability of the antibodies to form disulfide bonds in the reducing conditions of the cytoplasm [46], and would explain the low expression observed in this study for cytoplasmic VLR_{M1}. Some proteins can still form disulfide bonds in the cytosol, especially under oxidative stress conditions [47], so it is still possible for a fraction of the VLR_{M1} proteins to fold properly in the cytoplasm.

In contrast to cytoplasmic VLR, VLR_{M1} targeted to the apoplast or the TGN expressed well (Fig. 3a, c). HopM1 has been observed to be localized to the TGN when the

effector was expressed in transgenic plants [32]. In the future, plants with resistance against HopM1-expressing *P. syringae* strains could be developed by attaching SYP61 to VLR_{M1} and using VLR_{M1}-ubiquitin ligase or NBS–VLR_{M1} fusions (see below). However, currently it is unknown if HopM1 localizes to the lumen of the TGN or to the surrounding cytoplasm. Based on this study, we predict that VLR-based binding with HopM1 or other TGN-targeted plant proteins would probably work only if these proteins are localized on the lumen side of the TGN, since disulfide bond formation is not as efficient in the cytoplasm.

Antibodies with immunoglobulin domains have been used in the past to target plant- or plant pathogen-expressed proteins. For example, immunoglobulins have been used to modulate *in planta* abscisic and gibberellic acid availability [48, 49]. Plant viruses have also been the target of antibodies, as targeting the coat protein, the RNA-dependent RNA polymerase, and the protease that cleaves the viral polyprotein precursor reduced *in planta* viral accumulation and symptom development [50–52]. Immunoglobulins against cell wall proteins of fungal pathogens have even been engineered to be linked to antifungal peptides, which ultimately lead to reduced symptom production in transgenic plants carrying the antibody fusion [53].

We anticipate several ways in which the LRR modules from VLRBs could be used for targeting proteins and/or modifying protein function in plants (Additional file 1: Figure S4). Firstly, since LRRs are used as modules for interaction in numerous plant proteins [14, 15], VLRBs could replace the binding domains of these proteins to generate chimeric VLRB-proteins with new binding specificities. The potential for using VLR technology is such that one can conceive creating plants with a pseudo-adaptive immune system, in which pattern-recognition receptors (PRR) and disease resistance proteins with new specificities against invading pathogens may be tailored as needed. VLRBs could replace the binding modules of receptor-like proteins and receptor-like kinases and the LRR domains of NBS–LRR proteins. Functional chimeric PRRs in which the LRR domains were swapped with those from a different PRR have already been characterized [54]. Chimeric proteins responding to new stimuli and causing developmental changes could also be created (e.g., VLR–BRI1 chimera, BRI1 is the brassinosteroid hormone receptor) [5]. So far, an unsuccessful attempt at constructing a functional PRR–VLR chimera, in which no plant responses against lysozyme (the antigen recognized by the VLRA) were observed, has been described [55]. The newly characterized structure of the PRR bound to its ligand [6] might help in the future in constructing a proper functional chimera.

Antigen-specific VLRBs could also be used to explore a phenotype of interest by inhibiting the activity of a protein by direct binding (as has been observed for enzyme inhibitors carrying LRR domains) [2, 13] or by targeting a protein for degradation and observing the change in the phenotype. Direct inhibition would require a VLRB with a much higher affinity than that of the enzyme for its substrate. Since VLRBs with binding affinities in the picomolar range have been observed [41], this would be theoretically possible. For targeting proteins for degradation, VLRBs may be incorporated into LRR-containing E3 ubiquitin ligases that target proteins for proteasomal degradation in the plant cell. Multiple possibilities exist for the future use of LRR-containing VLRs for targeting plant-expressed proteins.

Conclusions

In this study, we have developed an original methodology for *in planta* targeting of proteins. This is achieved by immunizing lampreys with our target of interest, selecting VLRs with high-affinity for this protein target using flow cytometry and yeast surface display, and finally, expressing the target-specific VLR *in planta*. We found that VLR accumulation was higher when directed to the secretory pathway, although fusing the VLR to certain proteins, e.g., SYP61, might help stabilize them. With few systems available for *in planta* protein targeting, the VLR-based methodology offers the opportunity to bind and inhibit the function of specific plant proteins, and to construct chimeric proteins with new specificities in which the endogenous interacting domains are replaced by those of VLRs. This ultimately might facilitate the exploration and discovery of new phenotypes and mechanisms in plant biology.

Methods
Strains and antibiotics
Agrobacterium tumefaciens and *E. coli* strains (Additional file 2: Table S1) were grown on LB (Lennox) medium at 28–30 and 37 °C, respectively. Antibiotics were used at the following concentrations: 10 μg mL^{-1} gentamycin,

50 μg mL^{-1} kanamycin, 100 μg mL^{-1} rifampicin, and 50 μg mL^{-1} spectinomycin.

Saccharomyces cerevisiae strains (Additional file 2: Table S1) were grown on YPD (yeast extract peptone dextrose), SD-CAA (synthetic dextrose supplemented with casamino acids; 20 g L^{-1} dextrose, 6.7 g L^{-1} yeast nitrogen base, 100 mM sodium phosphate buffer, pH 6.0, and 5 g L^{-1} acid-hydrolyzed casamino acids lacking tryptophan) or SG-CAA (synthetic galactose supplemented with casamino acids; similar to SD-CAA but dextrose concentration is reduced to 1 g L^{-1} and 19 g L^{-1} galactose is included) media at 28–30 °C.

Plant growth conditions
Nicotiana benthamiana plants were grown at 22–24 °C with a 12-h photoperiod. *Arabidopsis thaliana* plants were grown under a 12-h photoperiod, at 23 °C when the lights were on and at 21 °C when the lights were off.

Sea lamprey culture
Sea lamprey larvae (*Petromyzon marinus*) of 12–15 cm in length were captured by commercial fishermen (Lamprey Services, Ludington, MI) and maintained in sand-lined, aerated aquariums at 16–20 °C. Lampreys were fed with brewer's yeast. All lamprey experiments were approved by the Emory Institutional animal care and use committee (IACUC).

Expression and purification of the N-terminus of HopM1
Escherichia coli BL21(DE3) pET28::His$_6$-hopM1$_{1-900}$ strain was grown at 37 °C until the O.D.$_{600}$ of a 200-mL culture reached 0.5. Protein expression was induced with the addition of 0.1 mM isopropyl β-D-1-thiogalactopyranoside (IPTG) and the culture was grown for 6 h at 22 °C. Cells were lysed by sonication (using the VirSonic 600 ultrasonic homogenizer from VirTis), centrifuged, and the supernatant was incubated with Ni–NTA agarose resin (QIAGEN) to capture polyhistidine-tagged proteins. Proteins were eluted from the resin with 0.5 M imidazole, and the sample diluted with 3 volumes of 30 mM Tris–HCl pH 8.3. A second-step purification of HopM1$_{1-300}$ used the UNO S-1 ion exchange chromatographic column (Bio-Rad) and the BioLogic DuoFlow™ chromatography system (Bio-Rad). HopM1$_{1-300}$ was eluted from the ion exchange column with 433 mM NaCl, and then desalted and resuspended in phosphate-buffered saline (PBS), pH 7.6, by dialysis.

Lamprey immunization
Lampreys respond to particulate antigens, such as intact viruses, bacteria and mammalian cells, but soluble proteins are weakly immunogenic on their own. Several adjuvants have been developed for vertebrates to enhance the

immune response, most of which are ineffective in lampreys. Although complete Freund's adjuvant can enhance the VLRB response, in our hands, it is toxic to lamprey larvae resulting in a high mortality rate. Given that mammalian cells are immunogenic, we determined that protein antigens or haptenated proteins covalently coupled to human Jurkat T cells by amine linkage reproducibly induced VLRB responses to both protein and hapten epitopes without toxicity. Accordingly, $HopM1_{1-300}$ was conjugated to Jurkat T cells before lamprey immunization.

For $HopM1_{1-300}$ conjugation, 10^8 Jurkat T cells were fixed overnight in 4% paraformaldehyde. The fixed Jurkat T cells were washed in 20 mM MES, pH 5.5, and then activated for amine conjugation with EDC/NHS (1-ethyl-3-(-3-dimethylaminopropyl) carbodiimide hydrochloride/N-hydroxysuccinimide) for 20 min at room temperature. Cells were briefly washed in PBS, and then 0.2 mg of $HopM1_{1-300}$ was added to the pelleted EDC/NHS-activated cells for 3 h at room temperature. After conjugation, $HopM1_{1-300}$-conjugated cells were washed once with PBS containing 10 mM Tris–HCl, pH 7.5; and stored at 4 °C until needed for lamprey immunization.

Sea lamprey larvae were sedated with 0.1 g L^{-1} of tricainemethanesulfonate (Tricaine-S; Western Chemical, Inc.) before injection into the coelomic cavity with 20 µg of recombinant $HopM1_{1-300}$ covalently conjugated to formaldehyde-fixed Jurkat T cells. Three lampreys were immunized for a total of 3 times at 2-week intervals. Two weeks after the final immunization, the lampreys were euthanized with 1 g L^{-1} of tricainemethanesulfonate and exsanguinated by tail severing. Blood was collected in a 30 mM EDTA solution (serving as an anticoagulant), and plasma and leukocytes were separated using a 55% Percoll gradient. The plasma samples were used to measure the lamprey VLRB response to immunization by ELISA, while the leukocytes were stored in RNAlater® (Thermo Fisher Scientific) at −80 °C until needed for VLRB cDNA library cloning.

Enzyme-linked immunosorbent assay (ELISA)

ELISA plates coated with 5 µg mL^{-1} of recombinant $HopM1_{1-300}$ were blocked with 2% skim milk in TBST (20 mM Tris–HCl, 150 mM NaCl, and 0.1% Tween-20, pH 7.5), before incubation with serial dilutions of plasma from HopM1-immunized lampreys or control non-immunized plasma. VLRB binding was detected with an α-VLRB mouse monoclonal antibody (4C4) [23] and an alkaline phosphatase (AP)-conjugated goat α-mouse IgG polyclonal antibody (SouthernBiotech; this secondary antibody binds to the α-VLRB antibody). In between each incubation period, five washes with TBST were performed. Enzyme activity was detected after addition of an AP substrate (p-nitrophenyl phosphate, SIGMA-Aldrich), after which plates were incubated for 30 min at room temperature, followed by AP enzyme inactivation with 0.1 M NaOH. Absorbance readings at 405 nm were collected and the data was graphed using GraphPad PRISM software.

VLRB library construction

RNA was isolated from total leukocytes samples collected from lampreys immunized with $HopM1_{1-300}$ using the RNeasy kit (QIAGEN). RNA was reverse transcribed into cDNA using SuperScript® III reverse transcriptase (Invitrogen™) and oligo-dT primers. VLRB transcripts were amplified from the leukocyte cDNA by nested PCR using high-fidelity DNA polymerase (Novagen®). The first round of PCR used primers that annealed to the 5′ and 3′ untranslated regions of VLRB, AVL001 and AVL002 (Additional file 2: Table S2), respectively. The second round of PCR used primers that amplified only the VLRB antigen-binding domain, from the N-terminal LRR to the C-terminal LRR (primers AVL003 and AVL004, respectively). These primers had approximately 50 bp of sequence homology to the yeast surface display (YSD) vector (pCT-ESO) for cloning by in vivo homologous recombination in transfected yeast cells.

The pCT-ESO plasmid adds a c-Myc epitope at the end of the VLRB insert and anchors the VLRB to the yeast cell wall by fusing the protein to Aga2p. VLRB expression in this system is controlled under a galactose-inducible promoter. To clone the VLRB cDNAs, the BDNF gene from the pCT-ESO-BDNF plasmid [56] was removed by restriction digestion with NheI and BamHI, and NcoI digestion (New England BioLabs® Inc.) to eliminate the BDNF insert.

For VLR library transformation, tryptophan-auxotroph S. cerevisiae strain EBY100 was grown to the log phase in YPD media at 30 °C until the $O.D._{600}$ reached 1.0. Cells were washed with H_2O, and incubated in 10 mM Tris–HCl, 10 mM DTT, 100 mM lithium acetate, pH 7.6, at 225 rpm and 30 °C for 20 min. After incubation, yeast cells were washed with H_2O and resuspended in 1 M sorbitol to a concentration of 10^9 cell mL^{-1}. Two hundred µL of yeast cells, 1 µg of digested pCT-ESO vector and 2 µg of the purified VLRB PCR product were mixed and electroporated at 2.5 kV using a Micropulser™ electroporator (Bio-Rad). The total number of transformants was estimated to be 1.1×10^6 VLRB clones. Aliquots of the transformed yeast library were stored at −80 °C in 15% glycerol.

Yeast surface display

Two rounds of enrichment for $HopM1_{1-300}$-binding VLRBs using Fluorescence-activated and Magnetic-activated cell sorting (FACS and MACS, respectively) were performed. For HopM1$_{1-300}$ biotinylation, the EZ-link NHS-LC-LC-biotin kit (Thermo Fisher Scientific) was used.

For determination of binding to HopM1 of individual yeast colonies, an overnight yeast culture was diluted into fresh SD-CAA medium and grown at 30 °C for 3 h. Culture was centrifuged and resuspended in SG-CAA medium (to induce VLRB expression) and further grown for 48 h at 20 °C. Fifty μL of yeast cells were washed once in staining buffer (PBS pH 7.4 with 1% BSA) and then incubated for 30 min with 50–250 nM biotinylated HopM1$_{1-300}$. Cells were washed thrice with staining buffer, and then incubated for 30 min at 4 °C in staining buffer with 5 μg mL^{-1} mouse α-Myc-Alexa Fluor$^{®}$ 488 (clone 4A6; EMD Millipore) and 2.7 μg mL^{-1} streptavidin, R-phycoerythrin conjugate (Invitrogen$^{™}$). HopM1$_{1-300}$ binding was evaluated using the BD Accuri$^{™}$ C6 flow cytometer (BD BioSciences) and a 488 nM excitation laser. For detection of Alexa Fluor$^{®}$ 488 fluorescence, a 533/30 nm filter (FL-1 channel) was used. For detection of phycoerythrin, a 585/40 nm filter (FL-2 channel) was used. The FL-1 channel was corrected by subtracting 5% of the FL-2 channel, while the FL-2 channel was corrected by subtracting 6.2% of the FL-1 channel.

To determine which events captured by the flow cytometer corresponded to VLRB binding of HopM1, the fluorescence intensity in the Alexa Fluor$^{®}$ 488 versus phycoerythrin plot was divided into four quadrants. The quadrant in the left lowermost corner represents those events in which neither *VLRB* expression nor HopM1 binding occurred, and its limits were determined by using samples in which *VLRB* expression was not induced. The quadrant in the right uppermost corner represents those events in which both VLRB expression and HopM1 binding occurred.

Cloning

VLRBs were amplified from individual colonies of yeast surface display libraries using KOD hot start DNA polymerase (Novagen$^{®}$) and Zymolase (Zymo Research), primers AVL005 and AVL006 (Additional file 2: Table S2), and a 30 min incubation at 37 °C prior to PCR (for the Zymolase to degrade the yeast cell wall). Purified PCR products were used as DNA templates with Phusion$^{®}$ high-fidelity DNA polymerase (Thermo Fisher Scientific) and primers AVL007 and AVL008 to clone into the pCR$^{™}$8/GW/TOPO$^{®}$ entry vector (Invitrogen$^{™}$). Prior to cloning, addition of adenine overhangs was performed using GoTaq$^{®}$ DNA polymerase (Promega).

The nucleotide sequence corresponding to the N-terminus of HopM1 (*PSPTO_1375*; amino acids 1–300) was amplified from *P. syringae* pv. *tomato* (*Pst*) DC3000 genomic DNA using Phusion$^{®}$ high-fidelity DNA polymerase (Thermo Fisher Scientific) and primers AVL009 and AVL010. The signal peptide sequence of *AtPR1* (*At2g14610*; *SP*) was amplified from *A. thaliana* cDNA with primers AVL011 and AVL012. *SYP61* (*At1g28490*) was amplified from *A. thaliana* cDNA with primers AVL013 and AVL014. *mRFP1* was amplified from plasmid pGWB554 [57] with primers AVL015 and AVL016. All these DNA sequences were cloned into pCR8$^{™}$/GW/TOPO$^{®}$. *Pst* DC3000 *hopK1* (*PSPTO_0044*) was amplified with primers AVL017 and AVL018 from *Pst* DC3000 genomic DNA and cloned into pDONR207. The nucleotide sequence corresponding to the N-terminus of HopM1 was also amplified using primers AVL019 and AVL020, cloned into plasmid pGEM$^{®}$-T-Easy (Promega Corporation) and then cloned into pET28a by using the restriction enzymes NdeI and EcoRI (New England BioLabs$^{®}$ Inc.).

To create a fusion between the signal peptide of AtPR1 (SP) and VLR$_{M1}$, *SP* was amplified using primers AVL021 and AVL022 such that the amplicon had overhangs that were identical in sequence to the pCR8 vector on the 5′ end and VLR$_{M1}$ on the 3′ end. The PCR product was purified and used in a second round of PCR with pCR8::*VLR$_{M1}$*; both templates had overlapping sequences, so that after the second PCR a single plasmid containing the signal peptide fused to the VLR would be produced. After amplification, removal of the original template plasmid was performed using restriction enzyme DpnI (New England BioLabs$^{®}$ Inc.).

Overlap-extension PCR (OE-PCR) was used to create a fusion between *SP* and *hopM1* using primers AVL011 and AVL023 to amplify *SP* with overlaps, and AVL024 and AVL010 to amplify *hopM1* with overlaps. The purified PCR products were used on a second PCR with primers AVL011 and AVL010 to create *SP-hopM1$_{1-900}$*, which was then cloned into pCR8$^{™}$/GW/TOPO$^{®}$. OE-PCR was also used to create fusions between *SP* and *VLR$_{TLR5}$* (using primers AVL011 and AVL022, and AVL025 and AVL008), and *SYP61* and *VLR$_{M1}$* (using primers AVL007 and AVL026, and AVL027 and AVL014).

VLR$_{M1}$, *VLR$_{TLR5}$*, *mRFP1*, *A. thaliana SYP61* and *SP*, *Pst* DC3000 *hopM1* and *hopK1*, and fusion proteins were cloned into destination vectors pGWB514, pGWB517, and pGWB554 [57]; and pDest-35S-X-YFP-6xHis [58] using Gateway$^{®}$ recombination technology (Invitrogen$^{™}$).

Alignment and 3-D structure modeling

Amino acid alignment was performed using MegAlign$^{™}$ (DNASTAR$^{®}$), and the alignment was graphed using BoxShade (Hofmann and Baron). Protein domains were predicted using the SMART tool [59].

3-D structure modeling was performed using SWISS-MODEL (Swiss Institute of Bioinformatics) and the structure of a VLRB specific for α-hen egg white lysozyme (VLR$_{HEL}$; 3g3a) [16].

Transient *in planta* expression of VLRBs in *Nicotiana benthamiana*

Agrobacterium tumefaciens strains were grown overnight in LB with appropriate antibiotics, washed twice with 10 mM MgCl$_2$, 10 mM MES (pH 5.6); and resuspended in the same buffer containing 200 μM acetosyringone to an O.D.$_{600}$ of 0.2, except for the YFP culture, whose O.D.$_{600}$ was adjusted between 0.010 and 0.025. Cultures were incubated in the dark for 3 h at room temperature, after which 5- to 6-week old *N. benthamiana* plants were infiltrated using a needleless syringe. Forty-eight hours post-infiltration, samples were collected for protein extraction or visualization on the microscope.

Stable expression of VLRBs in *Arabidopsis thaliana*

To generate transgenic *A. thaliana* plants, the floral dip method [60] was used. After seeds were collected, transformants were selected in ½ concentration Linsmaier and Skoog (LS) medium with 25 μg mL^{-1} hygromycin. Genomic DNA from putative transformants was extracted using the method of Edwards et al. [61] and the presence of the transgene confirmed by PCR using primers AVL007 and AVL014.

Protein extraction

Frozen leaf tissue was ground using 3-mm zirconium oxide beads (Glen Mills Inc.) and the TissueLyser II homogenizer (QIAGEN) or, for larger quantities, using a mortar and a pestle. Ground tissue was incubated with 3 volumes (μL) of extraction buffer (0.5–1.0% Triton X-100, 150 mM NaCl, 100 mM Tris–HCl pH 7.5, 10 mM dithiothreitol [DTT], 5 mM EDTA, and protease inhibitor cocktail for plant cell and tissue extracts from SIGMA-Aldrich) per mg of tissue for 10 min at 4 °C, after which the sample was centrifuged to remove the tissue debris. Protein concentration was determined using the Bradford method (Bio-Rad protein assay), so that every sample within an experiment was adjusted to have the same concentration.

Electrophoresis and Western blotting

Polyacrylamide gel electrophoresis was performed using the NuPAGE® electrophoresis system (Thermo Fisher Scientific) and NuPAGE® Novex® 4–12% Bis–Tris gels following manufacturer's recommendations (45 min at 200 V and 120 mA [maximum]). Protein transfer was confirmed by staining the PVDF membrane with Ponceau S stain (0.1% Ponceau S in 5% acetic acid). Western blotting was performed with the following antibodies: α-c-Myc and α-GFP (abcam®), α-HA-HRP (3F10; Roche), and α-rabbit IgG-HRP (Thermo Fisher Scientific). For chemiluminescent detection, the SuperSignal™ West Dura extended duration substrate (Thermo Fisher Scientific) and Blue Ultra Autoradiography film (GeneMate) were used.

Staining of gels during HopM1$_{1-300}$ purification was performed with Denville Blue™ protein stain (Denville Scientific Inc.) following manufacturer's recommendations.

Co-immunoprecipitation

Proteins were extracted by incubating ground tissue in extraction buffer (0.5% Triton X-100, 150 mM NaCl, 50 mM Tris–HCl pH 7.5, 0.5 mM EDTA, and protease inhibitor cocktail for plant cell and tissue extracts from SIGMA-Aldrich) for 1 h at 4 °C. No reducing agent (e.g., DTT) was included in the extraction buffer. Total protein immunoprecipitated was adjusted to be the same for every sample within an experiment. Extracted proteins (diluted to have a Triton X-100 concentration of 0.2%) were incubated with 20 μL of GFP-nAb™ (Allele Biotechnology), α-c-Myc (SIGMA-Aldrich), or α-HA (clone HA-7, SIGMA-Aldrich) agarose beads for 1 h at 4 °C. Beads were centrifuged and washed 4 times, after which immunoprecipitated proteins were released from the beads by resuspending them in 75 μL of LDS buffer (Thermo Fisher Scientific) and incubating for 10 min at 70 °C.

Confocal and epifluorescent microscopy

Confocal images were taken with the Olympus FluoView™ FV1000 confocal microscope. For YFP detection, the excitation used a 515 nm argon gas laser (10 mW, at 10% intensity), while the emission was collected between 530 and 569 nm. For mRFP1 detection, the excitation used the 559 nm solid-state diode laser (10 mW, at 10% intensity), coupled with an emission collected between 570 and 600 nm. Images were visualized with a 40×-magnification oil-immersion objective that had a numerical aperture (NA) of 1.3. Images were acquired at a voltage (HV) lower than one that gave fluorescence signal for an untransformed control and at which very few pixels for the image were starting to saturate. In all images, the offset parameter was adjusted to 10% and the line Kalman integration to 3.

Epifluorescent images were acquired with the Olympus IX71 inverted microscope equipped with a 120-W metal halide lamp and a YFP filter (Semrock). The filter had an excitation of 500/24 nm and an emission of 542/27 nm. Images were visualized with a 10×-magnification objective and were acquired at an exposure time at which the untransformed control did not show any autofluorescence.

Additional files

Additional file 1: Figure S1. Production of VLRB antibodies after HopM1$_{1-300}$ immunization in lampreys. ELISA results for VLRB production from dilutions of plasma from three lampreys immunized with HopM1$_{1-300}$ conjugated to Jurkat T cells and a control non-immunized lamprey (naïve). Binding of VLRBs to HopM1$_{1-300}$-coated plates was detected with a mouse monoclonal antibody and an alkaline peroxidase-conjugated goat α-mouse IgG polyclonal antibody. Absorbance at 405 nm (A$_{405}$) was measured 30 min after addition of an alkaline peroxidase substrate. Lamprey-1 showed the highest response to HopM1$_{1-300}$. **Figure S2**. VLRBs can be targeted to intracellular compartments. Visualization of intracellular accumulation of YFP, syntaxin SYP61 (*At1g28490*), and VLR$_{M1}$ fused to SYP61 in *N. benthamiana*. Images were taken with the Olympus IX71 inverted microscope using the YFP filter (excitation 500/24, emission 542/27). White bar length represents 50 μm. Image brightness increased 15% for YFP, and 20% for the other 2 images. Notice how the YFP fluorescence pattern is similar for SYP61 (which localizes to the early endosome/*trans*-Golgi network) [34, 37] and for VLR$_{M1}$-SYP61. **Figure S3**. *In planta* interaction of HopM1 with VLR$_{M1}$. Co-immunoprecipitation (co-IP) of HopM1 and its corresponding VLR in *Nicotiana benthamiana*. Interactions between HopM1 and VLR$_{M1}$ were tested with both proteins fused to 2 different epitope tags (HA and c-Myc). Highlighted in orange are those proteins detected in the Western blot, while in black are those proteins also expressed but not detected. As negative controls for the co-immunoprecipitations, different proteins that had low or no expression were co-expressed with HopM1 or VLR$_{M1}$ (data not shown). No reducing agents were used in the buffers. Abbreviations used: VLR$_{M1}$ = SP-VLR$_{M1}$, and M$_{1-300}$ = SP-HopM1$_{1-300}$. **a** Total protein input of HA and c-Myc tagged proteins. Proteins were detected with α-HA and α-c-Myc antibodies, respectively. Ponceau S staining of the PVDF membrane is shown below the Western blot image. **b** Immunoprecipitation (IP) using α-HA agarose beads. The IP (α-HA antibodies) and co-IP (α-c-Myc antibodies) Western blots are shown. **c** Reciprocal immunoprecipitation using α-c-Myc agarose beads. The IP (α-c-Myc antibodies) and co-IP (α-HA antibodies) Western blots are shown. **Figure S4**. Hypothetical modifications to VLRs to diversify their *in planta* use. Abbreviations used: NBS = nucleotide-binding site, RLK = receptor-like kinase, RLP = receptor-like protein, and VLR = variable lymphocyte receptor.

Additional file 2: Table S1. Strains used in this study. **Table S2.** Primers used in this study.

Authors' contributions
ACV designed and performed most of the experiments, and wrote the manuscript; KN performed experiments regarding HopM1$_{1-300}$ purification; MDC supervised experiments; BRH designed and performed experiments regarding lamprey immunizations and the yeast surface display library screens, and wrote the manuscript; and SYH designed and supervised the experiments, and wrote the manuscript. All authors read and approved the final manuscript.

Author details
[1] DOE Plant Research Laboratory, Michigan State University, East Lansing, MI 48824, USA. [2] Department of Pathology and Laboratory Medicine, Emory University, Atlanta, GA 30322, USA. [3] Department of Plant Biology, Michigan State University, East Lansing, MI 48824, USA. [4] Plant Resilience Institute, Michigan State University, East Lansing, MI 48824, USA. [5] Howard Hughes Medical Institute, Gordon and Betty Moore Foundation, Michigan State University, East Lansing, MI 48824, USA.

Acknowledgements
We would like to thank Dr. Alexandre Brutus for suggesting the VLR project idea, Dr. Kyaw Aung for constructing the *hopK1* constructs and for his help in confocal microscopy, Dr. John Scott-Craig for his help with protein purification, Dr. Melinda Frame for her help in confocal microscopy, James Kremer for his help using the flow cytometer, and Dr. Jian Yao for constructing the pEarleyGate104 vector without the Gateway cassette.

Competing interests
The authors declare that they have no competing interests.

Funding
Funding was provided by the Gordon and Betty Moore Foundation (GBMF3037 [S.Y.H.]), the National Institutes of Health (Grant GM109928 [S.Y.H.] and AI072435 [M.D.C.]), and the US Department of Energy (the Chemical Sciences, Geosciences, and Biosciences Division, Office of Basic Energy Sciences, Office of Science; DE-FG02-91ER20021 for infrastructural support [S.Y.H.]).

References
1. Kobe B, Kajava AV. The leucine-rich repeat as a protein recognition motif. Curr Opin Struct Biol. 2001;11:725–32.
2. Kobe B, Deisenhofer J. A structural basis of the interactions between leucine-rich repeats and protein ligands. Nature. 1995;374:183–6.
3. Jeong S, Trotochaud AE, Clark SE. The Arabidopsis *CLAVATA2* gene encodes a receptor-like protein required for the stability of the CLAVATA1 receptor-like kinase. Plant Cell. 1999;11:1925–34.
4. Stergiopoulos I, van den Burg HA, Okmen B, Beenen HG, van Liere S, Kema GH, et al. Tomato Cf resistance proteins mediate recognition of cognate homologous effectors from fungi pathogenic on dicots and monocots. Proc Natl Acad Sci USA. 2010;107:7610–5.
5. She J, Han Z, Kim TW, Wang J, Cheng W, Chang J, et al. Structural insight into brassinosteroid perception by BRI1. Nature. 2011;474:472–6.
6. Sun Y, Li L, Macho AP, Han Z, Hu Z, Zipfel C. Structural basis for flg22-induced activation of the Arabidopsis FLS2-BAK1 immune complex. Science. 2013;342:624–8.
7. Takken FL, Goverse A. How to build a pathogen detector: structural basis of NB-LRR function. Curr Opin Plant Biol. 2012;15:375–84.
8. Tan X, Calderon-Villalobos LI, Sharon M, Zheng C, Robinson CV, Estelle M, et al. Mechanism of auxin perception by the TIR1 ubiquitin ligase. Nature. 2007;446:640–5.
9. Sheard LB, Tan X, Mao H, Withers J, Ben-Nissan G, Hinds TR, et al. Jasmonate perception by inositol-phosphate-potentiated COI1-JAZ co-receptor. Nature. 2010;468:400–5.
10. Boruc J, Griffis AH, Rodrigo-Peiris T, Zhou X, Tilford B, van Damme D, et al. GAP activity, but not subcellular targeting, is required for Arabidopsis RanGAP cellular and developmental functions. Plant Cell. 2015;27:1985–98.
11. Forsthoefel NR, Dao TP, Vernon DM. *PIRL1* and *PIRL9*, encoding members of a novel plant-specific family of leucine-rich repeat proteins, are essential for differentiation of microspores into pollen. Planta. 2009;232:1101–14.
12. Baumberger N, Ringli C, Keller B. The chimeric leucine-rich repeat/extensin cell wall protein LRX1 is required for root hair morphogenesis in *Arabidopsis thaliana*. Genes Dev. 2001;15:1128–39.
13. Di Matteo A, Federici L, Mattei B, Salvi G, Johnson KA, Savino C, et al. The crystal structure of polygalacturonase-inhibiting protein (PGIP), a leucine-rich repeat protein involved in plant defense. Proc Natl Acad Sci USA. 2003;100:10124–8.
14. Shiu S-H, Bleecker AB. Receptor-like kinases from *Arabidopsis* form a monophyletic gene family related to animal receptor kinases. Proc Natl Acad Sci USA. 2001;98:10763–8.
15. Meyers BC, Kozik A, Griego A, Kuang H, Michelmore RW. Genome-wide analysis of NBS–LRR-encoding genes in Arabidopsis. Plant Cell. 2003;15:809–34.
16. Velikovsky CA, Deng L, Tasumi S, Iyer LM, Kerzic MC, Aravind L, et al. Structure of a lamprey variable lymphocyte receptor in complex with a protein antigen. Nat Struct Mol Biol. 2009;16:725–30.
17. Boehm T, McCurley N, Sutoh Y, Schorpp M, Kasahara M, Cooper MD. VLR-based adaptive immunity. Annu Rev Immunol. 2012;30:203–20.
18. Kirchdoerfer RN, Herrin BR, Han BW, Turnbough CL Jr, Cooper MD, Wilson IA. Variable lymphocyte receptor recognition of the immunodominant glycoprotein of *Bacillus anthracis* spores. Structure. 2012;20:479–86.

19. Luo M, Velikovsky CA, Yang X, Siddiqui MA, Hong X, Barchi JJ Jr, et al. Recognition of the Thomsen–Friedenreich pancarcinoma carbohydrate antigen by a lamprey variable lymphocyte receptor. J Biol Chem. 2013;288:23597–606.

20. Han BW, Herrin BR, Cooper MD, Wilson IA. Antigen recognition by variable lymphocyte receptors. Science. 2008;321:1834–7.

21. Pancer Z, Amemiya CT, Ehrhardt GR, Ceitlin J, Gartland GL, Cooper MD. Somatic diversification of variable lymphocyte receptors in the agnathan sea lamprey. Nature. 2004;430:174–80.

22. Hirano M, Guo P, McCurley N, Schorpp M, Das S, Boehm T, et al. Evolutionary implications of a third lymphocyte lineage in lampreys. Nature. 2013;501:435–8.

23. Alder MN, Rogozin IB, Iyer LM, Glazko GV, Cooper MD, Pancer Z. Diversity and function of adaptive immune receptors in a jawless vertebrate. Science. 2005;310:1970–3.

24. Kasamatsu J, Sutoh Y, Fugo K, Otsuka N, Iwabuchi K, Kasahara M. Identification of a third variable lymphocyte receptor in the lamprey. Proc Natl Acad Sci USA. 2010;107:14304–8.

25. Guo P, Hirano M, Herrin BR, Li J, Yu C, Sadlonova A, et al. Dual nature of the adaptive immune system in lampreys. Nature. 2009;459:796–801.

26. Alder MN, Herrin BR, Sadlonova A, Stockard CR, Grizzle WE, Gartland LA, et al. Antibody responses of variable lymphocyte receptors in the lamprey. Nat Immunol. 2008;9:319–27.

27. Herrin BR, Alder MN, Roux KH, Sina C, Ehrhardt GR, Boydston JA, et al. Structure and specificity of lamprey monoclonal antibodies. Proc Natl Acad Sci USA. 2008;105:2040–5.

28. Chao G, Lau WL, Hackel BJ, Sazinsky SL, Lippow SM, Wittrup KD. Isolating and engineering human antibodies using yeast surface display. Nat Protoc. 2006;1:755–68.

29. Alfano JR, Charkowski AO, Deng W-L, Badel JL, Petnicki-Ocwieja T, van Dijk K, et al. The Pseudomonas syringae Hrp pathogenicity island has a tripartite mosaic structure composed of a cluster of type III secretion genes bounded by exchangeable effector and conserved effector loci that contribute to parasitic fitness and pathogenicity in plants. Proc Natl Acad Sci USA. 2000;97:4856–61.

30. Baltrus DA, Nishimura MT, Romanchuk A, Chang JH, Mukhtar MS, Cherkis K, et al. Dynamic evolution of pathogenicity revealed by sequencing and comparative genomics of 19 Pseudomonas syringae isolates. PLoS Pathog. 2011;7:e1002132.

31. Nomura K, DebRoy S, Lee YH, Pumplin N, Jones J, He SY. A bacterial virulence protein suppresses host innate immunity to cause plant disease. Science. 2006;313:220–3.

32. Nomura K, Mecey C, Lee Y-N, Imboden LA, Chang JH, He SY. Effector-triggered immunity blocks pathogen degradation of an immunity-associated vesicle traffic regulator in Arabidopsis. Proc Natl Acad Sci USA. 2011;108:10774–9.

33. Onda Y. Oxidative protein-folding systems in plant cells. Int J Cell Biol. 2013;2013:585431.

34. Sanderfoot AA, Kovaleva V, Bassham DC, Raikhel NV. Interactions between syntaxins identify at least five SNARE complexes within the Golgi/prevacuolar system of the Arabidopsis cell. Mol Biol Cell. 2001;12:3733–43.

35. Pedrazzini E. Tail-anchored proteins in plants. J Plant Biol. 2009;52:88–101.

36. Luo Y, Scholl S, Doering A, Zhang Y, Irani NG, Di Rubbo S, et al. V-ATPase activity in the TGN/EE is required for exocytosis and recycling in Arabidopsis. Nat Plants. 2015;1:15094.

37. Stefano G, Renna L, Rossi M, Azzarello E, Pollastri S, Brandizzi F. AGD5 is a GTPase-activating protein at the trans-Golgi network. Plant J. 2010;64:790–9.

38. Gjetting KSK, Ytting CK, Schulz A, Fuglsang AT. Live imaging of intra- and extracellular pH in plants using pHusion, a novel genetically encoded biosensor. J Exp Bot. 2012;63:3207–18.

39. Yoon SI, Kurnasov O, Natarajan V, Hong M, Gudkov AV, Osterman AL, et al. Structural basis of TLR5-flagellin recognition and signaling. Science. 2012;335:859–64.

40. Gentile F, Amodeo P, Febbraio F, Picaro F, Motta A, Formisano S, et al. SDS-resistant active and thermostable dimers are obtained from the dissociation of homotetrameric beta-glycosidase from hyperthermophilic Sulfolobus solfataricus in SDS. Stabilizing role of the A-C intermonomeric interface. J Biol Chem. 2002;277:44050–60.

41. Tasumi S, Velikovsky CA, Xu G, Gai SA, Wittrup KD, Flajnik MF, et al. High-affinity lamprey VLRA and VLRB monoclonal antibodies. Proc Natl Acad Sci USA. 2009;106:12891–6.

42. Zimmermann S, Schillberg S, Liao Y-C, Fisher R. Intracellular expression of TMV-specific single-chain Fv fragments leads to improved virus resistance in Nicotiana tabacum. Mol Breed. 1998;4:369–79.

43. Harmsen MM, De Haard HJ. Properties, production, and applications of camelid single-domain antibody fragments. Appl Microbiol Biotechnol. 2007;77:13–22.

44. Teh YH, Kavanagh TA. High-level expression of Camelid nanobodies in Nicotiana benthamiana. Transgenic Res. 2010;19:575–86.

45. Jobling SA, Jarman C, Teh MM, Holmberg N, Blake C, Verhoeyen ME. Immunomodulation of enzyme function in plants by single-domain antibody fragments. Nat Biotechnol. 2003;21:77–80.

46. Biocca S, Ruberti F, Tafani M, Pierandrei-Amaldi P, Cattaneo A. Redox state of single chain Fv antibody targeted to the endoplasmic reticulum, cytosol and mitochondria. Biotechnology. 1995;13:1110–5.

47. Cumming RC, Andon NL, Haynes PA, Park M, Fischer WH, Schubert D. Protein disulfide bond formation in the cytoplasm during oxidative stress. J Biol Chem. 2004;279:21749–58.

48. Artsaenko O, Peisker M, zur Nieden U, Fiedler U, Weiler EW, Müntz K, et al. Expression of a single-chain Fv antibody against abscisic acid creates a wilty phenotype in transgenic tobacco. Plant J. 1995;8:745–50.

49. Shimada N, Suzuki Y, Nakajima M, Conrad U, Murofushi N, Yamaguchi I. Expression of a functional single-chain antibody against GA$_{24/19}$ in transgenic tobacco. Biosci Biotechnol Biochem. 1999;63:779–83.

50. Tavladoraki P, Benvenuto E, Trinca S, De Martinis D, Cattaneo A, Galeffi P. Transgenic plants expressing a functional single-chain Fv antibody are specifically protected from virus attack. Nature. 1993;366:469–72.

51. Boonrod K, Galetzka D, Nagy PD, Conrad U, Krczal G. Single-chain antibodies against a plant viral RNA-dependent RNA polymerase confer virus resistance. Nat Biotechnol. 2004;22:856–62.

52. Gargouri-Bouzid R, Jaoua L, Rouis S, Saïdi MN, Bouaziz D, Ellouz R. PVY-resistant transgenic potato plants expressing an anti-NIa protein scFv antibody. Mol Biotechnol. 2006;33:133–40.

53. Peschen D, Li HP, Fischer R, Kreuzaler F, Liao YC. Fusion proteins comprising a Fusarium-specific antibody linked to antifungal peptides protect plants against a fungal pathogen. Nat Biotechnol. 2004;22:732–8.

54. Brutus A, Sicilia F, Macone A, Cervone F, De Lorenzo G. A domain swap approach reveals a role of the plant wall-associated kinase 1 (WAK1) as a receptor of oligogalacturonides. Proc Natl Acad Sci USA. 2010;107:9452–7.

55. Koller T. Plant pattern-recognition receptor activation and exploration of techniques to engineer novel ligand specificities. Ph.D. dissertation, University of Wisconsin-Madison, Madison, Wisconsin, USA; 2014.

56. Burns ML, Malott TM, Metcalf KJ, Hackel BJ, Chan JR, Shusta EV. Directed evolution of brain-derived neurotrophic factor for improved folding and expression in Saccharomyces cerevisiae. Appl Environ Microbiol. 2014;80:5732–42.

57. Nakagawa T, Suzuki T, Murata S, Nakamura S, Hino T, Maeo K, et al. Improved Gateway binary vectors: high-performance vectors for creation of fusion constructs in transgenic analysis of plants. Biosci Biotechnol Biochem. 2007;71:2095–100.

58. Reumann S, Quan S, Aung K, Yang P, Manandhar-Shrestha K, Holbrook D, et al. In-depth proteome analysis of Arabidopsis leaf peroxisomes combined with in vivo subcellular targeting verification indicates novel metabolic and regulatory functions of peroxisomes. Plant Physiol. 2009;150:125–43.

59. Letunic I, Doerks T, Bork P. SMART: recent updates, new developments and status in 2015. Nucleic Acids Res. 2015;43:D257–60.

60. Bent A. Arabidopsis thaliana floral dip transformation method. Methods Mol Biol. 2006;343:87–103.

61. Edwards K, Johnstone C, Thompson C. A simple and rapid method for the preparation of plant genomic DNA for PCR analysis. Nucleic Acids Res. 1991;19:1349.

11

Growth curve registration for evaluating salinity tolerance in barley

Rui Meng[1], Stephanie Saade[2], Sebastian Kurtek[5], Bettina Berger[3], Chris Brien[3,4], Klaus Pillen[6], Mark Tester[2] and Ying Sun[1*]

Abstract

Background: Smarthouses capable of non-destructive, high-throughput plant phenotyping collect large amounts of data that can be used to understand plant growth and productivity in extreme environments. The challenge is to apply the statistical tool that best analyzes the data to study plant traits, such as salinity tolerance, or plant-growth-related traits.

Results: We derive family-wise salinity sensitivity (FSS) growth curves and use registration techniques to summarize growth patterns of HEB-25 barley families and the commercial variety, Navigator. We account for the spatial variation in smarthouse microclimates and in temporal variation across phenotyping runs using a functional ANOVA model to derive corrected FSS curves. From FSS, we derive corrected values for family-wise salinity tolerance, which are strongly negatively correlated with Na but not significantly with K, indicating that Na content is an important factor affecting salinity tolerance in these families, at least for plants of this age and grown in these conditions.

Conclusions: Our family-wise methodology is suitable for analyzing the growth curves of a large number of plants from multiple families. The corrected curves accurately account for the spatial and temporal variations among plants that are inherent to high-throughput experiments.

Keywords: Functional ANOVA model, High-throughput phenotyping, Nested association mapping, Plant growth, Spatial variation, Temporal variation

Background

Analysis of salinity tolerance in plants is necessary for our understanding of plant growth and productivity under saline conditions. Generally, high salinity has a negative effect on plant growth, causing decreases in productivity. High levels of salts in the soil reduce the ability of plant root cells to absorb water, and high levels of salts inside a plant lead to toxicity. A comprehensive review on the physiological and molecular mechanisms of salinity tolerance at cellular, organ, and whole-plant levels is written by Munns and Tester [1]. To understand how plants cope with salinity, Rajendran et al. [2] quantified three mechanisms that wheat uses to increase its salinity tolerance: osmotic tolerance, ion exclusion, and tissue tolerance.

Nowadays, advanced technologies and equipment allow the collection of large and reliable datasets related to plant growth variables, such as daily shoot growth and elemental concentration. These datasets allow us to explore salt tolerance in plants with sophisticated statistical tools. Hunt [3] proposed plant growth analyses using exponential curves to describe the relative growth rate, which they derived from the absolute growth rate, correcting for initial plant sizes. The maximum potential relative growth rate was then applied to analyze the growth of a wide range of plant species [4]. Golzarian et al. [5] showed that shoot biomass can be accurately inferred from projected shoot area, which is the total sum of pixels collected via high-throughput imaging at The Plant Accelerator®. These techniques can be used to capture

*Correspondence: ying.sun@kaust.edu.sa
[1] Computer, Electrical and Mathematical Science and Engineering Division, King Abdullah University of Science and Technology, Thuwal 23955-6900, Saudi Arabia
Full list of author information is available at the end of the article

large amounts of data that can help explain how plants respond under abiotic stresses; for example, the effects of drought on barley introgression lines [6] and the effects of salinity on rice diversity panels [7]. In fact, Al-Tamimi et al. [7] fitted cubic smoothing splines to estimate the daily growth of rice plants under saline conditions grown at The Plant Accelerator®.

In this paper, we use a functional data analysis approach to study the effects of salinity on growth patterns of barley. The field of functional data analysis is a branch of statistics that is concerned with analyzing datasets involving continuous curves and surfaces. In this work, we restrict ourselves to statistical analysis of temporal growth curves of barley plants from a nested association mapping population that consists of 25 diverse inbred families called HEB-25 [8]. For further details about the HEB-25 population, refer to Maurer et al. [8]. An important challenge in this approach is to resolve the intra- and inter-family misalignment or misregistration of the important growth patterns (peaks and valleys) of the plants. There exists a large amount of literature on statistical analysis of 1D functions, namely the pioneering work of Ramsay and Silverman [9], Kneip and Gasser [10], and Tang and Müller [11]. Some specific applications include disease classification using cyclostationary biomedical signals

[12], principal component analysis (PCA) for sparse longitudinal data [13], and classification of gene expression data [14]. When narrowing our focus to the analysis of functions that require temporal alignment, the literature is more limited [15–19]. The recent work of Srivastava et al. [20] and Kurtek et al. [21] provide a mathematically and statistically elegant approach for functional data registration (also referred to as amplitude-phase separation). The approach is based on the extension of the nonparametric Fisher–Rao metric and a convenient transformation called the square-root slope function. We use this method in conjunction with other functional data analysis tools to study family-wise salinity tolerance (FST) in the HEB-25 family.

Methods

The experiment was conducted in The Plant Accelerator®, a high-throughput phenotyping facility in Adelaide, Australia that includes northwestern (NW) and northeastern (NE) smarthouses. Each smarthouse has 24 lanes with 22 positions, and each four consecutive lanes are grouped as one zone due to homogeneous plant growth variability [22], dividing each smarthouse into a total of six zones. This setup is shown in Fig. 1. At each position, there is a cart that contains a pot with a single plant.

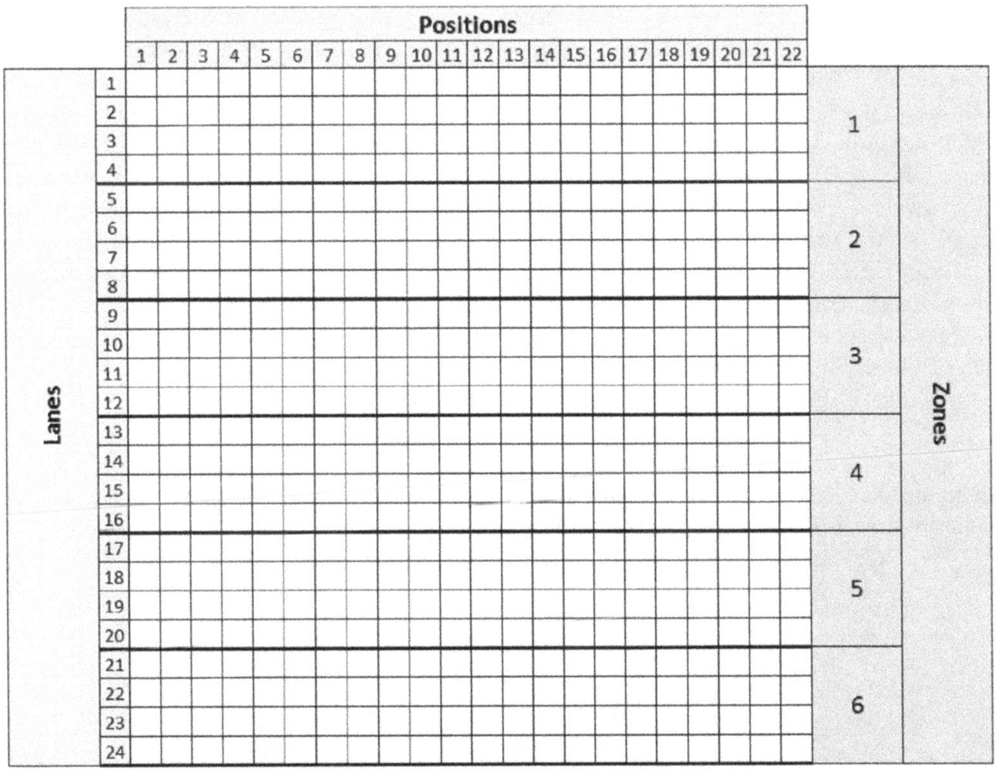

Fig. 1 Design of the smarthouse. The smarthouse includes 24 lanes where each lane contains 22 positions, and four consecutive lanes are grouped as one zone

To minimize spatial variation, plant lines are allocated to main plots, which are pairs of positions with randomly assigned plants, and designated to be part of either the control (plants watered with rain water) or the treatment (plants watered with saline water) group. The lines are assigned to three runs throughout the year and two smarthouses. Table 1 summarizes the family allocation information and the number of lines for each family of the HEB-25 population. In addition, 36 main plots were allocated to Navigator, a local Australian line used as a check line. Because only the Navigator is replicated, spatial variations between and within smarthouses and temporal variation are estimated based on this check line.

First, four seeds per accession for each condition (control and saline) were sown, and watered to a gravimetric water content of 17%. At the two-leaf stage, the seedlings were thinned down to one plant per pot, while ensuring that the plant in the control pot is similar in growth and development to that in the saline treatment pot. Marble chips were added to the surface of the pot to reduce soil evaporation. The plants were loaded on to the conveyor belts in The Plant Accelerator® at the time of emergence of the third leaf, about 16 days after sowing. After the appearance of the fourth leaf, about 20 days after sowing, we marked the third leaf and initiated the salt treatment by applying 200 mM NaCl to the treatment pots. After the stress imposition, daily images of the plants were taken for 14 days, using the LemnaTec Scanalyzer 3D, and the shoot biomass was inferred from the daily projected shoot area [5]. Fourteen days after salt imposition, the fully expanded fourth leaf was harvested and the sodium (Na) and potassium (K) contents per gram of leaf dry mass (μmol/g DM) were measured by flame photometer to provide a measure of ion exclusion (Na) and retention (K). At the end of the experiment, a large dataset with 17 daily measurements, including Na and K contents of more than 3000 plants from 25 families and two experimental conditions were recorded. The phenotypic data is available as part of the Additional files 1, 2 and 3 for the three runs, respectively.

Salinity sensitivity curves

In this section, we describe how to preprocess the data and define the salinity sensitivity (SS) curves. Let $x_{m\ell}(t)$ denote the number of pixels of the projected shoot area of the ℓth line in the mth family at time t, where $m = 1, \ldots, 26$, $\ell = 1, \cdots, n_m$, and n_m is the total number of lines in the mth family. For m, 1 to 25 refer to the HEB-25 families and 26 refers to the Navigator. First, for each line, $x_{m\ell}(t)$ was smoothed by cubic splines [23] over the common time interval, $t \in [16, 32]$ days. To account for the lines' differing initial sizes, we scaled each growth curve by its initial size: $y_{m\ell}(t) = x_{m\ell}(t)/x_{m\ell}(16)$. Then, for each pair under control and saline conditions, we took the difference in plant size between the two conditions and divided it by the size of the line in the control condition, $z_{m\ell} = (y_{m\ell,c} - y_{m\ell,s})/y_{m\ell,c}$. After smoothing the ratio by cubic splines, we predicted the first derivative denoted as $d_{m\ell} = z'_{m\ell}$. We then defined $d_{m\ell}$ to be the SS curve because it indicates how fast the relative difference, $z_{m\ell}$, changes over time. Oscillation values in $d_{m\ell}$ close to 0 suggest higher salinity tolerance, because this indicates that the growth of the plant under saline conditions was close to that under control conditions.

Pairwise and multiple registration of salinity sensitivity curves

To align SS curves temporally, we used the general framework proposed by Srivastava et al. [20] and Kurtek et al. [21] due to its theoretical and practical advantages over other methods. We provide some details of this framework next. Let f denote an absolutely continuous, real-valued function defined on the temporal domain [16, 32] (i.e., a single observation of an SS curve). Let $\mathcal{F}$ denote the set of all such functions. Also, let $\Gamma = \{\gamma : [16, 32] \to [16, 32] | \gamma(16) = 16, \gamma(32) = 32, 0 < \dot{\gamma} < \infty\}$ denote the set of temporal warping functions of the interval [16, 32], where $\dot{\gamma} = \frac{d\gamma}{dt}$. A temporal warping of an SS curve of f using $\gamma \in \Gamma$ is given by composition: $f \circ \gamma$. We seek a proper metric on $\mathcal{F}$ that provides tools for pairwise and multiple function alignment.

Table 1 Summary of family allocation and number of lines per family, where 25 families (F01–F25) are randomly allocated to two smarthouses (NW and NE), and three runs

Smarthouse	NW			NE		
Run	1	2	3	1	2	3
Family	F09 (43)	F07 (55)	F10 (54)	F03 (66)	F02 (46)	F01 (52)
	F18 (22)	F11 (55)	F17 (54)	F04 (36)	F05 (54)	F06 (54)
	F20 (47)	F15 (55)	F24 (55)	F12 (65)	F16 (53)	F08 (51)
	F21 (46)	F19 (49)	F25 (54)	F13 (49)	F23 (52)	F14 (55)
	F22 (40)					

The number of lines per family is shown in the parentheses. There are two plants (control and saline) per line

The simplest idea is to use the standard $\mathbb{L}^2$ metric. In fact, this is the most common approach in the literature on function registration. Unfortunately, such an approach is not well suited to the function registration problem because $\|f_1 - f_2\| \neq \|f_1 \circ \gamma - f_2 \circ \gamma\|$ for $f_1, f_2 \in \mathcal{F}$ and $\gamma \in \Gamma$. In other words, the action of Γ on $\mathcal{F}$ is not an isometry under the $\mathbb{L}^2$ metric. This theoretical deficiency has severe practical implications, including the pinching effect [24].

To overcome the previously described limitation, Srivastava et al. [20] used a different metric on $\mathcal{F}$ such that $d(f_1, f_2) = d(f_1 \circ \gamma, f_2 \circ \gamma)$ which is known as the Fisher–Rao distance. This metric has many fundamental advantages, including the fact that it is invariant under temporal warping [25]; however, it is difficult to compute in practice. Therefore, we used a different representation of the original SS curves called the square-root slope function (SRSF), defined as $q = \text{sign}(\dot{f})\sqrt{|\dot{f}|}$. It can be shown that if the SS curve of f is absolutely continuous, then the resulting SRSF is square-integrable (an element of $\mathbb{L}^2([16, 32], \mathbb{R})$). Furthermore, if we temporally warp an SS curve f using a $\gamma \in \Gamma$, the SRSF of $f \circ \gamma$ is given by $(q, \gamma) = (q \circ \gamma)\sqrt{\dot{\gamma}}$. The main motivation for using the SRSF representation for SS curves is that the complicated Fisher–Rao metric becomes the standard $\mathbb{L}^2$ metric and retains all of its desired properties, including isometry under the action of Γ. This result can be used to simply compute the Fisher–Rao distance d_{FR} between any two SS curves as follows: $d_{FR}(f_1, f_2) = \|q_1 - q_2\|$, where q_1 and q_2 are the SRSFs of f_1 and f_2, respectively. Let $\mathcal{C} = \mathbb{L}^2([16, 32], \mathbb{R})$ denote the space of all SRSFs. Then, for every $q \in \mathcal{C}$, there exists a unique SS curve of f such that $f(t) = f(16) + \int_{16}^{t} q(s)|q(s)|ds$. Thus, the representation $f \leftrightarrow (f(16), q)$ is invertible. Note that because we use SRSFs (defined using the derivative of the SS curve), the temporal registration will be independent of the baseline (or vertical) variability of SS curves.

Our general approach to multiple registration of SS curves will be to jointly search for an average SS curve as well as the pairwise alignment of each SS curve in the sample to this mean. Thus, we begin by describing the pairwise registration approach. Define the equivalence class of an SRSF $q \in \mathcal{C}$ under the action of Γ as $[q] = \{(q, \gamma) | \gamma \in \Gamma\}$. Each equivalence class represents the set of SRSFs associated with all possible time warpings of a given SS curve. Similarly, any two SS curves in the set $[q]$ differ only in their temporal alignment. Let $\mathcal{S}$ denote the set of all such equivalence classes (i.e., the quotient space $\mathcal{C}/\Gamma$). To compare any two equivalence classes, we will use the metric imposed on $\mathcal{C}$; given two SS curves f_1 and f_2, we register them using the $\mathbb{L}^2$ metric on the quotient space $\mathcal{S}$ using

$d_{\mathcal{S}}([q_1], [q_2]) = \inf_{\gamma \in \Gamma} \|q_1 - (q_2 \circ \gamma)\sqrt{\dot{\gamma}}\|$. Note that this is a proper distance on this space (symmetric, positive-semidefinite, and satisfies triangle inequality). The minimizer of d is denoted by γ^* and represents the warping function that achieves optimal temporal alignment of f_2 to f_1. We also let q_2^* denote $(q_2 \circ \gamma^*)\sqrt{\dot{\gamma}^*}$ and f_2^* denote $f_2 \circ \gamma^*$.

Next, we focus on mean estimation and multiple temporal alignment of SS curves. For a given collection of SS curves $f_1, f_2, \ldots, f_n$, let $q_1, q_2, \ldots, q_n$ denote their SRSFs, respectively. Then, the Karcher mean of the given SS curves is defined as $[\hat{\mu}] = \arg\min_{[q] \in \mathcal{S}} \sum_{i=1}^{n} d_{\mathcal{S}}([q], [q_i])^2$. We emphasize that the Karcher mean is actually an equivalence class $[\hat{\mu}]$ rather than an individual function. We choose a representative element of this equivalence class as follows. Select the element $\hat{\mu} \in [\hat{\mu}]$, which ensures that the mean of $\{\gamma_i^*\}$, the optimal warping functions aligning each SS curve in the given data to the Karcher mean, is the identity element of Γ given by $\gamma_{id}(t) = t$. This is called the orbit-centering step. The full algorithm for computing the Karcher mean of functions is given in Srivastava et al. [20] and Kurterk et al. [21]. This procedure results in three items: (1) $\hat{\mu}$, the preferred element of the Karcher mean equivalence class $[\hat{\mu}]$; (2) $\{f_i^*\}$, the set of optimally registered SS curves; and (3) $\{\gamma_i^*\}$, the set of optimal temporal warping functions with mean γ_{id}.

As a motivating example, we consider the 16 functions shown in Fig. 2. We suppose that these functions represent SS curves from one arbitrary family over the course of the experiment. Due to the natural variability in the response of plants to salinity stress, these functions clearly differ in relative heights and in the positions of their peaks and valleys. The time-warping method separates the amplitude and phase variabilities in Fig. 2 based on the Fisher-Rao Riemannian metric and using the square-root slope function representation to simplify the computation. The aligned functions display the relative heights of peaks and valleys, while the warping functions indicate their relative positions.

Figure 3 shows the distributions of the original and the aligned functions. The point-wise means ±2 standard deviations are shown in the top panels, and the functional boxplots [26] are displayed in the bottom panels. In the functional boxplot, the black line is the functional median, which is the most representative function, and the box contains 50% of the most central functions. Both approaches demonstrate that the mean or the median of the aligned functions summarizes the patterns of the peaks and valleys with smaller variability than in the original functions.

In our analysis, we apply the time-warping technique to the available lines within each barley family and choose

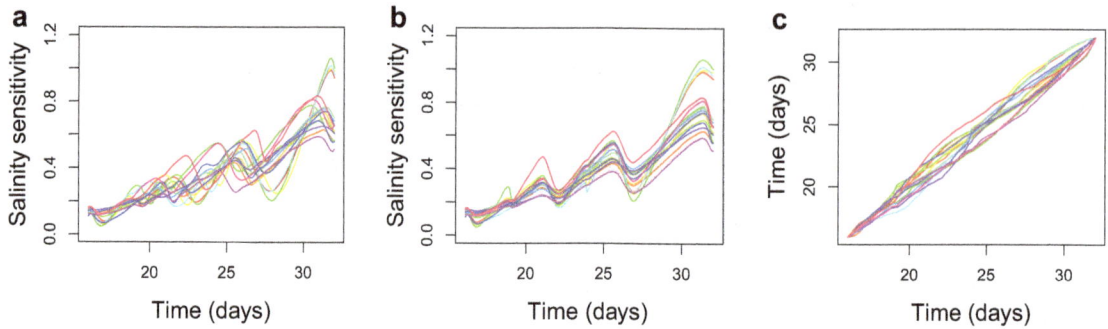

Fig. 2 An example of curve registration. **a** The salinity sensitivity (SS) curves of the 16 functions from an arbitrary family, **b** SS curves after the curve registration, and **c** the corresponding time-warping functions. The salinity sensitivity on the y-axis of **a** and **b** refers to the derivative of the relative decrease in plant biomass

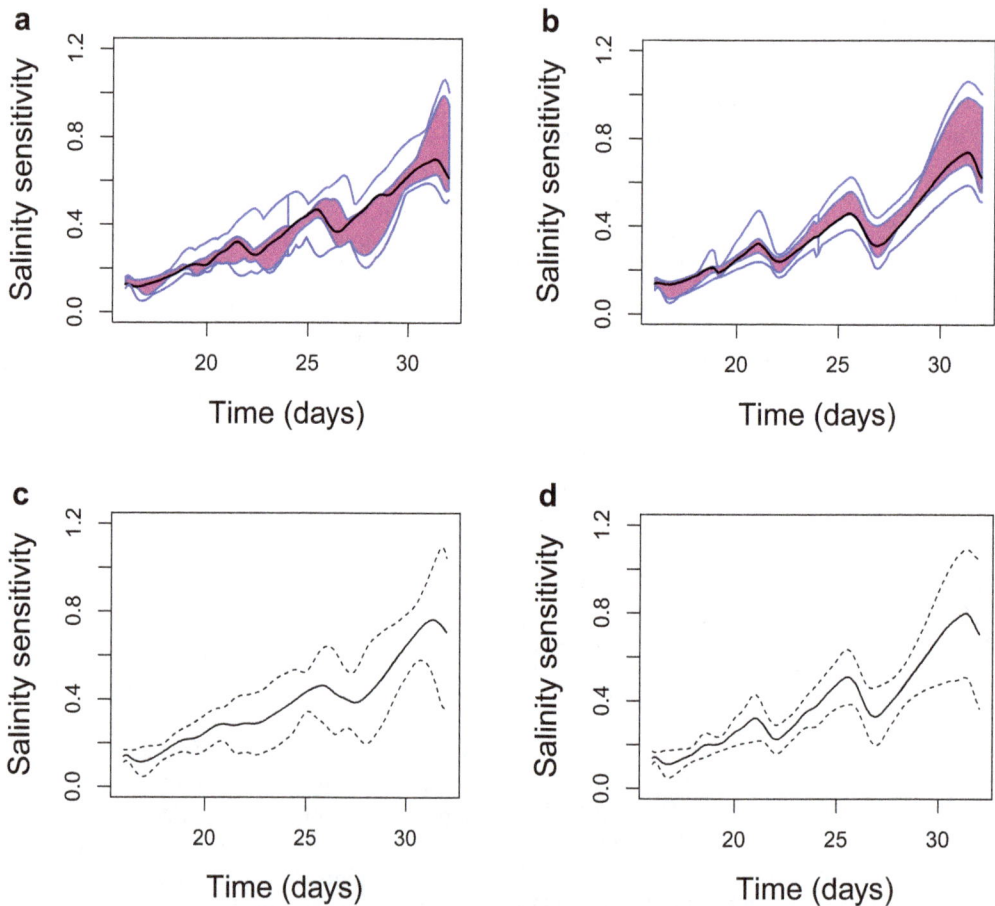

Fig. 3 Summaries of the 16 salinity sensitivity (SS) curves before and after alignment. The plots show the functional boxplots of **a** the original curves and **b** the aligned curves, where the *solid black lines* in the middle represent the functional median. The point-wise means ± 2 standard deviations before and after the alignment are shown in **c** and **d**, respectively. The salinity sensitivity on the y-axis refers to the derivative of the relative decrease in plant biomass

the aligned mean to represent the feature of a given family.

Correcting location and time effects

Plant growth can be considerably affected by differences in microclimate conditions across and within smart-houses. For example, air temperature and humidity differ in different areas of a smarthouse depending on proximity to an air conditioning unit, causing the spatial variation described by Brien et al. [22]. Moreover, since the three runs happen during different times of the year, we propose a functional ANOVA model involving the variability in both locations (spatial) and runs (temporal) effects.

Let $d_{ijk\ell}$ be the SS curve of the Navigator from the ith run, jth room, kth zone, and ℓth plant, where $i = 1, 2, 3$, $j = 1, 2$, $k = 1, \ldots, 6$, and $\ell = 1, \ldots, 6$. The model is

$$d_{ijk\ell} = \mu + \alpha_i + \beta_{jk} + \epsilon_{ijk\ell},$$

where μ represents the grand mean, α_i is the ith run effect, β_{jk} is the location effect in the kth zone of the jth room, and $\epsilon_{ijk\ell}$ is an independent error process with mean 0. We estimate each item as follows:

$$\hat{\mu} = \frac{\sum_{i=1}^3 \sum_{j=1}^2 \sum_{k=1}^6 \sum_{\ell=1}^6 d_{ijk\ell}}{216}, \tag{1}$$

$$\hat{\alpha}_i = \frac{\sum_{j=1}^2 \sum_{k=1}^6 \sum_{\ell=1}^6 d_{ijk\ell}}{108} - \hat{\mu}, \tag{2}$$

$$\hat{\beta}_{jk} = \frac{\sum_{i=1}^3 \sum_{\ell=1}^6 d_{ijk\ell}}{18} - \hat{\mu}. \tag{3}$$

Then, Fig. 4 shows the estimated grand effect, run effects and room effects after adding the salt in the time interval $t \in [21, 32]$, where the room effects are the averages of the zone effects within each smarthouse.

Although we used the available data from Day 16 to Day 32 for growth curve analyses, only the time interval [21, 32] is considered for the salinity tolerance analysis, because we are interested in comparing the treated and untreated families only after the salt was added. The addition of salt was performed on Day 20, for which we do not have images, so the first day after salting is 21. In Fig. 4, the mean curve is always around 0.028, suggesting that plants are increasingly sensitive to salinity with increasing length of time. The effect curve of Run 1 is overall greater than the others, indicating that plants in Run 1 have relatively lower salinity tolerance. This might be because Run 1 was conducted during the summer when plants were exposed to the sun for longer than during other runs. The location effects show that the difference between the NE and the NW smarthouse is significant. Overall, plants in the NE smarthouse were less sensitive to salinity than plants in the NW smarthouse.

For convenience, we redefine $d_{mijk\ell}$ as the SS curve for the mth family, ith run, jth room, kth zone, and ℓth line. The corrected salinity sensitivity (CSS) curve is $c_{mijk\ell} = d_{mijk\ell} - \hat{\alpha}_i - \hat{\beta}_{jk}$.

Family-wise salinity tolerance

To summarize the salinity tolerance of different families, we applied, within each family, the multiple registration described in subsection "Pairwise and multiple registration of salinity sensitivity curves" to SS curves and CSS curves, and took the aligned mean to represent the growth pattern. To compare across families, we aligned the aligned means again to obtain the family-wise salinity sensitivity (FSS) curves and the corrected family-wise salinity sensitivity (CFSS) curves denoted by f_m and g_m and showing the change of the relative growth difference based on salinity condition.

Taking the indefinite integral of f_m and g_m on time [21, 32] shows the growth relative difference directly. The

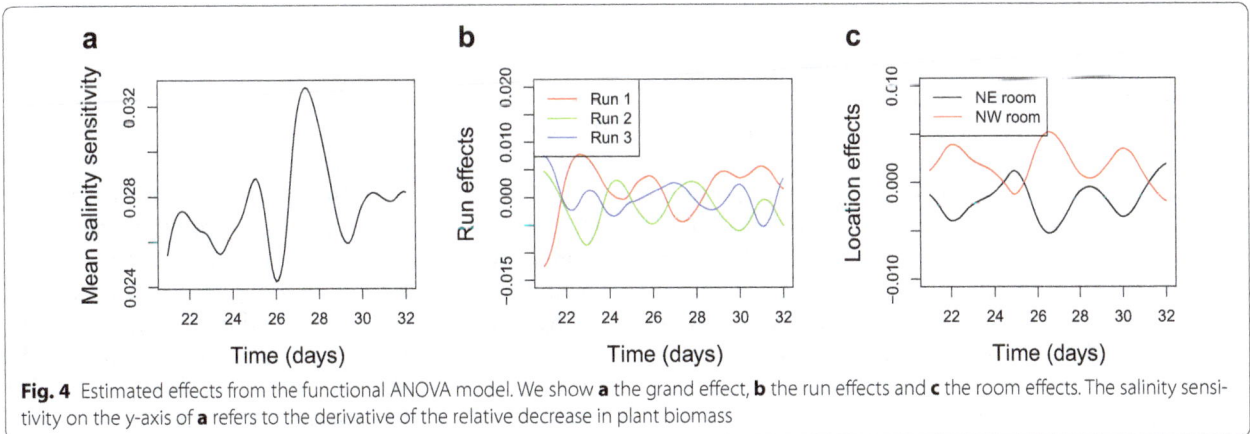

Fig. 4 Estimated effects from the functional ANOVA model. We show **a** the grand effect, **b** the run effects and **c** the room effects. The salinity sensitivity on the y-axis of **a** refers to the derivative of the relative decrease in plant biomass

resulting family-wise relative difference (FRD) curves and corrected family-wise relative difference (CFRD) curves are denoted with $F_m(t)$ and $G_m(t)$, $t \in [21, 32]$. The calculation of the integral is essentially computing the area under the curve. A similar technique, called the "area under the disease progress curve" (AUDPC), was used in the study of plant disease resistance. Details can be found in Gilligan [27]. The CFRD curves are shown in Fig. 5, showing the relative difference at different times for the 25 HEB-families and for the Navigator. Therefore, we can compare the salinity tolerance for different families based on these corrected curves. For example, if the CFRD curves for family A are overall higher than the CFRD curves for family B, it implies that family A has a lower salinity tolerance than family B.

The traditional salinity tolerance index only considers the ratio of projected shoot area between saline and control conditions at the last day. We propose the family-wise salinity tolerance (FST) by integrating the corrected ratio $1 - F_m(t)$ on $[21, 32]$, and we propose the corrected family-wise salinity tolerance (CFST) by integrating the corrected ratio $1 - G_m(t)$ on $[21, 32]$. Because a larger CFST suggests higher salinity tolerance, we evaluated the salinity tolerance of the 25-HEB families and the Navigator line by comparing their CFST values with their FST values.

Element content analysis

This section discusses the relationship between sodium and potassium contents, and the FST before and after correcting the location and time effects. Figure 6 shows the relationship between CFST and FST of each family with the within-family averaged Na and K contents, as well as the Na/K ratios. The scatter plots are color-coded according to their salinity tolerance. As can be seen in Fig. 6B, the CFST is strongly negatively correlated with the contents of Na, while the relationship to K is not significant. A similar negatively related pattern is also observed for Na/K ratios, which suggests that Na contents dominate salinity tolerance in all families. After fitting a linear regression line, as shown in Fig. 6a, the linear relationship between CFST and Na is stronger ($R^2 = 0.33$) than that for the FST ($R^2 = 0.21$). In addition, we use the t-test to test how significant the slope is below zero. After correcting for location and time effects, the increase of R^2 indicates a much stronger negative linear relationship between salinity tolerance and Na contents. Therefore, it is necessary to remove or adjust for these types of environmental effects when evaluating the plant growth. Table 2 summarizes the R^2 and p-values when both linear and nonlinear regression models are fitted to each of the six cases in Fig. 6. For the nonlinear model, we fit a linear regression model to the logarithm of these

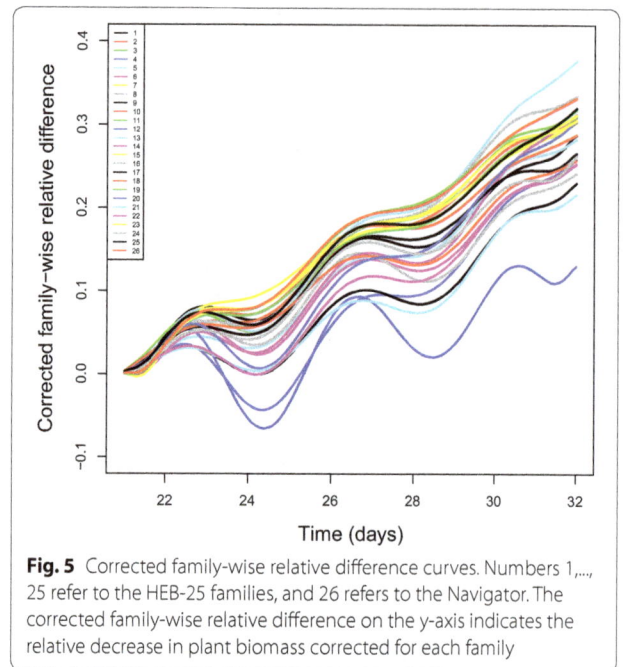

Fig. 5 Corrected family-wise relative difference curves. Numbers 1,...,25 refer to the HEB-25 families, and 26 refers to the Navigator. The corrected family-wise relative difference on the y-axis indicates the relative decrease in plant biomass corrected for each family

salinity tolerance indices, which is equivalent to fitting exponential curves for these six cases. We can see that in all cases, the relationship between salinity tolerance indices and element contents becomes stronger after correction for both models we have considered, but only slightly so, and not in all cases. Therefore, we prefer to use simpler, linear relationships, especially as there is no a priori reason biologically, to expect these relationships to be exponential. In addition, there appears, by eye, to be a difference in the relationship between Na/K and CFST for Na/K values below 0.6, apparent in plot (c) of Fig. 6B. There also appears to be a similarly distinct relationship between Na and CFST, as seen in plot (a)–differing at about 850 μmol/g DM. There may be a biological reason for this, where shoot Na is related to salinity tolerance at high values of Na, but not at low values of Na. Although this can make intuitive sense, at this stage we cannot take this further than noting it as a possible phenomenon.

Discussion

In this paper, we applied a set of advanced statistical tools for analysis of the barley growth curves in response to salinity. We used relative difference in growth rate between plants under control and saline conditions as an indicator of salinity tolerance. In addition, the FST values were corrected to account for spatial variation among plants in a smarthouse and for the temporal variation associated with high-throughput experiments. The growth pattern is summarized for the HEB-25 families and the Navigator line. Because different lines within the

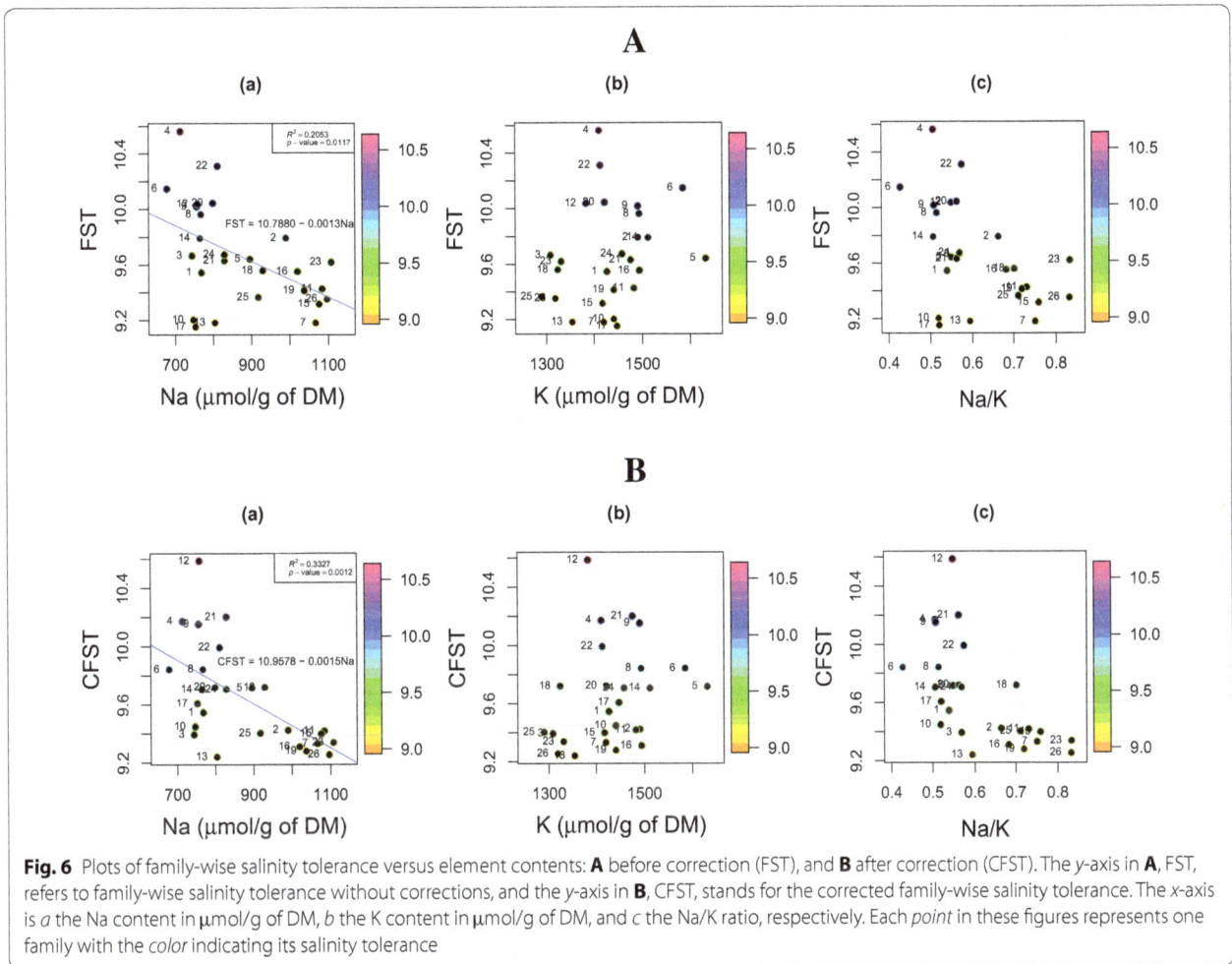

Fig. 6 Plots of family-wise salinity tolerance versus element contents: **A** before correction (FST), and **B** after correction (CFST). The y-axis in **A**, FST, refers to family-wise salinity tolerance without corrections, and the y-axis in **B**, CFST, stands for the corrected family-wise salinity tolerance. The x-axis is *a* the Na content in μmol/g of DM, *b* the K content in μmol/g of DM, and *c* the Na/K ratio, respectively. Each *point* in these figures represents one family with the *color* indicating its salinity tolerance

Table 2 Summary of the R^2 and p-values when both linear and nonlinear regression models are fitted to each of the six cases in Fig. 6

	Panel plots	Before correction			After correction		
		A(a)	A(b)	A(c)	B(a)	B(b)	B(c)
		FST-Na	FST-K	FST-Na/K	CFST-Na	CFST-K	CFST-Na/K
Linear regression	R^2	0.2371	0.0573	0.2525	0.3594	0.0587	0.3688
	p-value	0.0117	0.2385	0.0089	0.0012	0.2332	0.0010
Log scale (exponential)	R^2	0.2337	0.0593	0.2505	0.3665	0.0627	0.3781
	p-value	0.0124	0.2305	0.0092	0.0011	0.2172	0.0008

Log scale (exponential) indicates that linear regression models are fitted to the logarithm of the salinity tolerance indices

same family often do not respond to salinity at the same time, curve registration techniques were applied through time-warping, such that averaging aligned lines better display family-wise features. This method is suitable for analyzing growth curves of a large number of plants from multiple families, while accounting for the spatial and temporal variations inherent to high-throughput

experiments. It can also be used for experiments with similar designs but other stressors. In addition, our proposed CFST value allows a better understanding of the relationship between salinity tolerance and plant traits, such as the relationship between plant growth and Na and K contents, and the Na/K ratio. Although we proposed the CFST in our analysis, the curve registration technique can be used for any other functional indices of salinity tolerance as well if misalignment is an issue.

Authors' contributions

RM developed the statistical model. RM, SS and SK performed data analyses and wrote the manuscript. BB collected and performed the phenotypic analyses. CB designed the spatial allocation of plants to the smarthouses. KP developed the HEB-25 population and provided the genotypic data. MT and YS contributed to the original concept of the project and supervised the study. All authors read and approved the final manuscript.

Author details

[1] Computer, Electrical and Mathematical Science and Engineering Division, King Abdullah University of Science and Technology, Thuwal 23955-6900, Saudi Arabia. [2] Biological and Environmental Science and Engineering Division, King Abdullah University of Science and Technology, Thuwal 23955-6900, Saudi Arabia. [3] Australian Plant Phenomics Facility, The Plant Accelerator, University of Adelaide, Urrbrae, South Australia 5064, Australia. [4] Phenomics and Bioinformatics Research Centre, University of South Australia, Adelaide, South Australia 5001, Australia. [5] Department of Statistics, The Ohio State University, Columbus, OH, USA. [6] Institute of Agricultural and Nutritional Sciences, Martin Luther University Halle-Wittenberg, Betty-Heimann-Str. 3, 06120 Halle, Germany.

Acknowledgements

The authors would like to thank all members at The Plant Accelerator®for providing technical support in the phenotypic data collection. The Plant Accelerator, Australian Plant Phenomics Facility, is supported under the Collaborative Research Infrastructure Strategy (NCRIS). We thank Andreas Maurer for his scientific advice on the HEB-25 population.

Competing interests

The authors declare that they have no competing interests.

Funding

The research reported in this publication was supported by funding from King Abdullah University of Science and Technology (KAUST). This research was also partially supported by NSF DMS 1613054 (to SK).

References

1. Munns R, Tester M. Mechanisms of salinity tolerance. Annu Rev Plant Biol. 2008;59(1):651–81. doi:10.1146/annurev.arplant.59.032607.092911 PMID: 18444910.
2. Rajendran K, Tester M, Roy SJ. Quantifying the three main components of salinity tolerance in cereals. Plant Cell Environ. 2009;32(3):237–49. doi:10.1111/j.1365-3040.2008.01916.x.
3. Hunt R. Plant growth analysis. London: Edward Arnold; 1978.
4. Grime JP, Hunt R. Relative growth-rate: its range and adaptive significance in a local flora. J Ecol. 1975;63(2):393–422.
5. Golzarian MR, Frick RA, Rajendran K, Berger B, Roy S, Tester M, Lun DS. Accurate inference of shoot biomass from high-throughput images of cereal plants. Plant Methods. 2011;7(1):1–11. doi:10.1186/1746-4811-7-2.
6. Honsdorf N, March TJ, Berger B, Tester M, Pillen K. High-throughput phenotyping to detect drought tolerance QTL in wild barley introgression lines. PLoS ONE. 2014;9(5):1–13. doi:10.1371/journal.pone.0097047.
7. Al-Tamimi N, Brien C, Oakey H, Berger B, Saade S, Ho YS, Schmöckel SM, Tester M, Negrão S. Salinity tolerance loci revealed in rice using high-throughput non-invasive phenotyping. Nat Commun. 2016;7:13342.
8. Maurer A, Draba V, Jiang Y, Schnaithmann F, Sharma R, Schumann E, Kilian B, Reif JC, Pillen K. Modelling the genetic architecture of flowering time control in barley through nested association mapping. BMC Genomics. 2015;16(1):1–12. doi:10.1186/s12864-015-1459-7.
9. Ramsay JO, Silverman BW. Functional data analysis. 2nd ed. New York: Springer; 2005.
10. Kneip A, Gasser T. Statistical tools to analyze data representing a sample of curves. Ann Stat. 1992;20:1266–305.
11. Tang R, Müller HG. Pairwise curve synchronization for functional data. Biometrika. 2008;95(4):875–89.
12. Kurtek S, Wu W, Christensen GE, Srivastava A. Segmentation, alignment and statistical analysis of biosignals with application to disease classification. J Appl Stat. 2013;40(6):1270–88.
13. Yao F, Müller HG, Wang J-L. Functional data analysis for sparse longitudinal data. J Am Stat Assoc. 2005;100(470):577–90. doi:10.1198/016214504000001745.
14. Leng X, Müller HG. Classification using functional data analysis for temporal gene expression data. Bioinformatics 2006;22(1):68–76. doi:10.1093/bioinformatics/bti742.
15. Ramsay JO, Li X. Curve registration. J R Stat Soc Ser B. 1998;60:351–63.
16. Gervini D, Gasser T. Self-modeling warping functions. J R Stat Soc Ser B. 2004;66:959–71.
17. Liu X, Müller HG. Functional convex averaging and synchronization for time-warped random curves. J Am Stat Assoc. 2004;99:687–99.
18. James G. Curve alignment by moments. Ann Appl Stat. 2007;1(2):480–501.
19. Kneip A, Ramsay JO. Combining registration and fitting for functional models. J Am Stat Assoc. 2008;103(483):1155–65.
20. Srivastava A, Wu W, Kurtek S, Klassen E, Marron JS, Registration of functional data using fisher-rao metric. 2011. arXiv:1103.3817v2
21. Kurtek S, Srivastava A, Wu W, Signal estimation under random time-warpings and nonlinear signal alignment. In: Neural information processing systems (NIPS); 2011. p. 675–83
22. Brien CJ, Berger B, Rabie H, Tester M. Accounting for variation in designing greenhouse experiments with special reference to greenhouses containing plants on conveyor systems. Plant Methods. 2013;9(1):1–22. doi:10.1186/1746-4811-9-5.
23. Eilers PHC, Marx BD. Flexible smoothing with B-splines and penalties. Stat Sci. 1996;11(2):89–102.
24. Marron JS, Ramsay JO, Sangalli LM, Srivastava A, Functional data analysis of amplitude and phase variation. ArXiv e-prints 2015. 1512.03216
25. Čencov NN. Statistical decision rules and optimal inferences. Translations of mathematical monographs, vol. 53. AMS, Providence; 1982.
26. Sun Y, Genton MG. Functional boxplots. J Comput Graph Stat. 2011;20(2):316–34.
27. Gilligan CA. Comparison of disease progress curves. New Phytol. 1990;115:223–42.

12

Quantifying pruning impacts on olive tree architecture and annual canopy growth by using UAV-based 3D modelling

F. M. Jiménez-Brenes[1], F. López-Granados[1], A. I. de Castro[1], J. Torres-Sánchez[1], N. Serrano[2] and J. M. Peña[3]* [iD]

Abstract

Background: Tree pruning is a costly practice with important implications for crop harvest and nutrition, pest and disease control, soil protection and irrigation strategies. Investigations on tree pruning usually involve tedious on-ground measurements of the primary tree crown dimensions, which also might generate inconsistent results due to the irregular geometry of the trees. As an alternative to intensive field-work, this study shows a innovative procedure based on combining unmanned aerial vehicle (UAV) technology and advanced object-based image analysis (OBIA) methodology for multi-temporal three-dimensional (3D) monitoring of hundreds of olive trees that were pruned with three different strategies (traditional, adapted and mechanical pruning). The UAV images were collected before pruning, after pruning and a year after pruning, and the impacts of each pruning treatment on the projected canopy area, tree height and crown volume of every tree were quantified and analyzed over time.

Results: The full procedure described here automatically identified every olive tree on the orchard and computed their primary 3D dimensions on the three study dates with high accuracy in the most cases. Adapted pruning was generally the most aggressive treatment in terms of the area and volume (the trees decreased by 38.95 and 42.05% on average, respectively), followed by trees under traditional pruning (33.02 and 35.72% on average, respectively). Regarding the tree heights, mechanical pruning produced a greater decrease (12.15%), and these values were minimal for the other two treatments. The tree growth over one year was affected by the pruning severity and by the type of pruning treatment, i.e., the adapted-pruning trees experienced higher growth than the trees from the other two treatments when pruning intensity was low (<10%), similar to the traditionally pruned trees at moderate intensity (10–30%), and lower than the other trees when the pruning intensity was higher than 30% of the crown volume.

Conclusions: Combining UAV-based images and an OBIA procedure allowed measuring tree dimensions and quantifying the impacts of three different pruning treatments on hundreds of trees with minimal field work. Tree foliage losses and annual canopy growth showed different trends as affected by the type and severity of the pruning treatments. Additionally, this technology offers valuable geo-spatial information for designing site-specific crop management strategies in the context of precision agriculture, with the consequent economic and environmental benefits.

Keywords: Crown volume, Remote sensing, Unmanned aerial vehicle, Object-based image analysis, Precision agriculture

*Correspondence: jose.pena@fulbrightmail.org;
jmpena@ica.csic.es; mediciona@gmail.com
[3] Institute of Agricultural Sciences, CSIC, 28006 Madrid, Spain
Full list of author information is available at the end of the article

Background

Crop viability essentially relies on the management strategy adopted by the farmer. Among the tasks that impact orchard production, tree pruning remains as a costly practice with important implications for harvest [1], nutrition, pest and disease control, and irrigation strategies [2]. The pruning type and its intensity modify the tree crown to differing degrees of severity, which notably affects the tree physiology and, consequently, the fruit quantity and quality [3, 4]. Investigations on tree pruning usually involve the characterization of the tree architecture by measuring several geometric features of the crown. The conventional method consists in using a ruler to measure the primary dimensions of the tree (e.g., the tree height and its primary axis) and, next, estimating the canopy area and the crown volume either by applying equations that treat the trees as regular polygons or by applying empirical models [5]. Obviously, this task is very tedious; it requires intensive fieldwork and usually generates inconsistent results due to the irregular geometry of the tree crown [6].

Current advances in sensors and geo-spatial technologies offer an alternative to hands-on measurement tasks. Rosell and Sanz [7] described the following techniques: ultrasound, digital photographic techniques, light sensors, high-resolution radar images, high-resolution X-ray computed tomography, stereovision and LiDAR sensors. However, although some of these techniques, primarily terrestrial LiDAR laser scanning and stereovision systems, are very precise at measuring crop architecture [8, 9], they still pose some limitations under real agricultural scenarios that are usually characterized by large spaces and rugged areas. In these cases, unmanned aerial vehicles (UAVs) or drones have become a cost-effective tool for collecting continuous crop information at the field scale. The advantages of the UAVs in comparison to the traditional remote-sensing platforms are attributed to their lower cost, greater flexibility in flight scheduling and their capacity to collect remote images with much higher spatial resolution [10–12]. In addition, because the UAVs can fly at low altitude and acquire images with high overlaps, these images can be processed with automatic photo-reconstruction software and be used to produce a Digital Surface Model (DSM) of the flight area, i.e., the three dimensions (3D) of the topography and all the elements (e.g., trees) over the surface [13]. As a consequence, recent investigations have focused on evaluating the quality of UAV-based 3D models of tree plantations, and they have reported satisfactory results for olive trees [14–16], palm trees [17] and *Pinus pinea* [18]. For example, by comparing UAV-based estimations of olive trees to on-ground measurements, Torres Sánchez et al.

[15] obtained coefficients of determination (R^2) of 0.94, 0.90 and 0.65 for projected canopy area, tree height and crown volume, respectively.

To seize on all the benefits of the UAV capacity for collecting detailed information over large areas at a spatial resolution of a few centimeters, it is essential to develop and apply robust and automatic image analysis tools that are capable of computing a huge amount of crop data to produce useful maps for crop monitoring or other agronomic objectives. The object-based image analysis (OBIA) paradigm includes a wide array of techniques that offer a high level of automation and adaptability, improving on some of the limitations of pixel-based methods [19]. OBIA is based on two primary stages, called segmentation and classification. In the first stage, adjacent pixels with homogenous digital values are grouped as "objects", which are used as the basic elements of analysis and classification [20]. In the second stage, OBIA combines the spectral, topological and contextual features of these objects to successfully address complicated classification issues, e.g., in rangelands [21, 22], or urban areas [23].

The combined use of UAV images, 3D models and OBIA procedures offers new opportunities for the high-throughput monitoring of crop conditions at the level of individual plants or trees [24]. Therefore, this study takes advantage of this geo-spatial technology to compute the 3D geometric features of hundreds of olive trees with the ultimate objective of quantifying the pruning impact on the tree architecture and tree growth. Three different pruning treatments were evaluated by comparing a multi-temporal UAV-based dataset that was collected before tree pruning, after tree pruning and one year after tree pruning. The specific objectives of this research were separated in two linked sections (Fig. 1) as follows: (a) technological objectives, which involved the description and evaluation of the full procedure to acquire remote images with the UAV and to generate the image-based 3D tree models. These objectives included the development and implementation of an innovative OBIA algorithm with the capacity to automatically classify and identify every tree of the olive grove and to compute their position, projected area, height, and volume; and (b) agronomic objectives, which aimed to explore and interpret the temporal variability that was measured within the olive grove as affected by every applied pruning treatment. Additionally, the potential uses of the valuable dataset and maps obtained with this technology were also discussed, including applications for physiological and agronomical studies as well for designing site-specific crop management strategies in the context of precision agriculture.

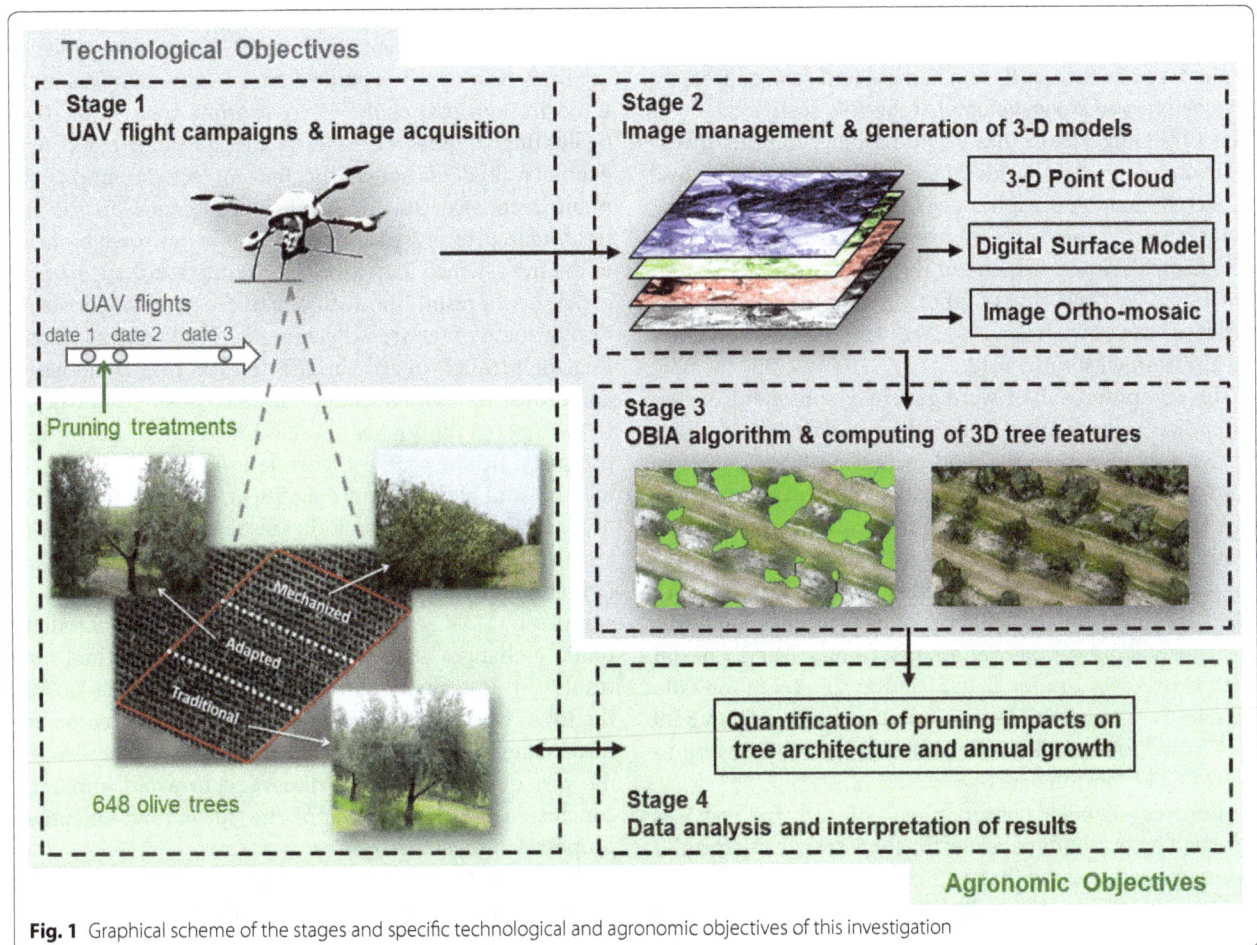

Fig. 1 Graphical scheme of the stages and specific technological and agronomic objectives of this investigation

Table 1 A sample of the output dataset delivered by the customized OBIA algorithm

Tree ID	Row	Column	Central coordinates[a]		Pruning treatment	Projected canopy area (m²)			Tree height (m)			Crown volume (m³)		
			X	Y		Date 1	Date 2	Date 3	Date 1	Date 2	Date 3	Date 1	Date 2	Date 3
1	1	1	485,726	4,217,762	Traditional	15.38	12.44	17.24	3.81	3.72	3.75	44.09	34.74	43.16
2	1	2	485,732	4,217,760	Traditional	15.88	9.61	16.69	4.10	4.01	4.14	47.75	27.52	44.68
3	1	3	485,738	4,217,757	Traditional	13.98	10.77	14.58	3.91	3.62	3.70	35.42	26.68	31.68
...	...	...	...	...	...	...	...	...	...	...	...	...	...	...
217	10	1	485,786	4,217,808	Adapted	15.34	5.16	11.13	4.37	4.04	4.43	43.81	13.28	31.43
218	10	2	485,789	4,217,806	Adapted	14.96	5.46	12.26	3.81	3.80	3.97	43.81	15.15	35.45
219	10	3	485,795	4,217,803	Adapted	12.64	7.62	13.02	4.53	3.77	3.45	31.73	18.99	27.43
...	...	...	...	...	...	...	...	...	...	...	...	...	...	...
433	19	1	485,840	4,217,852	Mechanical	7.00	5.75	7.20	3.85	3.48	3.83	18.78	15.67	17.68
434	19	2	485,846	4,217,851	Mechanical	10.11	10.00	13.64	4.31	3.58	3.79	27.28	26.73	31.97
435	19	3	485,850	4,217,848	Mechanical	13.55	13.46	14.93	4.17	3.54	3.61	43.87	39.38	42.49
...	...	...	...	...	...	...	...	...	...	...	...	...	...	...
648	27	24	486,017	4,217,841	Mechanical	6.06	5.93	6.30	3.48	3.34	3.54	15.27	14.52	15.46

[a] Coordinate system: UTM, zone 30 N, datum WGS84

Results

Technological objectives: multi-temporal quantification of the tree 3D features (location, projected canopy area, tree height and crown volume) at the field scale

The OBIA algorithm that was developed for this investigation automatically identified all the olive trees and reported their geographic coordinates, projected areas, heights and volumes on the three study dates (Table 1). The algorithm also accounted for the relative position of the trees, i.e. their row number (from 1 to 27) and their order in the row (from 1 to 24), which facilitated their localization within the field.

The computed values were generally consistent on the three dates, decreasing from date 1 to date 2 as a result of the pruning operation, and increasing from date 2 to date 3 due to the tree development that occurred over one year. On average, the projected canopy areas varied from a range of 11.6–14.7 m^2 on date 1 (before pruning) to 8.4–10.7 m^2 on date 2 (after pruning) and to 13.7–15.1 m^2 on date 3 (1-year after pruning). Similarly, the tree heights varied over a range from 3.9 to 4.1 m (on date 1) to 3.4–4.1 m (on date 2) and to 3.3–4.2 m (on date 3), and the crown volumes varied over a range from 31.9 to 42.7 m^3 (on date 1) to 22.7–28.1 m^3 (on date 2) and to of 32.8–41.0 m^3 (on date 3).

However, detailed observations of the full dataset revealed incorrect dimensions for some trees, which could be attributed to errors that occurred during the DSM generation. Therefore, the quality of the DSM created on the three dates was evaluated by visually comparing every tree perimeter that was defined by the OBIA algorithm (which was based on the DSM information) and the real tree perimeter observed in the orthomosaicked image. The accuracy that was achieved for the full 3D tree photo-reconstruction procedure varied as affected by the type of pruning treatment and the flight date (Table 2).

When the three dates were jointly considered, the most accurate olive tree photo-reconstruction was obtained for mechanical pruning (MP). However, for individual dates, the worst and best results for the three treatments were all obtained on date 2, after pruning. These values varied from 75.5% at the adapted pruning (AP) parcel, to 83.3% at the traditional pruning (TP) parcel and to 96.3% at the

MP parcel. This finding could be due to the specific characteristics of each treatment. The MP removed protruding branches, which produced a tree shape that was more uniform than that of the other pruning treatments, thus facilitating the task of building the 3D point cloud field geometry and consequently, the correct definition and photo-reconstruction of the MP tree edges. In the AP treatment, pruning drastically cut back the crown biomass of the trees, which increased the overall tree shape heterogeneity. As a result, the accuracy of the photo-reconstruction of some of the trees decreased. These results indicate that the pruning treatment affected the tree architecture and crown size, and it might have also positively or negatively affected the quality of the UAV-based 3D geo-spatial products. Additionally, no correlation between every tree location and photo-reconstruction errors was found at the orchard scale, which also indicates that other factors with respect to the weather conditions (e.g., wind or clouds) or operational issues (e.g., flight altitude, orientation of sensor axes or UAV velocity) could apparently produce slight random changes at the moment of image shooting. As a result of this evaluation, 512 trees (approximately 80% of the total) that were correctly photo-reconstructed on the three dates were studied in the subsequent analysis, and the rest of the trees were discarded to avoid imprecise conclusions in the context of the agronomic objectives proposed in this investigation.

The detailed information reported in Table 1 was ranked as four levels of projected canopy area (Fig. 2), tree height (Fig. 3) and crown volume (Fig. 4), which allowed for the observation of all the tree variability at the orchard scale.

A view of these figures on date 1 (before pruning) revealed the distribution of the tree sizes throughout the orchard, showing a higher number of small trees located at the three upper rows of the graph, which corresponded to the top zone of the field. On date 2 (after pruning), the homogeneity of the tree heights was clearly notable at the MP parcel (Fig. 3), as shown by a coefficient of variation (CV) of only 6%. On this date, a severe reduction of the projected canopy areas and crown volumes was also observed on the trees located at the AP and TP zones. Next, the geometric features on date 3 (1-year after pruning) of a majority of the trees showed similar values to

Table 2 Number and percentage of trees correctly photo-reconstructed on one of the three study dates (columns Date 1, Date 2 and Date 3) and on all the three study dates (column 1–2–3)

Pruning treatment	Number of trees	Date 1	Date 2	Date 3	Dates 1–2–3
Mechanical	216	185 (85.7%)	208 (96.3%)	191 (88.4%)	177 (81.9%)
Adapted	216	200 (92.6%)	163 (75.5%)	177 (81.9%)	161 (74.5%)
Traditional	216	200 (92.6%)	180 (83.3%)	190 (88.0%)	174 (80.6%)

Flight dates: Date 1 (Before pruning); Date 2 (After pruning); Date 3 (1-year after pruning)

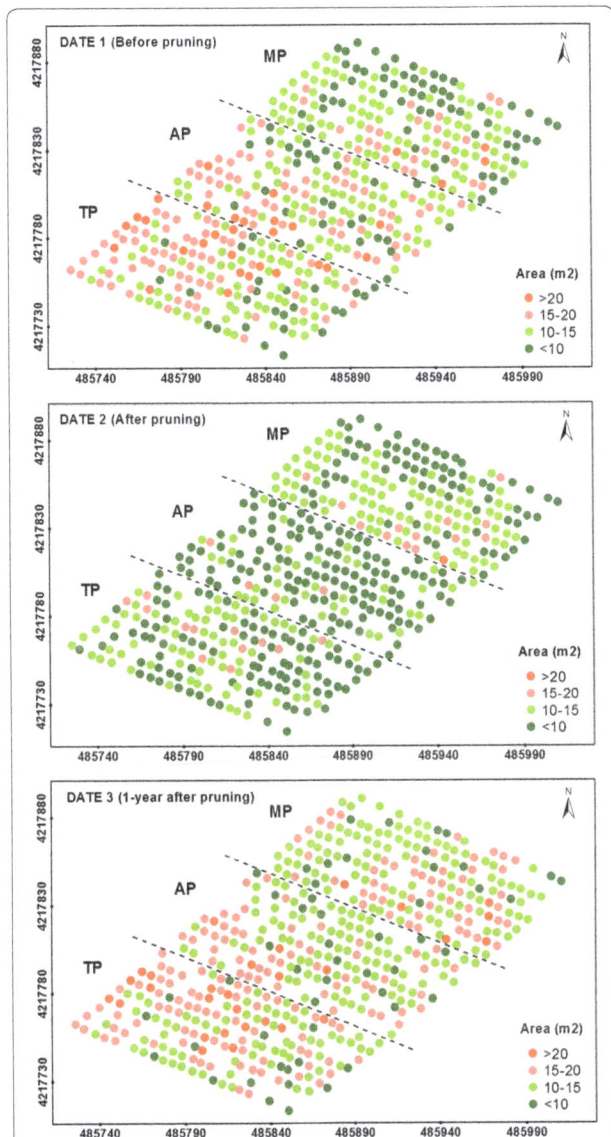

Fig. 2 Four-level representation of the projected tree canopy areas as computed on the three study dates. The *letters* indicate the pruning treatments as follows: TP (traditional), AP (adapted), and MP (mechanical). In the axes, coordinate system UTM zone 30 N, datum WGS84

on date 2 (after pruning) and date 1 (before pruning) (Fig. 5; Table 3). The AP treatment was generally the most aggressive for the olive trees, producing an average decrease of approximately 40% in canopy area and crown volume. However, the heights of the AP trees only decreased by an average of 0.05 m. These values were slightly greater than the ones obtained for the TP treatment. Regarding the MP treatment, the heights of these trees decreased 0.51 m on average, being up to ten times the overall decrease relative to the other two treatments, although the average reductions in the projected canopy area and crown volume were approximately five and four times lower than they were for the AP trees, respectively.

Regarding the impact of the pruning treatments on the annual tree growth (date 3–date 2), the type of pruning treatment might have a major influence on the vegetative response of the trees over time, mostly with respect to the crown volume growth (Fig. 6; Table 4). In comparing the tree data computed on date 2 and date 3 (after 1 year), the TP trees showed the greatest growth rates in projected canopy area (5.39 m^2, 64.27%), tree height (0.06 cm, 1.66%), and crown volume (13.90 m^3, 61.49%), although the projected canopy area of the AP trees showed higher growth in their percentage values (5.34 m^2, or 71.14%). By contrast, the MP showed the lowest growth rates for the three variables (3.25 m^2, −0.08 m and 4.66 m^3, respectively).

These results are linked to the ability of the trees to return to their initial dimensions from before the pruning task (Fig. 7; Table 5). Of the three treatments applied here, most MP trees were totally restored in terms of canopy area and volume in comparison to the original tree dimensions before the pruning task (date 1), but not in terms of the tree height. On average, the trees exceeded the canopy area and crown volume by 2.35 m^2 and by 0.87 m^3, respectively, although these trees were 0.59 m smaller in comparison to their heights on date 1. In the case of TP, these trees generally grew back to their original dimensions regarding the canopy area (0.39 m^2 of average excess) and height (0.03 m of average excess), but not in terms of crown volume (1.68 m^3 of average shortage). Finally, the AP trees did not generally reach their initial canopy area, tree height or crown volume in most of the trees.

A detailed analysis of the data by grouping the trees according to pruning intensity revealed that tree restoration might be affected not only by the pruning severity but also by the type of pruning treatment (Fig. 8). In general, the trees that were subjected to more aggressive pruning experienced much more vegetative development for the three studied treatments. Moreover, it was determined that the trees that were pruned to less than 30% of their crown volume grew approximately 20–40% over one year,

the one observed on date 1 for the projected areas and crown volumes (Figs. 2, 4, respectively) and on date 2 for the tree heights (Fig. 3). This finding indicates that the olive trees grew in area and volume over the duration of this experiment, but not in height.

Agronomic objectives: Impact of every pruning treatment on the tree architecture, annual tree growth and tree restoration

The impact of the pruning treatments on the tree architecture was evaluated by comparing tree dimensions

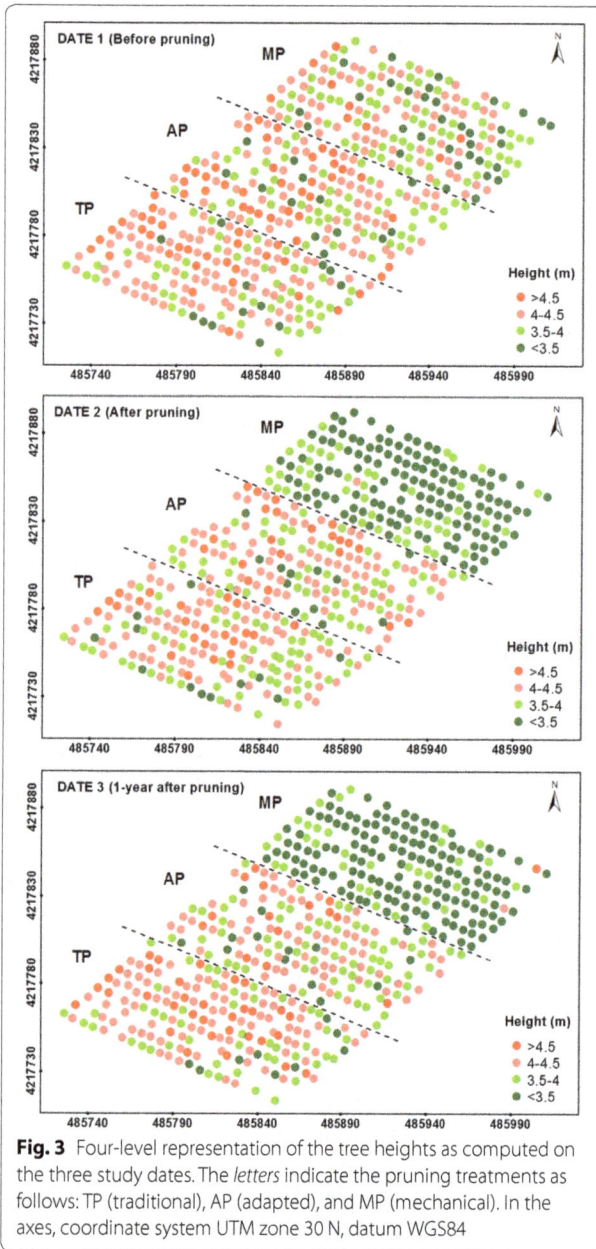

Fig. 3 Four-level representation of the tree heights as computed on the three study dates. The *letters* indicate the pruning treatments as follows: TP (traditional), AP (adapted), and MP (mechanical). In the axes, coordinate system UTM zone 30 N, datum WGS84

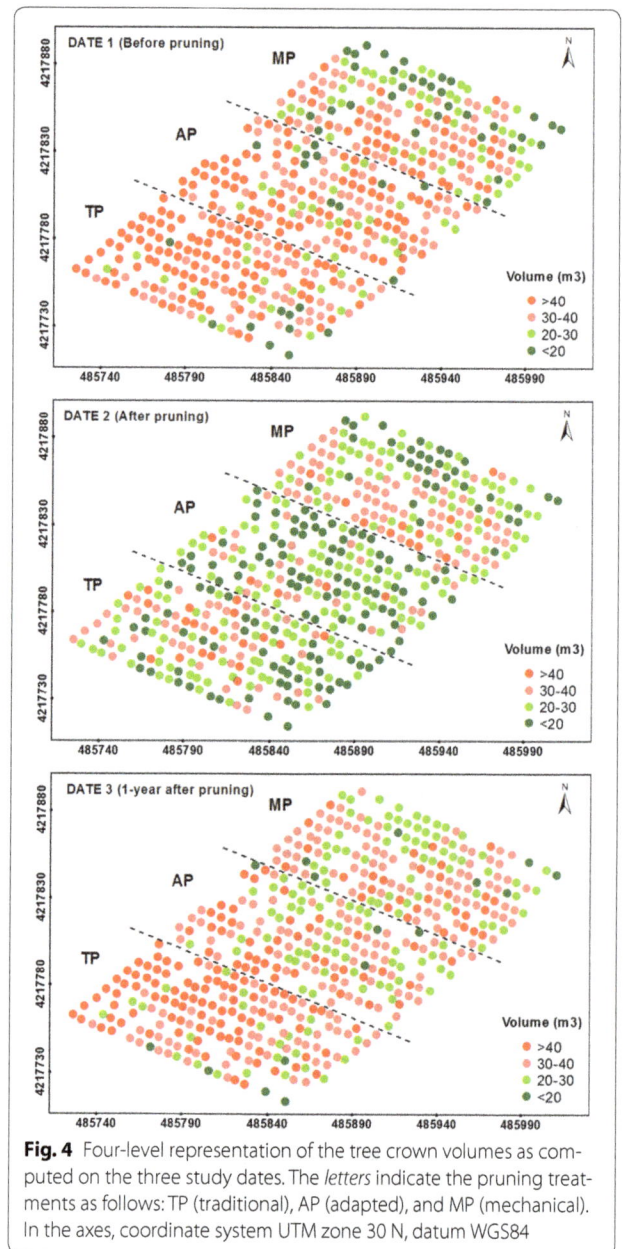

Fig. 4 Four-level representation of the tree crown volumes as computed on the three study dates. The *letters* indicate the pruning treatments as follows: TP (traditional), AP (adapted), and MP (mechanical). In the axes, coordinate system UTM zone 30 N, datum WGS84

while the trees that were pruned by up to 50% of their crown volume grew more than 75% for the same period. However, differences in tree growth were also observed among pruning treatments. Our results showed that the AP trees were relatively more productive in terms of vegetative growth than the other two treatments when the pruning intensity was low (<10%) and similar to that of the TP trees when the intensity was moderate (10–30%). By contrast, the TP generated more vegetative growth when the pruning intensity was very high (>50%), and similar to the MP when the intensity was high (30–50%).

Discussion

The direct assessment of the primary dimensions and certain geometric features of the olive trees such as the projected canopy area, height and crown volume is currently possible with the combined use of UAV imagery and advanced image processing and analysis procedures. This technology opens new opportunities to monitor tree status and progress at the field scale, as an efficient, objective and accurate alternative to the arduous and frequently inconsistent manual measurements on the ground [5]. This investigation takes advantage of this

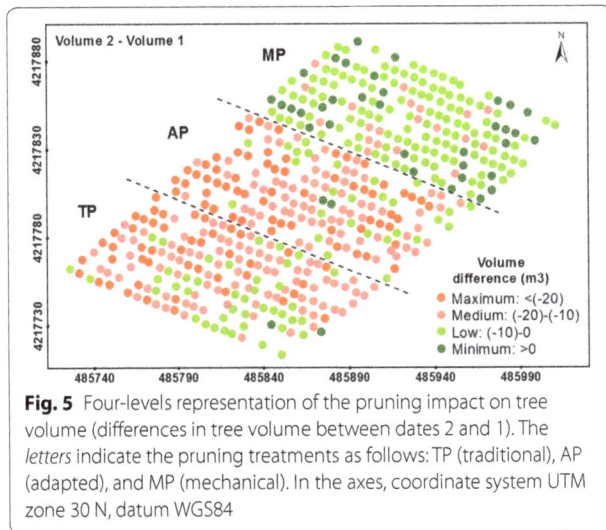

Fig. 5 Four-levels representation of the pruning impact on tree volume (differences in tree volume between dates 2 and 1). The *letters* indicate the pruning treatments as follows: TP (traditional), AP (adapted), and MP (mechanical). In the axes, coordinate system UTM zone 30 N, datum WGS84

innovative UAV-based technology to evaluate the impact of pruning treatments in an olive plantation, reporting quantitative data for an unprecedented number of trees.

From the multi-temporal analysis of the location and the 3D models of every tree in the olive orchard, the effect of the pruning intensity on the tree architecture and annual tree growth after pruning was quantified. AP and TP reduced the crown volume by approximately 42 and 36% on average, respectively, which is four and five-fold more compared to that of MP (approximately 10% on average). However, the trees that were subjected to TP grew slightly more than the AP trees, accounting for crown volume increases of approximately 62 and 59% on average, respectively. Regarding the effect of pruning on tree heights, important differences were only reported in the MP trees, which were homogenously cut to 3.5–4 m over the surface, as observed in Fig. 3. The data observed in this figure also confirmed the capacity of this UAV-based technology to accurately measure the tree heights. The trees also experienced minimum changes in terms of vertical growth after the pruning treatments, suggesting that the tree primarily grew along the horizontal axes. As expected, the trees that were subjected to more severe pruning generally experienced higher growth over time, although the magnitude of this event was affected by the given pruning treatment. In comparing the three pruning treatments, it was observed that AP benefitted tree development if the pruning intensity was lower than 10% of the crown volume, although this treatment was relatively worse in terms of tree growth when the pruning intensity was higher than 30%. These overall results were obtained in adult irrigated trees, so further analysis are needed to support such agronomic conclusions in other scenarios, e.g., under rain-fed conditions or with younger plantations.

Together with the evaluation of the pruning treatments, remote sensed information about the tree architecture at the orchard scale also has multiple implications for tree physiology, agronomy and field management [25] with potential applications for investigations about tree growth and yield [4], crown porosity [26], or the interception of solar irradiation [27, 28], among others. For example, the results presented here would improve the prediction models that connect the tree crown volume

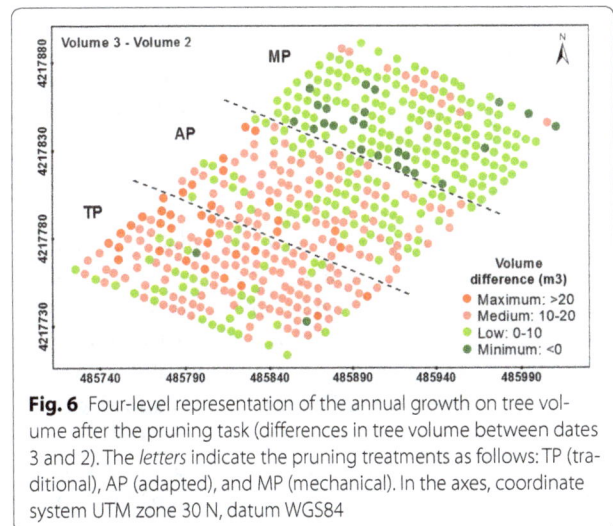

Fig. 6 Four-level representation of the annual growth on tree volume after the pruning task (differences in tree volume between dates 3 and 2). The *letters* indicate the pruning treatments as follows: TP (traditional), AP (adapted), and MP (mechanical). In the axes, coordinate system UTM zone 30 N, datum WGS84

Table 3 Impact of tree pruning treatment on the tree architecture, when computed as the differences in the projected canopy area, tree height, and crown volume between dates 2 (after pruning) and 1 (before pruning)

Pruning treatments	Area (Date 2 − Date 1)		Height (Date 2 − Date 1)		Volume (Date 2 − Date 1)	
	Mean ± SD (m²)	%[a]	Mean ± SD (m)	%[a]	Mean ± SD (m³)	%[a]
Mechanical	−0.89 ± 1.40	−6.97	−0.51 ± 0.40	−12.15	−3.79 ± 4.93	−9.89
Adapted	−5.56 ± 2.68	−38.95	−0.05 ± 0.38	−0.65	−17.46 ± 8.73	−42.05
Traditional	−5.00 ± 2.87	−33.02	−0.03 ± 0.31	−0.37	−15.58 ± 8.19	−35.72

[a] Average percentage of increase (+) or decrease (−) of each tree geometric feature between the date 2 and the date 1, as follow: % = (feature_date2 − feature_date1)/(feature_date1)

Table 4 Impact of tree pruning treatment on annual tree growth, computed as the differences of projected canopy area, tree height, and crown volume between the date 3 (1-year after pruning) and the date 2 (after pruning)

Pruning treatments	Area (Date 3 − Date 2)		Height (Date 3 − Date 2)		Volume (Date 3 − Date 2)	
	Mean ± SD (m²)	%[a]	Mean ± SD (m)	%[a]	Mean ± SD (m³)	%[a]
Mechanical	3.25 ± 1.50	37.37	−0.08 ± 0.19	−2.32	4.66 ± 4.60	23.65
Adapted	5.34 ± 1.71	71.14	−0.05 ± 0.24	−1.14	11.58 ± 4.98	58.51
Traditional	5.39 ± 2.03	64.27	0.06 ± 0.20	1.66	13.90 ± 6.06	61.49

[a] Average percentage of increase (+) or decrease (−) of each tree geometric feature between the date 3 and the date 2, as follow: % = (feature_date3 − feature_date2)/(feature_date2)

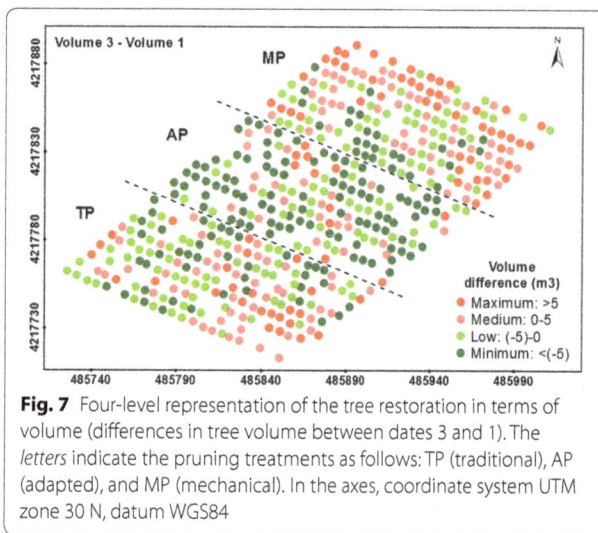

Fig. 7 Four-level representation of the tree restoration in terms of volume (differences in tree volume between dates 3 and 1). The *letters* indicate the pruning treatments as follows: TP (traditional), AP (adapted), and MP (mechanical). In the axes, coordinate system UTM zone 30 N, datum WGS84

harvesting. By contrast, a lack of pruning caused the crown to grow upward and away from the primary branches which resulted in defoliation due to a lack of light and from parasitic attack.

Additionally, quantifying the impact of pruning on the tree volume gives an estimated value of the available residual biomass [32], which could serve to calculate the potential energy from this raw material [33] or, furthermore, to evaluate the site-specific effects of the application of these by-products on the soil in no-till systems to prevent land degradation and improve the organic matter content [34–36]. The use of pruned residues as mulch is growing [37] and can help prevent pollutant dispersion in olive groves [38].

Geo-referenced maps with the locations and dimensions of every tree could also be the basis for designing a programme for a variable rate application of plant protection products [39], and in combination with on-ground equipment [40, 41], contribute to help fulfill the requirements of the European Directive for a Sustainable Use of Pesticides [42].

and tree canopy density with the tree yields, which is a complex issue since these models depend on a large number of factors [29]. Some investigations have also addressed the relationship of pruning treatments to tree productivity and mechanical harvesting [30]. Tombesi et al. [31] studied the influence of the canopy density on the efficiency of a trunk shaker after applying several pruning intensities, and they concluded that moderate and heavy annual pruning assisted mechanical

Conclusions

This investigation combined aerial images that were collected with an UAV on three different dates (before pruning, after pruning and 1-year after pruning), 3D models of the olive tree field that were created by photo-reconstruction procedures and an original OBIA algorithm

Table 5 Impact of the tree pruning treatment on tree restoration, when computed as the differences in the canopy area, tree height, and crown volume between date 3 (1-year after pruning) and date 1 (before pruning)

Pruning treatments	Area (Date 3 − Date 1)		Height (Date 3 − Date 1)		Volume (Date 3 − Date 1)	
	Mean ± SD (m²)	%[a]	Mean ± SD (m)	%[a]	Mean ± SD (m³)	%[a]
Mechanical	2.35 ± 1.76	26.14	−0.59 ± 0.42	−14.26	0.87 ± 6.47	10.26
Adapted	−0.23 ± 2.50	1.95	−0.10 ± 0.36	−1.99	−5.88 ± 7.81	−10.76
Traditional	0.39 ± 2.32	6.68	0.03 ± 0.28	1.01	−1.68 ± 6.80	0.29

[a] Average percentage of increase (+) or decrease (−) of each tree geometric feature between the date 3 and the date 1, as follow: % = (feature_date3 − feature_date1)/(feature_date1)

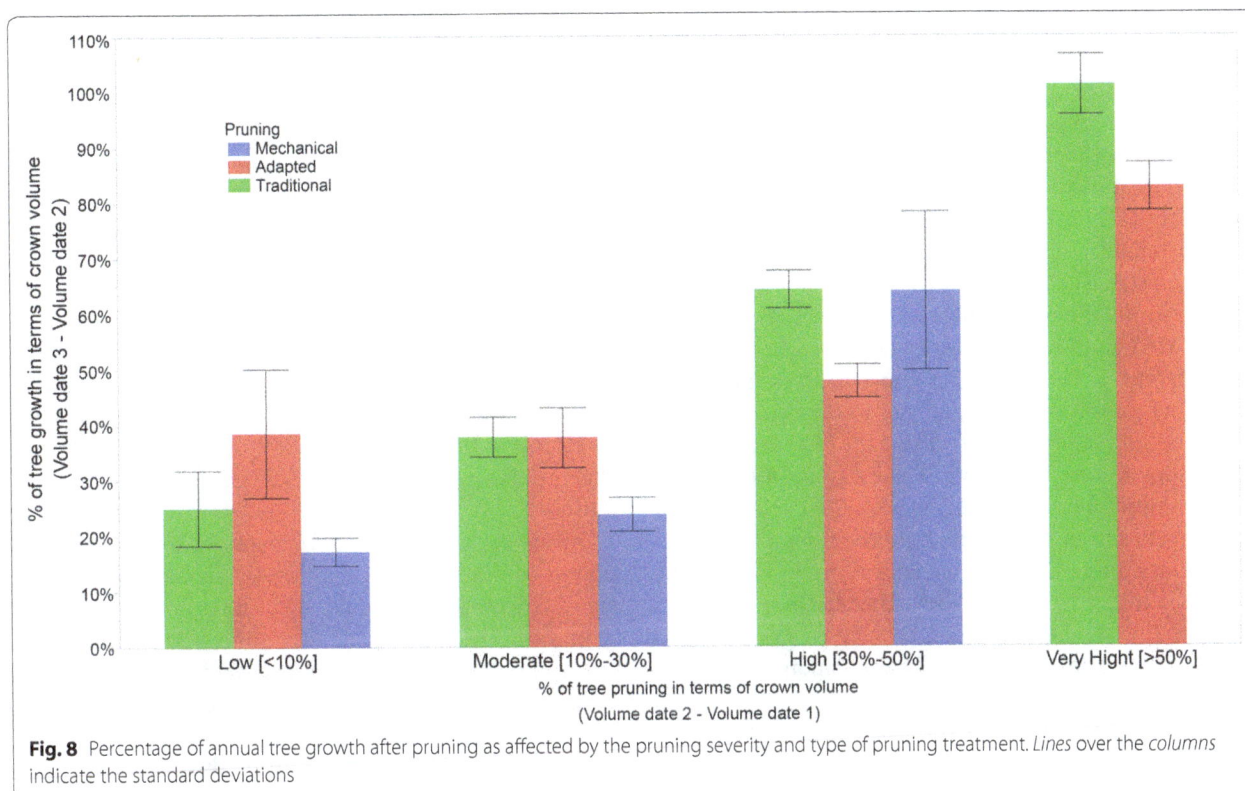

Fig. 8 Percentage of annual tree growth after pruning as affected by the pruning severity and type of pruning treatment. *Lines* over the *columns* indicate the standard deviations

Methods

Study area and description of the pruning treatments

to evaluate the impacts of three different pruning treatments (traditional, adapted and mechanical) on hundreds of irrigated trees. The projected canopy area, tree height and crown volume were quantified and compared among pruning treatments and flight dates.

The full procedure had high accuracy, and it correctly identified and measured every olive tree on the three dates with the exception of some cases when the 3D point cloud was incorrectly generated. As a general trend, the trees that were subjected to AP showed the highest foliage losses after pruning, followed by trees under TP. However, trees under TP experienced higher growths than the other trees for the quantification of this vegetative response one year after pruning. Due to the typical MP typology, the trees under this treatment maintained a more constant vegetative growth during this study.

This research offers valuable information for designing site-specific olive tree management strategies in the context of precision agriculture, which allows for the optimized application of agronomic tasks such as pruning, fertilization, pesticide use or irrigation, with the consequent economic and environmental benefits. The technology presented here can be made adaptable and transferable with corresponding adjustments to other woody crops such as vineyards or fruit orchards.

This research was performed in a commercial 20-year-old olive grove located in Villacarrillo, in the province of Jaen (southern Spain, central coordinates 485,885 m X, 4,217,810 m Y, system UTM zone 30 N, datum WGS84). A rectangular field of approximately 3 ha that was under drip-irrigation was selected for the experiment. This field was made up of 648 olive trees of the Arbequina variety, and it was laid out as 27 trees long and 24 across the field, with an intensive single-tree pattern of 8×4 m tree spacing. The field soil was loamy and silty clay loam, with at least 1.5 m deep without stoniness, and no limitations for crop production. The irrigation was controlled at a dose of approximately 800 m^3 per hectare, which corresponded to 3200 l per tree. A natural cover crop, 1.5 m wide and composed of grass and legume species, covered the soil among the tree lines. The cover crop was controlled with a brush-cutter, meanwhile herbicides were applied in early autumn and spring to control weeds under the olive trees. Fertigation was applied at 100 units of fertilizer per hectare.

The pruning strategy was part of a broader research program with the aim of studying the efficiency of different olive pruning and mechanical harvesting systems. Therefore, a simple demonstration strip design was selected in order to prioritize viability of mechanical pruning, which relies on continuous work at large areas, instead of designing a complex field experiment. The study field was divided in three sub-plots of 9 × 24 trees each, in which traditional, adapted and mechanical pruning treatments were separately performed on March 4th and 5th, 2015 (Fig. 9). In the TP, the highest branches, the crossed ones, and the ones below the base of the canopy and established at 60 cm over the soil were pruned. In the AP, the inner branches were totally removed, plus the crossed and low branches as described in the previous treatment. Thus, a large number of trees under this treatment presented a sizeable gap in the central crown part. The AP mainly aimed to adapt the olive architecture for canopy shaker harvesting. Finally, in the MP, a tractor with opposite mechanical cuts at a 30° angle removed the branches from 3.5 to 4 m above the terrain. This tractor also used a horizontal mechanical cut to remove the branches at less than 70 cm above the terrain.

Multi-temporal UAV flights and the generation of geo-spatial products

A set of remote images of the experimental field were acquired through UAV flights that were performed on the following three dates: (1) before tree pruning (date 1: December 9th, 2014), (2) a short time after tree pruning (date 2: April 14th, 2015), and (3) almost a year after tree pruning (date 3: February 1st, 2016). The flight equipment was a quadcopter UAV with vertical take-off and landing model MD4-1000 (Microdrones GmbH, Siegen, Germany), with a still point-and-shoot camera of model Olympus PEN E-PM1 (Olympus Corporation, Tokyo, Japan) (Fig. 10). This camera took 12.2 megapixel images in true color (Red, R; Green, G; and Blue, B, bands) with 8-bit radiometric resolution and at a 14 mm focal length. The camera's sensor size was 17.3 × 13.0 mm and the pixel size was 0.0043 mm. These parameters are needed to calculate the image resolution on the ground or, i.e., the ground sample distance as affected by the flight altitude.

During each flight, the UAV route was designed to take photos continuously at 1-s intervals, resulting in a forward lap of at least 90%. In addition, a side lap of 60% was programmed. The flight speed was 3 m/s and the flight altitudes were 100 m on the first and third dates and 50 meters on the second date. Due to the strong windy conditions, the 100-m flight was aborted on the second date. The UAV used a total of 24 and 15 min to fly the experiment field

Fig. 9 Description of the three pruning treatments evaluated in this investigation

Fig. 10 The quadcopter UAV and the Red–Green–Blue (RGB) camera used to acquire the remote images of the olive trees

Fig. 11 A partial view of the 3-D Point Cloud for the olive grove studied in this investigation, which was produced by the photogrammetric processing of the remote images taken with the UAV

at 50 and 100 m altitude, respectively. The area covered in each flight was 3.5 ha. To fully cover the experimental field, the camera collected 840 and 420 RGB images at ground sample distances of 1.90 and 3.81 cm/pixel for 50- and 100-m flight altitudes, respectively. The flight operations fulfilled the list of requirements established by the Spanish National Agency of Aerial Security, including the pilot license, safety regulations and limited flight distance [43]. The UAV flights were authorized by the person in charge of the olive grove as well.

In processing the set of UAV aerial images, the following three geo-spatial products of the olive grove were produced: (1) the 3D point cloud file, by applying the structure-from-motion technique (Fig. 11), (2) the DSM,

with height information, which was created from the 3D point cloud, and (3) the ortho-mosaicked image, with RGB information on every pixel. In this research, Agisoft PhotoScan Professional software, version 1.2.4 build 2399 (Agisoft LLC, St. Petersburg, Russia) was used for this task. The mosaicking process was fully automatic with the exception of the manual localization of six ground control points that were used to georeference the products. These ground control points were located in the corners and the center of the olive orchard, and their coordinates were taken with a GPS device after the flight operations. The automatic process involved the following three phases: (1) aligning images, (2) building field geometry, and (3) ortho-photo generation. The common points and the camera position for each image were located and matched, which facilitated the refinement of the camera calibration parameters. Once the images were aligned, the point cloud was generated. Next, the DSM was built on the basis of the estimated camera position and the images themselves. This process requires high computational resources and it can usually take approximately 5–6 h due to the use of many high-resolution images. Finally, the images were projected over the DSM, and the ortho-mosaicked image was generated. The DSM is a 3D polygon mesh that represents the overflown area and reflects the irregular geometry of the ground and the tree crowns. The DSM was joined to the ortho-mosaic in the form of TIFF files, which produced a 4-band multi-layer file (Red, Green, Blue and DSM). More information about the PhotoScan function is described in Dandois and Ellis [44].

Object-based image analysis algorithm for computing the 3D tree features

An innovative algorithm based on the OBIA paradigm was applied to the 4-band multi-layer file that was created during the previous stage to classify and identify every individual tree in the olive grove and to compute the tree geographic position and primary 3D geometric features, including the projected canopy area, the tree height and the crown volume. The OBIA algorithm was developed using eCognition Developer 9 software (Trimble GeoSpatial, Munich, Germany) and it was adapted from the basic version described in Torres-Sánchez et al. [15]. However, the procedure presented here was original and included improvements and variations related to the specifications of this research (Fig. 12).

The algorithm was specifically programmed to run in a fully automatic manner without the need for user intervention, and with the ability to be auto-adaptive to any olive grove, independent of the plantation pattern and of the given pruning treatment. The full procedure was composed of a sequence of phases (Fig. 13), which is described as follows:

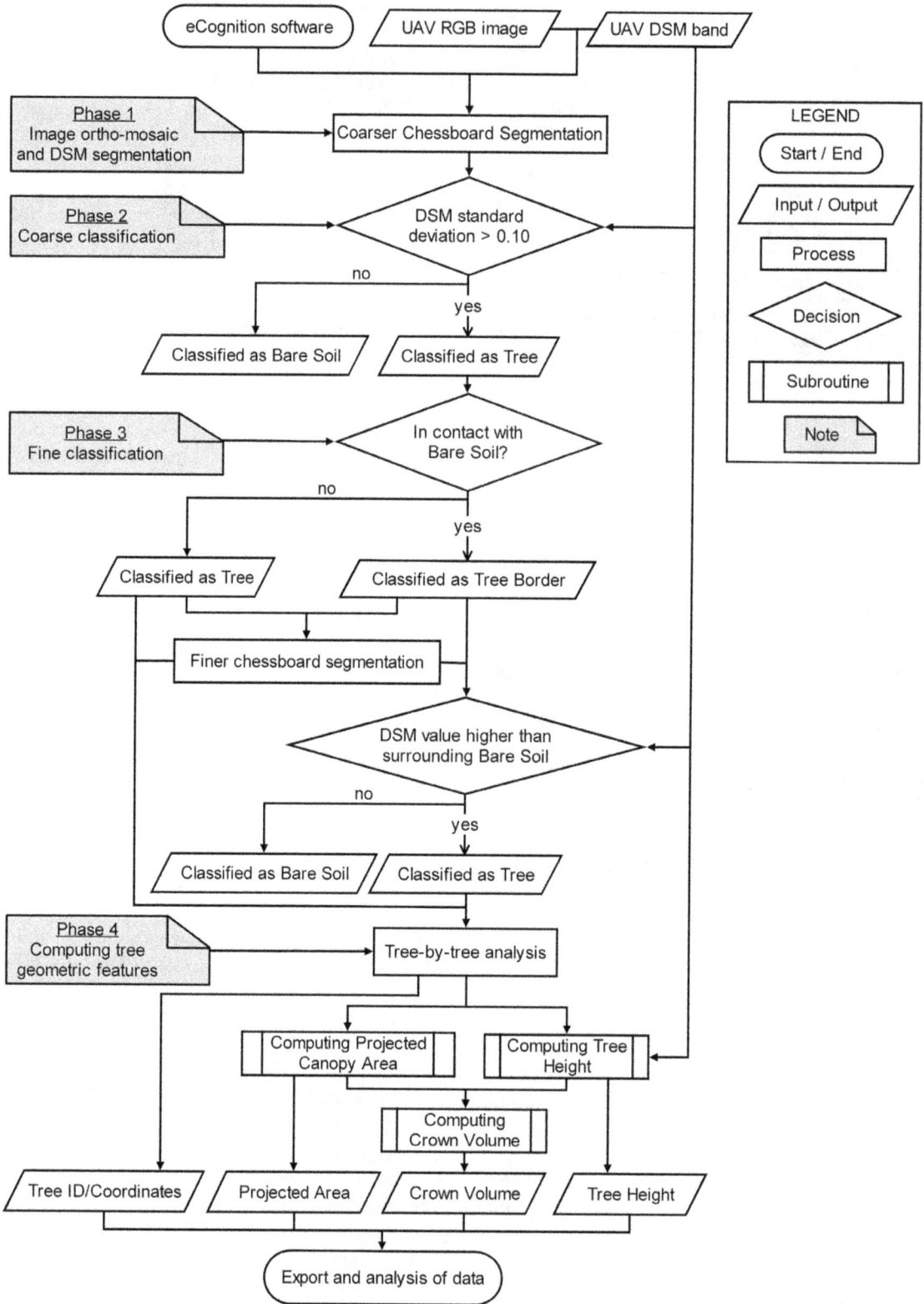

Fig. 12 Flowchart of the OBIA algorithm developed in this investigation

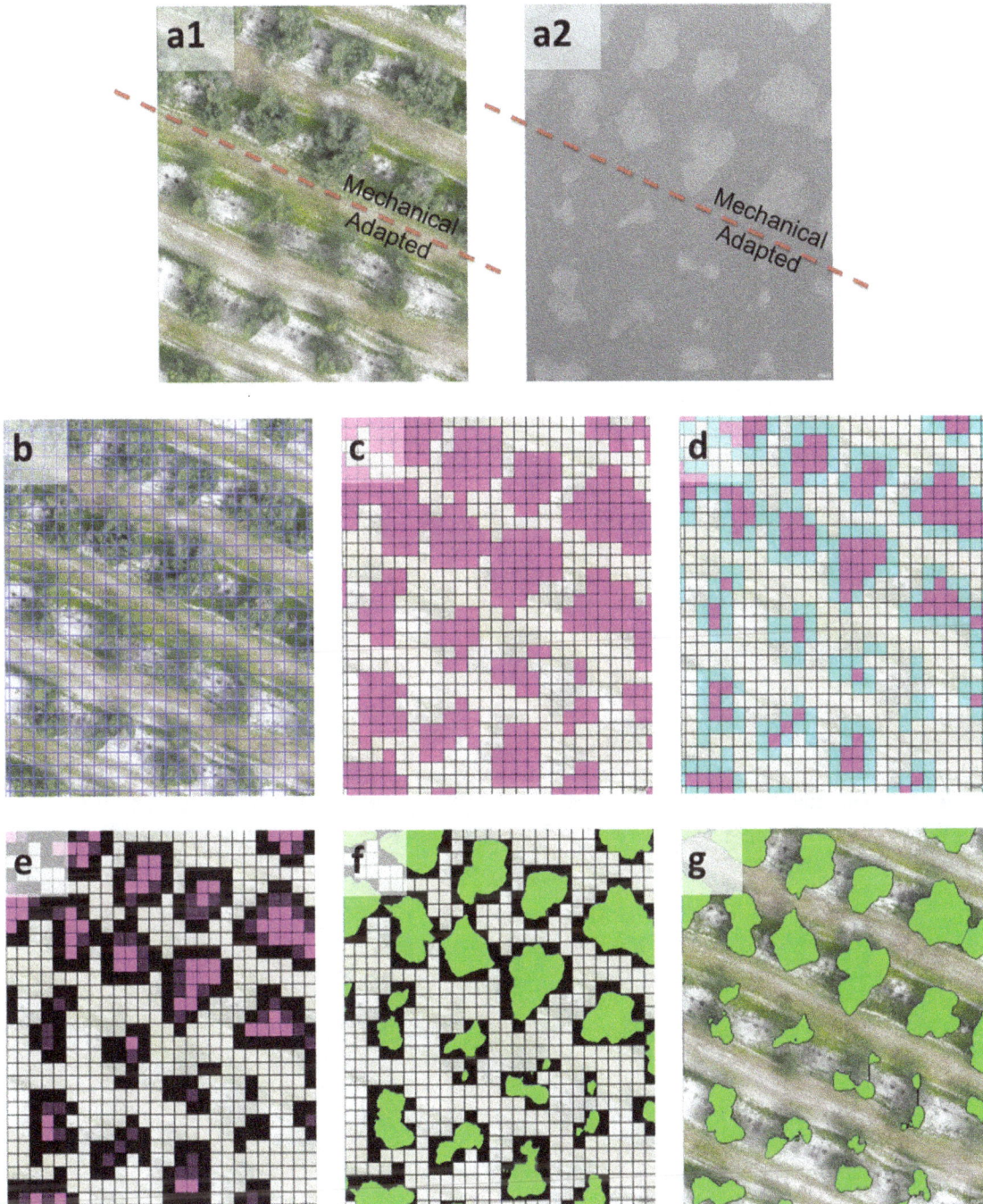

Fig. 13 Partial views of the primary OBIA algorithm outputs: **a** the 4-band multi-layer file with the RGB (**a1**) and the DSM (**a2**) layers, showing the results of mechanical and adapted pruning as applied to the trees on the two *top* and *bottom rows*, respectively; **b** chessboard segmentation output; **c** the coarse classification of the tree (*pink colour*) and bare soil (*white colour*) objects based on the difference in DSM (height) values; **d** the coarse classification of the tree borders (*blue color*); **e** pixel-based segmentation of the tree borders; **f** the fine classification of the tree (*green colour*) objects; and **g** the tree and the bare-soil objects were joined with separately. As a result of the whole procedure, the algorithm computed the 3D tree geometric features (projected area, height and volume) and exported the values as vector and table files for further analysis

Phase 1, image ortho-mosaic and DSM segmentation The 4-band (B, G, R, and DSM) multi-layer file (Fig. 13a1–a2) was segmented into 1-m² square objects by using the chessboard segmentation process (Fig. 13b). Because segmentation is by far the slowest task of the full OBIA procedure, the algorithm used chessboard segmentation instead of the multi-resolution option. In addition, it was programmed to only use the DSM band as the reference for the segmentation, which weighted the variable "height" instead of the spectral information. This configuration produced a notable decrease in computational demand, and consequently, an increase in the processing speed, without penalizing the segmentation accuracy.

Phase 2, coarse classification of trees and bare soil The segmented objects, whose standard deviation (SD) value of the DSM (height) layer was greater than 0.10 m, were classified as trees. The remaining objects were classified as bare soil (Fig. 13c).

Phase 3, fine classification of trees: To refine the tree delineation, the tree objects were analyzed at the pixel level. Firstly, the tree objects that were in contact with bare soil were classified as tree border objects (Fig. 13d), and next, they were segmented at the pixel size (Fig. 13e). Then, the algorithm classified every tree border object as a tree or bare soil by comparing their DSM value to the surrounding bare soil and tree DSM values (Fig. 13f). Finally, the objects classified as trees were joined into single objects and identified as individual trees (Fig. 13g).

Phase 4, computing the tree geometric features The algorithm automatically calculated the geometric features (projected canopy area, tree height and crown volume) of all the tree objects by applying a looping process in which every tree was individually identified and analyzed. During this sequential process, the height of every tree was obtained by comparing its maximum DSM value to the average DSM values of a bare soil area with a 1 m buffer surrounding each tree. Simultaneously, the crown volume was calculated by adding up the volumes (by multiplying the pixel areas and heights) of all the pixels corresponding to every tree. Finally, the OBIA algorithm automatically exported the identification, location and the three primary geometric features of every tree as vector (e.g., shapefile format) and table (e.g., Excel or ASCII format) files for further analysis.

Data analysis

The outputs delivered by the OBIA algorithm for the three study dates were subjected to descriptive analysis with JMP version 10 software (SAS Institute Inc., Cary, NC, USA). The impacts of pruning on the tree architecture and annual tree growth was separately evaluated at each pruning zone by analyzing tree-by-tree variability over time, i.e., by quantifying the differences in the three primary dimensions (projected area, tree height and crown volume) between dates 1 and 2 (the impact on the tree architecture) and between dates 2 and 3 (the impact on the annual tree growth). In addition, the ability of the trees to return to their original dimensions before the pruning task was quantified by comparing the data on

dates 1 and 3. The field experimental design, which prioritized mechanical pruning viability rather than testing statistical hypothesis, allowed for ranking pruning treatments according to averaged values obtained in every study date and to trends observed over time.

Abbreviations
3D: three dimensions; AP: adapted pruning; DSM: digital surface model; GPS: global position system; LiDAR: light detection and ranging; MP: mechanical pruning; OBIA: object-based image analysis; RGB: red–green–blue bands; SD: standard deviation; TP: traditional pruning; UAV: unmanned aerial vehicles.

Authors' contributions
FLG, NS and JMP conceived and designed the experiments. All authors performed the experiments. FMJB, JTS and JMP analyzed the data. FLG and JMP provided equipment, materials and analysis tools. FMJB, FLG and JMP wrote the manuscript. All authors read and approved the final manuscript.

Author details
[1] Institute for Sustainable Agriculture, CSIC, 14004 Córdoba, Spain. [2] Institute of Agricultural Research and Training (IFAPA), 14004 Córdoba, Spain. [3] Institute of Agricultural Sciences, CSIC, 28006 Madrid, Spain.

Acknowledgements
The authors thank Ms. Irene Borra-Serrano and Mr. Juan José Caballero-Novella for their very helpful assistance during the field work.

Competing interests
The authors declare that they have no competing interests.

Funding
This research was funded by the AGL2014-52465-C4-4R Project (The Spanish Ministry of Economy, Industry and Competitiveness, MINECO) and the CSIC-Intramural-Project (Ref. 201640E034). The research performed by Dr. de Castro, Dr. Torres-Sánchez and Dr. Peña was financed by the Juan de la Cierva, FPI and Ramón y Cajal Programmes (Spanish MINECO funds), respectively. We acknowledge support of the publication fee by the CSIC Open Access Publication Support Initiative through its Unit of Information Resources for Research (URICI).

References
1. Ferguson L, Glozer K, Crisosto C, Rosa UA, Castro-García S, Fichtner EJ, et al. Improving canopy contact olive harvester efficiency with mechanical pruning. Acta Hortic. 2012;965:83–7.
2. Connor DJ, Gómez-del-Campo M, Rousseaux MC, Searles PS. Structure, management and productivity of hedgerow olive orchards: a review. Sci Hortic. 2014;169:71–93.
3. Castillo-Ruiz FJ, Jiménez-Jiménez F, Blanco-Roldán GL, Sola-Guirado RR, Agüera-Vega J, Castro-García S. Analysis of fruit and oil quantity and quality distribution in high-density olive trees in order to improve the mechanical harvesting process. Span J Agric Res. 2015;13:e0209.
4. Villalobos FJ, Testi L, Hidalgo J, Pastor M, Orgaz F. Modelling potential growth and yield of olive (*Olea europaea* L.) canopies. Eur J Agron. 2006;24:296–303.

5. Miranda-Fuentes A, Llorens J, Gamarra-Diezma JL, Gil-Ribes JA, Gil E. Towards an optimized method of olive tree crown volume measurement. Sensors. 2015;15:3671–87.

6. West PW. Tree and forest measurement. Berlin: Springer; 2009. doi:10.1007/978-3-540-95966-3.

7. Rosell JR, Sanz R. A review of methods and applications of the geometric characterization of tree crops in agricultural activities. Comput Electron Agric. 2012;81:124–41.

8. Friedli M, Kirchgessner N, Grieder C, Liebisch F, Mannale M, Walter A. Terrestrial 3D laser scanning to track the increase in canopy height of both monocot and dicot crop species under field conditions. Plant Methods. 2016;12:9.

9. Rovira-Más F, Zhang Q, Reid JF. Stereo vision three-dimensional terrain maps for precision agriculture. Comput Electron Agric. 2008;60:133–43.

10. Torres-Sánchez J, Peña JM, de Castro AI, López-Granados F. Multi-temporal mapping of the vegetation fraction in early-season wheat fields using images from UAV. Comput Electron Agric. 2014;103:104–13.

11. Xiang H, Tian L. Development of a low-cost agricultural remote sensing system based on an autonomous unmanned aerial vehicle (UAV). Biosyst Eng. 2011;108:174–90.

12. Zhang C, Kovacs JM. The application of small unmanned aerial systems for precision agriculture: a review. Precis Agric. 2012;13:693–712.

13. Nex F, Remondino F. UAV for 3D mapping applications: a review. Appl Geomat. 2014;6:1–15.

14. Díaz-Varela RA, de la Rosa R, León L, Zarco-Tejada PJ. High-resolution airborne UAV imagery to assess olive tree crown parameters using 3D photo reconstruction: application in breeding trials. Remote Sens. 2015;7:4213–32.

15. Torres-Sánchez J, López-Granados F, Serrano N, Arquero O, Peña JM. High-throughput 3-D monitoring of agricultural-tree plantations with unmanned aerial vehicle (UAV) technology. PLoS ONE. 2015;10:e0130479.

16. Zarco-Tejada PJ, Díaz-Varela R, Angileri V, Loudjani P. Tree height quantification using very high resolution imagery acquired from an unmanned aerial vehicle (UAV) and automatic 3D photo-reconstruction methods. Eur J Agron. 2014;55:89–99.

17. Kattenborn T, Sperlich M, Bataua K, Koch B. Automatic single tree detection in plantations using UAV-based photogrammetric point clouds. In: ISPRS—the international archives of the photogrammetry, remote sensing and spatial information sciences. Copernicus GmbH; 2014 [cited 2017 Feb 10]. p. 139–44. http://www.int-arch-photogramm-remote-sens-spatial-inf-sci.net/XL-3/139/2014/.

18. Guerra-Hernández JG, González-Ferreiro E, Sarmento A, Silva J, Nunes A, Correia AC, et al. Short Communication. Using high resolution UAV imagery to estimate tree variables in *Pinus pinea* plantation in Portugal. For Syst. 2016;25:09.

19. Castillejo-González IL, Peña-Barragán JM, Jurado-Expósito M, Mesas-Carrascosa FJ, López-Granados F. Evaluation of pixel- and object-based approaches for mapping wild oat (*Avena sterilis*) weed patches in wheat fields using QuickBird imagery for site-specific management. Eur J Agron. 2014;59:57–66.

20. Blaschke T, Hay GJ, Kelly M, Lang S, Hofmann P, Addink E, et al. Geographic object-based image analysis—towards a new paradigm. ISPRS J Photogramm Remote Sens. 2014;87:180–91.

21. López-Granados F, Torres-Sánchez J, Castro A-ID, Serrano-Pérez A, Mesas-Carrascosa F-J, Peña J-M. Object-based early monitoring of a grass weed in a grass crop using high resolution UAV imagery. Agron Sustain Dev. 2016;36:67.

22. Peña JM, Torres-Sánchez J, de Castro AI, Kelly M, López-Granados F. Weed mapping in early-season maize fields using object-based analysis of unmanned aerial vehicle (UAV) images. PLoS ONE. 2013;8:e77151.

23. Qin R. An object-based hierarchical method for change detection using unmanned aerial vehicle images. Remote Sens. 2014;6:7911–32.

24. Rousseau D, Chéné Y, Belin E, Semaan G, Trigui G, Boudehri K, et al. Multiscale imaging of plants: current approaches and challenges. Plant Methods. 2015;11:6.

25. Usha K, Singh B. Potential applications of remote sensing in horticulture—a review. Sci Hortic. 2013;153:71–83.

26. Castillo-Ruiz FJ, Castro-García S, Blanco-Roldán GL, Sola-Guirado RR, Gil-Ribes JA. Olive crown porosity measurement based on radiation transmittance: an assessment of pruning effect. Sensors. 2016;16:723.

27. Cherbiy-Hoffmann SU, Searles PS, Hall AJ, Rousseaux MC. Influence of light environment on yield determinants and components in large olive hedgerows following mechanical pruning in the subtropics of the Southern Hemisphere. Sci Hortic. 2012;137:36–42.

28. Mariscal MJ, Orgaz F, Villalobos FJ. Modelling and measurement of radiation interception by olive canopies. Agric For Meteorol. 2000;100:183–97.

29. Álamo S, Ramos MI, Feito FR, Cañas JA. Precision techniques for improving the management of the olive groves of southern Spain. Span J Agric Res. 2012;10:583–95.

30. Farinelli D, Onorati L, Ruffolo M, Tombesi A. Mechanical pruning of adult olive trees and influence on yield and on efficiency of mechanical harvesting. Acta Hortic. 2011;924:203–9.

31. Tombesi A, Boco M, Pilli M, Farinelli D. Influence of canopy density on efficiency of trunk shaker on olive mechanical harvesting. Acta Hortic. 2002;586:291–4.

32. Velázquez-Martí B, Fernández-González E, López-Cortés I, Salazar-Hernández DM. Quantification of the residual biomass obtained from pruning of trees in Mediterranean olive groves. Biomass Bioenergy. 2011;35:3208–17.

33. Bilandzija N, Voca N, Kricka T, Matin A, Jurisic V. Energy potential of fruit tree pruned biomass in Croatia. Span J Agric Res. 2012;10:292–8.

34. Gómez-Muñoz B, Valero-Valenzuela JD, Hinojosa MB, Garcia-Ruiz R. Management of tree pruning residues to improve soil organic carbon in olive groves. Eur J Soil Biol. 2016;74:104–13.

35. Repullo MA, Carbonell R, Hidalgo J, Rodríguez-Lizana A, Ordóñez R. Using olive pruning residues to cover soil and improve fertility. Soil Tillage Res. 2012;124:36–46.

36. Rodríguez-Lizana A, Pereira MJ, Ribeiro MC, Soares A, Márquez-García F, Ramos A, et al. Assessing local uncertainty of soil protection in an olive grove area with pruning residues cover: a geostatistical cosimulation approach: assessing soil protection uncertainty through stochastic simulations. Land Degrad Dev. 2017. doi:10.1002/ldr.2734.

37. Calatrava J, Franco JA. Using pruning residues as mulch: analysis of its adoption and process of diffusion in Southern Spain olive orchards. J Environ Manag. 2011;92:620–9. doi:10.1016/j.jenvman.2010.09.023.

38. Rodríguez-Lizana A, Espejo-Pérez AJ, González-Fernández P, Ordóñez-Fernández R. Pruning residues as an alternative to traditional tillage to reduce erosion and pollutant dispersion in olive groves. Water Air Soil Pollut. 2008;193:165–73. doi:10.1007/s11270-008-9680-5.

39. Miranda-Fuentes A, Llorens J, Rodríguez-Lizana A, Cuenca A, Gil E, Blanco-Roldán GL, et al. Assessing the optimal liquid volume to be sprayed on isolated olive trees according to their canopy volumes. Sci Total Environ. 2016;568:296–305.

40. Pérez-Ruiz M, González-de-Santos P, Ribeiro A, Fernández-Quintanilla C, Peruzzi A, Vieri M, et al. Highlights and preliminary results for autonomous crop protection. Comput Electron Agric. 2015;110:150–61.

41. Rosell-Polo JR, Auat Cheein F, Gregorio E, Andújar D, Puigdomènech L, Masip J, et al. Chapter Three—Advances in structured light sensors applications in precision agriculture and livestock farming. In: Sparks DL, editor. Advances in agronomics. London: Academic Press; 2015. p. 71–112.

42. European Commission. Directive 2009/128/EC of the European Parliament and of the Council establishing a framework for Community action to achieve the sustainable use of pesticides. Cited 2017 Jun 2. http://www.fao.org/faolex/results/details/en/?details=LEX-FAOC113943.

43. AESA. Aerial work—legal framework. Cited 2017 Jun 2. http://www.seguridadaerea.gob.es/LANG_EN/cias_empresas/trabajos/rpas/marco/default.aspx.

44. Dandois JP, Ellis EC. High spatial resolution three-dimensional mapping of vegetation spectral dynamics using computer vision. Remote Sens Environ. 2013;136:259–76.

13

Use of ultrasonication to increase germination rates of Arabidopsis seeds

Ignacio López-Ribera and Carlos M. Vicient[*] (iD)

Abstract

Background: *Arabidopsis thaliana* is widely used as model organism in plant biology. Although not of agronomic significance, it offers important advantages for basic research in genetics and molecular biology including the availability of a large number of mutants and genetically modified lines. However, Arabidopsis seed longevity is limited and seeds stored for more than 10 years usually show a very low capacity for germination.

Results: The influence of ultrasonic stimulation was investigated on the germination of *A. thaliana* L. seeds. All experiments have been performed using a frequency of 45 kHz at constant temperature (24 °C). No germination rate differences were observed when using freshly collected seeds. However, using artificially deteriorated seeds, our results show that short ultrasonic stimulation (<1 min) significantly increased germination. Ultrasonic stimulation application of 30 s is the optimal treatment. A significant increase in the germination rate was also verified in naturally aged seeds after ultrasonic stimulation. Scanning electron microscopy observations showed an increase in the presence of pores in the seed coat after sonication that may be the cause, at least in part, of the increase in germination. The ultrasound treated seeds developed normally to mature fertile plants.

Conclusions: Ultrasound technology can be used to enhance the germination process of old Arabidopsis seeds without negatively affecting seedling development. This effect seems to be, at least in part, due to the opening of pores in the seed coat. The use of ultrasonic stimulation in Arabidopsis seeds may contribute to the recovering of long time stored lines.

Keywords: Ultrasound, Arabidopsis, Germination, Seed, Ageing

Background

Arabidopsis thaliana L. is an annual weed from the *Brassicaceae* family that lives in mild to cold climates and is widely distributed in Europe, North America and Asia. *A. thaliana* L. is universally recognized as a model for plant molecular biology and genetic studies. Although it is a non-commercial plant, it is favored among basic scientists because its short lifetime cycle, it is easy and inexpensive to grow, produces many seeds and contains a comparatively small genome. This allows large genetic experiments often involving thousands of plants [1]. Unfortunately, Arabidopsis as a model plant has also some disadvantages, being probably one of the worst the

difficulties in the long-term seed conservation without loss of viability [2]. When the *A. thaliana* seeds are stored for more than 10 years usually they retain low capacity for germination.

Many invigoration treatments of seeds, referred to as seed priming, have been used to increase and/or accelerate germination [3], as, for example, the addition of chemicals, plant hormones or by controlled hydration. The addition of gibberellic acid is probably the most effective method, but is time consuming and relatively expensive. Other chemical methods may add undesirable residues to the culture. Germination may also be stimulated by physical methods as, for example, heat treatments, ionizing radiation [4] or vacuum [5].

Sound is a vibration that propagates as a mechanical wave of pressure through a medium such as air or water. Sound that is perceptible by humans has frequencies

*Correspondence: Carlos.vicient@cragenomica.es
Centre for Research in Agricultural Genomics (CRAG) CSIC-IRTA-UAB-UB, Campus UAB Bellaterra, 08193 Barcelona, Spain

from 20 to 20,000 Hz. Ultrasounds are acoustic waves at frequencies higher than 20 kHz. Ultrasounds are often used in the agro-industry in order to enhance processes such as drying, extraction, emulsification and defoaming [6]. Ultrasound treatments have been reported to stimulate germination in different types of plants, such as *Calanthe* hybrids, bean [7], corn [8], barley [5], fern spores [9], alfalfa and broccoli [10], chickpea [11, 12], sorghum [13], navy beans [14], wheat, watermelon and pepper [15, 16], but not in Arabidopsis. This research investigates the effects of ultrasound treatments on germination of Arabidopsis seeds.

Results

To determine the effect of ultrasounds on the germinability of Arabidopsis seeds, recently collected seeds were subjected to different times of sonication, and germination (radicle protrusion) was assessed at different times (1, 1.5, 2, 2.5, 3, 3.5, 7 days) (Fig. 1). Short sonication (1 min) has no significant effect on germination compared to the control, but 2 min or longer sonication treatments led to a decline of germination. Only 2 min of ultrasounds reduced germination in about 10%. Surprisingly, about 25% of the seeds treated during 1 h were able to germinate after 7 days, however, none of these seedlings were able to develop a mature plant. On the contrary, all the seeds treated during one minute developed normal and fertile plants.

Next, we tested if ultrasounds increases the germination rate of aged seeds. First, we used artificially aged Arabidopsis seeds that were subjected to aging treatment for different periods of time (6 h–4 days). The time courses of germination varied depending on the length of the ageing treatment (Fig. 2). The 6 h treatment reduced germination at 7 days only about 5% respect to the control non-aged seeds. The 1 and 4 days ageing treatments produced a much more dramatic effect on seed germination with reductions of approximately 30 and 50%, respectively, after 7 days (Fig. 2). The seeds subjected to 4 days of artificially ageing treatment were selected for the subsequent assays.

We performed ultrasounds treatments on artificially aged and non-aged seeds using three different sonication times: 30 s, 1 and 2 min. As we previously noted, no significant differences were observed in recently collected seeds except for the 2 min treated seeds in which an approximately 10% reduction in germination was observed (Fig. 3). Different results were observed for the artificially aged seeds. First, the seeds subjected to the artificial aged treatment showed a strong reduction in their germinability compared to the non-treated seeds. Second, short ultrasound treatments (30 and 60 s) produced significant ($p < 0.05$) increases of approximately 10% in the germinability compared to the non-ultrasound treated seeds (Fig. 3). Aged seeds treated for two minutes showed a similar germination percentage as aged seed controls.

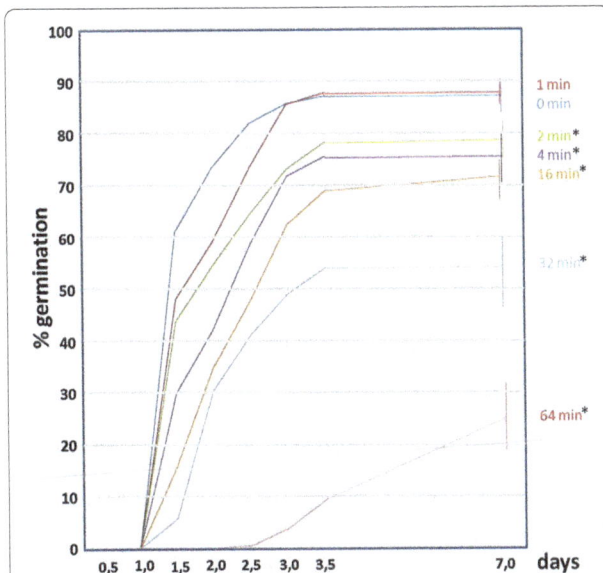

Fig. 1 Effect of ultrasounds on the germination percentage of seeds of *A. thaliana*. 30 seeds were subjected to 45 kHz sonication at 24 °C during the indicated periods of time. *Each point* shows a mean of ten independent samples. *Vertical bars* indicate standard error. For clarity, SD is only shown for the 7 days data. *Asterisks* indicate significant differences respect to the control (untreated seeds) according to the *t* test (p < 0.05)

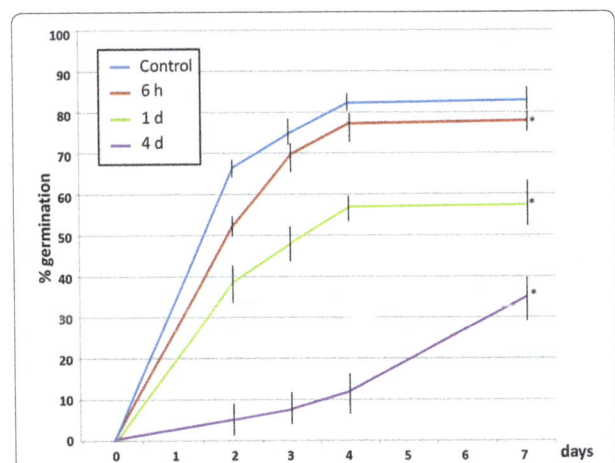

Fig. 2 Effect of ageing treatment (40 °C and 100% relative humidity) on the germination percentage of seeds of *A. thaliana*. 30 seeds were subjected to the ageing treatment during the indicated periods of time. *Each point* shows a mean of ten independent samples. *Vertical bars* indicate standard error. *Asterisks* indicate significant differences respect to the control (untreated seeds) according to the *t* test (p < 0.05)

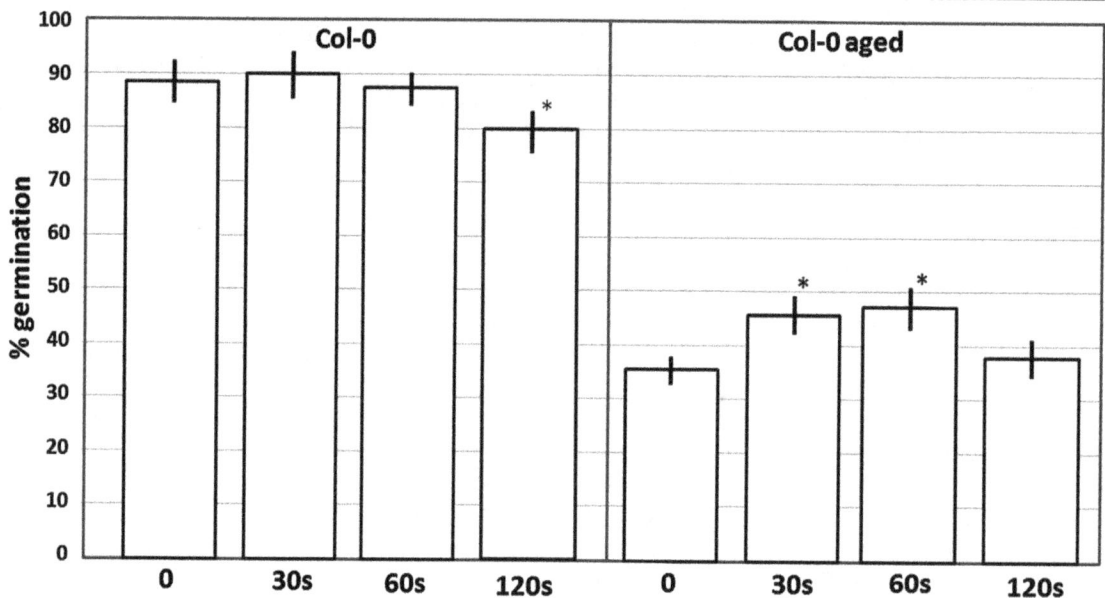

Fig. 3 Effect of ultrasounds on the germination percentage of seeds of *A. thaliana* (Col-0) and on the same seeds that have been artificially aged. Artificial ageing treatment consisted in 4 days at 40 °C and 100% relative humidity (Col-0 aged). 30 seeds were subjected to 45 kHz sonication at 24 °C during the indicated periods of time. *Each point* shows a mean of ten independent samples. *Vertical bars* indicate standard error. *Asterisks* indicate significant differences respect to the control (the corresponding ultrasound untreated seeds) according to the *t* test (p < 0.05)

Next, we tested if ultrasound treatment also induces an increase in the germination rate of "naturally" aged Arabidopsis seeds. Arabidopsis seeds collected 1, 9, 11, 13 and 16 years ago and keep at room temperature in 1.5 ml plastic tubes were tested. In this case we chose a treatment time of 30 s because although 60 s gave a slightly better result in artificially aged seeds the difference was minimal and not significant, and in that way we tried to minimize any possible side effect that the ultrasound treatment could produce. As expected, germination rates decreased with age (Fig. 4). Except for the 1-year old seeds, in all the other cases the sonicated seeds showed a significant increase in the germination rate, being the differences higher in the 11 and 13 years old seeds.

We closely examined the control and ultrasound treated Arabidopsis seeds using scanning electron microscopy (SEM) (Fig. 5). The surface of the ultrasound treated seeds presented many small pores (Fig. 5a–c). These pores are not present in the surface of the dry seeds (Fig. 5d, e) or embedded non-germinating untreated seeds (Fig. 5f, g), and are only present in small number in the untreated germinating seeds (Fig. 5h, i).

Discussion

Seed ageing decreases the quality of seeds and results in agricultural and economic losses, and it is also a serious problem in the research laboratories. *A. thaliana*

is widely used in plant basic research. It was the first plant whose complete genome was sequenced [20]. Since Arabidopsis is relatively easy to mutagenize and genetically transform, thousands of mutants and genetically modified lines are available at seed repositories. However, storage of Arabidopsis seeds is problematic because, even in optimal conditions, they deteriorate relatively quickly, losing their germination ability. This is a serious problem to seed stock centers, but also to individual researchers in order to conserve their precious Arabidopsis lines.

A series of methods collectively known as priming have been developed in order to induce a faster seed germination, a higher seed germination rate or a better seedling growth. Priming can be used to promote germination of seeds after a long storage period [21]. The positive effects of priming treatments on seed performance have been demonstrated in many species. One of the problems of priming is that some of these treatments are expensive and/or time-consuming. On the contrary, ultrasound treatments are easy, cheap (basically only the price of the ultrasound generator), and quick. Ultrasound treatments have shown promising results on other species [19], but no information is available on the effects of ultrasounds on the aged seeds of the model plant *A. thaliana*. In this study we showed that ultrasound treatment significantly increases Arabidopsis seed germination in artificially and naturally aged seeds.

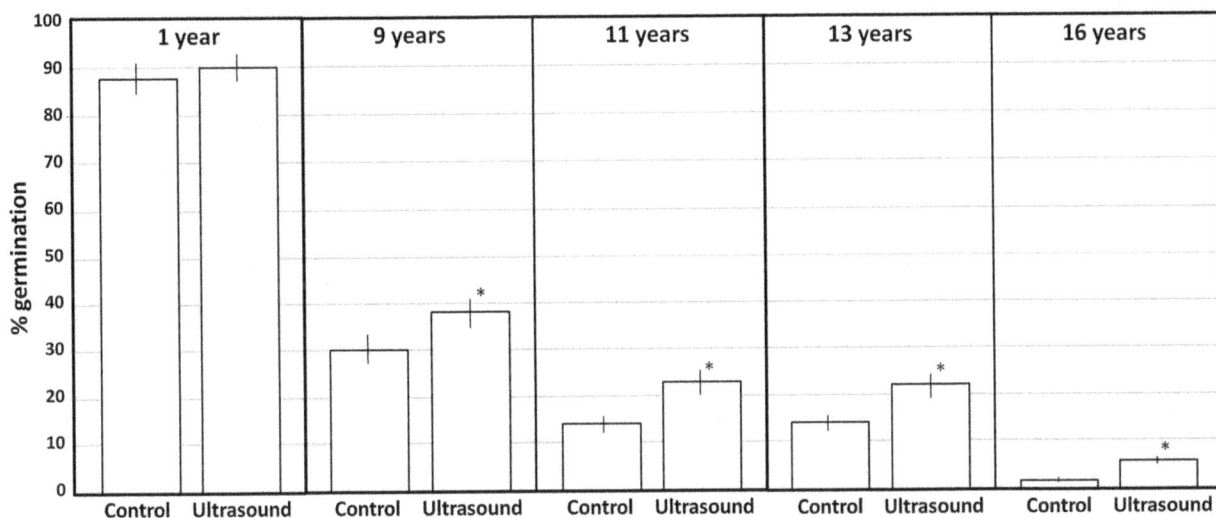

Fig. 4 Effect of ultrasounds on the germination percentage of seeds of *A. thaliana* (Col-0) collected at different years before testing. 30 seeds were subjected to 45 kHz sonication at 24 °C for 30 s. *Each point* shows a mean of 30 independent samples. *Vertical bars* indicate standard error. *Asterisks* indicate significant differences respect to the control (the corresponding ultrasound untreated seeds) according to the *t* test (p < 0.05)

The described possible effects of ultrasounds on plant tissues are multiple [19]: heat, inactivation of microorganisms and enzymes, acceleration of certain metabolic reactions, induced cavitation, etc. Cavitation consists in the formation of vapour cavities (bubbles) in a liquid that are the consequence of forces acting upon the liquid. The ultrasound induces in the liquid rapid changes of pressure that cause the formation of cavities where the pressure is relatively low. The collapse of the bubbles leads to a temperature increase and the differences in pressure may have mechanical consequences on the cellular and tissue structures. For example, if the bubbles collapse near the seed coat this may damage the surface, creating pores. Our results on SEM indicated that this is one of the effects of ultrasound in the Arabidopsis seeds. The seed coat acts as a physical barrier preventing the uptake of water and oxygen into the seed, both necessary for germination [22]. In consequence, an increase in seed coat porosity may produce an enhancement of water and oxygen intake, and, in consequence, in germination. Mechanical events such as vibration, cutting and brushing, can accelerate the germination process in *A. thaliana* L. seeds [23]. Therefore, a similar effect is expected for the opening of pores induced by ultrasounds on the seed coat of Arabidopsis seeds.

Conclusions

In this study we investigated the effects of ultrasound treatment on *A. thaliana* germination percentages in order to determine its possible application to promote germination, in special in aged seeds. No significant positive effects were observed in recently collected seeds, but a significant increase was observed in aged seeds. The ultrasound treated seeds present a greater number of pores in the seed coat than the control ones, perhaps being this the reason of the increase in the germination rate.

Methods
Biological materials
Arabidopsis thaliana plants (Col-0 ecotype) were grown in soil in controlled environmental chambers at 20 °C with a 16/8-h light–dark photoperiod. The harvested seeds were stored at room temperature in 1.5 ml tubes.

Ultrasound
Ultrasounds were generated with 45 kHz frequency and 0.028 W m^{-3} volumetric power in a USC-1400 ultrasonic bath (Unique, Brazil). Seed sonication was performed after cold imbibition for 4 days. Ultrasound treatments were performed in a water bath at constant temperature (24 °C) and different time periods (from 30 s to 64 min). Seeds were treated in 1.5 ml plastic tubes containing 50 μl of distillate water.

Germination test
Germination tests were performed basically as previously reported [17, 18] using 10 replicates of 30 seeds. Seeds were placed into rolled paper towels moistened with water at a proportion of 2.5-times the dry weight of the paper towels and placed at 24 °C with a photoperiod of 16 h light per day. Final germination rates were scored by counting seeds with radicle protrusion after 7 days.

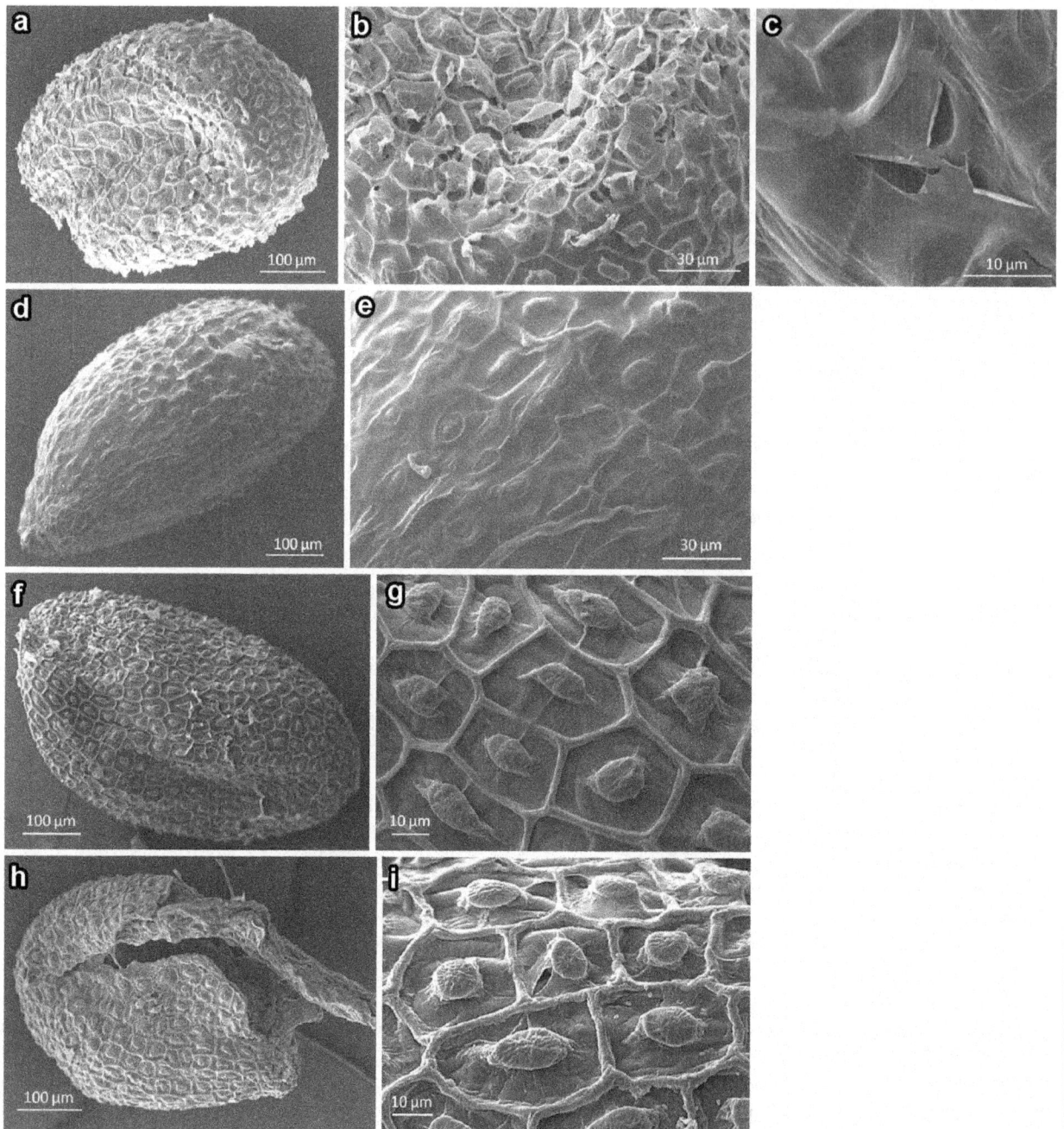

Fig. 5 Scanning electron microscopic view of Arabidopsis Col-0 seeds. **a–c** Seed subjected to 45 kHz sonication at 24 °C for 30 s. **d, e** dry seed, **f, g** seed placed in water for 4 days at 4 °C so they are embedded but they do not germinate, **h, i**, germinated seeds

Accelerated ageing

Artificial ageing of Arabidopsis seeds were performed according to previous reports [17, 18]. Basically, one layer of seeds were placed on a metal mesh into plastic boxes containing distilled water. These boxes were closed and placed in a chamber at 40 °C. Hence, within the boxes, the seeds were exposed to 100% relative humidity. The duration of the treatment depends on each experiment from 6 h to 4 days.

Data analysis

The statistical analyses were done using the t test for 2 Independent Means. Significance level were tested at $p < 0.05$.

Scanning electron microscopy

Arabidopsis seeds were dehydrated in an acetone series, critical point dried using carbon dioxide and mounted directly on stubs using double-side adhesive tape. Observations were made in a EVO MA-10 SEM.

Abbreviation

SEM: scanning electron microscopy.

Authors' contributions

IL carried out the microscopy experiments, CV contributed to conception, design and carried out the rest of the experiments. Both authors read and approved the final manuscript.

Acknowledgements

We would like to acknowledge the financial contribution to the research activities by MINECO (BFU2013-50058-EXP and SEV-2015-0533) and AGAUR (2014SGR-1434).

Competing interests

The authors declare that they have no competing interests.

Funding

MINECO (BFU2013-50058-EXP and SEV-2015-0533) and AGAUR (2014SGR-1434).

References

1. Weigel D, Glazebrook J. Arabidopsis: a laboratory manual. New York: Cold Spring Harbor Laboratory Press; 2002.
2. Nguyen TP, Cueff G, Hegedus DD, Rajjou L, Bentsink L. A role for seed storage proteins in Arabidopsis seed longevity. J Exp Bot. 2015;66:6399–413.
3. Rajjou L, Duval M, Gallardo K, Catusse J, Bally J, Job C, Job D. Seed germination and vigor. Ann Rev Plant Biol. 2012;63:507–33.
4. Ress P, Kiss I, Miltenyi G, Petro I, Farkas J, Biacs P, Kozma I, Debreczeny I. Process for controlling the germination of malting barley. Patent No. US4, 670,279. 1987.
5. Yaldagard M, Mortazavi SA, Tabatabaie F. Application of ultrasonic waves as a priming technique for accelerating and enhancing the germination of barley seed: optimization of method by the Taguchi approach. J Inst Brew. 2008;114:14–21.
6. Mason TJ, Riera E, Vercet A, Lopez-Buesa P. Application of ultrasound. In: Sun DW, editor. Emerging technologies for food processing. London: Academic Press; 2005. p. 323–51.
7. Rubtsova ID. Effect of ultrasonics on seed germination and productivity of fodder beans. Biofizika. 1967;12:489–92.
8. Karabascheff N, Metev V, Kutov T. On the effect of ultrasonic waves on the germination of corn. Compt Rendus Acad Bulg Sci Sci Math Nat. 1966;19:305–12.
9. Sossountzov L. Effect of ultrasonics on germination of fern spores; general observations. Compt Rend Seances Soc Biol et de ses Filial. 1954;148:293–6.
10. Kim HJ, Feng H, Kushad MM, Fan X. Effects of ultrasound, irradiation, and acidic electrolyzed water on germination of alfalfa and broccoli seeds and Escherichia coli O157:H7. J Food Sci. 2006;71:M168–73.
11. Yildirim A, Öner MD, Bayram M. Modeling of water absorption of ultrasound applied chickpeas (Cicer arietinum L.) using Peleg's equation. J Agric Sci. 2010;16:278–86.
12. Ranjbari A, Kashaninejad M, Aalami M, Khomeiri M, Gharekhani M. Effect of ultrasonic pre-treatment on water absorption characteristics of chickpeas (Cicer arietinum). Latin Am Appl Res. 2013;43:153–9.
13. Patero T, Augusto PED. Ultrasound (US) enhances the hydration of sorghum (Sorghum bicolor) grains. Ultrason Sonochem. 2015;23:11–5.
14. Ghafoor M, Misra NN, Mahadevan K, Tiwari BK. Ultrasound assisted hydration of navy beans (Phaseolus vulgaris). Ultrason Sonochem. 2014;21:409–14.
15. Shin YK, Baque MA, Elghamedi S, Lee EJ, Paek KY. Effects of activated charcoal, plant growth regulators and ultrasonic pre-treatments on in vitro germination and protocorm formation of Calanthe hybrids. Austr J Crop Sci. 2011;5:582–8.
16. Goussous SJ, Samarah NH, Alqudah AM, Othman MO. Enhancing seed germination of four crop species using an ultrasonic technique. Exp Agric. 2010;46:231–42.
17. Bentsink L, Alonso-Blanco C, Vreugdenhil D, Tesnier K, Groot SPC, Koornneef M. Genetic analysis of seed-soluble oligosaccharides in relation to seed storability of Arabidopsis. Plant Physiol. 2000;124:1595–604.
18. Clerkx EJ, El-Lithy ME, Vierling E, Ruys GJ, Blankestijn-De Vries H, Groot SP, Vreugdenhil D, Koornneef M. Analysis of natural allelic variation of Arabidopsis seed germination and seed longevity traits between the accessions Landsberg erecta and Shakdara, using a new recombinant inbred line population. Plant Physiol. 2004;135:432–3.
19. Liu J, Wang Q, Karagić D, Liu X, Cui J, Gui J, Gu M, Gao W. Effects of ultrasonication on increased germination and improved seedling growth of aged grass seeds of tall fescue and Russian wildrye. Sci Rep. 2016;6:22403.
20. The Arabidopsis Initiative. Analysis of the genome sequence of the flowering plant Arabidopsis thaliana. Nature. 2000;408:796–815.
21. De Sousa Araújo S, Paparella S, Dondi D, Bentivoglio A, Carbonera D, Balestrazzi A. Physical methods for seed invigoration: advantages and challenges in seed technology. Front Plant Sci. 2016;12:646.
22. Debeaujon I, Léon-Kloosterziel KM, Koornneef M. Influence of the testa on seed dormancy, germination, and longevity in Arabidopsis. Plant Physiol. 2000;122:403–14.
23. Uchida A, Yamamoto KT. Effects of mechanical vibration on seed germination of Arabidopsis thaliana (L.) Heynh. Plant Cell Physiol. 2002;43:647–51.

14

The combination of gas-phase fluorophore technology and automation to enable high-throughput analysis of plant respiration

Andrew P. Scafaro[1,2], A. Clarissa A. Negrini[1], Brendan O'Leary[1,4], F. Azzahra Ahmad Rashid[1], Lucy Hayes[1], Yuzhen Fan[1], You Zhang[1], Vincent Chochois[3], Murray R. Badger[3], A. Harvey Millar[4] and Owen K. Atkin[1*]

Abstract

Background: Mitochondrial respiration in the dark (R_{dark}) is a critical plant physiological process, and hence a reliable, efficient and high-throughput method of measuring variation in rates of R_{dark} is essential for agronomic and ecological studies. However, currently methods used to measure R_{dark} in plant tissues are typically low throughput. We assessed a high-throughput automated fluorophore system of detecting multiple O_2 consumption rates. The fluorophore technique was compared with O_2-electrodes, infrared gas analysers (IRGA), and membrane inlet mass spectrometry, to determine accuracy and speed of detecting respiratory fluxes.

Results: The high-throughput fluorophore system provided stable measurements of R_{dark} in detached leaf and root tissues over many hours. High-throughput potential was evident in that the fluorophore system was 10 to 26-fold faster per sample measurement than other conventional methods. The versatility of the technique was evident in its enabling: (1) rapid screening of R_{dark} in 138 genotypes of wheat; and, (2) quantification of rarely-assessed whole-plant R_{dark} through dissection and simultaneous measurements of above- and below-ground organs.

Discussion: Variation in absolute R_{dark} was observed between techniques, likely due to variation in sample conditions (i.e. liquid vs. gas-phase, open vs. closed systems), indicating that comparisons between studies using different measuring apparatus may not be feasible. However, the high-throughput protocol we present provided similar values of R_{dark} to the most commonly used IRGA instrument currently employed by plant scientists. Together with the greater than tenfold increase in sample processing speed, we conclude that the high-throughput protocol enables reliable, stable and reproducible measurements of R_{dark} on multiple samples simultaneously, irrespective of plant or tissue type.

Keywords: Dark respiration, Fluorophore, Gas-exchange, High-throughput, Oxygen consumption, Oxygen electrodes, Respiration, Respiratory flux, Respiratory quotient

Background

Mitochondrial respiration (R) is an essential physiological process in plants required for most energy-dependent metabolic processes. In mature leaves, R takes place in darkness (R_{dark}) and in the light, and is central to processing of carbon assimilates and nitrogen assimilation [1], while also supporting the energy requirements of phloem loading and maintenance processes (e.g. protein turnover and membrane transport) [2–6]. Respiration is also central to the functioning of roots, providing the energy needed for biosynthesis, nutrient uptake and assimilation, as well as maintenance processes [7]. As such, genotypic and/or environmentally-induced variations in leaf

---

*Correspondence: Owen.Atkin@anu.edu.au
[1] ARC Centre of Excellence in Plant Energy Biology, Research School of Biology, Building 134, The Australian National University, Canberra, ACT 2601, Australia
Full list of author information is available at the end of the article

and root R play a crucial role in determining growth/survival of individual plants, and productivity/functioning of terrestrial ecosystems [8–10]. Because of this, there is a growing need to describe and predict variability in rates of plant R, which in turn requires provision of large-scale data sets on leaf and root R. Recent studies reporting on expanded global data sets of leaf R_{dark} and its T-dependence [11–13]—compiled over several years using slow, low-throughput gas exchange protocols—are a step forward. However, our understanding of fine-scale temporal, spatial and developmental variation in plant R remains limited, both for natural and managed ecosystems. Addressing the need for new, large-scale datasets on plant R will require development of rapid, high-throughput methods capable of overcoming current bottlenecks in data provision.

One area where there is an urgent need for data on plant R is within the agriculture industry, where more energy-efficient crops are needed to improve global food security. For wheat (*Triticum aestivum*), only 10–15% of photosynthetic carbon gain contributes to yield [14], demonstrating the untapped potential for improving energy use efficiency. 30–80% of daily carbon gain by photosynthesis is subsequently respired [15–18], with respiratory costs increasing with increasing temperature [19]. Given that the efficiency of ATP synthesis per unit of CO_2 or O_2 equivalents respired varies (reflecting engagement of phosphorylating and non-phosphorylating pathways of mitochondrial electron transport [20, 21]), there is potential to improve crop yields via selecting for efficient genotypes with reduced rates of R [22, 23]. Indeed, there is growing evidence that physiological screening on a large scale assists crop breeders in identifying beneficial genetic material [24]. However, recombinant inbred line (RIL) populations, diversity panels and/or the structured genetic populations used in genome wide association studies (GWAS) typically include many hundreds of plant variants. Studying these for respiratory traits will require thousands of respiratory measurements to be routinely made on material at the same time of day and developmental stage.

Comprehensive R datasets are also needed to improve modelling of respiratory fluxes in terrestrial ecosystems [9, 25–27]. Using standard leaf gas exchange methods, recent surveys have greatly increased our understanding of biome-to-biome variation in leaf R_{dark} [11–13]; our understanding of how sustained changes in the environment affect respiratory rates is also improving [11, 28–31]. Yet, limitations in available data (e.g. documenting environmental, developmental and/or temporal variations) restrict our ability to fully describe the complexity of plant R that occurs in nature. Similarly, respiratory measurements have been conducted in only a small fraction of extant terrestrial plant species, limiting our ability to explore evolutionary changes in plant energy use efficiency. Addressing these challenges requires development of high-throughput methods for quantifying respiratory fluxes of plants growing in natural ecosystem across the globe.

Protocols using O_2-electrodes and infrared gas-analysers have dominated the measuring of plant R_{dark} for several decades (refer to Hunt [32] for a comprehensive review of each techniques application, advantages and disadvantages). The O_2-electrode technique was popularised in the form of Clark-type O_2-electrodes, being first applied to measure human blood O_2 levels [33]. O_2-electrodes are often used for measurements of root respiration [34–36] and to assess the impact of exogenous substrates, uncouplers and inhibitors on leaf slices, intact roots and isolated mitochondria [37–39]. While a series of O_2 electrodes can be set up in parallel to perform respiratory measurements, in most cases a single electrode is used and each measurement takes an estimated 25–50 min to complete (see Table 1 for a comparison of measurement times associated with this and other methods).

Infrared gas-analysers (IRGA) are also commonly used to measure rates of plant R (as respiratory CO_2 efflux), exploiting the infrared absorption properties of CO_2. The major benefit of the IRGA systems is that they can be portable and operate as a gas-phase/open system. Such systems have been extensively used in recent times for quantifying plant R_{dark} [12, 40–42], including specialised chambers for whole-plant R_{dark} [16, 19, 43]. While a few research teams have developed multiplex systems for single IRGA measurement of four to 12 samples [e.g. 44], most IRGA measurements are made individually, each requiring 10–20 min per sample (Table 1). Consequently, existing IRGA methods are unlikely to provide the high throughput capacity needed to screen for genetic variations in energy use efficiency and/or improved modelling of ecosystem gas exchange.

Less employed spectroscopy technology for detecting respiratory O_2 and/or CO_2 exchange include tuneable diode laser (TDL) spectroscopy [45] and cavity ring-down (CRDS) spectroscopy [46]. Mass spectrometry can also be used, with one example of a mass spectrometry technique being membrane inlet mass spectrometry (MIMS), a gas phase method that is used to discriminate between O_2 and CO_2 isotopes, enabling deeper insight into the photosynthesis/respiratory process [44, 47]. Although MIMS is beneficial in that it can discern gas isotopes, neither it nor the above spectroscopic approaches are high-throughput (Table 1). Similarly, calorimetry measurements of metabolic heat rate and respiratory fluxes [48, 49] while providing an opportunity to

Table 1 Measurement times required per sample for each of the R_{dark} techniques assessed

Technique	Step	Description	T (min)
Fluorophore	Calibration	Purge tubes of air using N_2 gas or sodium dithionite	0.02–0.05
	Sample preparation	Dissect tissue (e.g. scalpel, scissors or leaf punch) and place in measuring tube	0.5–1
	Measurements	In general, slopes taken from 1 to 2.5-h. 186 samples per run. Note: more than 186 samples can be simultaneously measured but cycle time between O_2 recordings will increase to >6-min, reducing resolution	0.8
	Total		1.3–1.9
O_2-electrode	Calibration	Prepare and assemble electrodes, including application of membrane and electrode solution. Aerate calibration solutions and obtain zero and saturated O_2 values after stabilisation of current	4–9
	Sample preparation	Dissect tissue and place inside cuvette and adjust plunger being careful not to introduce air pockets	1–2
	Measurements	Slopes taken after stabilisation of signal and before depletion of O2, usually within 10–40 min but dependent on sample	20–40
	Total		25–51
IRGA	Calibration	Change consumables (e.g. soda lime, desiccant, CO_2 canister) and zero IRGA chambers	1–2
	Sample preparation	Select and clip measuring chamber onto leaf	0.5–1
	Measurements	Allow steady-state gas-exchange to be reached	10–15
	Total		11.5–18
MIMS	Calibration	Apply membrane and test membrane stability. Purge tube and inject known volumes of O_2 and CO_2. Record background consumption	5–10
	Sample preparation	Dissect tissue and place inside cuvette and air-seal cuvette	1–3
	Measurements	Allow signal to stabilise (usually 5 min) and record slope between 5 and 20 min	20
	Total		26–33

T (min) represents the estimated time it takes to measure a single sample in minutes. For example, if 20 samples can be measured without recalibration and it takes 20-min to calibrate, then the calibration T is 1-min

explore relationships between respiration and growth—are also not high throughput.

Using O_2-sensitive fluorophores in combination with fibre-optic fluorescent detection mechanisms for measuring the O_2 evolution of photosynthesis of illuminated leaf disks was occurring by the late 1990s [50]. The technique works by exciting a fluorophore, in most cases a metal porphyrin, whose fluorescence is sensitive to O_2 quenching. The measured decay rate of the fluorescent emission is thus proportional to the partial pressure of O_2 present [51, 52]. This technology is becoming a more common technique for detecting respiratory O_2 consumption of biological samples ranging from bacterial plankton to benthic meiofauna [53, 54]. The power of this technology is that many tissue types of varying abundance can be simultaneously and accurately measured. For example, fluorophore technology has enabled multiple simultaneous measurements of leaf, root and seed respiratory rates [55]. The authors highlight the high-throughput and small tissue size capabilities of the technique, not achievable using conventional Clark-type electrodes, infrared gas-analyser, spectroscopy or calorimetry methods. Yet, take-up of fluorophore technology to facilitate high-throughput measures of plant R remains limited, reflecting the need for more straightforward

sample preparation than was possible using the liquid-phase approach of Sew et al. [55]. By contrast, using fluorophore technology in a gas-phase medium is likely to lead to faster processing times and avoid technical issues, such as floating tissues and air-pockets. To date, automated gas-phase measurements of O_2 consumption using fluorophore techniques for plants have primarily focused on large-scale analysis of seed germination [56, 57], with automated, high-throughput assessments of non-seed plant R yet to be attempted using gas-phase fluorophore approaches.

To address the urgent need for high-throughput measurements of plant R_{dark}, we have trialled an approach for measuring respiratory O_2 uptake which re-purposes equipment designed for seed germination assays and combines the advantages of: (1) fluorophore technology that can accurately measure changes in O_2 partial pressure in small measuring volumes that are easily calibrated; (2) closed, gas-phase measurements, which require minimal preparation time; and, (3) an automated sampling mechanism, relying on robotics to take measurements of multiple samples within a short period of time. As part of our study, we compare multiple O_2 consumption detection methodologies to ascertain the reliability and compatibility of the different approaches.

Further, to illustrate the potential of the high-throughput fluorophore technology to accelerate our understanding of plant R_{dark}, we report on: (1) a screen of R_{dark} in 138 genotypes of wheat (using >550 plants) that was conducted over a few days; and, (2) rapid assessments of respiration in leaf, stem and root tissues that enable whole-plant respiratory fluxes to be estimated by simultaneous analysis of individually dissected plants.

Methods

Plant material

The species used in this study were a grass (wheat— *Triticum aestivum*), a herb (thale cress—*Arabidopsis thaliana*) and an evergreen broadleaved tree (red river gum—*Eucalyptus camaldulensis*), enabling the method to be tested on a range of plant functional types. Considering its agricultural significance, *T. aestivum* was selected as the primary species of interest, and all experiments, including the high throughput practical applications, were undertaken on *T. aestivum*, with a sub-set of other experiments conducted using other tissues. All experiments took place at the Research School of Biology at the ANU, Canberra, Australia plants grown in organic potting mix, enriched with Osmocote® OSEX34 EXACT slow-release fertiliser, following manufacturer's instructions (Scotts Australia, Bella Vista, NSW) with an N/P/K ratio of 16:3.9:10. Plants were watered daily to field capacity. For experiments where roots were analysed, wheat plants were grown hydroponically in a nutrient solution consisting of 1.4 mM NH_4NO_3, 0.6 mM $NaH_2PO_4 \cdot 2H_2O$, 0.5 mM K_2SO_4, 0.2 mM $CaCl_2 \cdot 2H_2O$, 0.8 mM $MgSO_4 \cdot 7H_2O$, 0.07 mM Fe-EDTA, 0.037 mM H_3BO_3, 0.009 mM $MnCl_2 \cdot 4H_2O$, 0.00075 mM $ZnCl_2 \cdot 7H_2O$, 0.0003 mM $CuSO4 \cdot 5H_2O$, 0.0001 mM $(NH_4)_6Mo_7O_{24} \cdot 4H_2O$, 0.000138 mM NH_4VO_3, and 0.0012963 mM Na_2SiO_3. A pH ranging from 5 to 6 was maintained by adding concentrated sulphuric acid or sodium hydroxide, and monitoring of pH using a portable pH meter (Rowe Scientific Pty. Ltd., NSW, Australia). The hydroponic solution was aerated continuously using Infinity AP-950 aquatic air pumps (Kong's Pty Ltd, Ingleburn, Australia). Plants were grown at temperatures of 25/20 °C for *T. aestivum* and *E. camaldulensis*, in temperature controlled greenhouses with natural photosynthetically active radiation (PAR) of between 400 and 1200 µmol m^{-2} s^{-1}. *A. thaliana* was grown at 22/15 °C in temperature-controlled growth chambers (Thermoline, Wetherhill Park, Australia) with a PAR of 200 ± 30 µmol m^{-2} s^{-1} and a 12:12 h light/dark photoperiod. For leaf dissection samples, broad-leaved *A. thaliana* and *E. camaldulensis* leaf tissue was extracted using brass coring tools of known diameter and for *T. aestivum* a set distance of leaf blade was dissected with a

scalpel. Where sectioned, root segments were dissected transversely from base to tip.

High throughput fluorophore measurements

A Q2 O_2-sensor (Astec Global, Maarssen, The Netherlands) designed and marketed for seed germination assays was used to obtain automated, high-throughput fluorophore measurements of dark respiration from plant material. A custom-built frame covered in black cloth was used to maintain darkness during sample measurements. Plant material were freshly dissected and placed in empty tubes (1, 2 or 4 ml in volume) and hermetically sealed with specialised caps (Astec Global). The top surface of caps contained a fluorescent metal organic dye, sensitive to O_2 quenching. A blue-spectrum LED excitation pulse (approximately 480 nm) onto the surface of caps, followed by emission detection in the red spectrum (approximately 580 nm), enables the O_2 dependent decay in fluorescence signal to be quantified. The fibre optic fluorescence detection unit is attached to a robotic arm which sequentially measures vials placed in racks of 48 tubes each (or 24, 4 ml tubes). The machine can accommodate 16 racks allowing 768 samples (1 or 2 ml tubes) to be measured in a single run. The frequency of measurements was in most cases set to 4 min, enough time to measure approximately 180 samples (a minimum measurement frequency of 1-min is required). The Q2 O_2-sensor is calibrated before each set of measurements by measuring a designated tube containing ambient air (designated 100% O_2), and a tube purged of all O_2 using a sodium dithionite solution, or alternatively purging the tube of air using N_2 gas (designated 0% O_2). Output is given as an O_2 percentage, relative to the calibration readings.

Based on the ideal gas law, raw output as the % O_2 relative to the air calibration tube was converted to absolute values of dark respiration rates (R_{dark}) in moles of O_2 s^{-1} using Eq. 1.

$$R_{dark} = \frac{P_o VS}{RT} \qquad (1)$$

P_o equals 20.95, the partial pressure of ambient O_2 in kPa (i.e. 20.95% of atmospheric pressure), and V equals the volume of the sample tube (1, 2 or 4 ± 0.2 ml tubes were used in this study). S refers to the slope of sample tubes O_2 consumption, (as a % of air and subtracting the air calibration tube slope), from 1 to 2.5 h after the beginning of sample measurements, expressed as the % of O_2 per second. R is the gas constant (8314 cm^3 kPa K^{-1} mol^{-1}) and T is the temperature in Kelvin (K). The final calculation of O_2 consumption rates in moles s^{-1} were expressed on a leaf area (cm^2) basis, calculated from the diameter of the leaf corer (for leaf disks) or ruler measurements for grass leaf sections. Alternatively, for whole-plant developmental partitioning measurements respiration was expressed on a fresh mass basis. To test technical reproducibility of the instrument, a chemical oxidation assay consisting of 100 mM of cysteine in 600 μL of buffered solution (50 mM Hepes, 10 mM MES pH 6.5, 200 μM $CaCl_2$) was used, and the stabilised O_2 consumption rate over a 2-h run was measured repeatedly. To test fluorophore sensitivity to O_2 depletion, known volumes of pure CO_2 gas were injected into tubes through a pin-hole created on the side of tubes and sealed with blu-tack (Bostik, Paris, France) immediately after the gas-injection. All measurements were made at a room temperature of 21.5 ± 1.0 °C.

O_2-electrode measurements

Respiratory consumption of O_2 by leaves (3–42 mg fresh mass) or roots (56–214 mg fresh mass) were measured in the liquid-phase using Oxytherm Clark-type O_2-electrode (Hansatech Instruments, Pentney, UK) in a 2 ml measuring volume. Electrodes were calibrated by bubbling water with compressed air for approximately 2-h to reach saturation followed by adding sodium dithionite to record O_2 depleted signals. Leaf and root respiration was measured in a solution containing 20 mM Hepes (pH 7.2), 10 mM MES and 2 mM $CaCl_2$, at 21.5 ± 1 °C. All measurements were made by dark adapting tissue for >30 min, submerging tissue in the Clark-type electrode cuvettes below measuring solution, with no obvious air pockets and continually stirring, and recording O_2 consumption using Oxygraph Plus v1.02 software (Hansatech Instruments). The linear part of O_2 consumption (approximately 10–30 min into each run) was used to calculate respiration rates.

IRGA measurements

Infrared gas-analysis of CO_2 efflux by respiring leaves was measured using a Licor 6400XT with a 3×2 cm chamber head ((LI-COR, Lincoln, Nebraska, USA) on >30 min dark-adapted leaves. Attached whole leaves were placed across the measuring chamber and chamber gaskets and measurements recorded after CO_2 readings stabilised (~10–15 min). The flow rate was set to 300 μmol s^{-1}, the block temperature set to ambient air temperature of 22 °C and the CO_2 reference sample was set to 400 μmol mol^{-1}, to match ambient air. The light source was turned off.

Membrane inlet mass spectrometry

Dark-adapted wheat leaf disks (3×0.5 cm^2 or 6×0.5 cm^2) were placed in a 1 mL O-ring sealed cuvette containing only air and a polyethylene membrane sealed outlet attached to a mass spectrometer (MM6: VG, Winsford, UK). O_2 (m/z = 32) and CO_2 (m/z = 44) detection

over a 20-min period was recorded. Prior to leaf disk samples being placed in the cuvette, N_2 gas purging of the cuvette and injections of known volumes of O_2 and CO_2 allowed for conversion of mass detection signal to a gas concentration and the background consumption rate of O_2 and CO_2 by the mass spectrometer to be accounted for when determining leaf derived O_2 consumption and CO_2 evolution rates.

Replication and statistical analysis

For all experiments four to six biological replicates, with a biological replicate considered as plant material from individual plants grown in separate pots, or containers (when grown hydroponically) were measured. For the comparison of respiratory techniques, two or more samples from each biological replicate were analysed by each technique and sampling was standardised by selecting a 2 cm long mid-section of young, healthy, fully expanded leaves, or in relation to root samples, a longitudinal section from base to tip of the longest root segment. A one-way ANOVA was used to determine significance between leaf O_2 consumption techniques and Two-Sample t-tests for differences between leaf CO_2 evolution techniques and root O_2 uptake techniques.

Results
Technical and biological reliability and accuracy

The stability of the fluorescent oxygen concentration measurements performed using the Q2 is evident because control tubes containing either ambient air or no O_2, gave long-term stable readings at 100 ± 5 or $0 \pm 5\%$, respectively (Fig. 1a). The stability of O_2 in the purged tubes demonstrates that the sample tubes were hermetically sealed, providing a closed system, necessary for accurately measuring O_2 uptake. Nevertheless, we suggest periodically testing the accuracy of the calibration tubes, in case of drift over time, by placing 100% air and 0% O_2 tubes amongst samples during a run. Measuring the spontaneous chemical oxidation of a cysteine solution in replicate vials assessed the technical reproducibility of the O_2 consumption measurements. Analysis of 30 tubes in three separate experiments gave an average coefficient of variation of 8.1% (Additional file 1: Table S1).

When cut leaf material was placed inside the sample tubes, the fluorophore system was able to measure a consistent decline in O_2 over a greater than 7-h period following an initial 1 h period of stabilization (Fig. 1a). The decline was linear in all species and tissues tested. The 90-min O_2 consumption slope between 1 and 2.5-h had a mean r^2 of 0.99 across both species (Fig. 1b). Typically, the initial 0–30 min period of each run was associated with sharp declines in the O_2 consumption slope. R_{dark} calculated from a 1-h moving average of slopes over 7-h

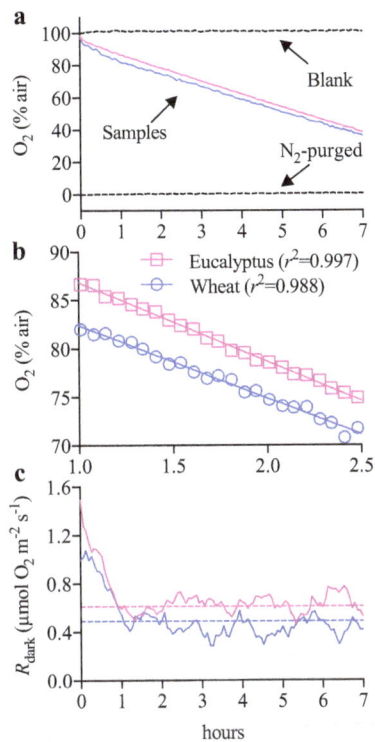

Fig. 1 O_2 consumption rates measured using the fluorophore technique. **a** O_2 is given as a percentage of O_2 in ambient air. A tube containing no sample (labelled as Blank) provided the baseline for no O_2 consumption, while a tube devoid of all air (labelled as N_2-purged) provided the baseline for total O_2 consumption, and samples (*magenta* for Eucalyptus; *blue line* for wheat) depleted O_2 within this range. **b** A higher resolution plot of individual data points over a 90-min period, from 1 to 2.5-h and linear regression analysis. **c** Respiration rates calculated from linear regression of O_2 consumption using Eq. 1. Presented are R_{dark} calculated from a 1-h moving slope (solid lines), and R_{dark} calculated from the 1–2.5 h slope as shown in Panel b (*dashed horizontal lines*). Values are the means of four biological replicates for each species with the % of O_2 measured every 4 min

was similar to the slope of O_2 consumption over a set 90 min period between 1 and 2.5-h (presented as dashed horizontal lines in Fig. 1c). The O_2 consumption slope between 1 and 2.5-h can therefore be used as a standard period for calculating R_{dark} across experiments.

Respiratory rates per unit leaf area were independent of the amount of leaf material placed within a given tube volume, apart from exceedingly small tissue abundance of below 0.1 cm^2 mL^{-1} (Fig. 2a). To test whether the signal was independent of CO_2 concentration and linearly related to O_2 concentration between 0 and 100% of atmospheric O2 known volumes of pure CO_2 gas were injected and sealed in measuring tubes. The measured percentage of O_2 in the tube declined linearly in close proximity to the expected values for the amount of

air displaced by CO_2 (Additional file 1: Fig. S1), validating that for the fluorophore in question, the O_2 dependent fluorescence quenching is linear and independent of CO_2 concentration. An increase in CO_2 concentration was not inhibitory to R_{dark}, evident in maintained R_{dark} when O_2 was depleted to less than 40% of ambient levels, equivalent to the gas volume being >8% CO_2, assuming a respiratory quotient of one. We provided further support of a lack of CO_2 inhibition of R_{dark} by purging tubes containing wheat leaf samples with various concentrations of pure CO_2 gas (Fig. 2b). Interestingly, replacing the volume of gas surrounding leaf material with as much as 90% CO_2 did not lead to a substantial decline in R_{dark}. When 100% of the air within a tube was replaced with CO_2, R_{dark} did essentially stop, understandable considering no O_2 would be available for respiration.

Although increased CO_2 concentration was not inhibitory to R_{dark}, heavy mechanical wounding of tissue resulted in higher R_{dark} (Fig. 2c). Intact wheat leaves versus a 2 × 0.5 cm transverse section from the middle of leaves (a ratio of 1:1, wounded boundary length to leaf area) did not exhibit significant differences in R_{dark} on an area basis (Fig. 2c). However, if the transverse section was further sliced into 20 smaller pieces (a 20-fold increase in the cut surface length to leaf area ratio), R_{dark} increased by as much as two-fold (Fig. 2c). Applying a buffered saline solution to the heavily wounded leaf partly mitigated the enhancement of R_{dark} by wounding. Thus it is important to reduce the amount of tissue exposed to mechanical damage when processing samples, to avoid the risk of artificially enhancing respiration rates.

Comparisons between leaf gas-exchange methods

Considering the many methods currently in use for determining plant respiratory gas-exchange, and the need to ensure that the fluorophore system was giving comparable rates, we compared R_{dark} values generated using the fluorophore technology, the more conventional Clark-type O_2-electrodes, Licor 6400 IRGA gas-exchange system, and membrane inlet mass spectrometry (MIMS). All of these techniques have varying degrees of difference in sample preparation and technical methodology that may influence the final respiratory rate recorded. For example, while we measured O_2 consumption in the gas-phase using the fluorophore technique, O_2-electrode measurements were made in aqueous-phase. Despite the IRGA measurements being made in gas-phase, measurements were of CO_2 rather than O_2 flux, and in an open gas-exchange system rather than the closed fluorophore system. Furthermore, IRGA measurements are made on intact not detached leaves. MIMS would be closest in methodology to the fluorophore technique in that both were measuring in the gas phase, in an essentially closed

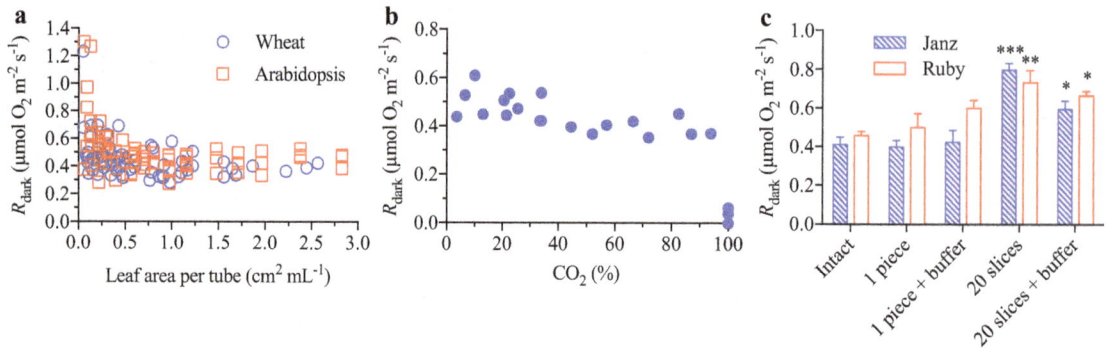

Fig. 2 The influence of leaf tissue amount, CO_2 concentrations, and mechanical wounding on the dark respiration rate (R_{dark}) of wheat and Arabidopsis leaf tissue. **a** Differing amounts of leaf area in measuring tube volumes (cm^2 mL^{-1}) plotted against corresponding R_{dark} for wheat (*open blue circles*) and Arabidopsis (*open red squares*). **b** R_{dark} of wheat leaves sealed in measuring tubes with varying CO_2 as a % of air. **c** The influence of mechanical wounding during sampling of leaf sections on the dark respiration rate (R_{dark}) of two wheat cultivars, Janz and Ruby. The R_{dark} of a 2×0.5 cm transverse section of leaf and a same sized leaf sliced a further 20 times was compared to an intact leaf, which was not mechanically damaged. Alternatively, leaves that were cut were washed with a wounding buffer prior to R_{dark} measurements. The values are mean $\pm$ SE of four biological replicates. *significance at $P < 0.05$, ** at $P < 0.01$ and *** at $P < 0.001$, for a one-way ANOVA and a Dunnett's multiple comparison test with the intact leaf set as the control

system. However, the MIMS system is not a completely closed system as the gradual leak of gasses through the semi-permeable membrane to the mass spectrometer would lead to changes in partial pressure and water vapour at the site of the leaf.

Understandably, due to the aforementioned differences in methodology, calculations of R_{dark} using matching leaf or root material were significantly different between methods (Fig. 3). On an O_2 basis, the conventional O_2-electrode technique gave lower values, MIMS gave higher values, and the fluorophore values were intermediate. On a CO_2 basis, MIMS measurements were significantly higher than IRGA measurements. MIMS, the only technique that can measure both O_2 and CO_2 concentrations, gave almost matching R_{dark} measurements on an O_2 and CO_2 basis, indicating a respiratory quotient near unity for darkened wheat leaf tissue. Root R_{dark} measurements in the gas-phase on the fluorophore system were significantly higher than in the liquid phase measured with O_2-electrodes. Thus, while the fluorophore and IRGA approaches provide similar estimates of leaf R_{dark}, both methods yield relatively lower estimated respiratory fluxes compared to MIMS; by contrast, the fluorophore approach yields relatively high values compared to liquid-phase Clark-type O_2 electrode measurements.

High-throughput analysis of respiration

Two studies were undertaken to verify the capabilities and versatility of automated O_2 fluorophore technology for measuring high-throughput plant respiration in leaves and other plant tissues.

For the first study we undertook a fully replicated experiment of leaf respiration in 138 wheat cultivars (Fig. 4). There were clear differences in R_{dark} among many genotypes, with a two-fold variation between the lowest and highest respiring cultivars (Fig. 4a). The wheat

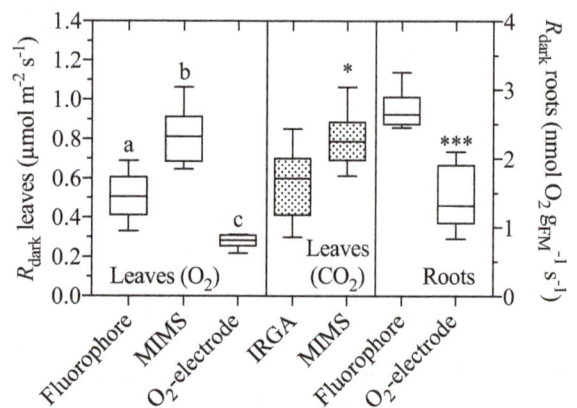

Fig. 3 Comparisons of wheat dark respiration (R_{dark}) measurements made using different experimental techniques. Leaf and root R_{dark} was calculated from O_2 consumption (*open boxes*) and CO_2 evolution rates (*hatched boxes*), measured using fluorophore quenching, membrane inlet mass spectrometry (MIMS), Clark-type O_2-electrodes, and infrared gas analysis (IRGA). Whiskers of *box-plots* represent the 5–95 percentile. For leaf O_2 analysis, letters indicate significant differences between techniques at $P < 0.05$, derived from a one-way ANOVA with a Tukey's multiple comparison test. For leaf CO_2 analysis and roots an unpaired t-test was performed and * indicates significance at $P < 0.01$, while ***significance at $P < 0.001$. All measurements were made at 21 ± 0.5 °C. The values are based on six biological replicates and greater than 12 technical replicates

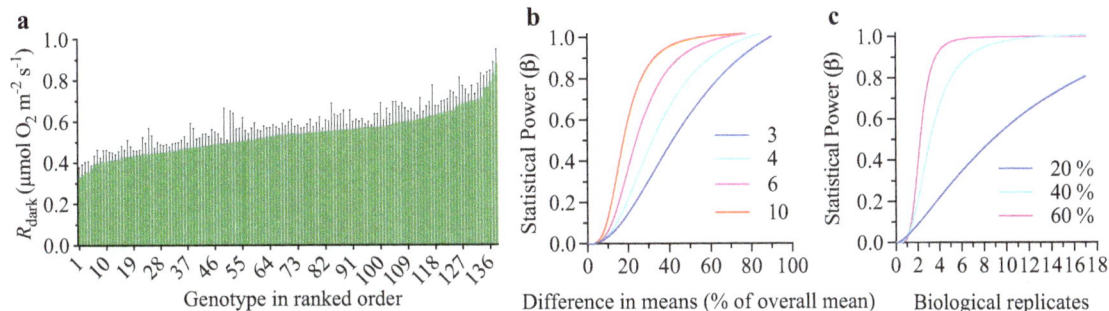

Fig. 4 A high-throughput genotype and statistical power analysis of wheat dark respiration (R_{dark}) of the youngest fully expanded leaf. **a** Leaf respiration in the dark (R_{dark}) across 138 wheat cultivars grown under common conditions in a controlled environment growth room and measured at 21.5 ± 1 °C. The experiment was replicated on four occasions and the means (*green columns*) ± SE (*black bars*) of four biological replicates are presented. **b** Statistical power analysis of the wheat dataset with lines representing different numbers of biological replicates (3, 4, 6, or 10) plotted as a function of statistical power versus difference in genotype mean values. **c** Statistical power analysis of the wheat dataset with *lines* representing differences in mean values between genotypes equal to 20, 40 or 60% of the overall mean (equal to effect sizes of 1, 2 and 3 times the standard deviation) plotted as a function of statistical power versus biological replicates

dataset was used to calculate the average standard deviation among biological replicates. As a proportion, the standard deviation was close to 20% of the overall mean R_{dark}. This coefficient of variation was used to estimate the statistical power for future t test comparisons of R_{dark} between wheat lines as a function of replicate number and difference in means, using a false discovery rate of 5% ($\alpha = 0.05$; not including corrections for multiple testing). Given the four biological replicates per genotype used in this 138-genotype study, there is sufficient statistical power [$(1 - \beta) > 0.8$] to consistently detect only large differences in R_{dark} between two lines equal to 50% of the mean (Fig. 4b). In a further example, to detect a 20% difference in mean respiration rates between any two wheat lines with the conventional statistical power target of $(1 - \beta) = 0.8$, 17 replicates would be appropriate (Fig. 4c). Of course, significant differences can still be detected with less replicates and less power, but given the high-throughput capacity of the fluorophore technique, appraisal of statistical power and appropriate biological replication can now be achieved, where previously, such high levels of replication were a barrier to experiments.

With the potential to run a single sample using the fluorophore system in less than 2 min (Table 1), a single replicate of all 138 genotypes could be processed in less than 4 h, and potentially, a fully replicated 138 genotype study could be achieved in a single day. The number of samples per day is limited by the capacity of the robotic system, and by the time taken to prepare samples. By comparison, the other techniques have significantly longer calibration, sampling and measurement times required to acquire a single measurement (Table 1). Hence, what can be undertaken in 8-h using the high-throughput fluorophore technique, would require a minimum of 83 equivalent hours,

or as much as 200-h for other commonly used procedures to measure R.

The second study looked at whole-plant developmental partitioning of R_{dark} between leaves, stems and roots of 46-day-old wheat plants, which had reached the tillering stage of development (Additional file 1: Fig. S2). This type of experiment enables the quantitative attribution of total plant R_{dark} to different parts of the plant at a specific stage of development. The simultaneous measurement of a whole dissected plant saves on the need to combine rates over time from measurements made on different plants. Plants were dissected and the individual leaves (including both leaf blade and sheath), the stem and roots were separated. The R_{dark} of all separated tissues was measured for six entire plants simultaneously. Relative to healthy fully expanded leaves of a tiller; R_{dark} was slightly higher in the oldest and much higher in the youngest leaves, on a fresh mass (i.e. nmol O_2 g^{-1} s^{-1}) basis (Fig. 5a, b). Leaves of intermediate ages exhibited similar rates of mass-based R_{dark}. The total R_{dark} for an entire leaf increased with age, presumably due to the increase in leaf size with stem and tiller developmental maturity. However, the total flux of O_2 for the youngest leaf of the main stem or tiller was low (Fig. 5b), due to the smaller leaf size (Fig. 5a). When considered together, the total respiratory output of wheat foliage is dominated by healthy, relatively young, fully-expanded leaves (Fig. 5b) despite the oldest and youngest leaf of a stem or tiller having greater rates of R_{dark} on a mass basis. When considering the partitioning of R_{dark} between all tissues of the entire plant, leaves accounted for 51% (Fig. 5c), roots 37% and stems 12% of the total respiratory flux. Although the stem accounted for 12% of total R flux, it was only 4% of the entire fresh mass of the plant; however, stems had the highest mass based fluxes,

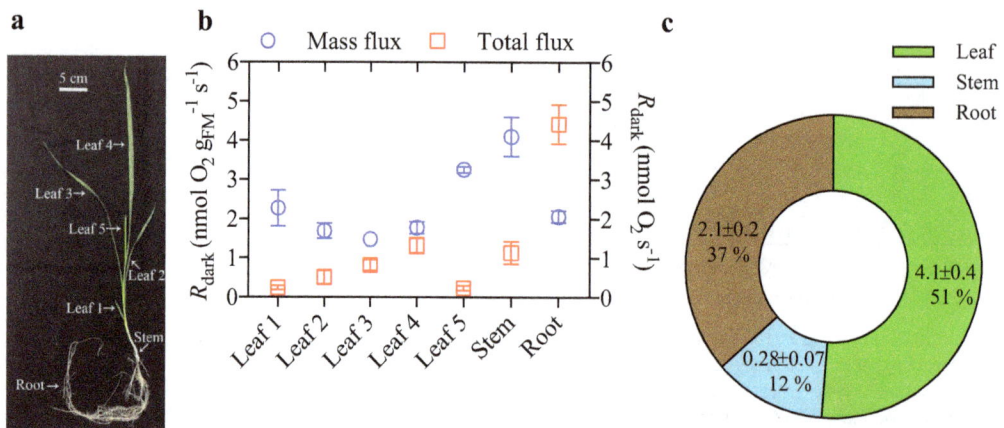

Fig. 5 Wheat plants were harvested and separated into individual leaves, stem and roots, enabling an analysis of whole-plant dark respiration (R_{dark}). **a** A representative image of a tiller with the roots, stem and leaf positions labelled. Leaf position 1 refers to the oldest leaf and position 5 refers to the newest emerging leaf. **b** Individual leaf, stem and root R_{dark} on a mass basis (Mass flux; plotted on the left-axis) and total flux basis (Total flux; plotted on the right-axis). **c** A pie diagram illustrating the partitioning of the total flux of R_{dark} between leaves as a whole, the stem and roots. The total plant leaf, stem and root fresh masses (g) are provided, as well as the % flux of total R_{dark}. Values are based on the mean ± SE of six biological and pot replicates

likely due to energy expensive processes of cell division and elongation at the site of the apical meristem.

Discussion

We demonstrate that using robotic fluorophore-based gas-phase measurements of O_2 consumption in sealed tubes provides a simple yet reliable and reproducible means of measuring R_{dark} for a diverse range of plant tissue types and species. The technique differentiates itself from other conventional methods in that it significantly reduces the time required for sample preparation and has substantial simultaneous measuring capabilities, making the technique a truly high-throughput means for measuring respiration. We demonstrate the potential capabilities of the method by measuring R_{dark} of 138 wheat genotypes, and by measuring R_{dark} of all tissues of six mid-vegetative stage plants simultaneously. A comparison of R_{dark} in absolute terms, generated by different methodologies suggests variation in respiratory rates depending on technique employed, which should be considered when making direct comparisons between methods.

Strengths and weaknesses of high-throughput fluorophore methods

There was an initial spike and rapid decline in respiratory activity within the first 30-min of measurements (Fig. 1b). We dark-adapted leaves for a minimum or 30-min prior to fluorophore analysis, so although it is common to find a spike in respiration of leaves following exposure to light within the initial 30-min post-illumination period [58], post-illumination bursts in respiration do not explain

the findings. Furthermore, while the O_2-electrode and MIMS measurements continuously recorded in a similar manner to the fluorophore system, neither approach showed the initial spike, followed by rapid decline in R_{dark} that was exhibited by the fluorophore approach (Fig. S3). Consequently, the first 60-min of each run were not used to calculate rates of R_{dark} in the genotypic and developmental studies; the initial stabilisation period, however, can be used as a dark-adaptation period if tissue is not dark-adapted prior to fluorophore experimentation.

CO_2 has previously been postulated to inhibit cytochrome c oxidase (COX) activity [59]. Reports initially suggested that a doubling of current atmospheric CO_2 (i.e. from 0.04% of atmospheric gas to 0.08%) reduced R_{dark} by 15–30% [60–62]. However, it was later discovered that CO_2 inhibition of R_{dark} was mostly likely an artefact of the measuring techniques used to quantify respiratory CO_2 release [63–65]. Our results show that CO_2 accumulation does not inhibit R_{dark}. In fact, even with CO_2 concentrations surrounding the sampled tissue reaching more than 90% of the gas volume (a 450-fold increase in concentration relative to previously reported measurements), no substantial inhibition in respiration occurred (Fig. 2b). We therefore conclude that leaf R_{dark} is highly insensitive to CO_2 accumulation over a course of several hours.

One factor that does seem to influence R_{dark} is mechanical wounding (Fig. 2c). Leaf wounding was thought to affect leaf respiration as far back as 1950 [66]. Increased R_{dark} with mechanical wounding is attributed to stimulation of the ATP/ADP ratio and activation of pyruvate kinase due to ion changes associated with wounding [67]. Pre-treatment by washing leaf samples with a buffered saline solution, the same as the measuring solution in liquid phase measurements, reduces any wounding effects on leaf R [38, 68]. We observed an increase in R_{dark} when a large proportion of the sample had a wounded edge, and a reduction in R_{dark} by applying wounding buffer, although not enough of a reduction to eliminate the wounded effect (Fig. 2c). However, minimal wounding did not significantly change R_{dark}. Considering the time required to wash the sample tissue with a wounding solution, we suggest minimising as much as possible the mechanical wounding of tissue, rather than applying a wounding solution, if high-throughput sampling is desired. However, minimising mechanical wounding may require using larger volume tubes (e.g. moving from 1 to 4 mL tubes) to adequately fit sample tissue. By running a preliminary experiment, one could initially check for wounding effects and use the appropriate tissue size thereon after.

The limited effect of leaf wounding and lack of any inhibition to R_{dark} from CO_2 accumulation resulted in respiration measurements being stable over a period of many hours (Fig. 1). The stability of R_{dark} for small leaf sections means that although the fluorophore technique we present is a closed-system that destroys the sampled tissue, a small sample of leaf collected in the field can be transported to the lab (making sure to keep detached leaves from desiccating), accurately representing in situ R_{dark}. Thus, the fluorophore method can be considered as a pseudo non-destructive technique for high-throughput analysis for field experiments, as demonstrated below in the 138 wheat genotypes study we present.

Comparisons between respiratory methods

Although Hunt [32] comprehensively compared the strengths and weaknesses of multiple photosynthesis and respiration measurement techniques, no study to our knowledge has directly compared the absolute values of R obtained from the same biological material but measured across multiple techniques. Determining if the fluorophore technique presented in our study is comparable with previously well-established methods is important. Firstly, if results are to be examined among studies that utilised different techniques, it must be established if the analysis is viable, or whether differences among studies are an artefact of measuring technique. Secondly, although in many cases only the relative differences in R between samples may be of interest (for example, the genotypic study we present here), in many circumstances, absolute R will be desired, such as for determining absolute photosynthesis, or modelling the impact of R on terrestrial carbon budgets. Hence, we directly compared the

fluorophore, O_2-electrode, IRGA, and MIMS output (Fig. 3). We found differences did exist between the techniques, suggesting that comparing results between studies utilising different R measuring apparatus may not be appropriate, or at least with the caveat that comparisons may require cross-calibration of method. Differences in measurements based on either O_2 consumption or CO_2 evolution may be expected considering the respiratory quotient (RQ) will not necessarily be equal to 1 (i.e. respiratory CO_2 release being equal to O_2 uptake) if pure carbohydrates were not the only source of respiratory substrate, or the oxidation state of respiratory products differed, although a RQ of 1 is usually assumed for higher plants under non-stressed conditions [69]. Indeed, the simultaneous measurement of R_{dark} derived from O_2 and CO_2 exchange by MIMS gave close to matching values, supporting a RQ of 1, in contrast to a study of wheat leaves measured in the dark, 6-h into the light period (similar conditions to this study), which gave a RQ value of 1.8 ± 0.21 [70]. However, the study by Azcón-Bieto, Lambers and Day [70] used values of R determined separately using O_2-electrode and IRGA systems, and since we found lower O_2 based O_2-electrode values relative to CO_2 IRGA values, we emphasise that caution must be taken when comparing R calculated from different methodologies. Of note, the widely used IRGA gas-exchange system on intact leaves gave similar rates to the fluorophore results, suggesting the two techniques may be complementary. We did not undertake subsequent experiments to determine the specific reasons for variations in R_{dark} between the techniques compared, and it will be of interest to further explore the reasons for why the techniques vary in future studies.

Genotypic and whole-plant analysis

Both a comprehensive genotype comparison and whole-plant respiratory balances were successfully obtained by use of the gas-phase automated fluorophore technique. Interestingly, a more than two-fold variation in R was observed between the 138 wheat genotypes (Fig. 4a). This demonstrates the inherent intra-specific divergence of R in Triticum aestivum, and a potential target for future yield improvements, if R not contributing to growth or yield can be minimised. Inherent differences in R_{dark} between species populations have previously been noted, such as in the ryegrass species Lolium perenne, attributed to adenylate limitations on glycolysis and varying ATP turnover rates between populations [71]. R was also highly variable among genotypes. This may not be considered surprising as leaf functional traits vary considerably among populations/genotypes within a given species. For example, a study of 13 common alpine species found that 30% of observable variance in measured traits, such as specific leaf area and leaf nitrogen content, was among populations/genotypes of a given species [72]. Similar results were found for species growing in a dry tropical forest [73]. Considering R is highly variable among genotypes within species, to gain sufficient statistical power a high level of replication is required (Fig. 4b, c), further supporting the benefit of the high-throughput fluorophore technique we present.

Our whole-plant respiratory analysis demonstrated the important effects of plant development on leaf R and partitioning of R between tissue types, as previously demonstrated in Arabidopsis by Sew et al. [55], which could be detrimentally ignored if the power of high-throughput respiratory analysis was not readily available. The results highlight the fact that, when measuring leaf, stem and root O_2 uptake in the gas phase, leaf R_{dark} accounted for 51% of the entire R budget. In other words, close to half of all vegetative-stage wheat R occurs in non-leaf tissue, a finding reported for previous studies that quantified whole-plant CO_2 fluxes [15–19]. Yet, we tentatively suggest that the majority of plant R reports would focus entirely, or predominantly on leaf R. Furthermore, the oldest and newest emerging leaves had considerably higher mass-based rates of R_{dark} than intermediate aged leaves. In regards to the latter, this is presumably due to the added cost of growth R as well as maintenance R for newly emerging leaves [6]. The spike in R for the oldest leaves may reflect the costs associated with senescence, such as an energy expensive remobilisation of nutrients from the senescing leaf to other parts of the plant. For example, in oats (*Avena sativa*), promotion of senescence of leaves by withholding light leads to a greater than two-fold increase in O_2 consumption, attributed to decoupling of R_{dark} from oxidative phosphorylation, and amino-acid and soluble sugar liberation during senescence [74].

Conclusions

The high-throughput and tissue size versatility of the experiments we conducted highlight the comparative advantages of an automated gas-phase system, over other systems based on the same technology but reliant on aqueous-phase and limited sample tubes and volumes. Although aqueous-phase fluorophore systems may be relatively high-throughput when compared to the older technology of Clark-type O_2 electrodes, liquid-phase measurements still require extensive time in preparation of solutions, dispensing of solutions, and delicate sample positioning or sufficient stirring to facilitate O_2 movement to the sensor [e.g. 75]. We processed 138 samples, from tissue harvesting to initial O_2 uptake measurements, in a period of less than 2-h, which was possible due to the simple procedure of placing tissue in tubes, tightening

the caps and placing tubes in the designated instrument position. Such a fast turnaround for sample processing would not be possible in a non-fluorophore and/or aqueous-phase procedure. The speed at which samples can be processed and the versatility in sample size and tissue type enables respiratory analysis that simply would not be feasible using other established approaches. The simultaneous measurement of many genotypes and the construction of multiple whole-plant respiratory budgets emphasise the potential of this method and its wider application.

Additional file

Additional file 1: Table S1. Technical reproducibility of the fluorophore instrument in measuring the chemical oxidation of cysteine. **Figure S1.** The measured and expected depletion of O_2 though replacement of air with known volumes of pure CO_2 gas. **Figure S2.** A representative 46-day-old wheat plant harvested for whole-plant respiration analysis. **Figure S3.** Derived respiration rates in the dark (R_{dark}) calculated from measurements between 5 and 20 min from experiment initiation using the various techniques.

Authors' contributions
APS, ACAN, BO, AHM and OKA conceived the idea for the study. APS, ACAN, BO, FAAF, LH, YZ, VC and MRB conducted the experiments. APS wrote the first draft; all authors contributed significantly to subsequent versions. All authors read and approved the final manuscript.

Author details
[1] ARC Centre of Excellence in Plant Energy Biology, Research School of Biology, Building 134, The Australian National University, Canberra, ACT 2601, Australia. [2] Bayer CropScience SA-NV, Technologiepark 38, 9052 Gent (Zwijnaarde), Belgium. [3] ARC Centre of Excellence for Translational Photosynthesis, Building 134, The Australian National University, Canberra, ACT 2601, Australia. [4] Australian Research Council Centre of Excellence in Plant Energy Biology, University of Western Australia, 35 Stirling Highway, Crawley, WA 6009, Australia.

Acknowledgements
The support of the Australian Research Council (CE140100008) to OKA and AHM is acknowledged.

Competing interests
The authors declare that they have no competing interests.

Funding
Australian Research Council (CE140100008).

References
1. Tcherkez G, Boex-Fontvieille E, Mahe A, Hodges M. Respiratory carbon fluxes in leaves. Curr Opin Plant Biol. 2012;15:308–14.
2. Bouma TJ, Devisser R, Janssen JHJA, Dekock MJ, Vanleeuwen PH, Lambers H. Respiratory energy requirements and rate of protein turnover in vivo determined by the use of an inhibitor of protein synthesis and a probe to assess its effect. Physiol Plant. 1994;92:585–94.

3. Bouma TJ, De VR, Van LPH, De KMJ, Lambers H. The respiratory energy requirements involved in nocturnal carbohydrate export from starch-storing mature source leaves and their contribution to leaf dark respiration. J Exp Bot. 1995;46:1185–94.

4. Noguchi K, Yoshida K. Interaction between photosynthesis and respiration in illuminated leaves. Mitochondrion. 2008;8:87–99.

5. Lambers H. Respiration in intact plants and tissues: its regulation and dependence on environmental factors, metabolism and invaded organisms. In: Douce R, Day DA, editors. Encyclopedia of plant physiology, vol. 18. New York: Springer; 1985. p. 417–73.

6. Amthor JS. The McCree–de Wit–Penning de Vries–Thornley respiration paradigms: 30 years later. Ann Bot. 2000;86:1–20.

7. Lambers H, Atkin OK, Scheurwater I. Respiration patterns in roots in relation to their functioning. In: Waisel Y, Eshel A, Kafkaki U, editors. Plant roots the hidden half. New York: Marcel Dekker; 1996. p. 323–62.

8. Gifford RM. Plant respiration in productivity models: conceptualisation, representation and issues for global terrestrial carbon-cycle research. Funct Plant Biol. 2003;30:171–86.

9. Huntingford C, Zelazowski P, Galbraith D, Mercado LM, Sitch S, Fisher R, Lomas M, Walker AP, Jones CD, Booth BBB, et al. Simulated resilience of tropical rainforests to CO_2-induced climate change. Nat Geosci. 2013;6:268–73.

10. Amthor JS, Wilkinson RE: Plant respiratory responses to the environment and their effects on the carbon balance. In: Plant-environment interactions. volume 1. New York: Marcel Dekker; 1994. p. 501–54.

11. Atkin OK, Bloomfield KJ, Reich PB, Tjoelker MG, Asner GP, Bonal D, Bönisch G, Bradford MG, Cernusak LA, Cosio EG, et al. Global variability in leaf respiration in relation to climate, plant functional types and leaf traits. New Phytol. 2015;206:614–36.

12. Heskel MA, O'Sullivan OS, Reich PB, Tjoelker MG, Weerasinghe KWLK, Penillard A, Xiang J, Egerton JJG, Creek D, Bloomfield KJ, et al. Convergence in the temperature response of leaf respiration across biomes and plant functional types. Proc Natl Acad Sci USA. 2016;113:3832–7.

13. O'Sullivan OS, Heskel MA, Reich PB, Tjoelker MG, Weerasinghe KWLK, Penillard A, Zhu L, Egerton JJG, Bloomfield KJ, Creek D, et al. Thermal limits of leaf metabolism across biomes. Glob Change Biol. 2016;23(1):209–23.

14. Reynolds M, Foulkes J, Furbank R, Griffiths S, King J, Murchie E, Parry M, Slafer G. Achieving yield gains in wheat. Plant Cell Environ. 2012;35:1799–823.

15. Poorter H, Remkes C, Lambers H. Carbon and nitrogen economy of 24 wild species differing in relative growth rate. Plant Physiol. 1990;94:621–7.

16. Loveys BR, Scheurwater I, Pons TL, Fitter AH, Atkin OK. Growth temperature influences the underlying components of relative growth rate: an investigation using inherently fast- and slow-growing plant species. Plant Cell Environ. 2002;25:975–87.

17. Atkin OK, Botman B, Lambers H. The causes of inherently slow growth in alpine plants: an analysis based on the underlying carbon economies of alpine and lowland poa species. Funct Ecol. 1996;10:698–707.

18. Gifford RM. Whole plant respiration and photosynthesis of wheat under increased CO_2 concentration and temperature: long-term vs short-term distinctions for modelling. Glob Change Biol. 1995;1:385–96.

19. Atkin OK, Scheurwater I, Pons TL. Respiration as a percentage of daily photosynthesis in whole plants is homeostatic at moderate, but not high, growth temperatures. New Phytol. 2007;174:367–80.

20. Millar AH, Whelan J, Soole KL, Day DA. Organization and regulation of mitochondrial respiration in plants. Annu Rev Plant Biol. 2011;62:79–104.

21. Vanlerberghe GC, McIntosh I. Alternative oxidase. from gene to function. Annu Rev Plant Physiol Plant Mol Biol. 1997;48:703–34.

22. Hauben M, Haesendonckx B, Standaert E, Van Der Kelen K, Azmi A, Akpo H, Van Breusegem F, Guisez Y, Bots M, Lambert B, et al. Energy use efficiency is characterized by an epigenetic component that can be directed through artificial selection to increase yield. Proc Natl Acad Sci USA. 2009;106:20109–14.

23. Wilson D, Jones JG. Effect of selection for dark respiration rate of mature leaves on crop yields of Lolium perenne cv. S23. Ann Bot. 1982;49:313–20.

24. Reynolds M, Langridge P. Physiological breeding. Curr Opin Plant Biol. 2016;31:162–71.

25. King AW, Gunderson CA, Post WM, Weston DJ, Wullschleger SD. Plant respiration in a warmer world. Science. 2006;312:536–7.

26. Wythers KR, Reich PB, Bradford JB. Incorporating temperature-sensitive Q_{10} and foliar respiration acclimation algorithms modifies modeled ecosystem responses to global change. J Geophys Res: Biogeosci. 2013;118:77–90.

27. Atkin OK, Millar AH, Turnbull MH. Plant respiration in a changing world. New Phytol. 2010;187:268–72.

28. Reich PB, Sendall KM, Stefanski A, Wei X, Rich RL, Montgomery RA. Boreal and temperate trees show strong acclimation of respiration to warming. Nature. 2016;531:633–6.

29. McLaughlin BC, Xu CY, Rastetter EB, Griffin KL. Predicting ecosystem carbon balance in a warming Arctic: the importance of long-term thermal acclimation potential and inhibitory effects of light on respiration. Glob Change Biol. 2014;20:1901–12.

30. Slot M, Rey-Sánchez C, Gerber S, Lichstein JW, Winter K, Kitajima K. Thermal acclimation of leaf respiration of tropical trees and lianas: response to experimental canopy warming, and consequences for tropical forest carbon balance. Glob Change Biol. 2014;20:2915–26.

31. Atkin OK, Bruhn D, Hurry VM, Tjoelker MG. The hot and the cold: unraveling the variable response of plant respiration to temperature. Funct Plant Biol. 2005;32:87–105.

32. Hunt S. Measurements of photosynthesis and respiration in plants. Physiol Plant. 2003;117:314–25.

33. Clark LC, Wolf R, Granger D, Taylor Z. Continuous recording of blood oxygen tensions by polarography. J Appl Physiol. 1953;6:189–93.

34. Poorter H, Van Der Werf A, Atkin OK, Lambers H. Respiratory energy requirements of roots vary with the potential growth rate of a species. Physiol Plant. 1991;83:469–75.

35. Loveys BR, Atkinson LJ, Sherlock DJ, Roberts RL, Fitter AH, Atkin OK. Thermal acclimation of leaf and root respiration: an investigation comparing inherently fast- and slow-growing plant species. Glob Change Biol. 2003;9:895–910.

36. Kurimoto K, Day DA, Lambers H, Noguchi K. Effect of respiratory homeostasis on plant growth in cultivars of wheat and rice. Plant Cell Environ. 2004;27:853–62.

37. Lambers H, Day DA, Azcón-Bieto J. Cyanide-resistant respiration in roots and leaves: measurements with intact tissues and isolated mitochondria. Physiol Plant. 1983;58:148–54.

38. Azcón-Bieto J, Lambers H, Day DA. Respiratory properties of developing bean and pea leaves. Aust J Plant Physiol. 1983;10:237–45.

39. Jacoby RP, Millar AH, Taylor NL. Assessment of respiration in isolated plant mitochondria using Clark-type electrodes. In: Whelan J, Murcha WM, editors. Plant mitochondria: methods and protocols. New York: Springer; 2015. p. 165–85.

40. Vasseur F, Pantin F, Vile D. Changes in light intensity reveal a major role for carbon balance in Arabidopsis responses to high temperature. Plant Cell Environ. 2011;34:1563–76.

41. Atkin OK, Scheurwater I, Pons TL. High thermal acclimation potential of both photosynthesis and respiration in two lowland Plantago species in contrast to an alpine congener. Glob Change Biol. 2006;12:500–15.

42. Whitehead D, Griffin KL, Turnbull MH, Tissue DT, Engel VC, Brown KJ, Schuster WSF, Walcroft AS. Response of total night-time respiration to differences in total daily photosynthesis for leaves in a Quercus rubra L. canopy: implications for modelling canopy CO_2 exchange. Glob Change Biol. 2004;10:925–38.

43. den Hertog J, Stulen I, Lambers H. Assimilation, respiration and allocation of carbon in Plantago major as affected by atmospheric CO_2 levels: a case-study. Vegetatio. 1993;104:369–78.

44. Kolling K, George GM, Kunzli R, Flutsch P, Zeeman SC. A whole-plant chamber system for parallel gas exchange measurements of Arabidopsis and other herbaceous species. Plant Methods. 2015;11:48.

45. Barbour MM, McDowell NG, Tcherkez GUIL, Bickford CP, Hanson DT. A new measurement technique reveals rapid post-illumination changes in the carbon isotope composition of leaf-respired CO_2. Plant Cell Environ. 2007;30:469–82.

46. Nakaema WM, Hao Z-Q, Rohwetter P, Wöste L, Stelmaszczyk K. PCF-based cavity enhanced spectroscopic sensors for simultaneous multicomponent trace gas analysis. Sensors. 2011;11:1620.

47. Beckmann K, Messinger J, Badger MR, Wydrzynski T, Hillier W. On-line mass spectrometry: membrane inlet sampling. Photosynth Res. 2009;102:511–22.

48. Hansen LD, Smith BN, Criddle RS, Breidenbach RW. Calorimetry of plant respiration. J Therm Anal Calorim. 1998;51:757–63.

49. Macfarlane C, Adams MA, Hansen LD. Application of an enthalpy balance model of the relation between growth and respiration to temperature acclimation of *Eucalyptus globulus* seedlings. Proc R Soc B: Biol Sci. 2002;269:1499–507.

50. Tyystjärvi E, Karunen J, Lemmetyinen H. Measurement of photosynthetic oxygen evolution with a new type of oxygen sensor. Photosynth Res. 1998;56:223–7.

51. Ast C, Schmälzlin E, Löhmannsröben H-G, van Dongen JT. Optical oxygen micro- and nanosensors for plant applications. Sensors. 2012;12:7015.

52. Yang J, Wang Z, Li Y, Zhuang Q, Gu J. Real-time monitoring of dissolved oxygen with inherent oxygen-sensitive centers in metal–organic frameworks. Chem Mater. 2016;28:2652–8.

53. Warkentin M, Freese HM, Karsten U, Schumann R. New and fast method to quantify respiration rates of bacterial and plankton communities in freshwater ecosystems by using optical oxygen sensor spots. Appl Environ Microbiol. 2007;73:6722–9.

54. Moodley L, Steyaert M, Epping E, Middelburg JJ, Vincx M, van Avesaath P, Moens T, Soetaert K. Biomass-specific respiration rates of benthic meiofauna: demonstrating a novel oxygen micro-respiration system. J Exp Mar Biol Ecol. 2008;357:41–7.

55. Sew YS, Ströher E, Holzmann C, Huang S, Taylor NL, Jordana X, Millar AH. Multiplex micro-respiratory measurements of Arabidopsis tissues. New Phytol. 2013;200:922–32.

56. Zhao G-W, Zhong T-L. Improving the assessment method of seed vigor in *Cunninghamia lanceolata* and *Pinus massoniana* based on oxygen sensing technology. J For Res. 2012;23:95–101.

57. Guangwu Z, Xuwen J. Roles of gibberellin and auxin in promoting seed germination and seedling vigor in *Pinus massoniana*. For Sci. 2014;60:367–73.

58. Atkin OK, Evans JR, Siebke K. Relationship between the inhibition of leaf respiration by light and enhancement of leaf dark respiration following light treatment. Aust J Plant Physiol. 1998;25:437–43.

59. Gonzalez-Meler MA, Ribas-Carbo M, Siedow JN, Drake BG. Direct inhibition of plant mitochondrial respiration by elevated CO_2. Plant Physiol. 1996;112:1349–55.

60. Amthor JS, Koch GW, Bloom AJ. CO_2 inhibits respiration in leaves of *Rumex crispus* L. Plant Physiol. 1992;98:757–60.

61. Drake BG, Azcon-Bieto J, Berry J, Bunce J, Dijkstra P, Farrar J, Gifford RM, Gonzalez-Meler MA, Koch G, Lambers H, et al. Does elevated atmospheric CO_2 concentration inhibit mitochondrial respiration in green plants? Plant Cell Environ. 1999;22:649–57.

62. Curtis PS, Wang XZ. A meta-analysis of elevated CO_2 effects on woody plant mass, form, and physiology. Oecologia. 1998;113:299–313.

63. Jahnke S. Atmospheric CO_2 concentration does not directly affect leaf respiration in bean or poplar. Plant Cell Environ. 2001;24:1139–51.

64. Bruhn D, Mikkelsen TN, Atkin OK. Does the direct effect of atmospheric CO_2 concentration on leaf respiration vary with temperature? Responses in two species of *Plantago* that differ in relative growth rate. Physiol Plant. 2002;114:57–64.

65. Tjoelker MG, Oleksyn J, Lee TD, Reich PB. Direct inhibition of leaf dark respiration by elevated CO_2 is minor in 12 grassland species. New Phytol. 2001;150:419–24.

66. Klinker JE. A modification of the warburg respirometer to measure the respiration rate of tomato leaf discs. Plant Physiol. 1950;25:354–5.

67. Macnicol PK. Rapid metabolic changes in the wounding response of leaf discs following excision. Plant Physiol. 1976;57:80–4.

68. Azcón-Bieto J, Day DA, Lambers H. The regulation of respiration in the dark in wheat (*Triticum aestivum* cultivar Gabo) leaf slices. Plant Sci Lett. 1983;32:313–20.

69. Reich PB, Tjoelker MG, Pregitzer KS, Wright IJ, Oleksyn J, Machado JL. Scaling of respiration to nitrogen in leaves, stems and roots of higher land plants. Ecol Lett. 2008;11:793–801.

70. Azcón-Bieto J, Lambers H, Day DA. Effect of photosynthesis and carbohydrate status on respiratory rates and the involvement of the alternative pathway in leaf respiration. Plant Physiol. 1983;72:598–603.

71. Day DA, de Vos OC, Wilson D, Lambers H. Regulation of respiration in the leaves and roots of two *Lolium perenne* populations with contrasting mature leaf respiration rates and crop yields. Plant Physiol. 1985;78:678–83.

72. Albert CH, Thuiller W, Yoccoz NG, Douzet R, Aubert S, Lavorel S. A multi-trait approach reveals the structure and the relative importance of intra- vs. interspecific variability in plant traits. Funct Ecol. 2010;24:1192–201.

73. Hulshof CM, Swenson NG. Variation in leaf functional trait values within and across individuals and species: an example from a Costa Rican dry forest. Funct Ecol. 2010;24:217–23.

74. Tetley RM, Thimann KV. The metabolism of oat leaves during senescence: I. Respiration, carbohydrate metabolism, and the action of cytokinins. Plant Physiol. 1974;54:294–303.

75. Sew YS, Millar AH, Stroeher E. Micro-respiratory measurements in plants. In: Whelan J, Murcha WM, editors. Plant mitochondria: methods and protocols. New York: Springer; 2015. p. 187–96.

15

High spatio-temporal-resolution detection of chlorophyll fluorescence dynamics from a single chloroplast with confocal imaging fluorometer

Yi-Chin Tseng[1] and Shi-Wei Chu[1,2*] (iD)

Abstract

Background: Chlorophyll fluorescence (CF) is a key indicator to study plant physiology or photosynthesis efficiency. Conventionally, CF is characterized by fluorometers, which only allows ensemble measurement through wide-field detection. For imaging fluorometers, the typical spatial and temporal resolutions are on the order of millimeter and second, far from enough to study cellular/sub-cellular CF dynamics. In addition, due to the lack of optical sectioning capability, conventional imaging fluorometers cannot identify CF from a single cell or even a single chloroplast.

Results and discussion: Here we demonstrated a fluorometer based on confocal imaging, that not only provides high contrast images, but also allows CF measurement with spatiotemporal resolution as high as micrometer and millisecond. CF transient (the Kautsky curve) from a single chloroplast is successfully obtained, with both the temporal dynamics and the intensity dependences corresponding well to the ensemble measurement from conventional studies. The significance of confocal imaging fluorometer is to identify the variation among individual chloroplasts, e.g. the temporal position of the P–S–M phases, and the half-life period of P–T decay in the Kautsky curve, that are not possible to analyze with wide-field techniques. A linear relationship is found between excitation intensity and the temporal positions of P–S–M peaks/valleys in the Kautsky curve. Based on the CF transients, the photosynthetic quantum efficiency is derived with spatial resolution down to a single chloroplast. In addition, an interesting 6-order increase in excitation intensity is found between wide-field and confocal fluorometers, whose pixel integration time and optical sectioning may account for this substantial difference.

Conclusion: Confocal imaging fluorometers provide micrometer and millisecond CF characterization, opening up unprecedented possibilities toward detailed spatiotemporal analysis of CF transients and its propagation dynamics, as well as photosynthesis efficiency analysis, on the scale of organelles, in a living plant.

Keywords: Optical section, 3D microscopy, Kautsky curve, Chlorophyll fluorescence transient

Background

Chlorophyll fluorescence (CF) has been proven to be one of the most powerful and widely used techniques for plant physiologists [1–7]. Despite of its low quantum efficiency (2–10% of absorbed light [8]), CF detections are meaningful due to its intricate connection with numerous internal processes during photosynthesis, such as reduction of photosystem reaction centers, non-photochemical quenching, etc. [9, 10]. It is well known that the efficiency of photosynthesis can be derived from CF dynamics, thus providing noninvasive, fast and accurate characterization for photosynthesis. CF characterization has been widely adopted to study plant physiology, including stress tolerance, nitrogen balance, carbon fixation efficiency, etc. [11]. It is not too exaggerated to say

---

*Correspondence: swchu@phys.ntu.edu.tw
[1] Department of Physics, National Taiwan University, No. 1, Section 4, Roosevelt Rd, Da'an District, Taipei City 10617, Taiwan
Full list of author information is available at the end of the article

that nowadays, no investigation about photosynthetic process would be complete without CF analysis.

Conventionally, the tool of choice to study CF is a fluorometer. There are many different fluorometry techniques, such as plant efficiency analyzer (PEA) [12], pulse amplitude modulation (PAM) [13], the pump and probe (P&P) [14, 15] and the fast repetition rate (FRR) [16]. It is interesting to note that these various detection approaches are all based on the same principle, i.e. the Kautsky effect [7], or equivalent, CF transient when moving photosynthetic material from dark adaption to light environment.

Conventional imaging fluorometers (e.g. PAM and P&P fluorometers) are based on wide-field detection, and are routinely adopted to study ensemble of CF transients from a large area of a leaf, significantly limiting its spatiotemporal resolution. For example, to study stress propagation in a plant leaf [17], current imaging fluorometers only provide spatial resolution on the order of millimeter, with temporal resolution on the order of second. To unravel the more detailed propagation dynamics, the required spatial resolution should be at least on single cell or sub-cellular level, while the temporal resolution should be enhanced to millisecond scale.

The concept of introducing fluorescent microscope to study high-resolution CF dynamics has been realized two decades ago [18], but the drawback of the early microscopic fluorometer version is the lack of optical section capability due to its wide-field nature, and thus prevents study of CF transient on a truly single cell or even a single chloroplast level. Confocal microscopy, which is known to provide optical sectioning with exceptionally high axial contrast, has been extensively used for CF imaging with sub-micrometer resolution [19–23]. However, the high-speed time-lapsed imaging capability is less explored in earlier works.

Here we introduce a concept of confocal imaging fluorometer, which is the combination of confocal microscopy and CF transient detection. The technique not only detects CF signals with millisecond temporal resolution, but also attains micrometer spatial resolution in all three dimensions. The CF transient (Kautsky curve) within a single chloroplast is successfully retrieved. With statistical comparison, the CF transients of a group of palisade cells and the ensemble of single chloroplasts are found to be similar to each other, and both correspond well to the result of conventional imaging fluorometers, showing the reliability of our result. Nevertheless, the CF transient of individual chloroplast can be substantially different, manifesting the value of the unusual capability to study plant cell organelles. Furthermore, we found that the shape of transients is highly intensity-dependent, which is also shown in an earlier study [24]. We also found that

the short integration time and optical section characteristic of confocal image fluorometer make a significant difference of illumination intensity comparing to that of conventional fluorometers. Given CF transient from a single chloroplast, it is possible to investigate degree of influence from external or internal plant-stress with scale of organelle, and confocal imaging fluorometer has paved the way for this high spatiotemporal resolution CF detection.

Principle
Basic concept of confocal imaging fluorometer

The optical principle of confocal imaging fluorometer is basically the same as confocal laser-scanning microscopy [25], which is an optical imaging technique for increasing contrast and resolution. The essential components of a confocal imaging fluorometer is shown in Fig. 1, including a laser system, a dichroic mirror, a scanning mirror system, an objective lens, a pinhole and a photomultiplier tube (PMT).

The laser system in a confocal imaging fluorometer provides strong and monochromatic illumination, whose wavelength can be selected to meet sample request. The laser beam is sent to the objective after the scanning mirror system to achieve two-dimensional raster scanning

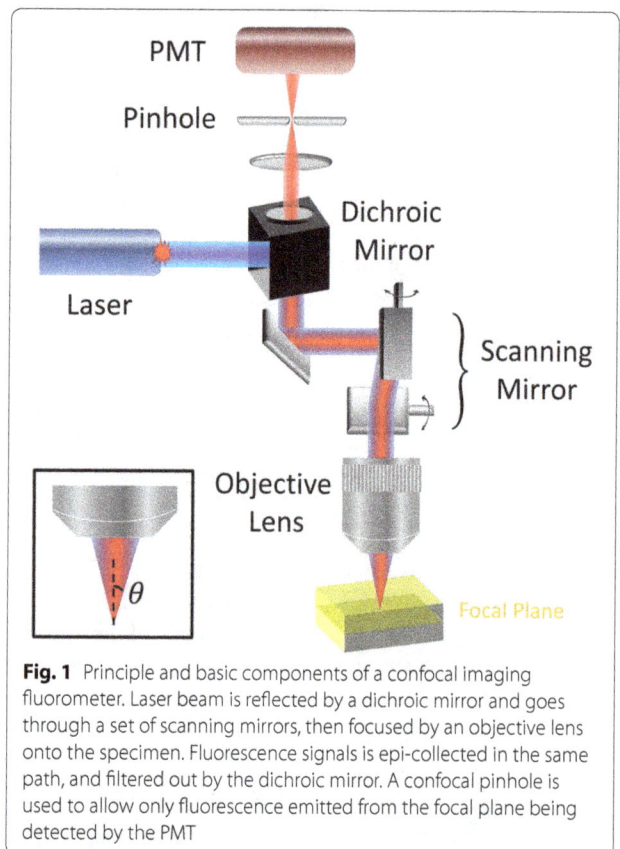

Fig. 1 Principle and basic components of a confocal imaging fluorometer. Laser beam is reflected by a dichroic mirror and goes through a set of scanning mirrors, then focused by an objective lens onto the specimen. Fluorescence signals is epi-collected in the same path, and filtered out by the dichroic mirror. A confocal pinhole is used to allow only fluorescence emitted from the focal plane being detected by the PMT

at the focal plane. The backward fluorescence signal is collected by the same objective, de-scanned through the scanning mirrors, and separated from residual laser by the dichroic mirror. The fluorescence signal then is focused onto the pinhole, which is placed at the conjugate plane of objective focus, to achieve optical sectioning by excluding out-of-focus signals. One or more PMTs are placed behind the pinhole to collect the in-focus fluorescence signals, which are reconstructed into images by synchronization with the scanning mirrors [25].

In general, a confocal imaging system is capable of collecting signal with a well-defined optical section on the order of 1 μm [26]. This high axial resolution makes confocal system an invaluable tool to observe single cell or sub-cellular organelles [27–29].

The objective lens is characterized by magnification and numerical aperture (NA). To enable large field-of-view observation, low magnification objectives are typically required. However, please note that resolution is determined by NA, which can be independent from magnification. NA describes the light acceptance cone of an objective lens and hence light gathering ability and resolution. The definition of NA is:

$$NA \equiv n \times \sin\theta, \tag{1}$$

where n is the index of refraction of the immersion medium, and θ is the half-angle of the maximum light acceptance cone. Both lateral (xy-direction) and axial (z-direction) resolutions for fluorescence imaging mode are defined by NA and the wavelength (λ) [30].

$$r_{lateral} = \frac{0.43 \times \lambda}{NA} \tag{2}$$

$$r_{axial} = \frac{0.67 \times \lambda}{n - \sqrt{n^2 - NA^2}} \tag{3}$$

To compare the actinic light illumination in a conventional fluorometer, e.g. PAM, and in a laser scanning confocal fluorometer, there are several aspects. First, in PAM the actinic light is provided by a lamp or an LED, which is an incoherent light source; while in a confocal system, the laser excitation is coherent. Second, the spectral bandwidth of a laser is in general much narrower than that of a lamp, which is typically tens of nanometers even after adding bandpass filters. Third, wide-field illumination is adopted in PAM, while point-scan is used in confocal.

Although there are many differences between the illumination method of the conventional fluorometer and the confocal one, in an early work [31], it has been shown that the actinic effect of using a Xe lamp or a laser is equivalent. In a more recent work [23], they have shown that frequency of scanning (~300 s^{-1}) does not seem to affect the response, even when compared to wide-field

illumination. In our current work, the scanning frequency on each chloroplast is about 10,000 s^{-1}. However, as we will show in the results, clear OPSMT transitions and similar intensity-dependent CF dynamics are all observable. Therefore, it seems that the high-frequency laser beam movement does not cause significant effects on CF dynamics.

Kautsky effect

Kautsky effect, discovered in 1931, describes the dynamics of CF when dark-adapted photosynthetic chlorophyll suddenly exposes to continuous light illumination [32]. After initial light absorption, chlorophyll becomes excited and soon releases its energy into one of the three internal decay pathways, including photosynthesis (photochemical quenching, qP), heat (non-photochemical quenching, NPQ) and light emission (CF). Owing to energy conservation, the sum of quantum efficiencies for these three pathways should be unity. Therefore, the yield of CF is strongly related to the efficiency of both qP and NPQ [33].

To be more specific, when transferring a photosynthetic material from dark adaption into light illumination, CF yield typically exhibits a fast rising phase (within 1 s) and a slow decay phase (few minute duration), as shown by the green curve in Fig. 2. The fast rising phase is labeled as O–P, where O is for origin, and P is the peak [24]. It is mainly caused by the reduction of qP; that is, depletion of electron acceptors, quinine (Qa) in the electron transport chain [34]. The slow decay phase is labeled as P–S–M–T, where S stands for semisteady state, M for a local maximum, and T for a terminal steady state level

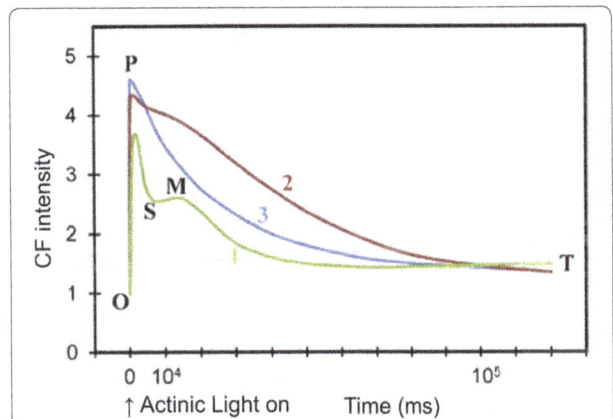

Fig. 2 The Kautsky effect, showing the CF transient as well as its intensity dependence. Wavelength of excitation: 650 nm. Excitation light intensity for curves labeled 1, 2 and 3 was 32, 320 and 3200 μmol/m^2/s, respectively. For definition of OPSMT, O is the origin, P is the peak, S stands for semi-steady state, M for a local maximum, and T for a terminal steady state level (Modified figure from [1], with copyright permission)

[24]. One very interesting phenomenon is the shape of this decay phase depends strongly on illumination intensity. At low intensity (32 μmol/m^2/s), the Kautsky curve is the green one. When the intensity grows one order larger, the amplitude of S–M rise in the transient is smaller, as shown by the red curve. At one more order higher intensity, the blue curve shows that the S–M section disappears completely, leaving an exponential decay in the P–T section. This is known as saturation state, which is critical to derive the quantum efficiency of photosynthesis. Such intensity-dependent curve transition is the result of photosynthetic state transition, and more detailed discussion can be found in the references [1, 10, 13, 35–37].

Methods

Plant sample

Brugmansia suaveolens (solanaceae), also known as Angel's Trumpet, was a woody plant usually 3–4 m in height with pendulous flowers and furry leaves distributed widely in Taiwan, especially in wet areas. Being interested in spatiotemporal dynamics of CF, we selected *B. suaveolens* as our target material since the CF of its cousin *Datura wrightii*, also known as Devil's Trumpet, had been studied in depth [17]. *B. suaveolens* leaves were collected from the Botanical Garden of National Taiwan University, Taipei, Taiwan (25°1′N, 121°31′E, 9 m a.s.l.). All sample leaves were picked as fully expanded leaves that had neither experienced detectable physical damage nor herbivory. In order to minimize the sampling error, three leaves were chosen within plants that grew in similar micro-climate. Furthermore, all the measurements were completed no longer than two hours after disleaving. Fresh leaves were sealed in slide glass (76 × 26 mm), and slide samples were dark-adapted under constant temperature and constant humidity dark environment (20 °C, 70%RH) for 20 min.

Experimental setup

A confocal microscope (Leica TCS SP5) in the Molecular Imaging Center of National Taiwan University was adopted. CF was excited by a HeNe laser, whose wavelength (633 nm) was the same as that used in popular conventional fluorometers, such as LI-6400 from LI-COR. A relatively low-NA objective (HC PL Apo 10×/0.4 CS) was selected to allow not only large field of view over a few millimeters, but also spatial resolution better than a single chloroplast. From Eqs. (2) and (3), the lateral and axial resolutions were 1 and 5 μm, respectively. Although this was not particularly high compared to common confocal imaging system, due to the low-NA objective here, the three-dimensional spatial resolution was much better than conventional imaging fluorometers.

To operate the confocal fluorometer, the initial step was to bring the sample to focus by weak excitation (~1 kW/cm^2, or equivalently 5.56 × 10^7 μmol/m^2/s for intensity conversion, please see "Discussion"), and then the leaf was left in dark again for 5 min. To observe the Kautsky effect, the 633-nm laser was focused on the sample, and the fluorescence emission was recorded in the spectral range of 670–690 nm. The intensity-dependent CF transient curves were obtained by taking time-lapsed images while varying the 633-nm excitation intensity from 1 to 55 kW/cm^2, at different sample regions. For experimental details, the scanning speed was 1400 Hz (1400 rows per second), the pinhole size was 52 μm (one Airy diameter), the built-in PMT voltage was set at 600 V, and a dichroic filter TD 488/543/633 was included in the optical path. With different number of total pixels, the temporal resolution of the CF transient varies from 10 ms (16 × 16 pixels) to about 200 ms (256 × 256 pixels). No significant photobleaching of CF was expected at this intensity range [38].

Results

Fluorescence dynamics from a single chloroplast

Conventional fluorometers observe CF dynamics over a large area on a leaf, and here we demonstrate that the confocal imaging fluorometer allows us to obtain CF transients from a precisely chosen cell or even a single chloroplast. Figure 3a shows the confocal images of a leaf sample. (a1) is the large-area view, showing the distribution of vascular bundles, while (a2) gives a zoom-in view of a group of palisade cells, showing clear distribution of chloroplasts in each cell. By further zooming in, the field of view is focused onto a single chloroplast, as given in (a3), showing the distribution of chlorophyll density inside the organelle [39].

Figure 3b presents the CF transients at low intensity illumination (3 kW/cm^2) from a group of palisade cells (b1) and a single chloroplast (b2). The latter is noisier due to less pixels involved. The characteristic P–S decay and S–M rise of Kautsky curve are obvious in both (b1) and (b2). In Fig. 3b1, based on the statistics of 30 chloroplasts, the averaged timing points for P–S–M states are 1.8, 5.9 and 10.4 s, respectively, corresponding well with the reported values in the literature (Fig. 2). On the other hand, box plots are embedded in Fig. 3b1 to show the variations of time and intensity in P–S–M states between the 30 individual chloroplasts. The bottom and top of the box are the first and third quartiles, while the ends of the whiskers represent the maximum and minimum values. This result not only confirms that the averaged Kautsky curves acquired by the confocal fluorometer are similar to the curves taken with conventional fluorometers, but

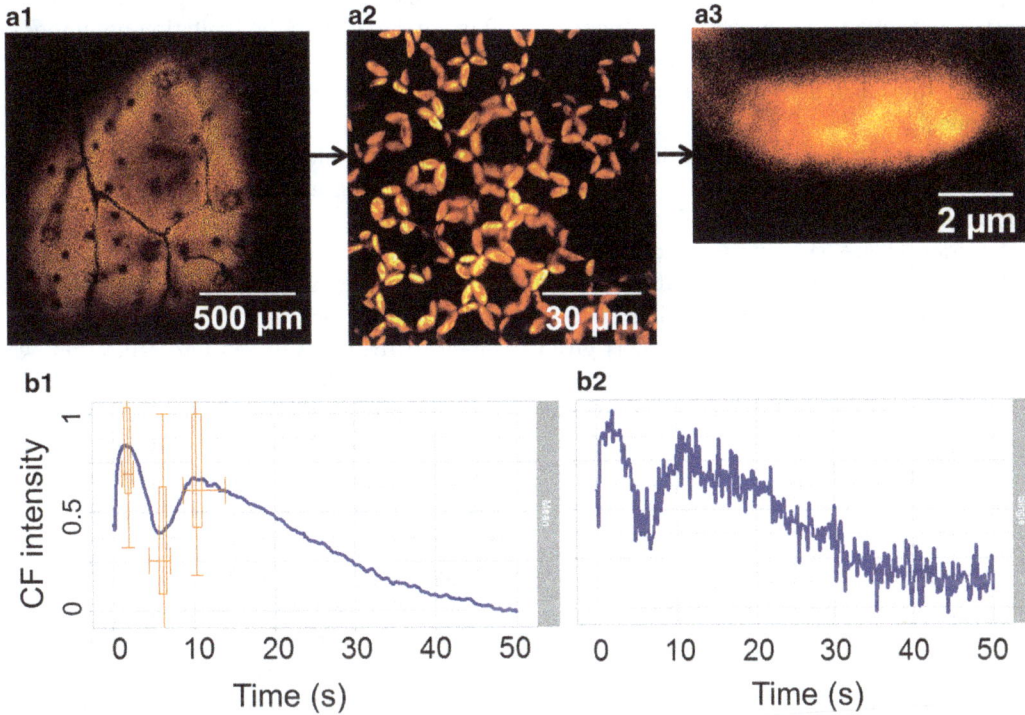

Fig. 3 Confocal images and CF transients on different spatial scales inside a living leaf. A 633-nm laser, with 3 kW/cm^2 intensity, is adopted for 50 s continuous confocal imaging. The sample leaf was kept in darkness for ~20 min before imaging. **a1** The image over a large area of the leaf, **a2** zoomed into show a group of palisade cells, and **a3** further zoomed into focus onto a single chloroplast. **b1**, **b2** CF transient from a group of palisade cells and a single chloroplast, respectively. **b2** Noisier since less pixels are involved

also shows that the variations between individual chloroplasts are indeed significant.

Intensity dependent fluorescence transient

As we have mentioned in Fig. 2, it is well known that the Kautsky curve changes with intensity. Figure 4 shows the intensity-dependent Kautsky curves from ~780 chloroplasts (colored lines) along with their standard error (gray lines), obtained by the confocal fluorometer. Note that to make the standard error visible on the same scale, it is multiplied by 16. Figure 4a is acquired with low laser intensity (3 kW/cm^2), and a temporal variation similar to curve 1 of Fig. 2 is found, i.e. a complete O–P–S–M–T curve. The CF intensity rises to its first peak within 1 s (O–P rise), quickly decreasing to a local minimum (P–S fall), rising again to a second peak (S–M rise) then slowly falling as exponential decay (M–T decay). At slightly higher intensity (10 kW/cm^2), a temporal variation similar to curve 2 of Fig. 2 is observed. The P–S fall and S–M rise still exist, but become much smaller, while the positions of P, S, and M appear earlier in the curve. At high intensity (55 kW/cm^2), the S–M part disappears completely, leaving a single exponential P–T decay, similar to

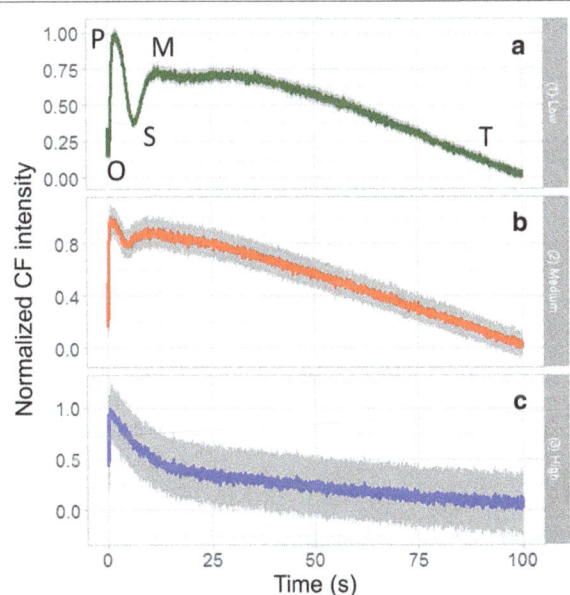

Fig. 4 Averaged CF transients from ~780 chloroplasts (*colored*) with standard error (×16, *gray*) under excitation intensity at **a** 3, **b** 10, and **c** 55 kW/cm2, respectively, showing clearly the intensity-dependent Kautsky curves

curve 3 of Fig. 2, i.e. saturation state. This result matches very well to the conventional wide-field fluorometer [1, 13, 36], but with much higher spatiotemporal resolution, manifesting again the reliability and usefulness of the confocal technique.

From Fig. 4, not only the curve shape is intensity-dependent, but the positions of local maxima and minimum (P, S, M points) are strongly dependent on excitation intensities. Figure 5a shows the detailed curve variation relative to intensity, in the range of 3–55 kW/cm^2, and the corresponding temporal position of local maximum of induced transients, i.e. point M, is given in Fig. 5b. Surprisingly, an almost perfect linear trend is observed. Similar linear results are found for the

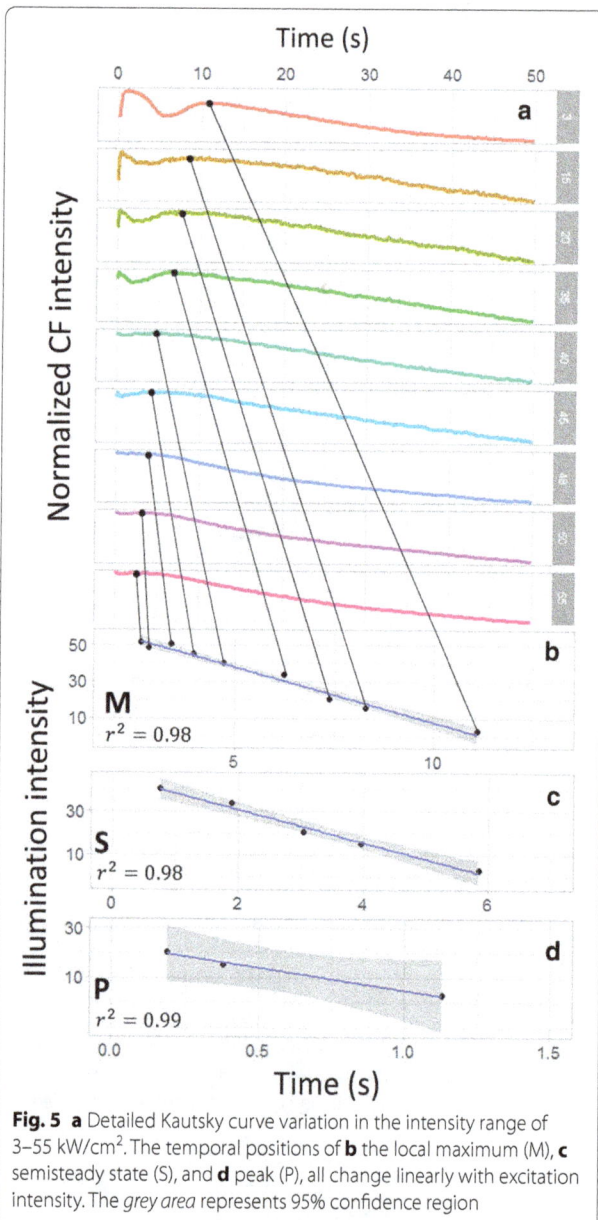

Fig. 5 a Detailed Kautsky curve variation in the intensity range of 3–55 kW/cm^2. The temporal positions of **b** the local maximum (M), **c** semisteady state (S), and **d** peak (P), all change linearly with excitation intensity. The *grey area* represents 95% confidence region

semi-steady state point S in Fig. 5c, and for the peak point P in Fig. 5d. Due to the limitation of temporal resolution (200 ms for 256 × 256 pixels), S and P points are analyzed with intensity range 3–40 kW/cm^2 and 3–20 kW/cm^2, respectively. The linear trends indicate that the state transition rate increases with higher excitation intensity. The underlying mechanism relies more investigation in the future.

Fluorescence dynamics under saturation intensity

In the last section, we have shown that at high excitation intensity, the CF is driven into saturation, which is very important for the quantum efficiency calculation. Thus, here we provide further characterization of the saturation states across individual chloroplasts. In Fig. 6a, the green color provides the spatial distribution of CF intensity over many living cells, and the red color shows the distribution of P–T phase decay time constant. For better identification, the two colors are shown separately in Fig. 6b, c. The statistical analysis for the P–T decay time constant of transients from individual chloroplasts is derived from Fig. 6c. The averaged decay time constant of a large area of leaf is 34.6 s, again matching well to the reported values in Fig. 2. Nevertheless, the standard variation of the decay time constant is 10.6 s, which reaches one-third of the average value, so significant divergence exists between each chloroplast. This decay time divergence is manifested by explicitly showing four Kautsky curves from individual chloroplasts in Fig. 6c1–c4.

Please note that error values in Fig. 6c1–c4 are the least square errors when fitting the curves with an exponential decay, different from the statistical standard deviation above. When analyzing data from a single chloroplast, the signal-to-noise ratio is relatively low, resulting in about 10% error in the time constant determination. Fortunately, the variation of time constants among chloroplasts is much larger than 10%, so this error is still tolerable. In the case where reduced error is necessary, the confocal system provides the flexibility to increase the integration time (reducing temporal resolution), so that higher signal-to-noise ratio can be achieved.

Since the laser intensity is relatively strong, it is necessary to confirm the reproducibility of the Kautsky curve in the same region of chloroplasts. Figure 7a1 shows the confocal CF image of a group of cells, and the corresponding averaged Kautsky curve is given in Fig. 7b1. The excitation intensity is 55 kW/cm^2, which is adequate to saturate the photosystem, so a curve similar to 3 in Fig. 2 is observed. The sample was then kept in dark for 5 min, before the same intensity was applied again. The results of second excitation is given in Fig. 7a2, b2. Apparently, the Kautsky curve is fully recoverable, even under relatively high illumination intensity.

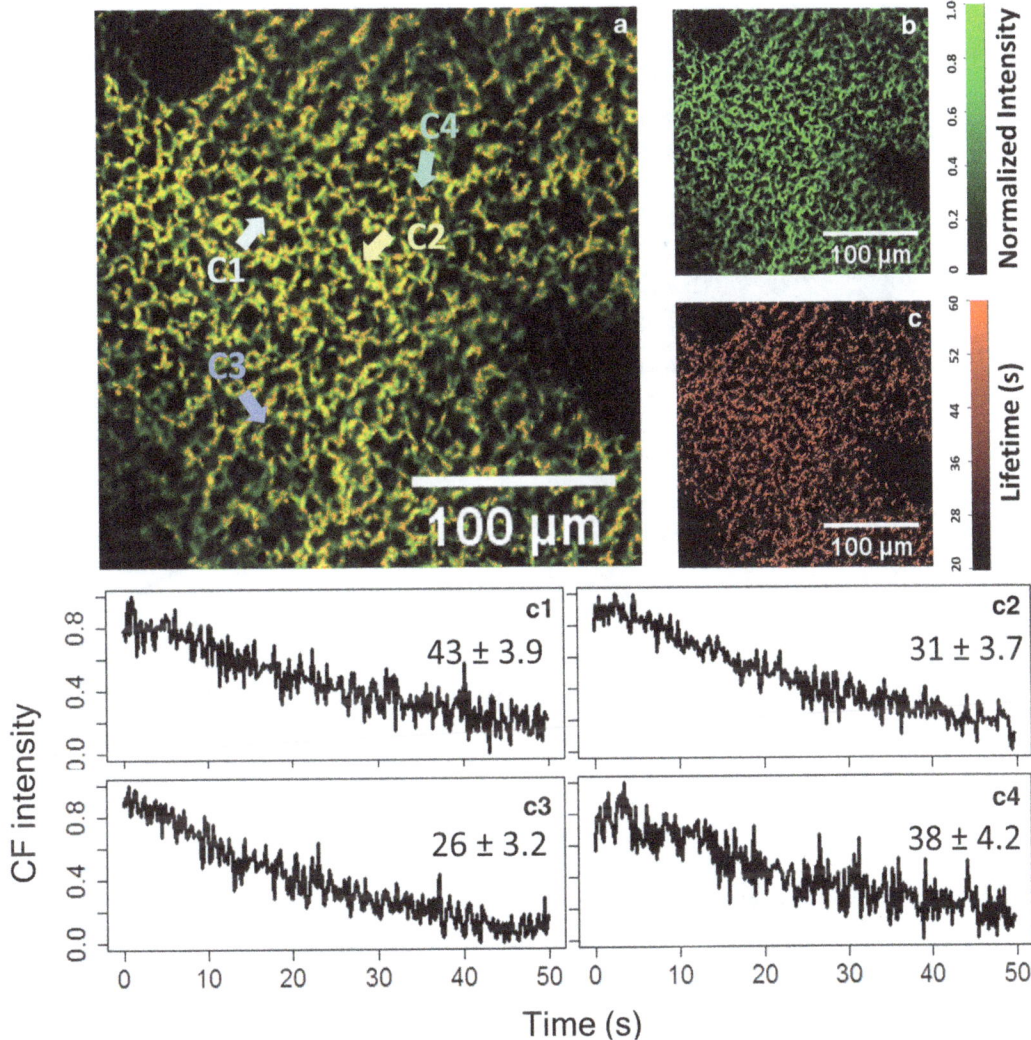

Fig. 6 High-resolution spatial distribution of CF intensity (*green* in **a**, **b**) and of the PT-phase time constant (*red* in **a**, **c**). The fluorescence transients of four selected chloroplasts within a living leaf are shown in the *bottom panels*, manifesting the significant difference in the time constants. The dataset is acquired at 40 kW/cm² with a HeNe laser (633 nm)

Deriving quantum efficiency of photosystem II

We have shown that intensity-dependent CF transient is found on the scale of cells and chloroplasts, it is then straightforward to derive the physiologically important factors, such as the maximal quantum efficiency of photosystem II (Φ_{PSII}). To derive maximal Φ_{PSII}, the first step is to quantify the fluorescence yield, which is the ratio between CF intensity and excitation intensity. In our work, the relative quantum yield values are obtained by normalizing the CF intensities to the fluorescence intensity of a commercial fluorescent slide (92001, Chroma Tech., VT) under the same excitation intensity. The values of relative quantum yield at low excitation intensity (Φ_{F0}, at 3 kW/cm²) and at saturation intensity (Φ_{Fm}, at 55 kW/cm²) are given in Table 1. Then the spatial distribution of Φ_{PSII} is obtained with pixel-by-pixel calculation of $(\Phi_{Fm} - \Phi_{F0})/\Phi_{Fm}$, as shown in Fig. 8. Apparently, the effect of the fluorescent slide is removed when calculating the quantum efficiency with the above equation. Numerical values of quantum efficiencies on different scales are also listed in Table 1. Similar to the results of Kautsky curves, the mean values of quantum efficiency are similar throughout a large area of leaf to a single chloroplast. On the other hand, from Table 1 and Fig. 8, the value of Φ_{PSII} can be very different among individual chloroplasts, once again manifesting the significance of high-resolution mapping of the CF dynamics inside a living plant.

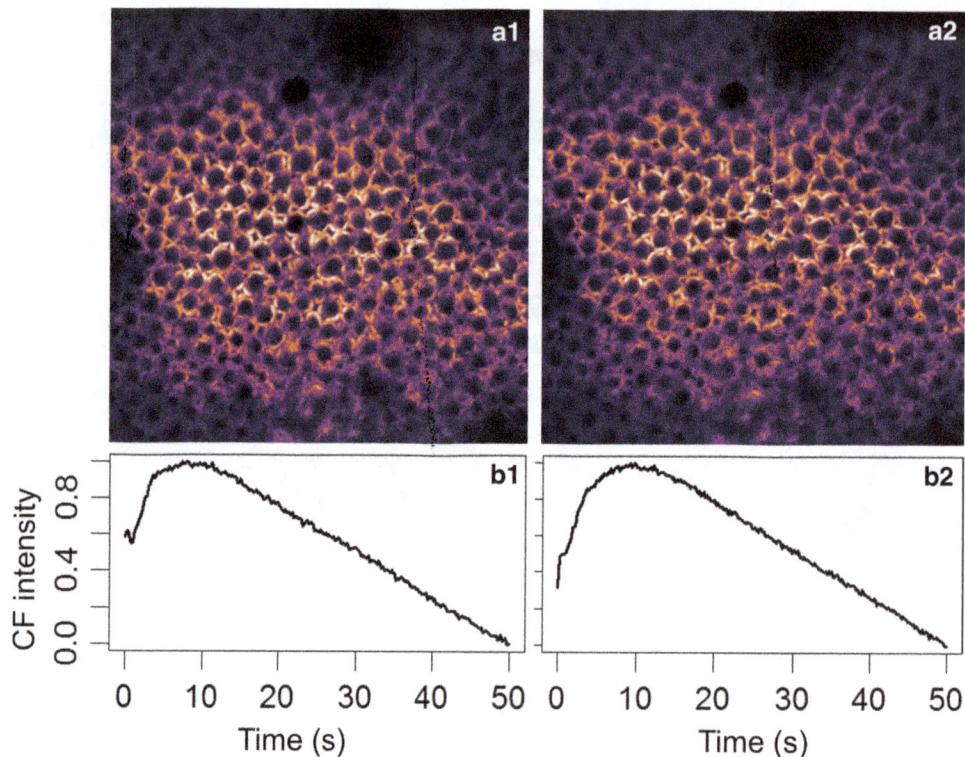

Fig. 7 The reproducibility of Kautsky curve under strong illumination intensity (55 kW/cm^2). **a1**, **b1** are confocal CF image and Kautsky curve for the first set of excitation. **a2**, **b2** are the corresponding results with the second set of excitation after 5 min in dark

Table 1 Relative quantum yields and quantum efficiencies at different spatial scales

	Large area of leaf (mean/SD)	Group of palisade cells (mean/SD)	Single chloroplast (mean/SD)
Φ_{Fm}	47.15/7.76	45.27/6.67	46.77/5.77
Φ_{F0}	13.44/3.66	12.46/3.19	11.94/2.44
Φ_{PSII}	0.71/0.22	0.72/0.20	0.74/0.16

Discussion

We have successfully obtained the Kautsky curve, as well as its intensity dependence, with the confocal imaging fluorometer. Comparing to conventional wide-field imaging fluorometers, the confocal technique allows much better spatial confinement due to optical sectioning capability, and thus observation from a single chloroplast becomes possible. With the statistical analyses for P, S, M, T states of the Kautsky curves, at low and high intensities in Figs. 3 and 6 respectively, it can be concluded that the behavior of individual chloroplasts under our confocal imaging fluorometer is indeed similar to a large area of leaf under a conventional wide-field fluorometer. However, the value of the confocal technique lies in the capability to unravel the significant difference between individual chloroplasts, as

highlighted by the box plot in Fig. 3 and the clear variation of P–T decay time constants in Fig. 6c.

In terms of the temporal resolution performance, the confocal and wide-field fluorometers should be similar in terms of a single pixel detection, which takes about 1–10 μs in both cases. As mentioned in [17], the wide-field fluorometer takes about 1 s to record one image. Nevertheless, the advantage of the confocal scheme is the freedom to select number of pixels, as well as the position of these pixels, significantly enhancing the temporal responses. By using more advanced scanning approaches, such as random-access microscopy [40], high-speed CF detection among distant chloroplasts is possible. In addition, by adopting a multi-focus scanning approach, such as being demonstrated by spinning disk confocal

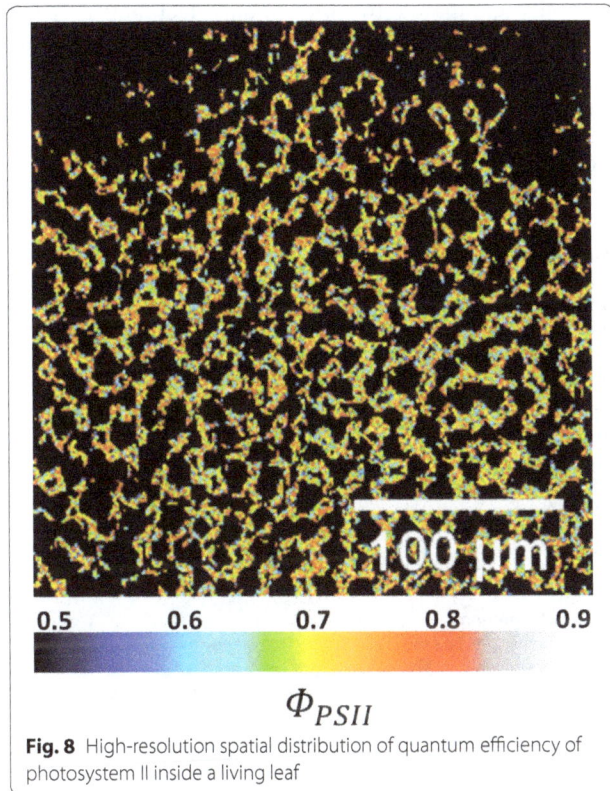

Fig. 8 High-resolution spatial distribution of quantum efficiency of photosystem II inside a living leaf

microscopy in 2009 [23], the frame rate of confocal fluorometer can be significantly improved.

Although spinning disk technique may potentially provide higher frame rate, there are several limitations that prevent it to be an ideal choice for fluorometry application [41]. First of all, due to the size limitation of camera, spinning disk confocal microscopy typically exhibits a small field of view, often only the size of a few cells, which is problematic when studying tissues. A good comparison is given in Fig. 8 of [41], where laser scanning confocal microscope provides much larger field of view.

Second, due to the existence of multiple pinholes on a pinhole array, the optical sectioning capability of spinning disk microscopy is in general less ideal than laser scanning confocal microscopy, especially when observing thick and scattering tissues. In addition, when using a low-magnification lens for large-area study, the spinning disk technique can significantly lose its optical sectioning ability. The reason is that most spinning disk system has a pinhole array comprising pinholes with a fixed size, which is designed for high-magnification and high-NA immersion objective lens, such as a $100\times$/NA 1.4 objective. However, for plant tissues, low magnification lens with moderate NA is preferred for large area observation. In our case, a $10\times$/NA 0.4 objective is employed, providing 1.5 mm $\times$ 1.5 mm field of view. If the $10\times$ objective

is used in a spinning disk system, whose pinhole diameter cannot be adjusted, both the axial sectioning capability and the lateral resolution shall be far less optimal than a confocal system. On the other hand, in a point-scanning confocal system, the pinhole size is easily adjustable, allowing observation for both high and low magnifications. Even with the low NA objective, as we mentioned in the main text, our confocal fluorometer still provides 1-micrometer lateral resolution and 5-micrometer axial resolution, adequate for single chloroplast imaging.

The third concern is image uniformity. In a spinning disk system, when using a Gaussian laser beam, the excitation intensity of the center region is larger than the edge, making it difficult to quantify the response from individual chloroplast. On the other hand, the image uniformity of laser scanning confocal fluorometer is much better than typical spinning disk one.

Last but not least, when comparing spinning disk and laser scanning confocal techniques, it is commonly accepted that the laser intensity at each focus is less for the former, so photobleaching is reduced. However, we would like to point out that the overall accumulated power/energy on the plant tissue is in fact higher, since the laser power is spread over hundreds of foci across the entire field of view. Therefore, more powerful lasers are required for the spinning disk system, and the issue of potential photothermal damage in the tissues has to be considered.

Another important aspect to notice is that the illumination intensity of the confocal fluorometer is much higher than that of the wide-field fluorometers. As shown in Figs. 5 and 6, to eliminate the semi-steady state S in the CF transient, about 55 kW/cm^2 is required for the confocal fluorometer. However, in the case of wide-field fluorometer, as shown in the example of Fig. 2 [1], to eliminate S, 3200 µmol/m^2/s is required. Considering the wavelength to be 650 nm in [1], the photon energy is $1240/650 = 1.9$ eV $= 3 \times 10^{-19}$ J. Therefore, the intensity unit (µmol/m^2/s) is equivalent to $[10^{-6} \times 6 \times 10^{23}$ (# of photons)] $\times$ [3×10^{-19} (J/photon)]/10^4 cm^2/s $= 18 \times 10^{-9}$ kW/cm^2. As a result, in the wide-field fluorometer, the required illumination intensity is $3200 \times 18 \times 10^{-9}$ kW/cm$^2 = 5.76 \times 10^{-5}$ kW/cm^2, six orders of magnitude smaller than that in the confocal one.

To explain this 6-order intensity difference, optical sectioning and illumination time of the confocal imaging fluorometer have to be considered. In a conventional fluorometer (wide-field detection), CF signals are emitted throughout the whole leaf in the axial direction, so the depth of field (i.e. signal collection depth) is equivalent to the thickness of a leaf, which is usually 100–1000 µm. On the other hand, for a confocal fluorometer, a pinhole is inserted before the detector to reject most out-of-focus fluorescence, and thus the total signal strength is significantly reduced. The typical depth of field in a confocal fluorometer is about 1–10 μm, which is 2-orders less than that of the wide-field one. Hence, the signal strength of the confocal fluorometer is expected to be 2-orders weaker than the wide-field counterpart.

In terms of the illumination time, in a conventional wide-field imaging fluorometer, the whole leaf sample is illuminated continuously, so the illumination time for each pixel is the same as the frame acquisition time. On the other hand, a small laser focus scans across the sample in the confocal scheme, making the illumination time for each pixel much shorter than the frame time. For example, in the case of Fig. 4a2, 1 frame takes about 1 s, and the frame is composed of 256×256 pixels, so the illuminating time for each pixel (1 pixel is roughly 1 μm^2 in this case) of the confocal imaging fluorometer is about 4-orders shorter than that of conventional wide-field imaging fluorometer.

Combining the above two reasons, it is reasonable that the illumination intensity in the confocal imaging fluorometer needs to be much higher than that in the wide-field fluorometer to achieve similar CF signal strength, as well as the Kautsky curves. The latter is somewhat surprising since it indicates that the physiological response of the chlorophyll remains the same with such high-intensity, yet short-period, illumination. One possible reason is that there is a slow reaction during photosynthesis and CF generation, so the chlorophyll only responses to the average intensity, not the instantaneous intensity. By looking into the electron transport chains in the photosystem, the bottleneck reaction might be the reduction of plastoquinone (PQ), which has a relatively slow reaction rate (100 molChl mmol^{-1} s^{-1}) [42]. Further studies are necessary to identify the underlying photochemical mechanism.

Conclusion

In this work, we demonstrated a confocal imaging fluorometer that can provide high spatiotemporal characterization of CF inside a living leaf. The three-dimensional spatial resolution is on the order of micrometer, and the temporal resolution reaches tens of milliseconds, allowing us to study CF transient, i.e. the Kautsky effect, from even a single chloroplast. Although the ensemble behavior of CF transient, as well as the intensity-dependent Kautsky curves, agree well with the results of conventional wide-field fluorometers, confocal imaging fluorometer provides valuable information toward the difference of CF dynamics among individual chloroplasts. The features of optical sectioning and laser focus scanning in the confocal fluorometer result in much higher illumination

intensity compared to conventional techniques, while maintaining normal cellular physiological responses. Our work not only opens up new possibilities to study CF dynamics on the level of organelles, but also is promising to unravel more spatial/temporal details in the associated photosynthetic processes.

Author's contributions
YCT designed the experiment, carried out signal analysis, and wrote most of the manuscript. SWC envisioned the idea, provided the experimental hardware, and helped to polish the manuscript. Both authors read and approved the final manuscript.

Author details
[1] Department of Physics, National Taiwan University, No. 1, Section 4, Roosevelt Rd, Da'an District, Taipei City 10617, Taiwan. [2] Molecular Imaging Center, National Taiwan University, No. 81, Changxing Street, Da'an District, Taipei 10672, Taiwan.

Acknowledgements
The authors appreciate the inspirational discussion with Prof. Govindjee from University of Illinois at Urbana-Champaign. SWC acknowledge the generous support from the Foundation for the Advancement of Outstanding Scholarship.

Competing interests
The authors declare that they have no competing interests.

Funding
This work is supported by the Molecular Imaging Center of NTU (105R8916, 105R7732), and by the Ministry of Science and Technology, Taiwan, under grant MOST-105-2628-M-002-010-MY4 and MOST-106-2321-B-002-020.

References
1. Strasser RJ, Srivastava A, Govindjee. Polyphasic chlorophyll a fluorescence transient in plants and cyanobacteria. Photochem Photobiol. 1995;61:32–42.
2. Zarco-Tejada PJ, Miller JR, Mohammed GH, Noland TL. Chlorophyll fluorescence effects on vegetation apparent reflectance: I. Leaf-level measurements and model simulation. Remote Sens Environ. 2000;74:582–95.
3. Krause GH, Weis E. Chlorophyll fluorescence and photosynthesis: the basics. Annu Rev Plant Biol. 1991;42:313–49.
4. Horton P, Ruban AV, Walters RG. Regulation of light harvesting in green plants (indication by nonphotochemical quenching of chlorophyll fluorescence). Plant Physiol. 1994;106:415.
5. Dobrowski SZ, Pushnik JC, Zarco-Tejada PJ, Ustin SL. Simple reflectance indices track heat and water stress-induced changes in steady-state chlorophyll fluorescence at the canopy scale. Remote Sens Environ. 2005;97:403–14.
6. Csintalan Z, Proctor MCF, Tuba Z. Chlorophyll fluorescence during drying and rehydration in the mosses *Rhytidiadelphus loreus* (hedw.) warnst., *Anomodon viticulosus* (hedw.) hook. & tayl. and *Grimmia pulvinata* (hedw.) sm. Ann Bot. 1999;84:235–44.
7. Bolhar-Nordenkampf HR, Long SP, Baker NR, Oquist G, Schreiber U, Lechner EG. Chlorophyll fluorescence as a probe of the photosynthetic competence of leaves in the field: A review of current instrumentation. Funct Ecol. 1989;3:497–514.

8. Trissl HW, Gao Y, Wulf K. Theoretical fluorescence induction curves derived from coupled differential equations describing the primary photochemistry of photosystem ii by an exciton-radical pair equilibrium. Biophys J. 1993;64:974.

9. Schansker G, Tóth SZ, Holzwarth AR, Garab G. Chlorophyll a fluorescence: beyond the limits of the QA model. Photosynth Res. 2014;120:43–58.

10. Papageorgiou GC. Photosystem ii fluorescence: slow changes–scaling from the past. J Photochem Photobiol B Biol. 2011;104:258–70.

11. DeEll JR, Toivonen PMA. Practical applications of chlorophyll fluorescence in plant biology. Berlin: Springer; 2012.

12. Lazár D, Ilik P. High-temperature induced chlorophyll fluorescence changes in barley leaves comparison of the critical temperatures determined from fluorescence induction and from fluorescence temperature curve. Plant Sci. 1997;124:159–64.

13. Schreiber U, Schliwa U, Bilger W. Continuous recording of photochemical and non-photochemical chlorophyll fluorescence quenching with a new type of modulation fluorometer. Photosynth Res. 1986;10:51–62.

14. Mauzerall D. Light-induced fluorescence changes in chlorella, and the primary photoreactions for the production of oxygen. Proc Nat Acad Sci. 1972;69:1358–62.

15. Falkowski PG, Wyman K, Ley AC, Mauzerall DC. Relationship of steady-state photosynthesis to fluorescence in eucaryotic algae. BBA-Bioenergetics. 1986;849:183–92.

16. Kolber ZS, Prášil O, Falkowski PG. Measurements of variable chlorophyll fluorescence using fast repetition rate techniques: defining methodology and experimental protocols. BBA-Bioenergetics. 1998;1367:88–106.

17. Barron-Gafford GA, Rascher U, Bronstein JL, Davidowitz G, Chaszar B, Huxman TE. Herbivory of wild manduca sexta causes fast down-regulation of photosynthetic efficiency in datura wrightii: an early signaling cascade visualized by chlorophyll fluorescence. Photosynth Res. 2012;113:249–60.

18. Oxborough K, Baker N. An instrument capable of imaging chlorophyll a fluorescence from intact leaves at very low irradiance and at cellular and subcellular levels of organization. Plant, Cell Environ. 1997;20:1473–83.

19. Scholes JD, Rolfe SA. Photosynthesis in localised regions of oat leaves infected with crown rust (Puccinia coronata): quantitative imaging of chlorophyll fluorescence. Planta. 1996;199:573–82.

20. Osmond B, Schwartz O, Gunning B. Photoinhibitory printing on leaves, visualised by chlorophyll fluorescence imaging and confocal microscopy, is due to diminished fluorescence from grana. Funct Plant Biol. 1999;26:717–24.

21. Hibberd JM, Quick WP. Characteristics of c4 photosynthesis in stems and petioles of c3 flowering plants. Nature. 2002;415:451–4.

22. Kim S-J, Hahn E-J, Heo J-W, Paek K-Y. Effects of leds on net photosynthetic rate, growth and leaf stomata of chrysanthemum plantlets in vitro. Sci Hortic. 2004;101:143–51.

23. Omasa K, Konishi A, Tamura H, Hosoi F. 3d confocal laser scanning microscopy for the analysis of chlorophyll fluorescence parameters of chloroplasts in intact leaf tissues. Plant Cell Physiol. 2009;50:90–105.

24. Papageorgiou G. Light-induced changes in the fluorescence yield of chlorophyll a in vivo: I. Anacystis nidulans. Biophys J. 1968;8:1299–315.

25. Paddock S, Fellers TJ, Davidson MW. Confocal microscopy. Berlin: Springer; 2001.

26. Shotton DM. Confocal scanning optical microscopy and its applications for biological specimens. J Cell Sci. 1989;94:175–206.

27. Roselli L, Paparella F, Stanca E, Basset A. New data-driven method from 3d confocal microscopy for calculating phytoplankton cell biovolume. J Microsc. 2015;258:200–11.

28. Matsumoto B. Cell biological applications of confocal microscopy. Cambridge: Academic Press; 2003.

29. Hibbs AR. Confocal microscopy for biologists. Berlin: Springer; 2004.

30. Pawley JB, Masters BR. Handbook of biological confocal microscopy. Opt Eng. 1996;35:2765–6.

31. Omasa K. Image instrumentation of chlorophyll a fluorescence. In SPIE Proc. 1998;3382:91–9.

32. Kautsky H, Appel W. Chlorophyllfluorescenz und kohlensaureassimilation. Biochemistry. 1960;322:279–92.

33. Bradbury M, Baker NR. Analysis of the slow phases of the in vivo chlorophyll fluorescence induction curve. Changes in the redox state of photosystem ii electron acceptors and fluorescence emission from photosystems i and ii. BBA-Bioenergetics. 1981;635:542–51.

34. Maxwell K, Johnson GN. Chlorophyll fluorescence—a practical guide. J Exp Bot. 2000;51:659–68.

35. Stirbet A, Govindjee. The slow phase of chlorophyll a fluorescence induction in silico: origin of the s–m fluorescence rise. Photosynth Res. 2016;130:193–213.

36. Omasa K, Shimazaki KI, Aiga I, Larcher W, Onoe M. Image analysis of chlorophyll fluorescence transients for diagnosing the photosynthetic system of attached leaves. Plant Physiol. 1987;84:748–52.

37. Kodru S, Malavath T, Devadasu E, Nellaepalli S, Stirbet A, Subramanyam R. The slow s to m rise of chlorophyll a fluorescence reflects transition from state 2 to state 1 in the green alga chlamydomonas reinhardtii. Photosynth Res. 2015;125:219–31.

38. Vermaas WF, Timlin JA, Jones HD, Sinclair MB, Nieman LT, Hamad SW, Melgaard DK, Haaland DM. In vivo hyperspectral confocal fluorescence imaging to determine pigment localization and distribution in cyanobacterial cells. Proc Nat Acad Sci USA. 2008;105:4050–5.

39. Chen M-Y, Zhuo G-Y, Chen K-C, Wu P-C, Hsieh T-Y, Liu T-M, Chu S-W. Multiphoton imaging to identify grana, stroma thylakoid, and starch inside an intact leaf. BMC Plant Biol. 2014;14:175.

40. Reddy GD, Kelleher K, Fink R, Saggau P. Three-dimensional random access multiphoton microscopy for functional imaging of neuronal activity. Nat Neurosci. 2008;11:713–20.

41. Jonkman J, Brown CM. Any way you slice it—a comparison of confocal microscopy techniques. J Biomol Tech. 2015;26:54–65.

42. de Wijn R, van Gorkom HJ. Kinetics of electron transfer from QA to QB in photosystem ii. Biochemistry. 2001;40:11912–22.

16

A real-time phenotyping framework using machine learning for plant stress severity rating in soybean

Hsiang Sing Naik[1], Jiaoping Zhang[2], Alec Lofquist[1], Teshale Assefa[2], Soumik Sarkar[1], David Ackerman[1], Arti Singh[2*], Asheesh K. Singh[2*] and Baskar Ganapathysubramanian[1*]

Abstract

Background: Phenotyping is a critical component of plant research. Accurate and precise trait collection, when integrated with genetic tools, can greatly accelerate the rate of genetic gain in crop improvement. However, efficient and automatic phenotyping of traits across large populations is a challenge; which is further exacerbated by the necessity of sampling multiple environments and growing replicated trials. A promising approach is to leverage current advances in imaging technology, data analytics and machine learning to enable automated and fast phenotyping and subsequent decision support. In this context, the workflow for phenotyping (image capture → data storage and curation → trait extraction → machine learning/classification → models/apps for decision support) has to be carefully designed and efficiently executed to minimize resource usage and maximize utility. We illustrate such an end-to-end phenotyping workflow for the case of plant stress severity phenotyping in soybean, with a specific focus on the rapid and automatic assessment of iron deficiency chlorosis (IDC) severity on thousands of field plots. We showcase this analytics framework by extracting IDC features from a set of ~4500 unique canopies representing a diverse germplasm base that have different levels of IDC, and subsequently training a variety of classification models to predict plant stress severity. The best classifier is then deployed as a smartphone app for rapid and real time severity rating in the field.

Results: We investigated 10 different classification approaches, with the best classifier being a hierarchical classifier with a mean per-class accuracy of ~96%. We construct a phenotypically meaningful 'population canopy graph', connecting the automatically extracted canopy trait features with plant stress severity rating. We incorporated this image capture → image processing → classification workflow into a smartphone app that enables automated real-time evaluation of IDC scores using digital images of the canopy.

Conclusion: We expect this high-throughput framework to help increase the rate of genetic gain by providing a robust extendable framework for other abiotic and biotic stresses. We further envision this workflow embedded onto a high throughput phenotyping ground vehicle and unmanned aerial system that will allow real-time, automated stress trait detection and quantification for plant research, breeding and stress scouting applications.

Keywords: High-throughput phenotyping, Image analysis, Machine learning, Plant stress, Smartphone

Background

Soybean (*Glycine max* (L.) Merr.) is a huge source of revenue for the United States, with production of approximately USD 40 billion in 2014 [1]. There are various factors that affect soybean yield, such as nutrient availability, weed management, genetics, row configuration, stress (biotic and abiotic) and soil fertility [2]. Iron deficiency chlorosis (IDC) is a yield-limiting abiotic stress which affects plants that usually grow on calcareous soil with high pH. Soybean plants growing in calcareous soils

*Correspondence: arti@iastate.edu; singhak@iastate.edu; baskarg@iastate.edu
[1] Department of Mechanical Engineering, Iowa State University, Ames, IA 50011, USA
[2] Department of Agronomy, Iowa State University, Ames, IA 50011, USA

(soils with free calcium carbonate and high pH) are unable to uptake iron from the soil leading to iron deficiency in plants. IDC causes reduced plant growth leading to a reduction in yield potential and quality of the crop. In the mid-west USA, IDC is one of the major problems reducing soybean yield, by as much as 20% for each visual rating point [3]. This causes an estimated economic loss of $ 260 million in 2012 alone [4]. IDC symptoms are observed at early plant growth stages on newly grown leaf tissue where chlorosis (yellowing) occurs in between the veins of the leaves, while the veins themselves remain green [5]. The extent of the problem varies depending on the cultivar, field and the year.

Soybean breeders in the US breed for genotypes with improved IDC tolerance by selecting for genes that help make the plant more iron uptake efficient [6]. Selection for desirable soybean genotype (with IDC tolerance) is done either through phenotyping in the field or in greenhouses [7], or genotyping with molecular markers linked to genes that improve IDC tolerance. More than 10 genes have been reported to be associated with improving IDC tolerance [8, 9] making genotyping approaches onerous where a breeding program may be working to select for several other traits. Phenotyping is most suitable as it allows identification of soybean genotypes that have an acceptable IDC tolerance. Furthermore, this method is cost effective and potentially requires little access to specialized labs.

Current methods for phenotypically measuring IDC are completely visual and labor-intensive. Rodriguez de Cianzio et al. [7] and Froechlich and Fehr [3] reported that visual scoring is the simplest, subjective measurement that requires relatively less labor. However, it has reduced accuracy if the evaluation is made in diverse environments and by different raters [10]. In addition, there can be intra-rater repeatability or inter-rater reliability [11] issues leading to incorrect visual scores. It also depends on the subjectivity (and its variability) of the IDC rater. Specifically, the human eye can get tired after long hours of scoring plants for various traits, which can produce large intra-rater variability in rating scores, thus resulting in diminished accuracy and reproducibility. In a breeding program, hundreds or thousands of plots are rated in a short time frame. A short time frame is crucial because one has to minimize plant stage variability, i.e., variability that is introduced if genotypes are rated over a longer time frame. *It is therefore essential to develop methods that allow for unbiased, accurate, cost effective and rapid assessment for IDC in particular, and plant biotic (e.g., diseases) and abiotic stresses in general.* There has been recent work in this regard to design, develop and deploy high efficiency methods/tools to quantify leaf surface damage [12] as well as plants response to pathogens

[13]. Additionally, a number of approaches using imaging methods for phenotyping, such as fluorescence and spectroscopic imaging have been successful for stress-based phenotyping [14], high throughput machine vision systems that use image analysis for phenotyping *Arabidopsis thaliana* seedlings [15] and barley [16], hyperspectral imaging for drought stress identification in cereal [17], and a combination of digital and thermal imaging for detecting regions in spinach canopies that respond to soil moisture deficit [18] which have proven to be successful. However, a simple, user friendly framework is unavailable for the public to phenotype for IDC in soybean plants. The availability of a simple modular approach could potentially be generalized for phenotyping of multiple stresses.

Motivated by these reasons, we developed a simple framework (image capture → data storage and curation → trait extraction → machine learning/classification → models/smartphone apps for decision support) that extracts features that are known to quantify the extent of IDC (amount of yellowing, amount of browning) from digital images. To determine a relationship between these features and their respective ratings, we evaluated a host of machine learning techniques, further elaborated in the latter stages of this paper, to perform supervised classification. Subsequently, using information obtained from these classifiers, a physically meaningful population canopy graph (PCG) connecting the features with the visual IDC rating was constructed for a diverse soybean germplasm. This complete framework, which is based on fast feature extraction and classification, can then be used as a high throughput phenotyping (HTP) system for real time classification of IDC. We enable real time phenotyping by implementing the software framework as a GUI-based, user-friendly software that is also deployed on smartphones. This step successfully abstracts the end-user from the mathematical intricacies involved, thus enabling widespread use. We showcase this software framework by extracting IDC features (amount of yellowing, amount of browning) from a set of 4366 plants that have different IDC resistances.

We envision our classifier based framework as a modular, extensible and accurate phenotyping platform for plant researchers including breeders and biologists.

Methods
Genetic material and field phenotyping
A total of 478 soybean genotypes, including 3 maturity checks and 475 soybean plant introduction (PI) lines acquired from the USDA soybean germplasm collection, were planted in the Bruner farm in Ames, IA, 2015, where soybean IDC was present in previous years. This set of PI lines exhibits a wide diversity in leaf and canopy shape [19]. The design for this field experiment follows a

randomized complete block design, with a total of four replications. Each PI line was planted once per replication, while the IDC checks (two) and maturity checks (three) were repeated at regular intervals in the field with four plots per replication. The plants were planted in 0.31 m length hill plot of five seeds per plot. At two soybean growth stages [20]: the second to third trifoliate (V2–V3) and fifth to sixth trifoliate (V5–V6) leaf stages, the soil pH was tested in the Soil and Plant Analysis Laboratory, Iowa State University. At each stage, eight soil samples were randomly collected from each replication and were mixed as one test sample. The soil pH values ranged at 7.80–7.95 and 7.75–7.85 at V2–V3 and V5–V6 growth stages, respectively. Field visual ratings (FVR) of IDC severity by experts were collected at V2–V3 and V5–V6 growth stages, as well as two weeks after the V5–V6 stage to obtain soybean canopies with a variety of IDC expression. FVR was done on a scale of 1–5 described by Lin et al. [21], where 1 indicates no chlorosis and plants were normal green; 2 indicates plants with modest yellowing of upper leaves; 3 indicates plants with interveinal chlorosis in the upper leaves but no stunting growth; 4 indicates plants are showing interveinal chlorosis with stunting growth; and 5 indicates plants show severe chlorosis plus stunted growth and necrosis in the new youngest leaves and growing points.

Image acquisition

We utilized a Canon EOS REBEL T5i camera for image acquisition. Images were stored in the native RAW format. Substantial effort was put in to develop a standard imaging protocol (SIP) (Additional file 1) to ensure imaging consistency and quality. The flash function was kept off and an umbrella was always used to shade the area under the camera view in order to minimize illumination discrepancies between images. A light/color calibration protocol was also followed. An image of a color calibration chart (X-Rite ColorChecker Color Rendition Chart) was taken at the beginning of imaging operations, and every 20 min thereafter or whenever light condition changes (cloud cover, etc.). When taking pictures, the whole canopy was fit in the field of view of the camera. Weed control was practiced consistent with research plots and commercial farms; however, due to the small size of the field weed removal was done manually. Weeds in the view of camera were removed for enhanced efficiency of subsequent image processing. Images were taken across several days (at several times of the day) under various illumination conditions. Finally, the imaging protocol was chosen so that the imaging window and the camera resolution resulted in images with at least 6 pixels/mm, ensuring that the approach is transferable to other cameras that use an appropriate imaging window to get this resolution.

Dataset description

A total of 5916 RGB (493 plots including PI accessions and checks × 4 replications × 3 time points) images were acquired, along with subsequent FVR. Each time point consists of four repetitions for a total of 1972 (493 × 4) images, with 493 images per repetition. Image acquisition at each of these time points was vital to obtain a large variety of IDC symptoms, as IDC symptoms progress in time. The idea was to develop a dataset with similar number of observations per IDC rating. This was, however, not possible simply due to the fact that a large fraction of plants remained healthy (FVR = 1) throughout the image acquisition period. Following image acquisition, for quality control, each image was inspected visually, and those that did not adhere to the Standard Imaging Protocol (SIP) were removed, which resulted in 4366 images in the remaining image set.

Preprocessing and feature extraction
Preprocessing

White balance and color calibration   As the appearance of color is affected by lighting conditions, using a calibration chart enables color correction to be applied to ensure that colors are uniform throughout all the plant canopy images collected. We primarily used the grey squares to identify the white balance, while the green, brown and yellow squares were used to calibrate the hue values of green, brown and yellow. Hue is defined as the color or tint of an object. Hue quantifies color in terms of angle around a circle (or more precisely around a color hexagon) with values ranging from 0° to 359° [22]. The red color axis is usually set as 0°. The hue of brown ranges from 21° to 50°, whereas yellow hue ranges from 51° to 80° [23]. Calibration is done by identifying how much the hue value of the green, brown and yellow squares on the color calibration chart has drifted from the defined hue values. This drift correction is then applied to the canopy images. This preprocessing resulted in an analysis pipeline that was robust to changes in illumination.

Segmentation   Each image was converted from native Red, Green, Blue (RGB) format to HSV (Hue, Saturation, Value) format [22] to efficiently perform background removal, leaving only the plant canopy (foreground). The background of an image (soil, debris) contains more gray pixels compared to the foreground (plant), and lacks green and yellow hue values; therefore, most of the background was removed by excluding pixels that had saturation value below a predefined threshold and hue values outside of a predefined range. The saturation threshold value was obtained by identifying the saturation values of the background in 148 diverse images. The hue range was simply obtained from the hue color wheel, removing pixels that were neither green nor brown.

This combined thresholding based on incorporating hue thresholding with saturation thresholding ensured a reliable and robust segmentation process.

Noise and outlier removal Once segmentation was done, the connected components method [24] was used on the processed image to remove spurious outliers and noise from the image, (for example, plant debris on soil). This was accomplished by identifying clusters of pixels which are connected to one another, labelling them, and identifying the largest connected component. Since the imaging protocol was designed to ensure that the plant was centered in the imaging window and in the foreground, it follows that the largest connected component is invariably the plant. Cleaning was done by removing any other connected components that contain fewer pixels than the largest connected component. Then, a mask of the isolated plant was applied onto the original RGB image in order to display the isolated plant in color. No significant pixel loss was observed which is common in other thresholding methods [25]. The use of the connected components approach to isolate plants from background is extremely fast and accurate. In conjunction with a SIP, using connected components for preprocessing is very promising, especially for near real time phenotyping applications. The preprocessing sequence is illustrated in Fig. 1.

Feature extraction from expert elicitation

Field visual ratings are assigned based on the extent of chlorosis (yellowing) and necrosis (browning) expressed on the canopy, as described earlier and illustrated in Fig. 2. Elicitation from domain knowledge experts (i.e., raters) suggested that color signatures (green to yellow to brown), specifically extent of (dis)coloration (chlorosis → yellowing, and necrosis → browning) were viable predictors to quantify IDC expression. Each pixel of the processed image belonging to the canopy was identified as either green, yellow, or brown through evaluating their respective hue values to identify which hue ranges they belong to, and the extent of discoloration from green was represented in the form of the percentage of canopy area that experience these visual changes (Y and B%), as seen in Fig. 3.

$$\frac{Area_{yellow}}{Area_{total}} \times 100\% = Percentage_{yellow},$$

$$\frac{Area_{brown}}{Area_{total}} \times 100\% = Percentage_{brown} \tag{1}$$

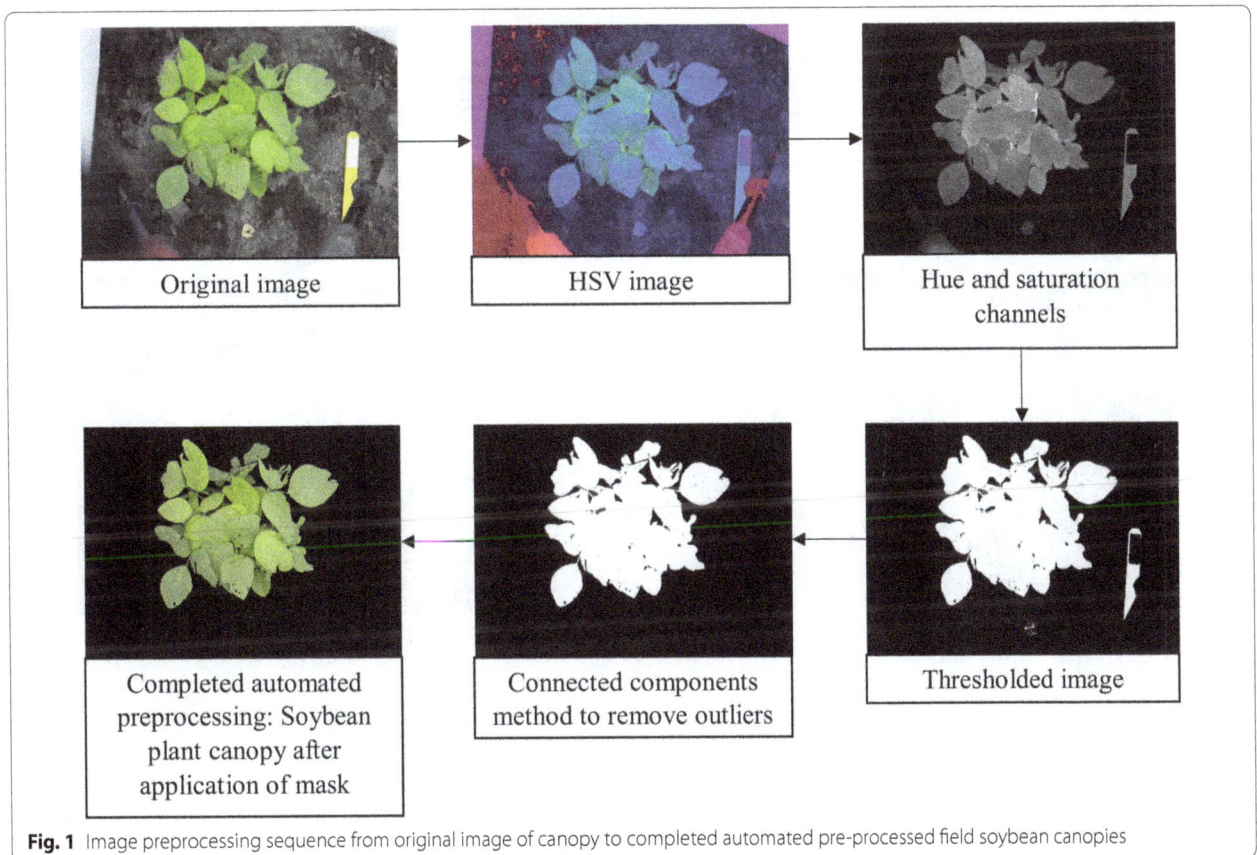

Fig. 1 Image preprocessing sequence from original image of canopy to completed automated pre-processed field soybean canopies

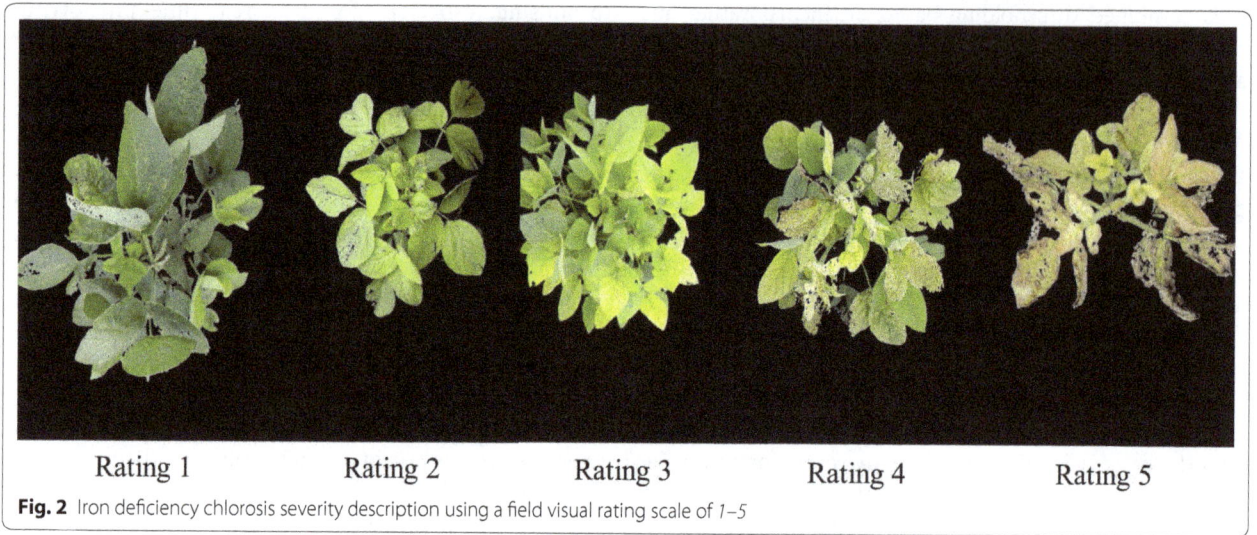

Fig. 2 Iron deficiency chlorosis severity description using a field visual rating scale of *1–5*

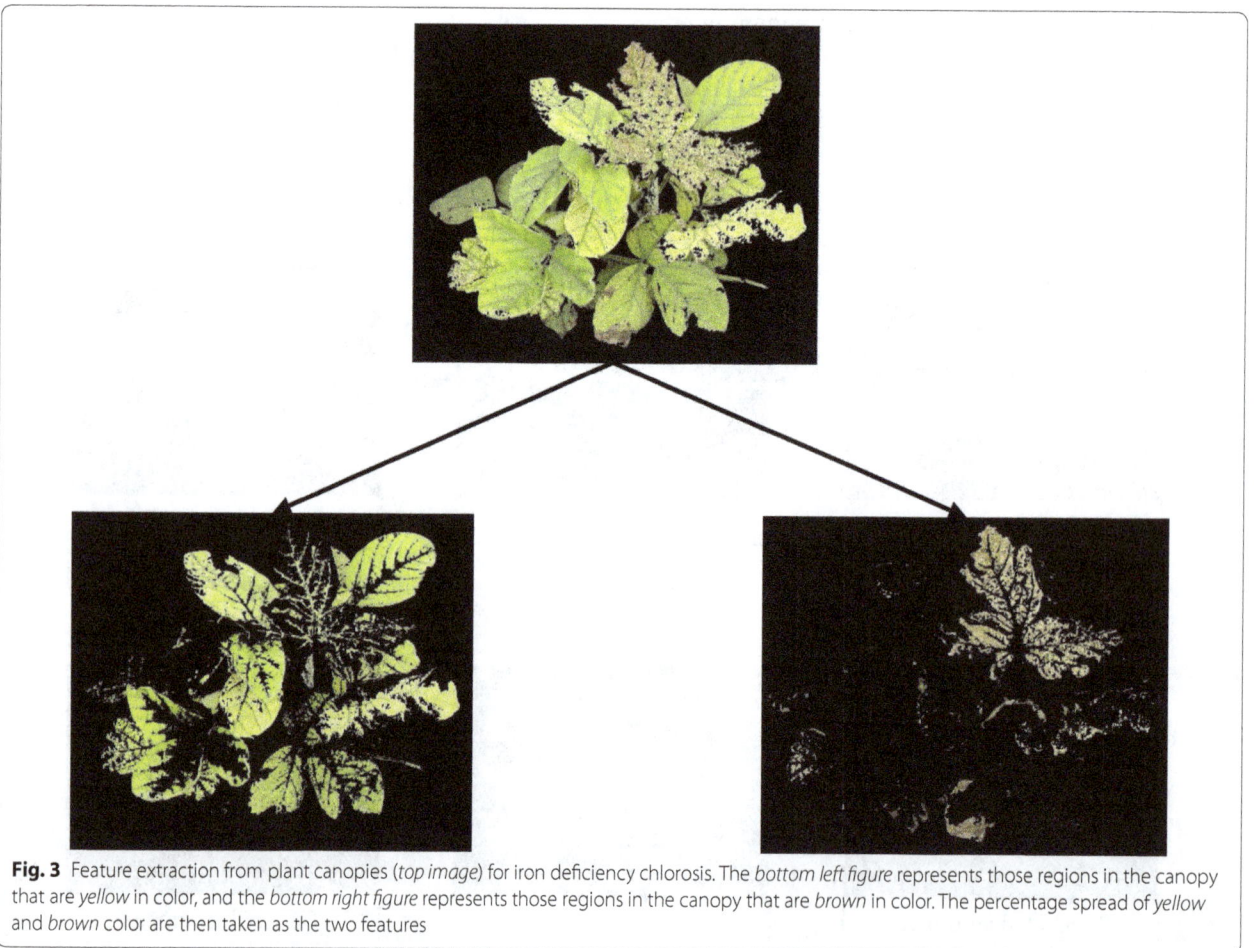

Fig. 3 Feature extraction from plant canopies (*top image*) for iron deficiency chlorosis. The *bottom left figure* represents those regions in the canopy that are *yellow* in color, and the *bottom right figure* represents those regions in the canopy that are *brown* in color. The percentage spread of *yellow* and *brown* color are then taken as the two features

This expert elicitation informed processes resulted in each image being represented by a quantitative measure of yellowing (Y%) and browning (B%), as shown in Eq. 1.

Classification

In order to map these quantitative variables to the visually rated IDC ratings, we utilize several state of the art machine learning algorithms to construct classification models. The field visual rating served as the categorical output variable (classes) while the inputs were the 2-tuple (Y, B%). The classification models are then eventually used to generate IDC ratings given different input variables.

The total dataset consisted of 4366 images following quality control as detailed in the "Dataset description" section. The images were sorted into 5 groups which correspond to their respective FVR, with majority of the observations falling into group 1 (FVR = 1). The remaining groups (FVR = 2/3/4/5) meanwhile contained a balanced distribution of observations amongst themselves.

Due to the imbalanced nature of the dataset with a preponderance of images belonging to FVR 1, two variations of the dataset were used to develop classification models: (a) Using observations from time point 2 and (b) for a combination of time point 1, 2, and 3. Time point 2 served as a standalone dataset due to the fact that it has the largest distribution of observations containing each of the FVRs. We utilized several classification algorithms, namely classification trees (CT), random forests (RF), Naïve Bayes (NB), linear discriminant analysis (LDA), quadratic discriminant analysis (QDA), multiclass support vector machines (SVM), k-nearest neighbors (KNN), and Gaussian mixture models (GMM). Building upon the results, we subsequently utilized the concept of hierarchical classification to develop two additional models using a combination of LDA and SVM algorithms.

The dataset was randomly sampled into two subsets in a 75–25% ratio. The larger subset (75%) served as the training set, while the remaining subset served as the testing dataset (25%). We additionally evaluated the performance of the classifier across additional datasets. One dataset consisted of images from completely different genetic material. Additionally, we repeated the field experiment in 2016 and used the trained classifier on images from this experiment [26]. The training dataset is used to train the classifier, by learning a mapping of the Y and B% with their expected IDC ratings. Subsequently, the testing dataset is used to estimate the performance of the classification model, by applying it on the testing dataset to classify each observation. The performance of the classifier can be interpreted from the confusion matrix (Table 1). The diagonals on a confusion matrix show the number of observations where the predicted rating is equal to the actual rating, whereas the off-diagonal elements are observations that have been misclassified.

An example confusion matrix for a *binary* classification problem is shown below:

Table 1 Confusion matrix

	Predicted positive (class 1)	Predicted negative (class 2)
Actual positive (class 1)	True positive (TP)	False negative (FN)
Actual negative (class 2)	False positive (FP)	True negative (TN)

Three measures of accuracy of the classifier are reported from the confusion matrix

(a) *Accuracy* which quantifies the fraction of the training dataset that is correctly predicted.

$$Accuracy = \frac{TP + TN}{TP + TN + FP + FN} \times 100 \qquad (2)$$

(b) *Per-class accuracy* is a more refined metric which calculates how the classifier performs for each of the classes. This is useful when the instances in each class vary a lot, i.e., when the classes are *imbalanced (as is the case in this work)*, since accuracy is usually overestimated due to the impact of the class with the most instances dominating the accuracy statistic.

Per class accuracy
$$= \frac{i-th \; observation \; of \; row \; i}{Sum \; of \; observations \; of \; row \; i}, \qquad (3)$$
$$i = 1, \ldots, n,$$

where n number of classes, *row* row on the confusion matrix

c) *Mean per-class accuracy (MPCA)* is the mean per-class accuracy over these classes.

$$Mean \; Per \; Class \; Accuracy = \frac{1}{n} \sum_{i=1}^{n} Per \; class \; accuracy \qquad (4)$$

In addition, we compute the misclassification costs in order to quantify the cost of the misclassification errors—i.e., if an observation in rating 1 were to be classified as rating 5, it would have a higher misclassification cost than if it were to be classified as rating 2. Essentially, calculating the misclassification cost enables us to know, if errors are made, how bad the errors are. To do so, we defined a misclassification cost matrix, as detailed in Table 2. The off-diagonals of the matrix are the misclassification cost for each of the ratings, which are finite, real values [27]. For example, if the actual rating of an observation is rating 1, the error of misclassifying the observation to rating 5 is 4 times as costly as misclassifying the observation to rating 2, and so on. Then, misclassification cost is computed using Eq. 5.

$$cost = \frac{1}{N} \sum_{i} \sum_{j} CM_{ij} * w_{ij}, \qquad (5)$$

Table 2 Cost matrix, w_{ij}

	Predicted ratings				
Actual ratings	0	1	2	3	4
	1	0	1	2	3
	2	1	0	1	2
	3	2	1	0	1
	4	3	2	1	0

CM_{ij} confusion matrix, w_{ij} cost matrix, N number of observations

Next, we employ cross-validation to estimate the average generalization error for each classifier. Cross-validation essentially is a method of assessing the accuracy and validity of a statistical model for generalization on future datasets. From a generalizability standpoint, the absolute accuracy of a classifier is less important as it could be subject to bias and overfitting. Hence, cross-validation is a method of performance estimation based on the variance. The ideal estimation method would have low bias and low variance [28]. We used k-fold cross-validation, with k = 10 which is a good compromise between variance and bias [28]. K-fold CV was repeated 10 times to compute the mean cross-validation misclassification error for each model. While accuracy and MPCA detail the performance of a classifier on essentially the same dataset, mean cross-validation misclassification error provides information on how well the classifier performs on other datasets.

A brief description of classification algorithms deployed

We briefly describe each of the classification algorithms [29]. We refer the interested reader to a more detailed description of these methods in [30–32].

Decision trees It is based on the construction of predictive models with a tree-like structure that correlates observations to their corresponding categories such as classes (for classification) and rewards (for decision-making problems). These observations are sorted down the tree from the root to a leaf node, which in turn classifies the observation. Decision trees [33] perform well on lower dimensional classification problems, but tend to falter when the dimension of the classes increases.

Random forests An ensemble method employed to regularize the greedy, heuristics nature of the decision tree training which sometimes causes overfitting. This method [34] combines results and structures from a number of trees prior to coming to a conclusion. Multiple trees are grown from random sampling of the data. Nodes and branch choices of a tree are also determined through a

non-deterministic manner. These models are more robust to uncertainties.

Naïve Bayes A supervised classification technique for constructing classifiers of a probabilistic graphical model. It is based on the assumption that each feature is independent of each other. Naïve Bayes [35] have been used in a variety of fields, and is a popular method for text categorization.

Linear discriminant analysis (LDA) A linear classification technique [36] based on the idea of Fisher's Metric, with an aim to maximize between class variance, while minimizing within-class variance. This allows the linear combination of features to improve separability among two or more classes. This requires an assumption of equal variance–covariance matrices of the classes.

Quadratic discriminant analysis (QDA) A modification of linear discriminant analysis, except a covariance matrix must be estimated for each class. This allows overcoming the problem where the variance–covariance differs substantially [36], where LDA will not perform well.

Support vector machine (SVM) The most popular among supervised, discriminative kernel-based methods for classification. SVM [37] uses kernel functions to project data into a higher dimensional space in order to separate data from different classes which cannot be linearly separated. A hyperplane is constructed to determine the bounds in which each class is separated, to maximize class separability.

K-nearest neighbors (KNN) A non-parametric classification method [38]. This algorithm assigns the same class label to data samples as its k nearest neighbors based on a similarity metric defined on the feature space, where k is an integer. This nonlinear algorithm works reasonably well for multi-class classification problems.

Gaussian mixture model (GMM) A generative, unsupervised data model that aims to identify a set of Gaussian distributions mixtures which best describe the data. GMM [39] is a probabilistic technique where every data example is expressed as a sample of the distribution which is a weighted sum of k Gaussian distribution. Once this model is created, a Bayes classifier is applied in attempt to solve classification problems.

Hierarchical classification

We subsequently pursued a hierarchical classification strategy that is motivated by expert elicitation of information about IDC susceptibility. Hierarchical classification is known to work well on datasets with a larger number of classes but with fewer observations. The IDC data set fell into this category. Also, the task of designing the hierarchy in this classification strategy enables the inclusion of expert knowledge. Here, the hierarchical structure is predefined, based on insight and existing knowledge of class hierarchies, which then contributes to improving classification accuracy.

In this case, the hierarchies were identified based on the susceptibility of the genotypes to IDC. Specifically, rating 1 and 2 are usually taken together as low susceptibility genotypes, while rating 4 and 5 are taken together as high susceptibility genotypes. We thus designed a two-step classification strategy: In Step A, a classifier is learnt that can separate the data into low, medium and high susceptibility groups. Step B then further classifies these groups into rating 1 or 2 (for the low susceptibility group), and rating 4 or 5 (for the high susceptibility group).

For Step A, we deploy both LDA and multi-class SVMs. The learnt classifier is called Model 0, and seeks to classify the dataset into three groups (low, medium and high susceptibility) based on their yellow and brown percentage. For Step B, we deploy Support Vector Machine as the classification is binary. Figure 4 displays a flowchart of this hierarchical classifier.

Results and discussion

A number of classification algorithms were capable of achieving high mean per class accuracy, more than 90%, for classification on the time point 2 data set. Hierarchical models performed relatively well, with a mean per class accuracy at 95.9%. More importantly, when the classifier made incorrect predictions, the results were predominantly within the same susceptibility class—i.e., an error in rating 1 typically falls to rating 2, and not into rating 5 etc. This is illustrated in the misclassification cost metric for each classifier, as detailed in Eq. 5. The best performing classifier, classification trees, were able to correctly predict new observations 100% of the time.

When data from all time points were used to train and test a classifier, the hierarchical model performed the best, with 91% accuracy. Other classifiers fell short of the 90% mark. The decrease in accuracy was expected simply because combining all three time points caused the data set to be more imbalanced that before.

While being able to have high classification accuracy is important, the capability of a classifier to produce an interpretable PCG was extremely vital. This is quantified by the *interpretability* of the PCG, and is further discussed in the "PCG" and "Model selection" sections.

The results of each of the classification models are displayed in Tables 3 and 4. Table 3 consists of the results from a classification model developed using a sub-set of the IDC data (which consists of 3 time points). Instead of developing a model using 3 time points, this model was developed using data from time point 2 as it has the largest distribution of observations containing each of the FVRs. Table 4 consists of results from a model developed using the data spanning across all 3 time points (the whole dataset).

Population canopy graph

It was interesting to note that the learnt classifier revealed insightful phenotypic intuition. Specifically, we queried the classifier to predict ratings for a uniform sampling of the Y and B% range. This information is used to construct a 2D plot that depicts decision boundaries that separate

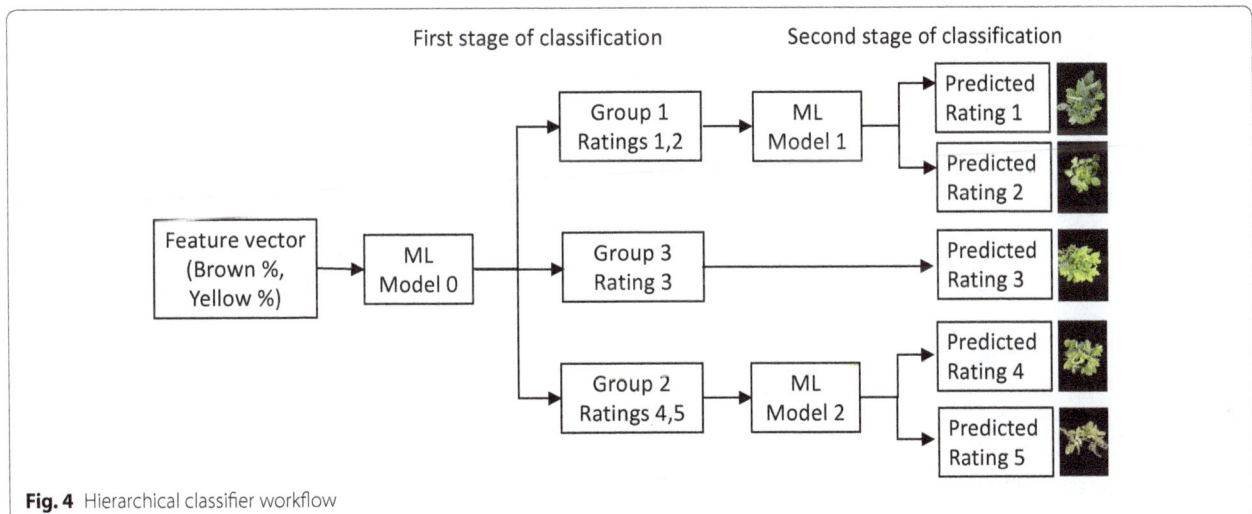

Fig. 4 Hierarchical classifier workflow

various IDC classes (as a function of Y and B%), which we refer to as a population canopy graph (PCG). This graph, shown in Fig. 5 which displays the PCG output from Hierarchy^2 classification results on the test set, correlates very well with expert intuition. Expert intuition suggests that ratings 1–3 exhibit low brown values (corresponding to minimal to no necrosis), which is clearly seen in the PCG in Fig. 5. Similarly, beyond a certain stage of necrosis, a plant is automatically rated as 5 irrespective of the amount of chlorosis. This trend is also exhibited by the nearly horizontal line marking the Rating 5 class in Fig. 5. Finally, the linear boundaries that allow a graceful transition from rating 1 through to 2 and 3, which is similar to how experts rate the transition of chlorosis.

Model selection

Several of the trained models exhibit good accuracy. We choose one of them as our best model based on a combination of a set of two objective measures and one subjective measure. The ideal model would have high MPCA and cross-validated MPCA as it illustrates the capability of the model to predict the IDC ratings of soybean through features extracted from images. We use MPCA instead of just accuracy due to the imbalanced nature of the dataset, as accuracy alone gives a distorted picture as the class with more examples will dominate the statistic. These two constitute the set of objective measures. Our subjective measure is based on a notion of interpretability—which we define as the ability of the end-user (plant researchers, breeders, and/or farmer) to interpret the PCG created

Table 3 Results for machine learning algorithm model accuracies developed using a sub-set of iron deficiency chlorosis data on a diverse set of soybean accessions

Algorithm	Accuracy	MPCA[a]	Cross validated MPCA	Interpretability	Cost metric
CT	100.0	100.0	96.0	Medium	0.0000
KNN	99.7	96.7	95.0	Low	0.0031
RF	99.7	96.0	85.0	Low	0.0031
Hierarchy[b]	99.4	95.9	79.8	High	0.0062
QDA	99.4	92.0	98.9	Medium	0.0620
Hierarchy[c]	98.5	86.6	70.8	High	0.0155
GMMB	99.1	82.0	87.0	Medium	0.0093
NB	99.1	82.0	93.8	Medium	0.0093
LDA	98.8	79.3	84.3	High	0.0124
SVM	93.8	39.8	50.0	Low	0.1084

[a] Mean per class accuracy

[b] SVM and SVM

[c] LDA and SVM

Table 4 Results for machine learning algorithm model accuracies developed using the complete set of iron deficiency chlorosis data on a diverse set of soybean accessions

Algorithm	Accuracy	MPCA[a]	Cross validated MPCA	Interpretability	Cost metric
CT	99.7	91.7	78.4	Low	0.0027
Hierarchy[b]	99.2	90.7	79.2	High	0.0082
Hierarchy[c]	98.3	84.0	79.0	High	0.0201
QDA	98.5	83.2	77.9	Medium	0.0201
NB	98.4	79.0	78.5	Medium	0.0284
KNN	99.5	75.8	84.3	Low	0.0073
RF	99.1	75.0	81.1	Low	0.0092
GMMB	99.4	74.2	82.7	Low	0.0064
LDA	98.5	71.7	76.9	High	0.0156
SVM	97.3	45.8	45.3	Low	0.0458

[a] Mean per class accuracy

[b] SVM: using SVM for both classifiers

[c] LDA and SVM

and link it to the visible rating characteristics that are currently used. Specifically, we check to see if the shape of the decision boundaries produced by the model makes physical sense—that the decision boundaries correlated with the physical aspect of IDC, e.g.: plants with IDC rating 4 and above display significantly more browning compared to ratings 3 and below. Interpretability was scored either 'Low', 'Medium', or 'High'; 'Low' for models that did not correlate with expert intuition (e.g.: individual islands, quadratic boundaries that appear to be biased), 'Medium' for models that partially correlates with expert intuition, and 'High' for models that correlated well with expert intuition. The hierarchical model Hierarchy[2] had the best trade-offs amongst these criterions, as shown in Tables 3 and 4, and was chosen as the best model.

Smartphone and PC software

To enable high throughput phenotyping using the developed classifier, we embed the preprocessing stage as well as the classifier into an easy to use GUI that is deployable as a smartphone app. This app is supported on all Android-based devices, such as tablets and smartphones and has the full functionality of the desktop-based version. The Android-based app allows users to take pictures with their devices and extract the IDC rating in real time. This allows for portability and instant acquisition of data. Figure 6 shows a flowchart of illustrating the app. When the app is launched, the user has a choice between taking a new picture, and analyzing a picture already contained in the device. The picture should be taken in the native RAW format (usually in the.dng format), and not using standard JPEG formats which use lossy compression that may cause color changes. Once a picture has been selected, it is processed and the IDC score evaluated and displayed on the screen. The user can export single or multiple images in tabular form through various methods, such as Dropbox, Bluetooth, Google Drive, and through email. This app allows untrained personnel and/or unmanned ground vehicles to extract and transmit IDC ratings without the need for a trained plant researcher/phenotyper looking at every plant. This is a tremendous enabler in terms of dramatically increasing the number of plants that can be accessed. In addition to the smartphone based app, a desktop based GUI will also be released to enable batch processing of a large number of images. This allows offline (or off site) analysis of images that are either captured manually or in an automated fashion.

Conclusion

We designed, developed and deployed an end-to-end integrated phenotyping work-flow that enables fast, accurate and efficient plant stress phenotyping. We show how image processing and machine learning can be deployed to construct classifiers that can automatically evaluate stress severity from image data. We emphasize that expert knowledge is crucial in designing appropriate classifiers. This is clearly seen in the markedly superior performance of the hierarchical classifier over single stage classifiers. The classifier is additionally used to produce a phenotypically meaningful population canopy graph. Subsequently, we deploy the developed classifier onto smartphones that serves as a high-throughput framework that can be utilized cross-platform for evaluating IDC ratings of soybean using only digital images. It is clear that image based analysis is more reliable and consistent than visual scoring as it removes the human error aspect involved in visual rating when repeated IDC measurements are needed at different growth stages. We compared the computed IDC ratings with provided visual scores from domain experts, and observed a close similarity, supporting accurate measurements and the accuracy of this HTP framework. We envision that such systems will help the plant researchers and breeders increase the efficiency and accuracy of selecting genotypes compared to visual scoring to enable fast phenotyping and reduce researcher bias. It is also relatively low cost and has the potential to speed up and improve crop development. The newly developed software framework is being embedded onto a high throughput phenotyping ground vehicle and unmanned aerial system (UAS) that will allow real-time, automated stress trait detection and quantification for plant breeding and stress scouting applications. This framework is also currently under further development by our group for numerous biotic stresses in soybean.

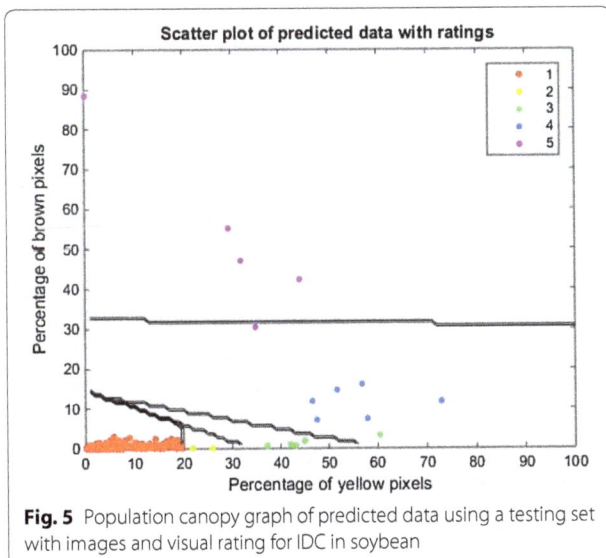

Fig. 5 Population canopy graph of predicted data using a testing set with images and visual rating for IDC in soybean

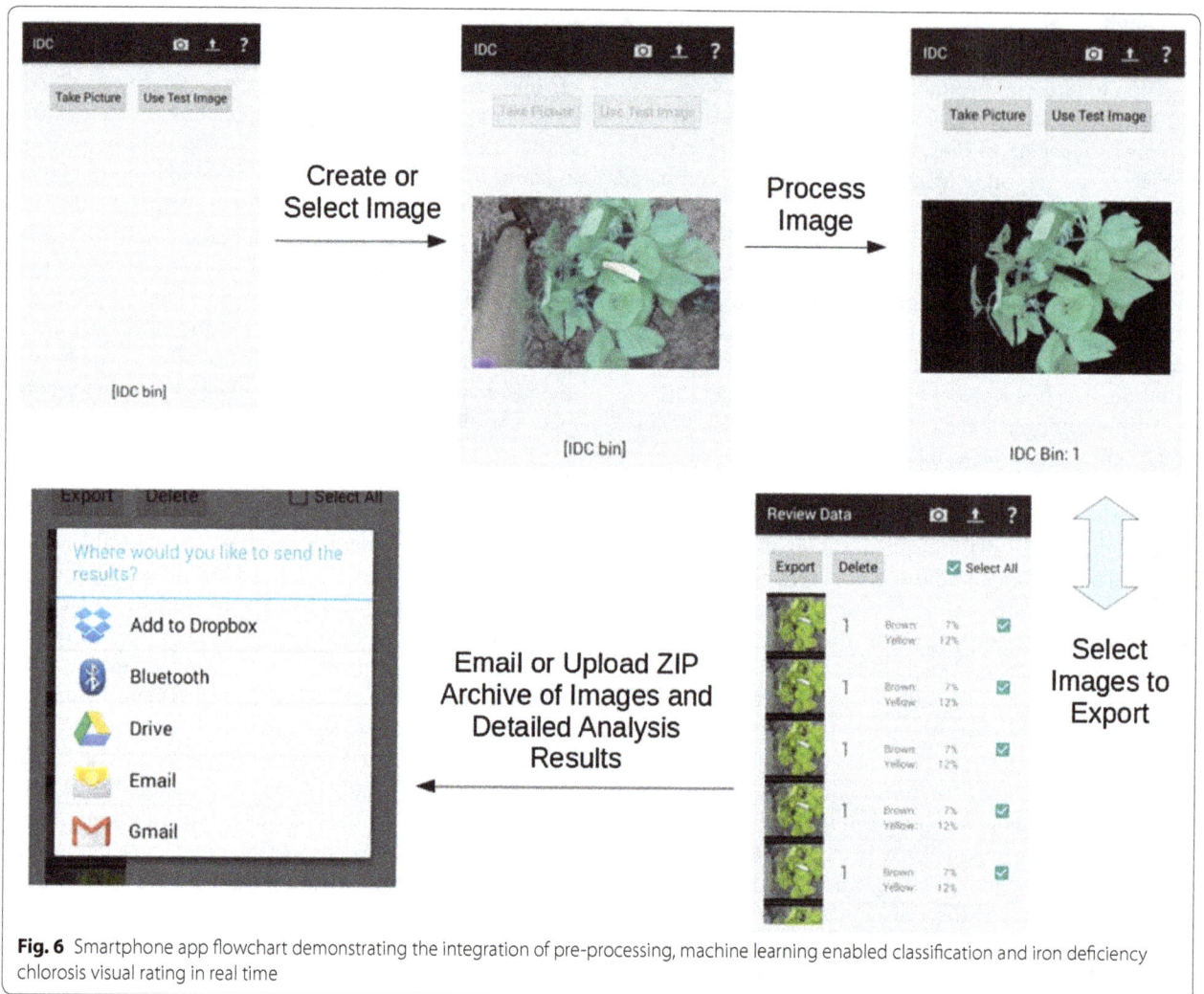

Fig. 6 Smartphone app flowchart demonstrating the integration of pre-processing, machine learning enabled classification and iron deficiency chlorosis visual rating in real time

Authors' contributions
AS, BG, AKS, SS formulated research problem and designed approaches. JZ, TA, AKS, AS directed field efforts and phenotyping. JZ, HSN, AS, BG, AKS, TA designed the standard imaging protocol. HSN, BG, AL developed processing workflow. DA suggested using hierarchical classification. HSN, BG, SS performed data analytics. AL designed smartphone app. HSN, JZ, AKS, AS, SS, BG contributed to the writing and development of the manuscript. All authors read and approved the final manuscript.

Acknowledgements
Undergraduate and graduate students of soybean groups of AKS and AS who participated in phenotyping. Brian Scott, Jae Brungardt, and Jennifer Hicks for support during field experiments.

Competing interests
The authors declare that they have no competing interests.

Funding
Funding for this work came from Iowa Soybean Association, ISU Presidential Initiative for Interdisciplinary Research in Data Driven Science, Iowa State University (ISU) PSI Faculty Fellow, Monsanto Chair in Soybean Breeding at ISU, Baker Center for Plant Breeding at ISU. CRIS Project (IOW04314).

References
1. Soybean production in 2014. http://quickstats.nass.usda.gov/results/65A32870-615A-3A90-85D0-330CD8A77361.
2. Systematic strategies to increasing yield. In: Illinois soybean production guide. Edited by Association IS. Illinois Soybean Association; 2012.
3. Froechlich DM, Fehr WR. Agronomic performance of soybeans with differing levels of iron deficiency chlorosis on calcareous soil. Crop Sci. 1981;21(3):438–41.
4. Peiffer GA, King KE, Severin AJ, May GD, Cianzio SR, Lin SF, Lauter NC, Shoemaker RC. Identification of candidate genes underlying an iron efficiency quantitative trait locus in soybean. Plant Physiol. 2012;158(4):1745–54.
5. Morgan J. Iron deficiency chlorosis in soybeans. Crops and Soils Magazine. American Society of Agronomy. 2012. p. 5–9.
6. Fehr WR. Control of iron-deficiency chlorosis in soybeans by plant breeding. J Plant Nutr. 1982;5(4–7):611–21.
7. Rodriguez de Cianzio S, de Fehr WR, Anderson IC. Genotypic evaluation for iron deficiency chlorosis in soybeans by visual scores and chlorophyll concentration. Crop Sci. 1979;19(5):644–6.
8. Mamidi S, Lee RK, Goos JR, McClean PE. Genome-wide association studies identifies seven major regions responsible for iron deficiency chlorosis in soybean (Glycine max). PLoS ONE. 2014;9(9):e107469.
9. Lauter ANM, Peiffer GA, Yin T, Whitham SA, Cook D, Shoemaker RC, Graham MA. Identification of candidate genes involved in early iron

deficiency chlorosis signaling in soybean (*Glycine max*) roots and leaves. BMC Genomics. 2014;15:702.

10. Wiersma JV. Chapter 2: Importance of seed [Fe] for improved agronomic performance and efficient genotype selection, in "Soybean - Genetics and Novel Techniques for Yield Enhancement". Croatia: INTECH Open Access Publisher; 2011. ISBN:978-953-307-721-5

11. Nutter FW Jr, Gleason ML, Jenco JH, Christians NC. Assessing the accuracy, intra-rater repeatability, and inter-rater reliability of disease assessment systems. Phytopathology. 1993;83(8):806–12.

12. Kruse OMO, Prats-Montalbán JM, Indahl UG, Kvaal K, Ferrer A, Futsaether CM. Pixel classification methods for identifying and quantifying leaf surface injury from digital images. Comput Electron Agric. 2014;108:155–65.

13. Baranowski P, Jedryczka M, Mazurek W, Babula-Skowronska D, Siedliska A, Kaczmarek J. Hyperspectral and thermal imaging of oilseed rape (*Brassica napus*) response to fungal species of the genus Alternaria. PloS ONE. 2015;10(3):e0122913.

14. Sindhuja S, Ashish M, Reza E, Cristina D. Review: A review of advanced techniques for detecting plant diseases. Comput Electron Agric. 2010;72(1):1–13.

15. Subramanian R, Spalding EP, Ferrier NJ. A high throughput robot system for machine vision based plant phenotype studies. Mach Vis Appl. 2013;24:619–36.

16. Chen D, Neumann K, Friedel S, Kilian B, Chen M, Altmann T, Klukas C. Dissecting the phenotypic components of crop plant growth and drought responses based on high-throughput image analysis. Plant Cell. 2014;26(12):4636–55.

17. Römer C, Wahabzada M, Ballvora A, Pinto F, Rossini M, Panigada C, Behmann J, Léon J, Thurau C, Bauckhage C. Early drought stress detection in cereals: simplex volume maximisation for hyperspectral image analysis. Funct Plant Biol. 2012;39(11):878–90.

18. Smith HK, Clarkson GJJ, Taylor G, Thompson AJ, Clarkson J, Rajpoot NM. Automatic detection of regions in spinach canopies responding to soil moisture deficit using combined visible and thermal imagery. PloS ONE. 2014;9(6):e97612.

19. Jubery TZ, Shook J, Parmley K, Zhang J, Naik HS, Higgins R, Sarkar S, Singh A, Singh AK, Ganapathysubramanian B. Deploying Fourier coefficients to unravel soybean canopy diversity. Front Plant Sci. 2017;7:2066–75.

20. Licht M. Soybean growth and development. Iowa State Univ. Ames: Ext Pub PM 1945 Iowa State Univ; 2014. p. 28.

21. Lin S, Cianzio S, Shoemaker R. Mapping genetic loci for iron deficiency chlorosis in soybean. Mol Breed. 1997;3(3):219–29.

22. Gonzalez RC, Woods RE, Eddins SL. Digital image processing using MATLAB®. New York City: McGraw Hill Education; 2010.

23. Color names by hue ranges and luminance. http://www.workwithcolor.com/orange-brown-color-hue-range-01.htm.

24. Connected components labeling. http://homepages.inf.ed.ac.uk/rbf/HIPR2/label.htm.

25. Lee H, Park RH. Comments on "An optimal multiple threshold scheme for image segmentation. IEEE Trans Syst Man Cybern. 1990;20(3):741–2.

26. Zhang J, Naik H, Assefa T, Sarkar S, Chowda-Reddy RV, Singh A, Ganapathysubramanian B, Singh AK. Computer vision and machine learning for robust phenotyping in genome-wide studies. Sci Rep. 2017;7.

27. Turney PD. Cost-sensitive classification: Empirical evaluation of a hybrid genetic decision tree induction algorithm. J Artif Intell Res. 1995;2:369–409.

28. Kohavi R. A study of cross-validation and bootstrap for accuracy estimation and model selection. International Joint Conference on Artificial Intelligence (IJCAI) 1995; pp. 1137–1145.

29. Singh A, Ganapathysubramanian B, Singh AK, Sarkar S. Machine learning for high-throughput stress phenotyping in plants. Trends Plant Sci. 2016;21(2):110–24.

30. James G, Witten D, Hastie T, Tibshirani R. An introduction to statistical learning. 6th ed. New York: Springer; 2015.

31. Rish I. An empirical study of the naive Bayes classifier. In: IJCAI 2001 workshop on empirical methods in artificial intelligence. Vol. 3, No. 22. New York: IBM; 2001. pp. 41–46.

32. Kamarainen J, Paalanen P: GMMBayes Toolbox. http://www.it.lut.fi/project/gmmbayes/ (2003). Accessed 20 Feb 2016.

33. Quinlan JR. Improved use of continuous attributes in C4.5. J Artif Intell Res. 1996;4(1):77–90.

34. Breiman L. Random forests. Mach Learn. 2001;45(1):5–32.

35. Rish I. An empirical study of the naive Bayes classifier. In: IJCAI workshop on empirical methods in AI, 2001.

36. McLachlan GJ. Discriminant analysis and statistical pattern recognition. Hoboken: Wiley; 2004.

37. Cortes C, Vapnik V. Support-vector networks. Mach Learn. 1995;20(3):273.

38. Cover T, Hart P. Nearest neighbor pattern classification. IEEE Trans Inf Theory. 1967;13(1):21–7.

39. Reynolds DA. Gaussian mixture models, encyclopedia of biometric recognition. Heidelberg: Springer; 2008.

17

Development of a phenotyping platform for high throughput screening of nodal root angle in sorghum

Dinesh C. Joshi[1]*, Vijaya Singh[2], Colleen Hunt[3], Emma Mace[3], Erik van Oosterom[2], Richard Sulman[4], David Jordan[5] and Graeme Hammer[2]

Abstract

Background: In sorghum, the growth angle of nodal roots is a major component of root system architecture. It strongly influences the spatial distribution of roots of mature plants in the soil profile, which can impact drought adaptation. However, selection for nodal root angle in sorghum breeding programs has been restricted by the absence of a suitable high throughput phenotyping platform. The aim of this study was to develop a phenotyping platform for the rapid, non-destructive and digital measurement of nodal root angle of sorghum at the seedling stage.

Results: The phenotyping platform comprises of 500 soil filled root chambers (50 × 45 × 0.3 cm in size), made of transparent perspex sheets that were placed in metal tubs and covered with polycarbonate sheets. Around 3 weeks after sowing, once the first flush of nodal roots was visible, roots were imaged in situ using an imaging box that included two digital cameras that were remotely controlled by two android tablets. Free software (*openGelPhoto. tcl*) allowed precise measurement of nodal root angle from the digital images. The reliability and efficiency of the platform was evaluated by screening a large nested association mapping population of sorghum and a set of hybrids in six independent experimental runs that included up to 500 plants each. The platform revealed extensive genetic variation and high heritability (repeatability) for nodal root angle. High genetic correlations and consistent ranking of genotypes across experimental runs confirmed the reproducibility of the platform.

Conclusion: This low cost, high throughput root phenotyping platform requires no sophisticated equipment, is adaptable to most glasshouse environments and is well suited to dissect the genetic control of nodal root angle of sorghum. The platform is suitable for use in sorghum breeding programs aiming to improve drought adaptation through root system architecture manipulation.

Keywords: Drought, High throughput phenotyping, Root system architecture, Nodal root angle, Sorghum

Background

Root system architecture (RSA) is a major factor determining the ability of plants to access soil moisture in drought prone environments, particularly in cereals like sorghum (*Sorghum bicolor* (L.) Moench), maize (*Zea mays* L.), and wheat (*Triticum aestivum* L.), which are frequently grown in such environments. Of the many traits constituting RSA, the growth angle of the seminal

and nodal roots at the seedling stage has important implications for drought adaptation of adult cereal plants [1–3], because this trait can influence both horizontal and vertical exploration of the soil by roots [4–7]. These spatial effects on the ability of plants to access water can be exploited through crop management [8, 9].

The root system of sorghum is characterized by a single seminal root originating directly from the embryo and by multiple postembryonic nodal roots that emerge from the below-ground nodes of the stem [10].The seminal root plays an important part only in initial water and nutrient uptake and hence is of little importance in

---

*Correspondence: dinesh.pbl@gmail.com
[1] ICAR- Indian Grassland and Fodder Research Institute, Jhansi, Uttar Pradesh 284003, India
Full list of author information is available at the end of the article

mature sorghum, for which the RSA is predominantly constituted by post embryonic nodal roots [11, 12]. The angle of the first flush of nodal roots, which appears when around five leaves have fully expanded [10], is associated with the spatial distribution of roots of mature sorghum plants and hence with their ability to extract soil water [7]. As a consequence, an association between nodal root angle and grain yield has been reported for sorghum [13]. Nodal root angle is thus an important selection trait in sorghum breeding programs for improving drought adaptation.

High throughput screening of agronomically relevant traits is often restricted by the availability of suitable phenotyping systems, rather than the availability of genetic information. Multiparental breeding populations such as nested association mapping (NAM) populations have emerged as an excellent mapping resource to dissect the genetic control of complex quantitative traits by combining the advantages of traditional linkage analysis with association mapping [14]. Genetic dissection of various complex agronomic traits has been investigated in maize by utilising abundant genetic diversity of a NAM population [15–22]. A comparable backcross (BC) nested association mapping (BC-NAM) population has been developed in sorghum [23] and has been used to dissect the genetic control of complex quantitative traits [24]. Despite the availability of mapping resources such as BC-NAM populations to tackle complex rooting traits, the lack of high throughput phenotyping methods remains a bottle neck to quantify the genetic control of nodal root angle and enable its use as a selection trait in breeding programs.

Various phenotyping methods have been proposed for screening of root traits, including root angle, in common bean (*Phaseolus vulgaris* L.) [25, 26], barley (*Hordeum vulgare* L. and *H. spontaneum* C. Koch) [1], wheat (*Triticum aestivum* L) [4], and maize (*Zea mays* L.) [2, 27]. However, all these methods are designed to measure root traits within a few days of germination and are not suitable for phenotyping of nodal roots in sorghum, which needs to grow for 3 weeks before the first flush of nodal roots starts to appear [10]. Field based phenotyping methods have been proposed for phenotyping postembryonic root architecture of adult plants [28–30]. Although these methods can provide good data on a range of traits associated with RSA, the throughput of this system is limited by the duration of the crop cycle, and the area of land required to screen large numbers of genotypes.

To exploit nodal root angle as a selection trait in sorghum breeding programs that target drought stressed environments, efforts need to be directed towards the development of robust root phenotyping platforms that

are capable of (1) supporting root and shoot growth up to 5th–6th leaf stage, (2) expressing high heritability (repeatability) for the trait, and (3) minimizing the genotype × environment interaction. Ideally, any platform should have a short phenotyping cycle, be suitable for continuous screening throughout the year, and not be labour intensive. Therefore, the objectives of this study were to (1) develop a simple, low cost and high throughput phenotyping platform that supports root development of sorghum up to the 5th–6th leaf stage, (2) develop a high throughput imaging system for the digital measurement of nodal root angle and (3) test its high throughput capacity, reproducibility and ability to identify sorghum genotypes with contrasting root angle by characterizing a large BC-NAM population and a large set of advanced hybrids.

Methods
Phenotyping platform setup
As a first step in the development of a high throughput phenotyping platform for nodal root angle in sorghum, we compared the suitability of three artificial growing media, specifically gel chambers, seed germination blotting paper (Anchor Paper Co, St Paul, MN, USA), and geotextile capillary mat with pore size of 60 microns (Global Synthetics, Virginia, QLD, Australia), with soil filled chambers that were developed previously [10]. The gel filled observation chambers were constructed from two plates (one black perspex and one clear glass) as described by Bengough et al. [1] and Manschadi et al. [4]. Sterilized agar (Sigma Type A; 2% w/v) was poured onto each plate and after the agar had set, the two plates were taped together. Two germinated seeds were placed between the agar layers of the vertically mounted chambers. The seeds were oriented vertically with the radicle facing downwards. These gel chambers have been used successfully for wheat [4], where seminal root angle can be phenotyped within days of germination. However, they did not support the growth of sorghum plants for the extended period until the appearance of nodal roots, because of microbial contamination, difficulties in maintaining a consistent density of gel and an inability to meet the nutrient requirements of the plants.

The germination paper and capillary mat were cut into 40 × 35 cm units. Both were sterilized in an autoclave before being hung into a large plastic tub with similar height and width as the units so that the lowest 3 cm of the units were submerged in 3.5 l of nutrient solution, which moved up through the capillary mat and germination paper through capillary action. Both the germination paper and capillary mat were always wet throughout the experimental period, suggesting adequate capillary rise in both systems. The nutrient solution consisted of quarter strength of Hoagland solution and was changed every 5 days. Sorghum seeds were sterilized by rinsing in 70% ethanol and washed three times with sterile distilled water. The sterilized seeds were pre-germinated for 3 days on filter paper in petri dishes with a day/night temperature regime of 30/22 °C and 60% relative humidity. The germinated seeds were fixed between two sheets of germination paper or capillary mat with two standard paper clips and each unit was attached to a wooden stick with two fold back clips, one on each side of the upper edge of the unit. Each unit was covered with black polythene foil to prevent penetration of light and had one 2 × 1 cm slit at the top to allow emergence of the shoot.

The soil filled chambers, which consisted of two perspex sheets, separated by a rubber strip and clamped together with fold back clips, have been described in detail by Singh et al. [10]. The growth parameters of the plants grown in the soil-filled chambers were compared with those of plants grown on the germination paper and capillary mat. Shoot dry weight, root dry weight, root length, and root diameter of 3 week old plants were measured, when around five leaves had fully expanded and nodal roots started to appear. At harvest, the shoots of five plants from each of the three media were cut off at the base of the stem and shoot dry weight was determined after drying at 60 °C for 3 days. After imaging, roots were carefully removed from the medium and washed. Washed roots were stored in 70% ethanol and scanned using an Epson scanner (Epson, Long Beach, CA, USA), after submerging in a water bath to facilitate separation of roots and to minimize overlap. Scanning was done at a resolution of 600 dpi, using both top and bottom lighting and a threshold setting of 68 was used to distinguish roots from the background. Scanned roots were analyzed using WinRHIZO Pro (Regent Instruments, Inc., Quebec City, QC, Canada) which was calibrated to obtain total root length and average root diameter. Root samples were then dried at 60 °C for 3 days and dry weight determined.

The germination paper and capillary mat did support plant growth until five leaves had fully expanded, but plant growth was suboptimal and non-representative

of soil grown plants. The rate of development was significantly slower than in soil filled chambers, as was evident from a slower leaf appearance rate and this delayed appearance of nodal roots at the five-leaf stage (Table 1). Despite this delayed harvest, plants had significantly shorter and thinner roots and less root and shoot biomass (Table 1). Moreover, nodal roots often penetrated the germination paper surface (Fig. 1a) or grew vertically along the capillary mat (Fig. 1b), making measurement of nodal root angle cumbersome or even impossible. In contrast, in the soil filled chambers, nodal roots became clearly visible against the transparent perspex sheets once they started appearing around the 5th leaf stage (Fig. 1c). The prolonged duration of the screens using germination paper and capillary mat made it more cumbersome to meet the nutrient requirements of the plants and to keep the surface of the growth medium free from microbial contamination. This required sterilization of the substrate and seeds, pre-germination of seeds, and preparation of Hoagland solution at regular intervals, making these two systems inefficient and more time consuming for screening large numbers of genotypes compared to the soil filled chambers. Hence, a phenotyping platform based on soil-filled chambers was deemed to be the most suitable for high throughput phenotyping of nodal root angle in sorghum.

Soil based phenotyping platform

The phenotyping platform consists of custom built root chambers (Fig. 2) and imaging equipment (Fig. 3) that were designed and constructed by Biosystems Engineering, Toowoomba, QLD, Australia (www.biosystem-eng.com). The platform was an improvement on a previously published system [31], which lacked automation for imaging and was not conducive for high throughput applications. The basic units of the phenotyping platform were custom made root chambers (Fig. 2a). Each chamber comprised of two 6 mm thick transparent perspex sheets of 50 cm high, 45 cm wide and 3 mm thick that were separated on three sides (two long sides and one of the short sides) by 3 mm thick rubber and held in place

Table 1 Comparison of seedling attributes using soil filled root chambers, germination paper and capillary mat

Trait	Soil filled chamber	Germination paper	P^a	Capillary mat	P^b
Days to emergence of nodal roots	25	36	0.001	36	0.001
Total root length (m)	312	245	0.003	232	0.002
Average root diameter (mm)	2	1.3	0.002	1.2	0.002
Root mass (g)	0.13	0.08	0.005	0.06	0.005
Shoot mass (g)	0.33	0.21	0.002	0.24	0.001

[a] P value for difference of germination paper with soil filled chamber

[b] P value for difference of capillary mat with soil filled chamber

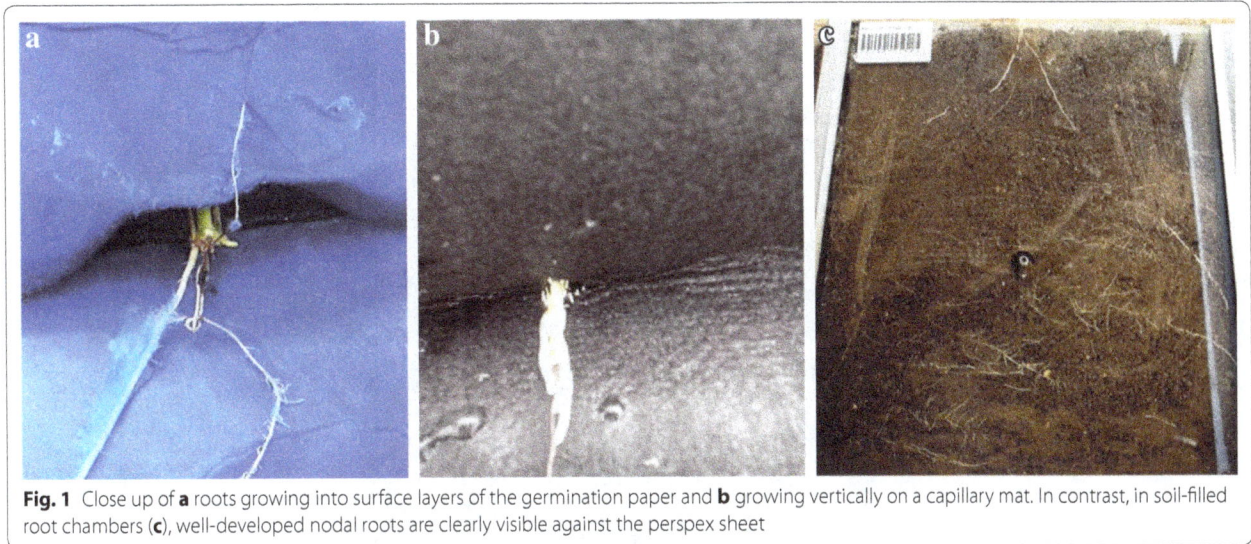

Fig. 1 Close up of **a** roots growing into surface layers of the germination paper and **b** growing vertically on a capillary mat. In contrast, in soil-filled root chambers (**c**), well-developed nodal roots are clearly visible against the perspex sheet

Fig. 2 High throughput phenotyping platform for screening genetic variation for nodal root angle in sorghum. **a** Purpose built root chamber filled with soil. **b** Metal tubs containing root chambers. **c** Polycarbonate sheet covering the top of the root chambers to exclude light. **d** Nodal roots visible through the transparent wall of perspex sheets at 6th leaf stage

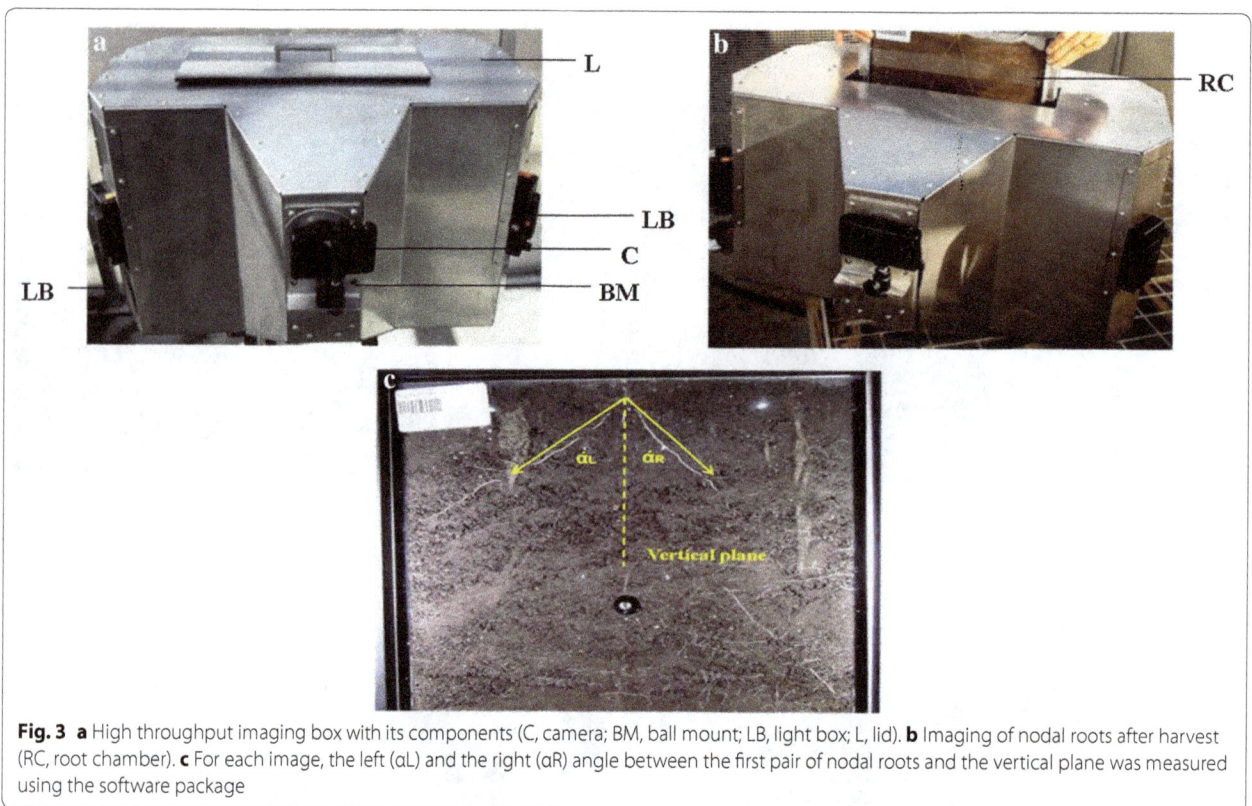

Fig. 3 **a** High throughput imaging box with its components (C, camera; BM, ball mount; LB, light box; L, lid). **b** Imaging of nodal roots after harvest (RC, root chamber). **c** For each image, the left (αL) and the right (αR) angle between the first pair of nodal roots and the vertical plane was measured using the software package

by three metal clamps. To minimize bulging of the chambers during soil filling and to maintain constant 3 mm spacing between the two perspex sheets, the sheets were connected at the centre with a nut and bolt. Each chamber was filled with 1100 g of black, clay-textured soil, which provided a suitable contrast with the roots for image analysis. Root chambers were placed vertically in 2 m long stainless steel tubs that had similar height and width as the chambers (Fig. 2b). Each tub had slots at the top and bottom to vertically position 50 root chambers, and had six holes in the bottom to allow drainage of excess water and nutrient solution. The entire platform contained ten tubs, giving a total capacity of 500 root chambers.

Before sowing, the filled chambers were watered to field capacity. Three seeds were sown per chamber. In order to ensure root growth along the transparent perspex sheets, seeds were placed vertically at a depth of 3 cm, with the embryo downwards and facing one of the perspex sheets. After sowing, the top surfaces of all chambers in each tub were covered with 2 m long black polycarbonate sheets to exclude light from the developing roots while leaving 5 cm long slits for the seedlings to emerge (Fig. 2c). Three days after emergence, seedlings were thinned to one per chamber. A complete hydroponic nutrient solution (Peters Professional Water Soluble Fertilizer Hydro-sol,

Scotts-Sierra Horticultural Products Co., Marysville, OH, USA) was applied once a week to ensure nutrients and water were not limiting growth and development of the plants. The solution comprised of the following nutrients (mg per liter of water); N, 200; P, 48; K, 210; Ca, 193; Mg, 40; Na, 3.6; S, 53.5; B, 0.50; Cl, 0.04; Cu, 0.15; Fe, 5.0; Mn, 0.50; Mo, 0.10; and Zn, 0.15.

Roots were imaged when 5–6 leaves had fully expanded and the first flush of nodal roots was visible. At that time, the shoot of each plant was removed by cutting the base of the stem. Previous studies have shown that differences in plant size and vigour do not affect root angle [31]. Both perspex sheets of each chamber were barcoded (Fig. 2d) to track the identity of individual plants during imaging. Root imaging was done in a metal box of 55 cm high, 62 cm wide, and 40 cm thick that contained a central imaging plane (Fig. 3). Two cameras (Canon PowerShot SX610 HS 20.2 MP Ultra-Zoom Digital Cameras) were positioned on versatile ball mounts (Universal 1/4"-20 Camera Accessory Mount) on either side of this plane (Fig. 3a). The imaging box also housed Glanz LED98A Camera LED light boxes (121 × 78 × 35 mm) that were mounted in four corners to uniformly illuminate the root chambers during imaging. These light boxes were fully dimmable and included one magnetic diffuser for soft day light balance light and a second one for the soft tungsten balance light. The imaging plane contained three metal clamps to allow easy insertion and removal of each root chamber, and to ensure a fixed position within the imaging plane. The cameras were equipped with in-built Wi-Fi® technology to connect directly to Android™ devices (Samsung Galaxy Tab 3 Lite-7.0 T-113 Wifi-Only 8 GB White Tablets) using camera connect (www.canon.com.au), a free app that controlled the imaging set up and synchronized the imaging of both sides of each root chamber (Fig. 3b). An exposure time of 1 s allowed for high quality images with limited background noise (Fig. 3c). Hence, there was no need for calibrating the cameras. Root images, which captured the entire perspex sheets on both sides, were downloaded manually from the camera to a computer and were saved as JPEG files.

Plant material

Two different sets of sorghum germplasm were used in the study: a backcross nested association mapping (BC-NAM) population and a set of advanced hybrids from the Australian sorghum pre-breeding program. The BC-NAM population was derived from crosses between a recurrent inbred line (R931945-2-2) and 23 diverse founder parents [23]. R931945-2-2 is an elite line adapted to Australian growing environments and the founder parents include exotic genotypes that represent the global diversity of sorghum, spanning countries of origin, racial type, and wild and weedy genotypes. In total, 976 BC-NAM progenies and 24 parents were included in the study. The advanced hybrid set included 628 hybrids that were based on three female and 395 male parent lines that represented a random sample from the set of hybrids used in the advanced trial series of the sorghum pre-breeding program.

Experimental details

The experiments were conducted at The University of Queensland, St. Lucia, Australia (27°23′S, 153°06′E) in a naturally lit, temperature controlled glasshouse with a day/night temperature regime of 30/22 °C. The BC-NAM population was phenotyped across four independent runs (Table 2) that each consisted of 10 tubs with 50 root chambers each. Experimental runs had two blocks of five tubs and were designed using a row column design where column was represented by the tubs and row by root chamber position within each tub. In each run, 240 genotypes were replicated twice and two check genotypes that were known to differ in root angle (R931945-2-2 and SC-170-6-8, [31]) were each replicated once in each tub. Across runs, the first two had 20% of genotypes in common, whereas run 3 and run 4 had 10% in common (Table 2). This design enabled each pair of runs to be combined in a single analysis.

The advanced hybrids were screened in two experimental runs that used a multi-site partially replicated row column design with two experiments that each contained two blocks of five tubs each. Each run included 339 unique hybrids with 50 of these in common across the two experiments, giving a total of 628 hybrids. Within each experimental run, approximately 50% of the hybrids were replicated twice and the other 50% had only a single replicate. The multi-site design enabled the two runs to be combined in a single analysis and replicated hybrids and their randomization to be optimised between the two runs.

Measurement of nodal root angle

The JPEG files that contained the images of the root systems were used to determine the encompassing angle, relative to the vertical plane, of the first flush of nodal roots. This was done using *openGelPhoto.tcl* (www.activestate.com/activetcl), free software that calculates the angle of individual roots relative to the vertical plane by identifying a point of origin (base of the plant) and an end point for each nodal root. In order to standardize observations, the end point of each root was taken at a distance of 3 cm from the base of the plant. The observed

Table 2 Broad sense heritability and genetic correlations for nodal root angle between the experimental runs of the BC-NAM population and the advanced hybrids

Pairs of experiments	Run number	Number of genotypes	Mean root angle (degree)	Heritability (%) (from spatial analysis)	Heritability (%) (from non-spatial analysis)	Percentage of common genotypes (genotypic overlap)	Correlation between runs
BC-NAM population	1	199	23.6	77	55	20	0.94
	2	242	25.9	93	89		
BC-NAM population	3	242	26.5	95	85	10	0.70
	4	242	26.0	91	78		
Advanced hybrids	5	339	27.3	94	93	15	0.96
	6	339	27.2	96	94		

root angle for each plant was the mean of four observations (left and right for both sides of each chamber).The identity of the genotype in each image was tracked using the barcodes that were attached to each individual perspex sheet.

Statistical analysis

The six experimental runs were analyzed as three independent multi-site trials that consisted of two pairs of runs with sufficient common genotypes to be analyzed together (Table 2). The observed values for nodal root angle (y) were analyzed using a linear mixed model, written as

$$y = X\tau + Zu + e,$$

where the fixed effect τ contained the mean root angle in each run and the random effects u contained genotypes within runs and the genetic correlation between runs. The design matrices for fixed and random effects were given by X and Z respectively and e was the random vector of residual effects. For each run, independent components associated with the structure of the experimental design were included as random effects. These were variation between and within tub.

Possible neighbour effects were allowed by fitting spatial correlations between tubs and root chambers within each run. Since there were 4 measurements within each tub and slot position, an equal variance AR1 model was used for the spatial interaction between tub and slot (AR1v). Despite the relative small footprint of each experimental run, significant spatial effects were detected in all runs and adjustment for these effects generally improved the accuracy of each run by 1–20%. Data were analyzed using ASReml-R [32] and R software Version 3.1.1 (R core team 2014).

In order to quantify the capability of the phenotyping platform to detect genotypic differences in nodal root angle in a repeatable and consistent manner, a broad sense heritability (H^2) was calculated firstly from a model without the spatial AR1v variation and then from a model that included the spatial variation.

Results

Genotypic variation for nodal root angle was detected

Genotypic variation in root angle was observed in all experimental runs. Nodal root angle ranged from 17.6° to 41.3° for BC-NAM population and from 16.0° to 42.0° for the advanced hybrids (Fig. 4). Across runs of the BC-NAM population, the minimum root angle ranged from 17.6° (Run 3) to 18.5 (Run 2), whereas the maximum varied more, ranging from 30.5° (Run 1) to 41.3° (Run 2) (Fig. 4a). As a consequence of these differences in maximum nodal root angle, the range in observed phenotypes

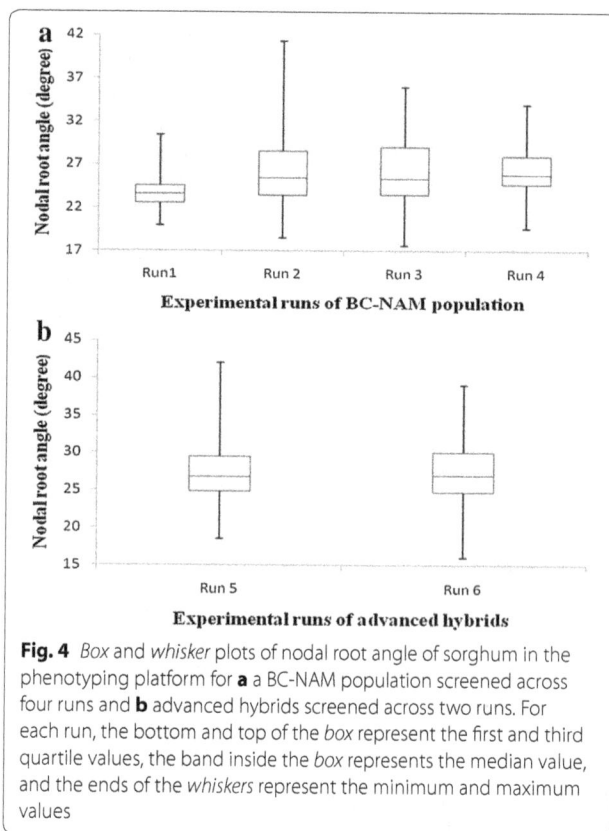

Fig. 4 *Box* and *whisker* plots of nodal root angle of sorghum in the phenotyping platform for **a** a BC-NAM population screened across four runs and **b** advanced hybrids screened across two runs. For each run, the bottom and top of the *box* represent the first and third quartile values, the band inside the *box* represents the median value, and the ends of the *whiskers* represent the minimum and maximum values

was lowest in Run 1 (10.5°) and greatest in Run 2 (22.8°). Across the two runs with hybrids, the mean root angle was almost identical (27.3° in Run 5 and 27.2° in Run 6). The greater differences in mean and range in the runs of BC-NAM genotypes was likely due to their increased genetic diversity, because the hybrids represent a random sample from advanced trials that have already undergone selection.

The nodal root angle of 24 parental genotypes of the BC-NAM population screened in Runs 1 and 2 varied from 23.1° to 38.8° (Fig. 5). Out of the 24 parents, 20 were previously screened for nodal root angle under slightly different experimental conditions using a prototype version of the phenotyping platform used in this study [31]. A strong correlation ($r^2 = 0.73$) was observed for nodal root angle recorded for these 20 parents common between both studies (Fig. 6).

Heritability and genetic correlations for nodal root angle were generally high

High estimates of H^2 (94 and 96%) for nodal root angle were obtained in both runs of advanced hybrids (Table 2). Without the spatial AR1v effects the H^2 were 93 and 94% for the advanced hybrids (Table 2). For runs of the BC-NAM population, when spatial effects were included in

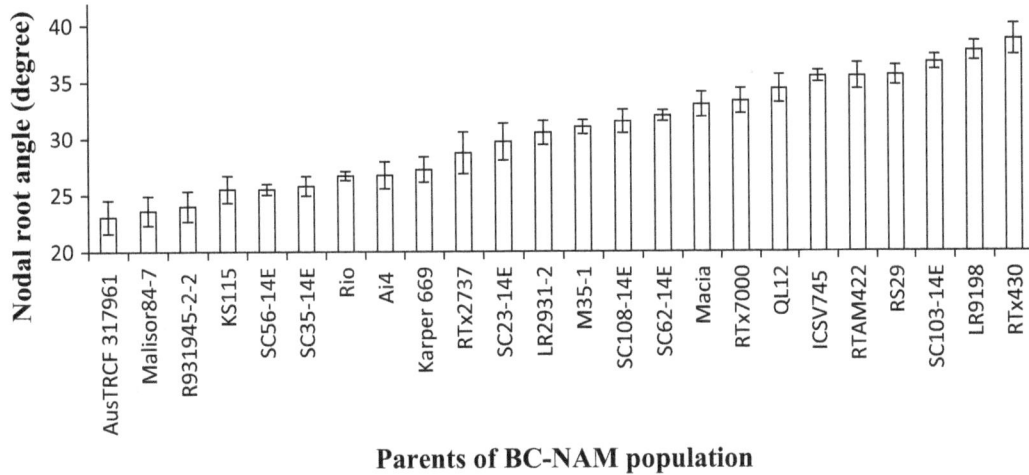

Fig. 5 Genetic variation for nodal root angle across 24 parents phenotyped in Run 1 and Run2 of the BC-NAM population. Parents have been sorted for average root angle. The *vertical bars* indicate the relevant standard error

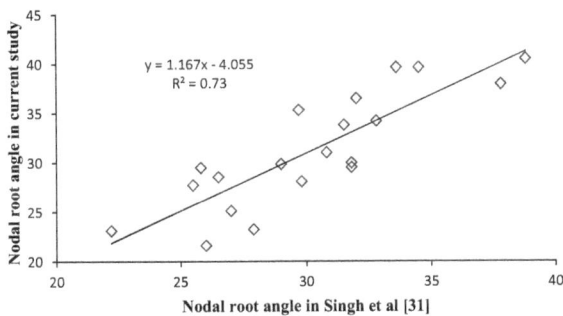

Fig. 6 Association between nodal root angle (in degrees) in the current study and that reported by Singh et al. [31] for 20 parents of the BC-NAM population

the analysis H^2 ranged from 77% (Run 1) to 95% (Run 3) and when spatial effects were not included H^2 ranged from 55% (Run 1) to 89% (Run 2) (Table 2). These differences in H^2 estimates of nodal root angle were likely to be a consequence of differences in the genotypes that were included in each run. In general, the high estimates of H^2 across runs indicated that for individual runs, differences between replications were smaller than those across the genotypes.

Strong genetic correlations were calculated for nodal root angle for all three pairs of experimental runs with common genotypes (Table 2). The genetic correlation between the two advanced hybrid runs was 0.96, whereas for the BC-NAM population it ranged from 0.70 for Runs 3 and 4 to 0.94 for Runs 1 and 2. These high correlations between the three pairs of runs confirmed that the genotypes displayed consistent rankings for nodal root angle across each pair of runs.

A single experimental run requires around 110 h of user time

The phenotyping platform demonstrated a capacity to identify genetic variation for nodal root angle in large numbers of sorghum lines for a relatively low requirement of user time of 100–120 h per run of 500 plants. Most of this time (nearly 80 h) was required to set up the root chambers, including cleaning and assembly, soil filling, and initial watering. Time requirements for other activities included up to 14 h for sowing, thinning, watering, and application of the nutrient solution and up to 8 h for barcoding, harvesting, and image acquisition. Image analysis and measurement of nodal root angle using the software package takes on average 1 min per image (around 15 h for 1000 images). Hence, the overall time requirement is around 100–120 h, or 13–15 min per plant.

Discussion

Late appearance of nodal roots in sorghum requires a soil-based phenotyping platform

Despite the proven utility of nodal root angle in improving drought adaptation, the difficulty of measuring the trait in situ under field conditions has hindered both the genetic dissection and its use as a selection criterion in sorghum breeding programs. Non-soil based phenotyping methods like gel chambers, which have been used in wheat [1, 4], barley [33] and rice [34], and growth pouches and blotting paper, which have been used in maize [27, 35], have been proposed to study root architecture, including root angle, in cereals. These methods were found suitable for root phenotyping in those species, because their root development allowed phenotyping within a few days after germination. Wheat and barley produce multiple seminal roots soon after germination that can be used for phenotyping root angle, whereas in rice and maize, nodal roots appear around the 2nd leaf stage, 1 week after sowing [3, 10]. For sorghum in contrast, the single seminal root and the relatively late appearance of nodal roots [10] require sorghum plants to grow for 3 weeks before root angle can be phenotyped. This can potentially restrict the utility of these non-soil based methods [1, 4, 27, 33–35] that are specifically designed for phenotyping within a few days after germination, before seed reserves used for growth run out.

Nonetheless, non-soil based phenotyping systems that support plant growth for longer periods of time have been developed. Le Marié et al. [36] developed Rhizoslides, a system based on germination paper and plexiglass that was used to phenotype root systems of 20 day old maize plants. Our preliminary studies (Table 1) also indicated that sorghum seedlings can be grown on germination paper and capillary mat for extended periods of time. However, a major problem of the germination paper and capillary mat was that the sorghum roots would penetrate the surface of the media, making acquisition of high quality images cumbersome. It is possible that this behavior of roots might differ across species, because a comparison between maize and sorghum root systems in soil filled rhizotrons suggested that maize and sorghum roots respond differently when hitting the sides of the rhizotrons, with sorghum roots generally growing vertically along the side of the chamber (Fig. 3), whereas maize roots were more likely to reflect off the side back towards the centre of the rhizotrons [10]. This contrasting response could potentially make sorghum roots more prone to penetrating the top layers of the germination paper than maize roots. In addition, the generally poorer growth of sorghum seedlings on germination paper and capillary mat compared to soil-based systems (Table 1) might reflect the high susceptibility of sorghum to water logging during early vegetative stages [37], which could be associated with the lack of aerenchyma in sorghum roots [38]. The specific nature of the root system development of sorghum and its sensitivity to water logging at the seedling stage make the crop not conducive to root phenotyping in non-soil based phenotyping platforms.

Soil-based phenotyping platforms for root architecture traits in general and nodal root angle in particular have been developed before. Richard et al. [39] used 4L transparent pots to image the angle and number of seminal roots in wheat 5 days after sowing. Hargreaves et al. [33] combined soil sacks and X-ray microtomography to measure root traits in barley. However, because of the plant size required for sorghum at the time of imaging and because the root angle observed from 2D gel based phenotyping platform in barley is representative of the 3D angular root spread measured in soil sacks [33], 2D root chambers are logistically the most attractive option. Nagel et al. [40] developed GROWSCREEN-Rhizo, which combined soil-filled rhizotrons for non-destructive 2D measurement of a range of root architecture traits with fully automated imaging at a rate of 60 rhizotrons per hour. Conceptually, this system is quite similar to the system we developed, with the main difference being that our system requires less investment in automating imaging as it requires no sophisticated equipment. This relative simplicity makes our platform adaptable to most glasshouse environments and a version of the system has already been implemented in Africa.

The two critical components of the platform are the root chambers (Fig. 2) and the imaging box (Fig. 3). To facilitate high quality images of roots, it is important to ensure that nodal roots are clearly visible against at least one of the perspex sheets of the root chamber. To achieve this, at sowing the seed needs to be positioned along one of the perspex sheets, because even minor bulging of the perspex sheets can sometimes render the roots invisible against both perspex sheets. Nagel et al. [40] addressed this issue in their setup by having an option to position root chambers at an angle, thus ensuring that a complete root system is visible against the transparent perspex on one side. Apart from the additional costs involved in this feature, the disadvantage compared to our setup is that this will likely result in getting only two estimates for root angle per chamber, whereas our setup generally resulted in four estimates, with increased accuracy and precision. In addition, it is important that the embryo of the seed faces downwards to ensure that the radicle and other roots emerge from the bottom of the seeds and thus avoid gravitropic responses affecting observed root angles. The root chambers also need to be covered to prevent penetration of light to the roots and to prevent algal growth on the perspex sheets. The black polycarbonate sheet we used to cover the top of the tubs that contain the root chambers satisfied these criteria and is readily available in the market. The important elements of the high throughput imaging box were (1) two software controlled digital cameras, (2) four LED light controlling boxes, and (3) two android tablets. The ability to remotely control the digital cameras to capture the image by android tablets through an in-built Wi-Fi® system ensured an image acquisition throughput of at least 60 chambers per hour, similar to the throughput reported by Nagel et al. [40] for GROWSCREEN-Rhizo. The four LED light boxes provided illumination that was uniform across the surface of both perspex sheets for the root chamber inside the imaging box (Fig. 3) and that was consistent across successive root chambers. This uniformity in space and time was key to maintain high resolution of the images obtained. It is likely that our soil-based phenotyping platform can be used for a range of crops. However, for crops for which root angle can be phenotyped within a week from germination, smaller systems that are based on gel, germination paper, or even soil [1, 4, 27, 33–35, 39] might be more appropriate.

The focus during development of our platform was on the phenotyping of nodal root angle. This was guided by observations that this trait is associated with the spatial distribution of roots of mature sorghum plants [7]. This will affect the spatial and temporal ability to extract water from the soil profile [7] and hence grain yield in field conditions [13]. Although other traits associated with root architecture, such as root length, could be measured in our platform, the soil could potentially obscure the fine detail of root architecture required for these measurements (Fig. 3). Although this issue can be resolved by placing the root chambers at an angle [40], the root length density required by sorghum to access all available water is only around 0.2 cm cm^{-3} [41]. Hence, small differences in branching may have only a minor effect on the ability of sorghum plants to extract water. However, branching might be relevant in the context of root system efficiency (RSE), the transpiration per unit leaf area per unit root mass, which represents functional mass allocation to roots to support water capture, relative to allocation to aerial mass that determines water demand [42].

Phenotypic platform revealed high genetic variation, high heritability, and reproducibility for nodal root angle

The wide range in observed nodal root angle (Fig. 4) and the high heritabilities (Table 2) indicated that the phenotyping platform has the power to detect genetic variation in nodal root angle. The observed range of 17.6°–41.3° (BC-NAM population) and 16.0°–42.0° (advanced hybrids) was comparable with the range of 14.5°–32.3° reported for a recombinant inbred line (RIL) population of sorghum [13]. Variability observed for the parents of the BC-NAM population (23.1°–38.8°) was comparable with that of 21.6°–40.5° in an earlier study [31]. The slightly wider range in nodal root angle observed in the BC-NAM progenies compared to the parental genotypes provided some evidence of minor transgressive segregation for the trait. This can possibly be explained by the recombination of additive alleles of the diverse male parents and a common female line (R931945-2-2), which may have influenced many allelic effects of the female inbred line. The range observed in root angle in the present study was quite different to the range of 60.1°–84.0° and 52.0°–88.0° (relative to the vertical plane) reported for wheat [39] and maize [30] respectively. These species differences were possibly associated with differences

in root types (seminal roots for wheat, crown roots for maize, nodal roots for sorghum) and stage of development at the time of phenotyping (seedling stage for wheat, early vegetative for sorghum, mature plants for maize). The overall broad sense heritability H^2 (repeatability) for nodal root angle observed in the present study (91% averaged across the six experimental runs from the spatial analysis and 82.4% averaged from non-spatial analysis, Table 2), was considerably greater than the H^2 of 46.6% observed for 44 diverse inbred lines in a previous study [31], but close to the H^2 value of 73.7% reported for a RIL population of sorghum [13]. The high H^2 in our studies indicated that differences in nodal root angle were predominately influenced by genotype, and that variation associated with random factors was minor compared to the genetic variation. The magnitude of the heritability for a trait measured in a phenotyping platform is an important factor in determining its efficiency and relevance to a large scale germplasm screening programs.

The accuracy and repeatability of the phenotyping platform was further highlighted by strong genetic correlations for the three pairs of experimental runs (Table 2) and by the high correlation with the results reported by Singh et al. [31] under different experimental conditions (Fig. 6). Overall, strong genetic correlations, high repeatability (heritability) and consistent ranking obtained across runs in different sets of genetic material indicate that the platform could provide a useful screen for nodal root angle across the diverse germplasm of a breeding program. This would make root angle an important selection trait in breeding programs that aim at improving drought adaptation of sorghum through genetic manipulation of RSA.

Value of phenotyping for nodal root angle to breeding programs

Higher efficiency in selecting for RSA can be achieved by designing phenotyping platforms that are capable of (1) screening root traits at the seedling stage to shorten the selection cycle and speed up the rate of genetic improvement and (2) establishing the genetic correlation between the root trait phenotyped and the ultimate breeding target (grain yield) [43]. The advanced hybrids screened for nodal root angle in the current study were also evaluated for grain yield in seven different environments across Australia. Analysis of the data revealed an association between narrow nodal root angle and grain yield across environments (D Jordan, unpublished data). This was consistent with the findings of Mace et al. [13], who reported a weak but significant association between the presence of QTL for narrow root angle and grain yield in a set of RILs of sorghum evaluated in hybrid combination in yield trials. This association reflects the observation that in sorghum, QTL for nodal root angle co-locate with QTL for stay-green [13], the ability of a crop to retain green leaf area during grain filling [44]. A possible mechanism for this would be that narrow root angle could increase the ability of plants to access water from deeper soil layers [7, 45], which can prolong maintenance of photosynthesis and remobilization activities during grain filling [44] under drought. In this context, it is interesting to note that genotypes SC56-14E, SC35-14E, Rio, and R931945-2-2, which are the main sources of stay green in various breeding program across the world [23], each had narrow nodal root angle (Fig. 5). Moreover, four near isogenic lines (NILs) that each contained a single introgression of a stay green QTL (NIL6078-1, *Stg1*; NIL2219-3, *Stg2*; NIL2290-19, *Stg3*;NIL6085–9, *Stg4*) displayed narrow root angle as well (data not shown). These results indicate that the sorghum breeding program in Australia has likely indirectly selected for narrow root angle when pursuing improved adaptation to post- flowering drought stress as one of its most important breeding objectives. Hence, nodal root angle at the seeding stage can be predictive of grain yield across environments.

Conclusion

The platform presented in this paper provides a high throughput, low cost, easy to implement screen for phenotyping nodal root angle of sorghum. The setup requires no sophisticated instruments, has a relatively small foot print and allows rapid, non-destructive, two dimensional analysis of root angle with minimal disturbance to plant growth. Integration of the high throughput phenotyping platform with advanced genomic approaches allows identification of QTLs governing nodal root angle and mining of alleles to tailor RSA of genotypes to predominant management and environmental conditions to exploit specific adaption to drought stress. Apart from screening large breeding populations for genetic mapping, this platform is equally applicable to enhancing the efficiency of breeding programs. For instance, high throughput screening of elite inbred lines and a large number of experimental hybrids at early seedling stage will assist in their selection, release and adoption for a particular production environment (terminal drought stress environment, skip row management system). In addition, rapid screening of early segregating generations (F_2 and F_3) will enhance the selection efficiency and enrich the gene pool with the alleles governing desirable root phenotype (narrow or wide angle). The platform would also be suitable to evaluate the post-embryonic root architecture of any other graminaceous species for which nodal roots appear at similar development stages as sorghum.

Authors' contributions
DCJ and VS phenotyped the breeding material and were the major contributors in writing the manuscript. CH designed the experiments and conducted statistical analysis. EM and DJ conceived the study, participated in germplasm development and selection, and helped to draft the manuscript. EvO participated in the development of the phenotyping platform and editing of the manuscript. RS constructed the root chambers and imaging box. GH conceived the phenotyping platform and assisted in design of the study and editing of the manuscript. All authors discussed the results and commented on the manuscript. All authors read and approved the final manuscript.

Author details
[1] ICAR- Indian Grassland and Fodder Research Institute, Jhansi, Uttar Pradesh 284003, India. [2] Queensland Alliance for Agriculture and Food Innovation, The University of Queensland, Brisbane, QLD 4072, Australia. [3] Department of Agriculture and Fisheries, Hermitage Research Facility, Warwick, QLD 4370, Australia. [4] Biosystems Engineering, 323 Margaret Street, Toowoomba, QLD 4350, Australia. [5] Queensland Alliance for Agriculture and Food Innovation, Hermitage Research Facility, The University of Queensland, Warwick, QLD 4370, Australia.

Acknowledgements
The first Author is thankful to Department of Biotechnology, Government of India for providing the financial support through Indo-Australia Career Boosting Gold Fellowship. The authors thank the Grains Research and Development Corporation and the Queensland Department of Agriculture and Fisheries for access the germplasm used in this study.

Competing interests
The authors declare that they have no competing interests.

Funding
This work was funded by the Bill and Melinda Gates Foundation.

References
1. Bengough AG, Gordon DC, Al-Menaie H, Ellis RP, Allan D, Keith R, Thomas WB, Forster BP. Gel observation chamber for rapid screening of root traits in cereal seedlings. Plant Soil. 2004;262:63–70.
2. Hochholdinger F, Katrin W, Sauer M, Dembonsky D. Genetic dissection of root formation in maize reveals root-type specific development programmes. Ann Bot. 2004;93:359–68.
3. Kato Y, Abe J, Kamoshita A, Yamagishi J. Genotypic variation in root growth angle in rice (*Oryza sativa* L.) and its association with deep root development in upland fields with different water regimes. Plant Soil. 2006;287:117–29.
4. Manschadi AM, Hammer GL, Christopher JT, deVoil P. Genotypic variation in seedling root architectural traits and implications for drought adaptation in wheat (*Triticum aestivum* L.). Plant Soil. 2008;303:115–29.
5. Hammer GL, Dong ZS, McLean G, Doherty A, Messina C, Schussler J, Zinselmeier C, Paszkiewicz S, Cooper M. Can changes in canopy and/or root system architecture explain historical maize yield trends in the U.S. Corn Belt? Crop Sci. 2009;49:299–312.
6. Uga Y, Okuno K, Yano M. Dro1, a major QTL involved in deep rooting of rice under upland field conditions. J Exp Bot. 2011;62:2485–94.
7. Singh V, van Oosterom EJ, Jordan R, Hammer GL. Genetic control of root angle in sorghum and its implication in water extraction. Eur J Agron. 2012;42:3–10.
8. McLean G, Whish J, Routley R, Broad I, Hammer GL. 2003. The effect of row configuration on yield reliability in grain sorghum: II. Modelling the effects of row configuration. In: Proceedings of 11th Australian agronomy conference Geelong, VIC, Australia. 2–6 Feb 2003. The Regional Institute, Gosford, NSW, Australia. http://www.regional.org.au/au/asa/2003/c/9/mclean.htm. Assessed 24 Aug 2016.

9. Whish J, Butler G, Castor M, Cawthray S, Broad I, Carberry P, Hammer GL, McLean G, Routley R, Yeates S. Modelling the effects of row configuration on sorghum in north-eastern Australia. Aust J Agric Res. 2005;56:11–23.

10. Singh V, van Oosterom EJ, Jordan DR, Messina CD, Cooper M, Hammer GL. Morphological and architectural development of root systems in sorghum and maize. Plant Soil. 2010;333:287–99.

11. Salih AA, Ali IA, Lux A, Luxova M, Cohen Y, Sugimoto Y, Inanga S. Rooting, water uptake, and xylem structure adaptation to drought of two sorghum cultivars. Crop Sci. 1999;39:168–73.

12. Tsuji W, Inanaga S, Araki H, Morita S, An P, Sonobe K. Development and distribution of root system in two grain sorghum cultivars originated from Sudan under drought stress. Plant Prod Sci. 2005;8:553–62.

13. Mace E, Singh V, van Oosterom E, Hammer G, Hunt C, Jordan D. QTL for nodal root angle in sorghum (Sorghum bicolor L. Moench) co-locate with QTL for traits associated with drought adaptation. Theor Appl Genet. 2012;124:97–109.

14. Yu J, Holland JB, McMullen MD, Buckler ES. Genetic design and statistical power of nested association mapping in maize. Genetics. 2008;178:539–51.

15. Buckler ES, Holland JB, Bradbury PJ, Acharya CB, Brown PJ, et al. The genetic architecture of maize flowering time. Science. 2009;325:714–8.

16. Kump KL, Bradbury PJ, Wisser RJ, Buckler ES, Belcher AR, et al. Genome-wide association study of quantitative resistance to southern leaf blight in the maize nested association mapping population. Nat Genet. 2011;43:163–8.

17. Poland JA, Bradbury PJ, Buckler ES, Nelson RJ. Genome-wide nested association mapping of quantitative resistance to northern leaf blight in maize. Proc Natl Acad Sci. 2011;108:6893.

18. Tian F, Bradbury PJ, Brown PJ, Hung H, Sun Q, Flint-Garcia S, Rocheford TR, McMullen MD, Holland JB, Buckler ES. Genome-wide association study of leaf architecture in the maize nested association mapping population. Nat Genet. 2011;43:159–62.

19. Brown PJ, Upadyayula N, Mahone GS, Tian F, Bradbury PJ, et al. Distinct genetic architectures for male and female inflorescence traits of maize. PLoS Genet. 2011;7:e1002383.

20. Cook JP, McMullen MD, Holland JB, Tian F, Bradbury P, Ross-Ibarra J, Buckler ES, Flint-Garcia SA. Genetic architecture of maize kernel composition in the nested association mapping and inbred association panels. Plant Physiol. 2012;158:824–34.

21. Peiffer JA, Flint-Garcia SA, De Leon N, McMullen MD, Kaeppler SM, Buckler ES. The genetic architecture of maize stalk strength. PLoS ONE. 2013;8:e67066.

22. Peiffer JA, Romay MC, Gore MA, Flint-Garcia SA, Zhang Z, et al. The genetic architecture of maize height. Genetics. 2014;196:1337–56.

23. Jordan DR, Mace ES, Cruickshank AW, Hunt CH, Henzell RG. Exploring and exploiting genetic variation from unadapted sorghum germplasm in a breeding program. Crop Sci. 2011;51:1444–57.

24. Mace ES, Tai S, Gilding EK, Li Y, Prentis PJ, Bian L, Campbell BC, Hu W, Innes DJ, Han X, Cruickshank A, Dai C, Frère C, Zhang H, Hunt CH, Wang X, Shatte T, Wang M, Su Z, Li J, Lin X, Godwin ID, Jordan DR, Wang JL. Whole genome sequencing reveals untapped genetic potential in Africa's indigenous cereal crop sorghum. Nat Commun. 2013;4:2320.

25. Bonser AM, Lynch J, Snapp S. Effect of phosphorus deficiency on growth angle of basal roots in Phaseolus vulgaris. New Phytol. 1996;132:281–8.

26. Liao H, Yan XL, Rubio G, Beebe SE, Blair MW, Lynch J. Genetic mapping of basal root gravitropism and phosphorus acquisition efficiency in common bean. Funct Plant Biol. 2004;31:959–70.

27. Hund A, Trachsel S, Stamp P. Growth of axile and lateral roots of maize. I: development of a phenotyping platform. Plant Soil. 2009;325:335–49.

28. Trachsel S, Kaeppler SM, Brown KM, Lynch JP. Shovelomics: high throughput phenotyping of maize (Zea mays L.) root architecture in the field. Plant Soil. 2011;341:75–87.

29. Colombi T, Kirchgessner N, Le Marié CA, York LM, Lynch JP, Hund A. Next generation shovelomics: set up a tent and REST. Plant Soil. 2015;388:1–20.

30. Grift TE, Novais J, Bohn M. High-throughput phenotyping technology for maize roots. Biosyst Eng. 2011;110:40–8.

31. Singh V, van Oosterom EJ, Jordan DR, Hunt CH, Hammer GL. Genetic variability and control of nodal root angle in sorghum. Crop Sci. 2011;51:2011–20.

32. Butler D, Cullis BR, Gilmour AR, Gogel BJ. {ASReml}-R Reference Manual. 2007.

33. Hargreaves C, Gregory P, Bengough A. Measuring root traits in barley (Hordeum vulgare ssp. vulgare and ssp. spontaneum) seedlings using gel chambers, soil sacs and X-ray microtomogrphy. Plant Soil. 2009;316:285–97.

34. Clark RT, MacCurdy RB, Jung JK, Shaff JE, McCouch SR, Aneshansley DJ, Kochian LV. Three-dimensional root phenotyping with a novel imaging and software platform. Plant Physiol. 2011;156:455–65.

35. Trachsel S, Messmer R, Stamp P, Hund A. Mapping of QTLs for lateral and axile root growth of tropical maize. Theor Appl Genet. 2009;119:1413–24.

36. Le Marié C, Kirchgessner N, Marschall D, Walter A, Hund A. Rhizoslides: paper-based growth system for non-destructive, high throughput phenotyping of root development by means of image analysis. Plant Methods. 2014;10:13.

37. Promkhambut A, Polthanee A, Akkasaeng C, Younger A. Growth, yield and aerenchyma formation of sweet and multipurpose sorghum (Sorghum bicolor L. Moench) as affected by flooding at different growth stages. Aust J Crop Sci. 2011;5:954–65.

38. McDonald MP, Galwey NW, Colmer TD. Similarity and diversity in adventitious root anatomy as related to root aeration among a range of wetland and dryland grass species. Plant Cell Environ. 2002;25:441–51.

39. Richard C, Hickey L, Fletcher S, Jennings R, Chenu K, Christopher J. High-throughput phenotyping of seminal root traits in wheat. Plant Methods. 2015;11:13.

40. Nagel KA, Putz A, Gilmer F, Heinz K, Fischbach A, Pfeifer J, Faget M, Blossfeld S, Ernst M, Dimaki C, Kastenholz B, Kleinert AK, Galinski A, Scharr H, Fiorani F, Schurr U. GROWSCREEN-Rhizo is a novel phenotyping robot enabling simultaneous measurements of root and shoot growth for plants grown in soil-filled rhizotrons. Funct Plant Biol. 2012;39:891–904.

41. Robertson MJ, Fukai S, Ludlow MM, Hammer GL. Water extraction by grain sorghum in a sub-humid environment. I. Analysis of the water extraction pattern. Field Crops Res. 1993;33:81–97.

42. van Oosterom EJ, Yang Z, Zhang F, Deifel KS, Cooper M, Messina CD, Hammer GL. Hybrid variation for root system efficiency in maize: potential links to drought adaptation. Funct Plant Biol. 2016;43:502–11.

43. Kuijken RCP, van Eeuwijk FA, Marcelis LFM, Bouwmeester HJ. Root phenotyping: from component trait in the lab to breeding. J Exp Bot. 2015;66:5389–401.

44. Borrell AK, Mullet JE, George-Jaeggli B, van Oosterom EJ, Hammer GL, Klein PE, et al. Drought adaptation of stay-green sorghum is associated with canopy development, leaf anatomy, root growth, and water uptake. J Exp Bot. 2014;65:6251–63.

45. Manschadi AM, Christopher J, deVoil P, Hammer GL. The role of root architectural traits in adaptation of wheat to water-limited environments. Funct Plant Biol. 2006;33:823–37.

18

An improved butanol-HCl assay for quantification of water-soluble, acetone:methanol-soluble, and insoluble proanthocyanidins (condensed tannins)

Philip-Edouard Shay[1], J. A. Trofymow[1,2] and C. Peter Constabel[1*]

Abstract

Background: Condensed tannins (CT) are the most abundant secondary metabolite of land plants and can vary in abundance and structure according to tissue type, species, genotype, age, and environmental conditions. Recent improvements to the butanol-HCl assay have separately helped quantification of soluble and insoluble CTs, but have not yet been applied jointly. Our objectives were to combine previous assay improvements to allow for quantitative comparisons of different condensed tannin forms and to test protocols for analyses of condensed tannins in vegetative plant tissues. We also tested if the improved butanol-HCl assay can be used to quantify water-soluble forms of condensed tannins.

Results: Including ~50% acetone in both extraction solvents and final assay reagents greatly improved the extraction and quantification of soluble, insoluble and total condensed tannins. The acetone-based method also extended the linear portion of standard integration curves allowing for more accurate quantification of samples with a broader range of condensed tannin concentrations. Estimates of tannin concentrations determined using the protocol without acetone were lower, but correlated with values from acetone-based methods. With the improved assay, quantification of condensed tannins in water-soluble forms was highly replicable. The relative abundance of condensed tannins in soluble and insoluble forms differed substantially between tissue types.

Conclusions: The quantification of condensed tannins using the butanol-HCl assay was improved by adding acetone to both extraction and reagent solutions. These improvements will facilitate the quantification of total condensed tannin in tissues containing a range of concentrations, as well as to determine the amount in water-soluble, acetone:MeOH-soluble and insoluble forms. Accurate determination of these three condensed tannin forms is essential for careful investigations of their potentially different physiological and ecological functions.

Keywords: Tannin, Proanthocyanidins, Leaf litter, Nitrogen stress, Polyphenol, Flavonoid, Poplar, Douglas-fir

Background

Condensed tannins (CT), also known as proanthocyanidins, are the most abundant secondary metabolite of land plants. They can be found in many species but are most prevalent in woody plants, where they accumulate in most major tissues including leaves, bark, and roots [1].

Condensed tannins are polymers of flavan-3-ols, and end products of the well-characterized flavonoid pathway [2]. High concentrations of CTs have been shown to provide protection against vertebrate herbivores due to binding of dietary protein in the gut [2]. A defensive function against lepidopteran insect herbivores has also been proposed, but is more likely based on the pro-oxidant nature of tannins [2]. Condensed tannins demonstrate broad anti-microbial properties in vitro [3], and CT concentrations have been correlated with reduced rates of

*Correspondence: cpc@uvic.ca
[1] Department of Biology & Centre for Forest Biology, University of Victoria, P.O. Box 3020, STN CSC, Victoria, BC V8W 3N5, Canada
Full list of author information is available at the end of the article

pathogen infection in the field [4]. Other hypothesized functions of CTs include roles in roots as protection against metal toxicity, and in soil as mediators of decomposition and nutrient cycling by microbes [5].

Condensed tannin structures can show subtle variation in the degree of polymerization, hydroxylation of the flavonoid B-ring, and stereochemistry [6]. Both abundance and structure of CTs can vary depending on tissue type, species, genotype, age and environmental conditions [7–14]. Much of this variation still awaits functional characterization [2], but is likely to be associated with differences in biological activity [6, 15, 16]. Furthermore, CTs extracted from a given species and tissue will contain a mixture of CTs with subtle structural differences, for example a broad range of polymer lengths. The individual compounds are difficult to separate using chromatographic methods, which has made precise determination of individual structures difficult. Nevertheless, methods that rely on depolymerization in the presence of thiol or phloroglucinol give useful structural information and average subunit composition [17, 18]. In addition, sophisticated LC–MS/MS methods which rely on the in-source fragmentation of oligomeric and polymeric CTs have been developed recently. These also provide data on subunit composition and degree of polymerization in complex extracts [19].

The most common and straightforward method for quantifying CTs has been the butanol-HCl method developed by Swain and Hillis [20] and improved by Porter et al. [21]. This involves depolymerization of the polymer in acid and conversion of the monomers to anthocyanidin, which can be spectrophotometrically quantified. Structural differences in CTs from different sources can lead to differences in reactivity to this assay; therefore, a purified standard of CT isolated from the same species and tissues to be analyzed is essential if absolute quantitation is required. However, using the standard for quantifying CT across species and tissues, as done in this study, may be more workable and still allows for relative quantitation and comparisons between treatments. An advantage of the butanol-HCl method is that it permits a direct quantification of all CT fractions, i.e., water-soluble, organic solvent-extractable, and unextractable (functionally insoluble) CTs [17, 22]. Depending on the solvent used, these insoluble CTs typically make up between 10% and 50% of total CT content, and in some tissues can constitute 90% of total CTs [23]. Nevertheless, they are poorly investigated and often ignored [17, 22, 24]. What renders these fractions insoluble is not well-understood, but greater polymer length has been associated with a greater proportion of insoluble CT [17].

In leaf litter, cross-linking of CTs during cell death and senescence has been suggested [7, 25]. The proportion of insoluble CTs in foliar litter fall is thus dynamic, and likely also differs between species, but is rarely reported [26–28]. For example, the impact of CTs on decomposition and nitrogen mineralization in soil has been extensively investigated, but differential roles for soluble and insoluble components of the CTs are not easily defined [10, 15, 25, 27, 29–32]. Preliminary data suggest that these tannin fractions have different stabilities in soils [33], and a method that can efficiently assay and compare these is needed for ecological studies. The butanol-HCl assay can be used directly on plant material or on solvent-extracted residue, providing a measure of both soluble and insoluble CT fractions from the same sample. Associating potential biological functions to soluble and insoluble components of CT is difficult since these are defined by the solvents used for extraction. Furthermore, water soluble tannins may be ecologically more relevant than solvent soluble fraction, yet CTs are rarely quantified in water extracts using the butanol-HCl method [14]. One reason is likely the assay's sensitivity to the presence of water [21]. Typically, the analysis of CT using the butanol-HCl protocol is carried out on fractions in aqueous acetone [7, 26, 32, 34] or methanol [9, 35]. Where quantification of CT in insoluble forms is performed, available approaches have not applied the appropriate solvent concentrations needed for a direct comparison of soluble and insoluble fractions of CTs [36] and for exhaustive extraction [35, 37]. Consolidating disparate extraction and assay conditions was a major motivation for developing our modified method.

Recent improvements of the butanol-HCl protocol for CT quantification have separately enhanced the extraction of soluble CT forms by optimizing solvent concentrations and heating temperatures [38] and improving quantitation of total CT [i.e. soluble and insoluble forms; 36] by modifying reagent concentrations. However, these modifications have not yet been combined into one efficient method. The objectives of our study, therefore, were to combine improvements of the butanol-HCl assay described by Mané et al. [38] and Grabber et al. [36] into one efficient protocol. Our aim is to improve solvent extraction of CT, and facilitate direct quantitative comparison of soluble and insoluble fractions of CTs forms from different types of tissues and samples, in particular foliar litter. Additionally, we show how our methodological improvements allow for easy quantification of water-soluble CTs.

Methods

Condensed tannin standards

CT standards were purified from leaves of *Populus tremula* × *tremuloides* [INRA clone 353-38; 34] by the method of Fierer et al. [40] using chromatography on

Sephadex LH-20 resin with the following changes: sample pre-treatment with hexane was omitted, and the dried crude extract was resuspended in 50% EtOH, and filtered on 0.45 μm polyvinylidene difluoride membrane (EMD Millipore, Germany) rather than treating with ethyl acetate. Elution of purified CT from the Sephadex LH-20 column using 70% aqueous acetone was only carried out after successive washing of the column with 80% EtOH yielded fractions with a UV absorbance at 280 nm of less than 0.5 absorbance units (AU). The CT standard was characterized and checked for purity by NMR as described by Preston and Trofymow [14].

Plant material

Assay development and improvements were performed using naturally abscised poplar leaves (*Populus angustifolia*) and Douglas-fir (*Pseudotsuga menziesii*) needles. Naturally abscised Douglas-fir needle litter fall (hereafter litter) was collected from the Shawnigan Research Forest, Vancouver Island, British Columbia, Canada [39]. Naturally abscised poplar leaf litter was from a common garden at the Ogden Nature Center (Ogden, UT, USA) kindly provided by Dr. Thomas Whitham and the Cottonwood Research Group at Northern Arizona University. Poplar-leaf litter was pooled from trees with known leaf chemistries in order to set leaf litter treatments with low and high CT concentrations. To generate poplar or Douglas-fir foliar litter with high and low nitrogen concentrations, samples were sprayed with either a glutamine solution or distilled water.

Fresh leaf and root tissues for testing the assay improvements on samples with known CT content were obtained from the University of Victoria's Glover Greenhouse and the Constabel laboratory. Fresh leaf tissues consisted of greenhouse grown *Populus tremula* × *tremuloides* (INRA clone 353-38) and MYB134-overexpressing high CT line of the same hybrid [41]. Fresh roots and leaf tissues from untransformed (WT line 353) plants grown under N-limited conditions to induce CT synthesis were also tested. All fresh material was first flash-frozen in liquid N, ground using mortar and pestle, and lyophilized prior to analysis. For the experiment comparing tissue homogenization, a hammer mill model (Polymix PX-MFC 90D, Kinematica, Switzerland) was also used.

Extraction and assay conditions

The assay conditions and method are summarized in Fig. 1. Butanol-containing reagents were prepared on the same day they were used. The ratio of tissue weight to solvent (or assay reagent) depended on the tissue type and is described under Results, and sample amounts were adjusted to keep absorbance readings within the preferred range (Fig. 2). In some cases where CT concentrations in tissues were extremely high, solvent and reagent volumes were increased in order to avoid using too little sample.

To obtain solvent-extractable CTs, an appropriate volume (300 μl) of MeOH acidified with 0.05% trifluoroacetic acid (TFA) was added to dried tissue samples first. The slurry was briefly vortexed (<5 s) prior to adding 1.7 ml 60% aqueous acetone acidified with 0.05% TFA. Thus, the final solvent mixture consisted of 51% acetone, 34% MeOH, and 15% dH_2O, the whole acidified with 0.05% TFA. The slurry was briefly vortexed, sonicated for 10 min and extracted for a further 57 min at room temperature (67 min total). During the extraction period, sample tubes were mixed by vortexing periodically (3 times for 5 s). Extracts were then clarified by centrifugation (5 min at 4000 rpm) and supernatants (CT extracts) removed and placed into fresh tubes.

To assay soluble CTs, 0.5 ml of extract was mixed with 2 ml of butanol-containing reagent (51.5% acetone/43% butanol/5% 12 N concentrated HCl/0.5% H_2O) and 67 μl of Fe-reagent (2% w/v $FeNH_4(SO_4)_2$ in 2 N HCl). The final assay mixture, containing both sample and assay solutions in a 2.5 ml volume, was thus comprised (v/v) of 50.1% acetone, 33.5% butanol, 3.9% 12 N concentrated HCl, 7% dH_2O, 2.9% MeOH, and 2.6% Fe-reagent in a total volume of 2.5 ml (For simplicity, we do not include the additional H_2O found in concentrated HCl and in the Fe-reagent in this breakdown; with this, the actual total H_2O content approaches 12%). Aliquots (200 μl) of the final assay mixture of were removed to be read as non-heated controls. Assay samples were heated to 70 °C for 2.5 h, allowed to cool to RT, and the absorbance read at 550 nm using a Victor™ X5 Multi-label plate reader (PerkinElmer Inc.). To determine CT concentration, absorbances from unheated aliquots were subtracted from heated samples. When assaying water extracts for CTs, the proportion of assay reagent components, as well as volumes of extract and iron reagent, were adjusted to give the same water and solvent concentrations in the final assay mixture as for solvent extracted CTs (see Fig. 1). Maintaining a consistent proportion of water in the final assay mixture throughout is critical, as water is known to dramatically influence anthocyanidin formation in this assay [18, 21].

Insoluble CTs were determined directly on the centrifuged tissue pellets after extraction of soluble CTs. First, 75 μl MeOH (with 0.05% TFA) was added to pellets and vortexed, prior to adding 425 μl of 60% aqueous acetone (with 0.05% TFA). The remaining components of the assay reagent mix were then added, and the assay carried out directly on the suspended pellets (Fig. 1). The direct assay for total CTs was carried out in the identical manner. For assays performed directly on pellets or samples,

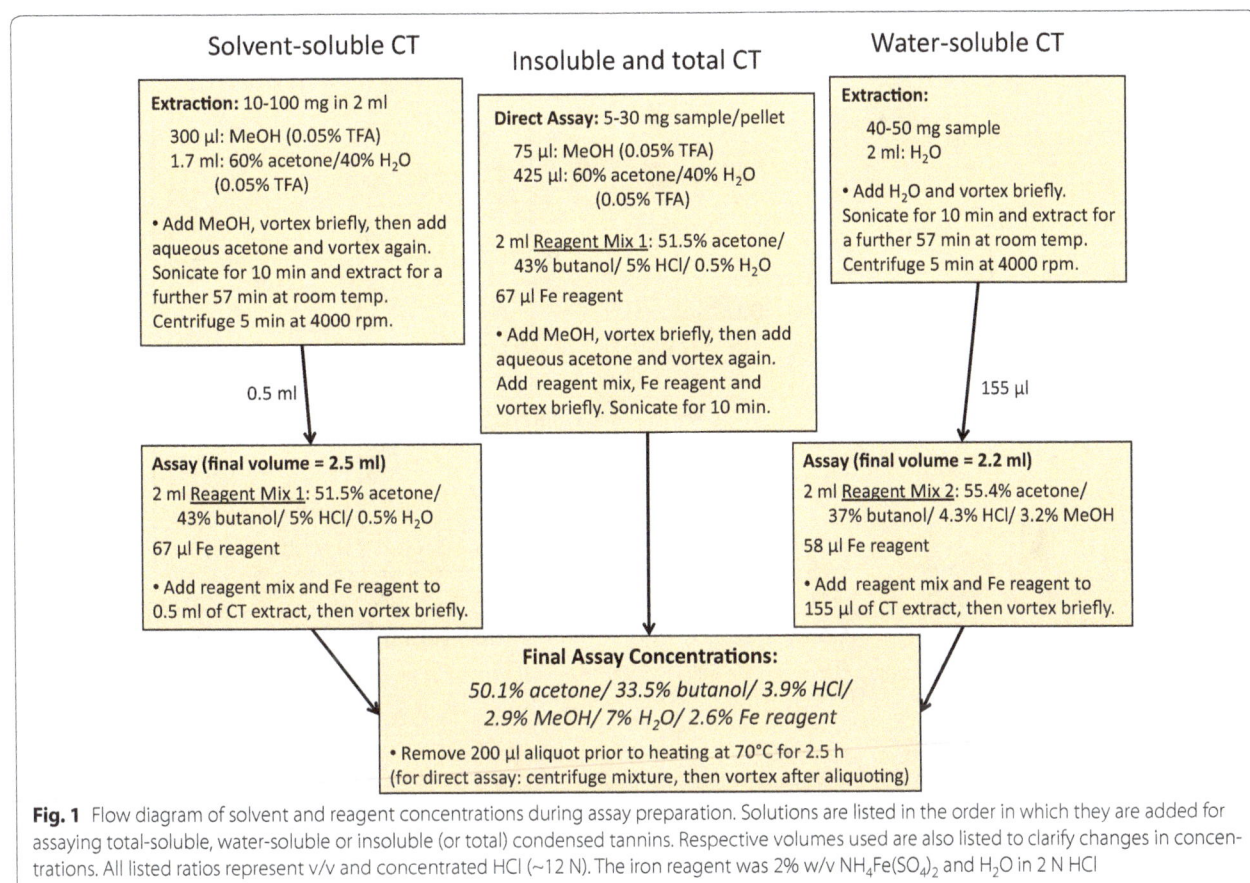

Fig. 1 Flow diagram of solvent and reagent concentrations during assay preparation. Solutions are listed in the order in which they are added for assaying total-soluble, water-soluble or insoluble (or total) condensed tannins. Respective volumes used are also listed to clarify changes in concentrations. All listed ratios represent v/v and concentrated HCl (~12 N). The iron reagent was 2% w/v $NH_4Fe(SO_4)_2$ and H_2O in 2 N HCl

mixtures were centrifuged 5 min at 4000 rpm prior to taking non-heated and heated aliquots for absorbance readings, and pellets re-suspended by vortex mixing before heating.

For comparison, samples were also assayed for CT concentration using 80% MeOH for soluble-CT extraction and the standard butanol-HCl assay reagent with methanol (74% butanol/3.9% 12 N concentrated HCl/15.6% MeOH/3.9% H_2O/2.6% Fe-reagent (v/v). CT analyses are often carried out on MeOH extracts, which are more compatible with other analytical methods such as HPLC. To our knowledge, only one study of butanol-HCl assays comparing solely MeOH extracts with assays of acetone extracts have been made [42].

Statistical analyses were performed using R-statistics version 3.1.2 [43]. Purified CT standard curves were analysed using linear regression and r^2. ANOVA and Kruskal–Wallis H-test were used to compare CT quantification in samples ground in liquid N_2 using mortar and pestle or mechanically milled. Pearson correlations and linear regressions were used to compare one- versus two-step assay approaches as well as acetone-based versus acetone-free methods for quantification of total CT. Pearson correlations and ANOVA

were used to assess differences in CT content between tissue types.

Results

In order to more effectively measure CTs in a diversity of plant samples, but foliar litter in particular, we incorporated several modifications into one method. We used the solvent ratios of 51:34:15 (acetone:water:methanol) previously optimized by Mané et al. [38] in order to maximize extraction of the soluble CTs. The inclusion of 50% acetone in the final assay reagent, as per Grabber et al. [36], also improved the assay by extending the linear range of the assay response of purified poplar CT (Fig. 2). It also reduced the slope of the standard curve. This is in contrasts with results shown by Grabber et al. [36], but could be due to trends associated with the different H_2O concentrations in our assays with and without acetone [21]. We also checked whether diluting a standard sample post-assay, i.e. after the heat treatment, would provide the same result as a dilution prior to the assay, but note that this leads to underestimates (Fig. 2 inset).

When we tested the length of the incubation time at 70 °C, we found that absorption values and the slope of integration curves continued to increase for at least 2.5 h.

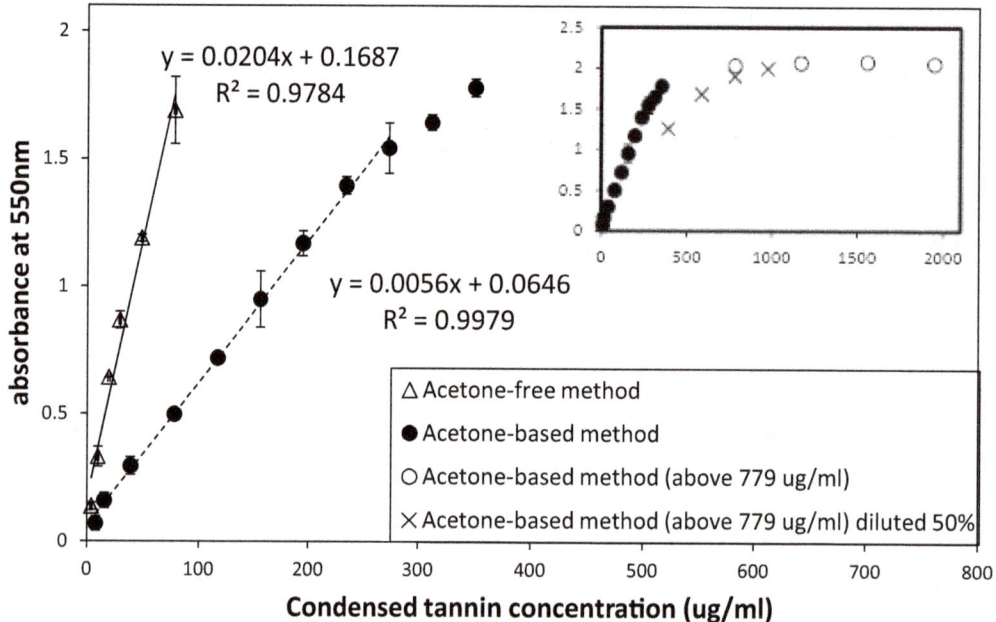

Fig. 2 Standard curves generated with purified poplar condensed tannin. *Each point* in the main panel represents the mean of at least two independent values (±SE). Best fit lines were drawn through all data when using acetone-free method (74% butanol/3.9% 12 N concentrated HCl/15.6% MeOH/3.9% H_2O/2.6% Fe-reagent) and between absorbance values of 0.07–1.54 for acetone-based method (50.1% acetone/33.5% butanol/3.9% 12 N concentrated HCl/2.9% MeOH/7% H_2O/2.6% Fe-reagent). The preferred absorbance range for CT quantification when using acetone-based method was from 0.158 to 1.247 (abs. value ± 3 × SE). In the inset, samples with the four most concentrated CT standards (*open circles*) were diluted by 50% (*crosses*) and absorbance readings repeated. The comparison to equivalent concentrations of undiluted extract (*solid circles*) indicates that a discrepancy in predicted and actual absorbance values, suggesting that post-assays dilution will lead to erroneous results

However, after 2.5 h of heating, the average absorbance increased by 2.7% in the last 30 min of heating, and less than one-half of the increase measured in the previous 30 min (data not shown). Therefore, 2.5 h was used as the standard heating time. We observed minimal to no color development in non-heated controls at room temperature over 2.5 h (data not shown). Nonetheless, we read the absorbance of unheated controls immediately (<5 min) after pipetting aliquots in order to reduce the potential for anthocyanidin production and evaporation losses. Homogenizing samples in liquid-N_2 by mortar and pestle did not improve the quantification of soluble and insoluble CT compared to hammer mill homogenization, suggesting access to solvent did not limit extraction (Fig. 3).

One of our goals was to adapt and test the method on the solvent-insoluble CT fraction, since insoluble CTs appear to be of particular ecological relevance in foliar litter. Total CT concentrations, comprising both soluble and insoluble fractions, were determined by using the direct assay on the ground tissue (Fig. 4). The resulting concentrations were compared to those obtained with the improved method and assaying acetone-extracted soluble tannins first, and then assaying the insoluble CTs in the remaining pellet (two-step assay). Tannin

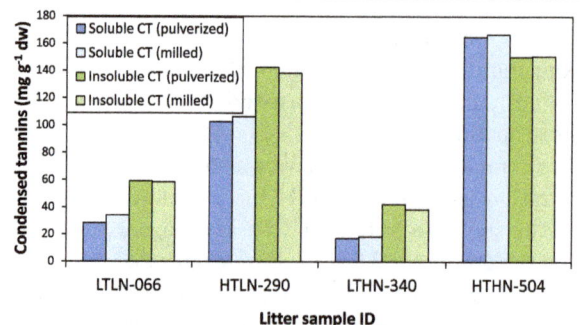

Fig. 3 Quantification of soluble and insoluble condensed tannin using hammer-milled or liquid N_2-pulverized samples. *Each column* represents an individual assay on a unique litter sample, from a larger-scale litter decay study; $p > 0.05$, ANOVA or Kruskal–Wallis H-test depending on data normality

concentrations determined by the direct assay method were highly correlated with concentrations determined using the two-step assay approach (Pearson correlation $r = 0.99$). However, the direct assay method generally led to slightly smaller estimates of CT concentration (Fig. 4).

We compared results obtained using the improved acetone-containing method with those measured using 80% MeOH as a solvent and a butanol-containing

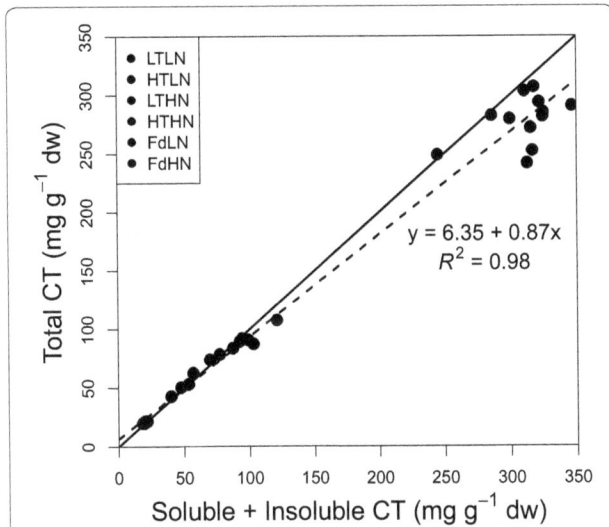

Fig. 4 Comparison of total condensed tannin quantification using one- or two-step assay approaches. Condensed tannin values obtained using the two-step assay (soluble + insoluble CT) and the one-step method (direct assay for total CTs) are compared using Pearson's correlation value and the best fit linear response (*dashed line*). *Colors* represent different leaf-litter chemistries, namely Douglas-fir (Fd) and poplar litter with low and high nitrogen concentrations (LN and HN) and CT concentrations (LT and HT, poplar only). The *black line* represents a 1:1 response

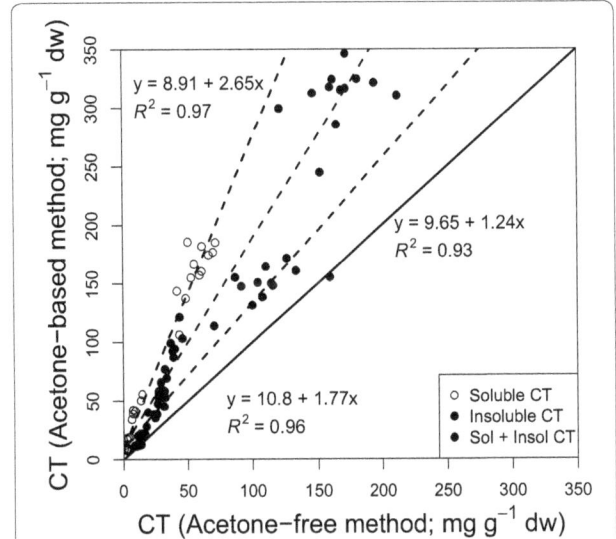

Fig. 5 Comparison of two-step condensed tannin quantification using butanol-HCl assays with or without acetone. *Points* represent concentrations of soluble or insoluble CTs, or the sum of both two forms (Sol + Insol CT). Sample to solvent concentrations were ~1 mg ml^{-1} for the protocol without acetone, and 5–20 mg ml^{-1} for the acetone containing protocol. *Top, middle and bottom* equations correspond to soluble, insoluble, and soluble + insoluble CT quantifications

reagent without acetone. Despite being highly correlated ($r = 0.96$ and 0.98 for soluble and insoluble fractions respectively), soluble and insoluble CT concentrations measured using the acetone-based assay were on average $3\times$ and $1.4\times$ greater, respectively, than concentrations measured using the former assay (Fig. 5). In other words, including acetone in both extraction solvent and final assay reagent appeared to lead to a more exhaustive extraction of CTs in foliar litter, while also improving the accuracy of CT standard curves across greater concentration ranges.

The quantification of water-soluble CT fractions was highly replicable (Fig. 6). However, the high sample to solute ratio required for the assay (25–200 mg ml^{-1}) prevented the quantification of residual CT in pellets after H$_2$O extraction, due to the high volume of butanol-containing reagent needed to maintain absorbance values within spectrophotometric detection limits.

We next tested the improved assay methods for total soluble (water- and acetone:MeOH-soluble forms) and insoluble CTs concentration on a set of samples consisting of high- and low-tannin litter (poplar and Douglas-fir) as well as poplar tissue samples. The latter included leaves from high-CT transgenic poplars [44] as well as leaves and roots from N-deficient plants, as these also show elevated tannins. Soluble and insoluble CTs were

correlated across both species when considering only litter samples (Fig. 7a; $r = 0.95$; $p < 0.001$), and across tissue types when considering only fresh poplar tissue (Fig. 7b; $r = 0.95$; $p < 0.001$). By contrast, the ratios of soluble CT to total CT differed significantly between naturally abscised and fresh tissues (0.43 and 0.80 on average, respectively; $p < 0.001$; Fig. 7). In fresh plant tissues, CT was mostly in soluble form, but the amount and proportion of soluble to total CT also varied with N status (Fig. 7b). By contrast, foliar litter from both species had a significant insoluble CT component, comprising almost 50% of total CTs (Fig. 7a). Water-soluble CT concentration in foliar litter correlated better with acetone:MeOH-soluble CTs than the insoluble CT concentration ($r = 0.94$ and 0.90, respectively). Water-soluble CTs concentrations in fresh tissue was not measured.

The improved method was validated with several different available poplar tissue types and different stages of decaying foliar litter, to determine appropriate ratios of sample to solvent and sample to reagent for assays of soluble, insoluble, and total CTs (Table 1). The complete assay was repeated on each sample, incrementally adjusting sample to solvent (or reagent) ratios prior to heating, until the AU of heated solutions and the difference between heated and unheated solutions fell within the linear portion of our standard curve (without the need for dilution). Appropriate amounts of poplar tissue

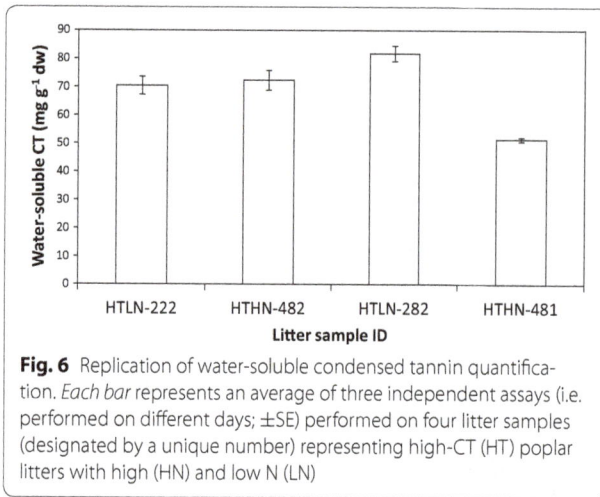

Fig. 6 Replication of water-soluble condensed tannin quantification. *Each bar* represents an average of three independent assays (i.e. performed on different days; ±SE) performed on four litter samples (designated by a unique number) representing high-CT (HT) poplar litters with high (HN) and low N (LN)

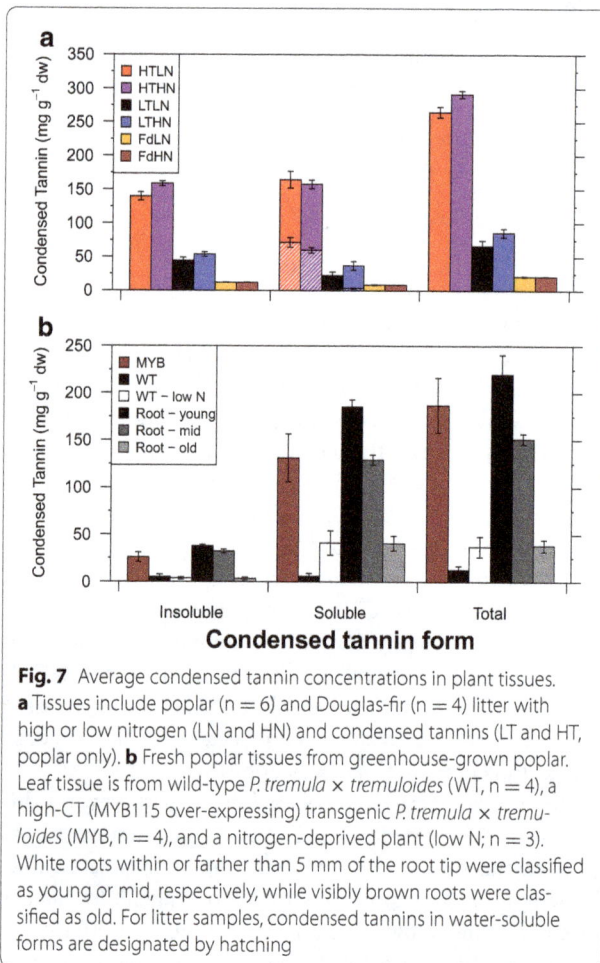

Fig. 7 Average condensed tannin concentrations in plant tissues. **a** Tissues include poplar (n = 6) and Douglas-fir (n = 4) litter with high or low nitrogen (LN and HN) and condensed tannins (LT and HT, poplar only). **b** Fresh poplar tissues from greenhouse-grown poplar. Leaf tissue is from wild-type *P. tremula* × *tremuloides* (WT, n = 4), a high-CT (MYB115 over-expressing) transgenic *P. tremula* × *tremuloides* (MYB, n = 4), and a nitrogen-deprived plant (low N; n = 3). White roots within or farther than 5 mm of the root tip were classified as young or mid, respectively, while visibly brown roots were classified as old. For litter samples, condensed tannins in water-soluble forms are designated by hatching

ranged from 5 to 10 mg ml^{-1} solvent for soluble CTs, and 20 mg ml^{-1} solvent for Douglas-fir litter. For the insoluble or direct methods, only small amounts of sample/

pellet are needed, in particular for high-tannin tissues such as roots, where 0.6 mg ml^{-1} was found to be appropriate. Therefore, before applying these methods on different plant tissues, preliminary analyses should be done to determine appropriate sample to solvent ratios.

Discussion

Quantification of condensed-tannin forms using improvements on the butanol-HCl assay

The butanol-HCl assay for condensed tannins was improved using solvent concentrations and heating temperatures from Mané et al. [38] for better extraction of soluble tannins and applying reagent concentrations from Grabber et al. [36] that allow for quantitative comparison between CT forms. Our results suggest that solvent mixtures and protocols derived by Mané et al. [38] and Grabber et al. [36] were transferable to poplar and Douglas-fir tissues, therefore indicating broader application potential. Incorporating trifluoroacetic acid (TFA), MeOH, water, and ~50% acetone into extraction solvents for soluble CT assays, as well as including ~50% acetone in final assay reagents for assaying total CT, allowed for a more thorough CT quantification in foliar litter. As noted earlier [36], the 50% acetone in the final assay solution also eliminated quantification issues associated with the commonly observed 'biphasic' standard curve seen with the classic butanol-HCl protocol [21]. This is an additional advantage of the improved method.

We did not attempt re-optimize the earlier protocols and solvent ratios for our plant species and tissues, based on the assumption that poplar CTs are sufficiently similar in structure to those analyzed in the previous studies. Mané et al. [38] optimized their solvent mixture for grape seeds pulp, and skin extracts, with CTs having a mean degree of polymerization (DPM) of 2.9–39, prodelphinidin content of 0–14.5%, and galloylation at 1.1–9.5% depending on the tissues. Grabber et al. [36] developed their reagent mixture for use on two *Lotus* species differing mainly in procyanidin to prodelphinidin ratios (60:40 and 21:79), with no galloyl groups and polymers with average DPM from 5 to 38, depending on species. Poplar CTs vary in DPM from 2 to 28, with up to 50% prodelphinidins [12], and our purified poplar CT standard contained approximately 10% prodelphinidin and with an average DPM of 5.6 as verified by NMR [14, C. Preston, and C. P. Constabel, unpublished data]. Since both previous reports suggest that acidified mixtures with ~50% acetone are effective at removing fibre-bound CTs [36, 38], we are confident that this concentration is effective across a range of CT and tissue types. The consistently higher concentrations of soluble CT measured with ~50% acetone in both extraction solvent and reagent solution (Fig. 7) is likely due to the greater extractability of bound

Table 1 Optimal sample concentrations (mg ml^{-1}) for quantification of condensed tannins using improved acetone-based butanol-HCl assay

Samples	Species	Growing conditions	Tissue type	Soluble CT (sample/ solvent)	Insoluble CT (pellet/ reagent)	Total CT (sample/ reagent)
Fresh leaf_ MYB115/353(4)	*Populus tremula × tremuloides*	Greenhouse grown, well fertilized	Green leaf	5.0	1.3	0.6
Fresh leaf_WT353	*Populus tremula × tremuloides*	Greenhouse grown, well fertilized	Green leaf	10.0	7.8	3.9–9.7
Fresh leaf_Low N_WT 353	*Populus tremula × tremuloides*	Greenhouse grown, nitrogen limited	Green leaf	10.0	7.8	3.9–9.7
Roots (young)_ Low N	*Populus tremula × tremuloides*	Greenhouse grown, nitrogen limited	Live root	5.0	1.3	0.6
Roots (mid-sized)_ Low N	*Populus tremula × tremuloides*	Greenhouse grown, nitrogen limited	Live root	5.0	1.3	0.6
Roots (older)_ Low N	*Populus tremula × tremuloides*	Greenhouse grown, nitrogen limited	Live root	10.0	2.6	1.3
Litter_LTLN	*Populus angustifolia*	Field grown	Abscised leaf	5–10	1.3–2.6	1.3
Litter_HTLN	*Populus angustifolia*	Field grown	Abscised leaf	5.0	1.3	0.6
Litter_LTHN	*Populus angustifolia*	Field grown	Abscised leaf	5–10	1.3–2.6	1.3
Litter_HTHN	*Populus angustifolia*	Field grown	Abscised leaf	5.0	1.3	0.6
Litter_FdLN	*Pseudotsuga menziesii*	Field grown	Abscised leaf	20.0	5.2	3.9
Litter_FdHN	*Pseudotsuga menziesii*	Field grown	Abscised leaf	20.0	5.2	3.9
Decayed litter_ Poplar	*Populus angustifolia*	Field grown and decayed	Decayed abscised leaf	Not detected	19.5–39	11.7
Decayed litter_ Douglas-fir	*Pseudotsuga menziesii*	Field grown and decayed	Decayed abscised leaf	39 to not detected	31–39	11.7

CT due to acetone/TFA, or the enhanced depolymerisation of CT complexes into flavan-3-ols. The increase in soluble CTs did not occur at the expense of insoluble CTs, which also increase with the new method. Therefore, the method appears to facilitate more efficient depolymerisation of CT bound to the litter matrix as well as in solution. Heating the assay tubes for 2.5 h was adequate for cleaving almost all of the CT polymers found in our samples, as additional incubation times had minimal effect. By contrast, quantification of CT using MeOH as an extraction solvent leads to underestimates but does provide reliable relative quantification of soluble CT concentration. It could thus be useful where relative quantification is the priority, and where methanolic extracts are preferred for additional downstream analyses such as HPLC or antioxidant tests.

The insoluble CT fraction may have been overestimated when performing the butanol-HCl assay directly on the residual pellet after extracting soluble CT (Fig. 4), due to remnant solvent in the pellet. Subtracting soluble from total CT values, with each assay done on separate subsamples, should lead to better quantification of the insoluble CT fraction. Washing pellets with MeOH prior to assaying insoluble CT [27] could also resolve this carry-over. In addition, we tested litter to solvent ratios for assaying both soluble and insoluble CT fractions

to ensure that coloured anthocyanidin solutions were directly within the linear range of our standard curve. We note that dilution of reaction media after heating is sometimes carried out if absorbance values are too high [e.g. 35], but this should be avoided as it can lead to underestimates of CT concentrations (Fig. 2, inset).

The more effective extraction and depolymerisation of CT tightly bound to proteins or cell wall polysaccharides is most likely responsible for the greater proportion of insoluble CTs (~50%) we measured in our foliar litter samples compared to previous studies on foliar litter [26–28]. This suggests that the input of insoluble CTs into soil systems is greater than previously thought. By contrast, in fresh leaves, the proportion of insoluble CTs was only 10–20%. While these leaf samples are from a distinct poplar species grown under different environmental conditions, the much higher insoluble CT proportion in our litter could suggest that during senescence, soluble CTs become cross-linked and insoluble. Lindroth et al. [7] had previously suggested CT in poplar species undergo a shift from soluble to insoluble forms during senescence, especially in N-limited trees. By contrast, studies of developmental trajectories of CT in mangrove species do not show such increase in insoluble CT prior to senescence [28, 45]. A change in CT form during senescence could have important implications for litter decay, nutrient-cycling and other below-ground processes. Our assay improvements, together with methods for distinguishing protein and fibre-bound CT [46], will help in understanding the dynamics of soluble and insoluble CTs during development and senescence.

Slight modifications of the reagent concentrations used for the butanol-HCl assay allowed us to quantify the amount of water-soluble CT in leaf litter. This variation of our procedure releases an ecologically relevant form of CT, since we made extracts using room-temperature rather than hot water [14]. Water can solubilise low molecular weight phenols such as flavan-3-ol monomers, yet these are not converted to anthocyanidins during the butanol-HCl assay since they lack the carbocations resulting from interflanavoid bond cleavage [18, 21].

Sources of error for the butanol-HCl assay

It is well established that the choice of a standard used for the butanol-HCl assay is critical, since structural differences of CTs from different species will influence reactivity and color formation. For example, differences in degrees of polymerization [6, 14, 17, 18, 37, 46] alters the ratio of extender to terminal subunits, which are not detected by the assay [21]. Our estimates of CT concentrations in *P. menziesii* needle litter could therefore be low, due to differences in degree of polymerization of CT from *P. menziesii* compared to *P. tremuloides* leaves [6, 13, 27], as we did not have access to a *P. menziesii* CT standard for this work. Choice of standards could also explain the slightly lower (2.1 vs. 4.3% dw) CT concentrations measured in our study compared to results by Preston et al. [27]. These authors used the same *P. menziesii* litter collection as we did, but quantified CT using purified CTs from *Abies balsamea*, a genus with shorter CT polymers than is typical for *P. tremuloides* [6, 13, 47]. Total CT concentrations in the *P. menziesii* litter used by Preston et al. [27] were indeed reduced to values similar to those obtained in our study when we recalculated concentrations using the poplar CT standard [14]. The absence of hydrolysable tannins in our poplar CT standard also avoids another common factor leading to overestimation of CTs [6, 35]. The high purity of our poplar CT standard [14] further avoided potential sources of error [36].

Estimates of CT concentrations in our poplar litter should be more accurate since we used a poplar CT standard; however, high variation in the degree of CT polymerization has been described in *Populus* [7, 12], and may have led to slight over-estimation of our comparatively more polymerized *P. angustifolia* species. We also note that discrepancies can arise when the CT not extracted from plant material using the same solvents

as for the sample analysis. The degree of CT polymerization can affect their solubility and binding affinity [6, 18, 48], and thus their extractability. As a result, during the purification process, some smaller oligomeric CTs may have been eliminated [40], biasing the standard towards larger polymers. This could be prevented via the development of CT purification protocols encompassing water-soluble, MeOH-soluble, acetone-soluble and insoluble CT fractions for better representation of the range of CT forms present. However this issue may be difficult to resolve for the insoluble CTs, since to our knowledge, pure insoluble CT cannot be isolated [22] and is thus not included in typical purified CT standards [for details about extracts with insoluble CT, see ref. 24].

Conclusions

Solvents have varying absorbance qualities and influence absorbance response curves to CTs. It is therefore important to maintain the same final reagent concentrations for all assays, whether analysing soluble, insoluble, or purified CTs so they can be directly compared. Our improvements on the butanol-HCl protocol allowed for a highly-replicable and more thorough quantification of both soluble and insoluble CT fractions in foliar litter and plant tissues. Our results show that the concentration of insoluble CTs in senesced foliar litter is greater than previously thought, and that water-soluble CT forms can make up a substantial proportion of foliar litter CT. Changes in the distribution of the CT forms during senescence may have important implications for above- and below-ground interactions.

Abbreviations
CT: condensed tannins; TFA: trifluoroacetic acid.

Authors' contributions
PES, CPC and JAT conceived the study. PES developed laboratory protocols and performed analyses. PES, CPC and JAT wrote the manuscript. All authors read and approved the final manuscript.

Author details
[1] Department of Biology & Centre for Forest Biology, University of Victoria, P.O. Box 3020, STN CSC, Victoria, BC V8W 3N5, Canada. [2] Pacific Forestry Centre, Canadian Forest Service, Natural Resources Canada, Victoria, BC, Canada.

Acknowledgements
We sincerely thank Professor Thomas G. Whitham and Zacchaeus Compson (Northern Arizona University) for supplying us with poplar litter from their common garden, and to Caroline Preston (Pacific Forestry Centre, Victoria) and Juha-Pekka Salminen (University of Turku, Finland) for help with the characterization of purified CT standards. We also acknowledge Nicholas von Wittgenstein for help in preparing the foliar litter samples, Rebecca Westley for providing the poplar material grown under nitrogen deficiency, and Gerry Gourlay for providing wild-type and transgenic poplar leaves.

An improved butanol-HCl assay for quantification of water-soluble, acetone:methanol-soluble, and insoluble...

209

Funding

Natural Sciences and Engineering Research Council of Canada (NSERC) (CREATE and Discovery grants to CPC).

References

1. Porter LJ. Flavans and proanthocyanidins. In: Harborne JB, editor. The flavanoids. London: Chapman & Hall; 1988. p. 21–62.
2. Barbehenn RV, Constabel CP. Tannins in plant-herbivore interactions. Phytochemistry. 2011;72(13):1551–65.
3. Scalbert A. Antimicrobial properties of tannins. Phytochemistry. 1991;30(12):3875–83.
4. Holeski LM, Vogelzang A, Stanosz G, Lindroth RL. Incidence of *Venturia* shoot blight in aspen (*Populus tremuloides* Michx.) varies with tree chemistry and genotype. Biochem Syst Ecol. 2009;37(3):139–45.
5. Constabel CP, Yoshida K, Walker V. Diverse ecological roles of plant tannins: plant defense and beyond. Recent Adv Polyphen Res. 2014;4(4):115–42.
6. Kraus TEC, Yu Z, Preston CM, Dahlgren RA, Zasoski RJ. Linking chemical reactivity and protein precipitation to structural characteristics of foliar tannins. J Chem Ecol. 2003;29(3):703–30.
7. Lindroth RL, Osier TL, Barnhill HRH, Wood SA. Effects of genotype and nutrient availability on phytochemistry of trembling aspen (*Populus tremuloides* Michx.) during leaf senescence. Biochem Syst Ecol. 2002;30(4):297–307.
8. Liu LL, King JS, Giardina CP. Effects of elevated concentrations of atmospheric CO_2 and tropospheric O_3 on leaf litter production and chemistry in trembling aspen and paper birch communities. Tree Physiol. 2005;25(12):1511–22.
9. Donaldson JR, Stevens MT, Barnhill HR, Lindroth RL. Age-related shifts in leaf chemistry of clonal aspen (*Populus tremuloides*). J Chem Ecol. 2006;32(7):1415–29.
10. Madritch M, Donaldson JR, Lindroth RL. Genetic identity of *Populus tremuloides* litter influences decomposition and nutrient release in a mixed forest stand. Ecosystems. 2006;9(4):528–37.
11. Osier TL, Lindroth RL. Genotype and environment determine allocation to and costs of resistance in quaking aspen. Oecologia. 2006;148(2):293–303.
12. Scioneaux AN, Schmidt MA, Moore MA, Lindroth RL, Wooley SC, Hagerman AE. Qualitative variation in proanthocyanidin composition of *Populus* species and hybrids: genetics is the key. J Chem Ecol. 2011;37(1):57–70.
13. Norris CE, Preston CM, Hogg KE, Titus BD. The influence of condensed tannin structure on rate of microbial mineralization and reactivity to chemical assays. J Chem Ecol. 2011;37(3):311–9.
14. Preston CM, Trofymow JA. The chemistry of some foliar litters and their sequential proximate analysis fractions. Biogeochemistry. 2015;126(1–2):197–209.
15. Kraus TEC, Dahlgren RA, Zasoski RJ. Tannins in nutrient dynamics of forest ecosystems—a review. Plant Soil. 2003;256(1):41–66.
16. Whitham TG, DiFazio SP, Schweitzer JA, Shuster SM, Allan GJ, Bailey JK, Woolbright SA. Extending genomics to natural communities and ecosystems. Science. 2008;320(5875):492–5.
17. Tarascou I, Souquet JM, Mazauric JP, Carrillo S, Coq S, Canon F, Fulcrand H, Cheynier V. The hidden face of food phenolic composition. Arch Biochem Biophys. 2010;501(1):16–22.
18. Schofield P, Mbugua DM, Pell AN. Analysis of condensed tannins: a review. Anim Feed Sci Technol. 2001;91(1–2):71–40.
19. Engström MT, Palijarvi M, Fryganas C, Grabber JH, Mueller-Harvey I, Salminen JP. Rapid qualitative and quantitative analyses of proanthocyanidin oligomers and polymers by UPLC-MS/MS. J Agric Food Chem. 2014;62(15):3390–9.
20. Swain T, Hillis WE. The phenolic constituents of *Prunus domestica* L.—the quantitative analysis of phenolic constituents. J Sci Food Agric. 1959;10(1):63–8.
21. Porter LJ, Hrstich LN, Chan BG. The conversion of procyanidins and prodelphinidins to cyanidin and delphinidin. Phytochemistry. 1986;25(1):223–30.
22. Perez-Jimenez J, Torres JL. Analysis of nonextractable phenolic compounds in foods: the current state of the art. J Agric Food Chem. 2011;59(24):12713–24.
23. Marles MAS, Gruber MY, Scoles GJ, Muir AD. Pigmentation in the developing seed coat and seedling leaves of *Brassica carinata* is controlled at the dihydroflavonol reductase locus. Phytochemistry. 2003;62(5):663–72.
24. Perez-Jimenez J, Arranz S, Saura-Calixto F. Proanthocyanidin content in foods is largely underestimated in the literature data: an approach to quantification of the missing proanthocyanidins. Food Res Int. 2009;42(10):1381–8.
25. Hattenschwiler S, Vitousek PM. The role of polyphenols in terrestrial ecosystem nutrient cycling. Trends Ecol Evol. 2000;15(6):238–43.
26. Parsons WFJ, Bockheim JG, Lindroth RL. Independent, interactive, and species-specific responses of leaf litter decomposition to elevated CO_2 and O_3 in a northern hardwood forest. Ecosystems. 2008;11(4):505–19.
27. Preston CM, Nault JR, Trofymow JA, Smyth C, Grp CW. Chemical changes during 6 years of decomposition of 11 litters in some Canadian forest sites. Part 1. Elemental composition, tannins, phenolics, and proximate fractions. Ecosystems. 2009;12(7):1053–77.
28. Lin YM, Liu JW, Xiang P, Lin P, Ding ZH, Sternberg LDL. Tannins and nitrogen dynamics in mangrove leaves at different age and decay stages (Jiulong river estuary, China). Hydrobiologia. 2007;583:285–95.
29. Driebe EM, Whitham TG. Cottonwood hybridization affects tannin and nitrogen content of leaf litter and alters decomposition. Oecologia. 2000;123(1):99–107.
30. Kraus TEC, Zasoski RJ, Dahlgren RA, Horwath WR, Preston CM. Carbon and nitrogen dynamics in a forest soil amended with purified tannins from different plant species. Soil Biol Biochem. 2004;36(2):309–21.
31. Schweitzer JA, Bailey JK, Rehill BJ, Martinsen GD, Hart SC, Lindroth RL, Keim P, Whitham TG. Genetically based trait in a dominant tree affects ecosystem processes. Ecol Lett. 2004;7(2):127–34.
32. Liu LL, King JS, Giardina CP, Booker FL. The influence of chemistry, production and community composition on leaf litter decomposition under elevated atmospheric CO_2 and tropospheric O_3 in a Northern hardwood ecosystem. Ecosystems. 2009;12(3):401–16.
33. Shay PE. The effects of condensed tannins, nitrogen and climate on decay, nitrogen mineralisation and microbial communities in forest tree leaf litter. Dissertation. Victoria, BC, Canada: University of Victoria; 2016.
34. Lindroth RL, Hwang SY. Clonal variation in foliar chemistry of quaking aspen (*Populus tremuloides* Michx). Biochem Syst Ecol. 1996;24(5):357–64.
35. Preston CM, Trofymow JA, Sayer BG, Niu JN. [13]C nuclear magnetic resonance spectroscopy with cross-polarization and magic-angle spinning investigation of the proximate-analysis fractions used to assess litter quality in decomposition studies. Can J Bot. 1997;75(9):1601–13.
36. Grabber JH, Zeller WE, Mueller-Harvey I. Acetone enhances the direct analysis of procyanidin- and prodelphinidin-based condensed tannins in *Lotus* species by the butanol-HCl-iron assay. J Agric Food Chem. 2013;61(11):2669–78.
37. Yu Z, Dahlgren RA. Evaluation of methods for measuring polyphenols in conifer foliage. J Chem Ecol. 2000;26(9):2119–40.
38. Mané C, Souquet JM, Olle D, Verries C, Veran F, Mazerolles G, Cheynier V, Fulcrand H. Optimization of simultaneous flavanol, phenolic acid, and anthocyanin extraction from grapes using an experimental design: application to the characterization of Champagne grape varieties. J Agric Food Chem. 2007;55(18):7224–33.
39. Trofymow JA, CIDET Working Group. The Canadian Intersite Decomposition Experiment (CIDET): Project and site establishment report. In: BC-X-378-126. Victoria: Pacific Forestry Centre; 1998.
40. Fierer N, Schimel JP, Cates RG, Zou JP. Influence of balsam poplar tannin fractions on carbon and nitrogen dynamics in Alaskan taiga floodplain soils. Soil Biol Biochem. 2001;33(12–13):1827–39.
41. Mellway RD, Tran LT, Prouse MB, Campbell MM, Constabel CP. The wound-, pathogen-, and ultraviolet B-responsive MYB134 gene encodes an R2R3 MYB transcription factor that regulates proanthocyanidin synthesis in poplar. Plant Physiol. 2009;150(2):924–41.
42. Cork SJ, Krockenberger AK. Methods and pitfalls of extracting condensed tannins and other phenolics from plants—insights from investigations on eucalyptus leaves. J Chem Ecol. 1991;17(1):123–34.

43. R Core Team. R: a language and environment for statistical computing. Vienna, Austria: R Foundation for Statistical Computing; 2014.

44. James AM, Ma D, Mellway R, Gesell A, Yoshida K, Walker V, Tran L, Stewart D, Reichelt M. Jussi Suvanto, Salminen J-P, Gershenzon J, Séguin A, Constabel CP. The poplar MYB115 and MYB134 transcription factors regulate proanthocyanidin synthesis and structure. Plant Physiol. 2017;174(1):154–71.

45. Lin YM, Liu JW, Xiang P, Lin P, Ye GF, Sternberg L. Tannin dynamics of propagules and leaves of *Kandelia candel* and *Bruguiera gymnorrhiza* in the Jiulong river estuary, Fujian, China. Biogeochemistry. 2006;78(3):343–59.

46. Terrill TH, Rowan AM, Douglas GB, Barry TN. Determination of extractable and bound condensed tannin concentrations in forage plants, protein-concentrate meals and cereal-grains. J Sci Food Agric. 1992;58(3):321–9.

47. Schweitzer JA, Madritch MD, Bailey JK, LeRoy CJ, Fischer DG, Rehill BJ, Lindroth RL, Hagerman AE, Wooley SC, Hart SC, et al. From genes to ecosystems: the genetic basis of condensed tannins and their role in nutrient regulation in a *Populus* model system. Ecosystems. 2008;11(6):1005–20.

48. Zeller WE, Sullivan ML, Mueller-Harvey I, Grabber JH, Ramsay A, Drake C, Brown RH. Protein precipitation behavior of condensed tannins from *Lotus pedunculatus* and *Trifolium repens* with different mean degrees of polymerization. J Agric Food Chem. 2015;63(4):1160–8.

19

Nano-mechanical characterization of the wood cell wall by AFM studies: comparison between AC- and QI™ mode

Kirstin Casdorff[1,2], Tobias Keplinger[1,2] and Ingo Burgert[1,2]*

Abstract

Background: Understanding the arrangement and mechanical properties of wood polymers within the plant cell wall is the basis for unravelling its underlying structure–property relationships. As state of the art Atomic Force Microscopy (AFM) has been used to visualize cell wall layers in contact resonance- and amplitude controlled mode (AC) on embedded samples. Most of the studies have focused on the structural arrangement of the S_2 layer and its lamellar structure.

Results: In this work, a protocol for AFM is proposed to characterize the entire cell wall mechanically by quantitative imaging (QI™) at the nanometer level, without embedding the samples. It is shown that the applied protocol allows for distinguishing between the cell wall layers of the compound middle lamella, S_1, and S_2 of spruce wood based on their Young's Moduli. In the transition zone, S_{12}, a stiffness gradient is measured.

Conclusions: The QI™ mode pushes the limit of resolution for mechanical characterization of the plant cell wall to the nanometer range. Comparing QI™- against AC images reveals that the mode of operation strongly influences the visualization of the cell wall.

Keywords: Atomic Force Microscopy, Wood, Spruce, Cell wall, Young's Modulus

Background

Wood is a unique biological material with several levels of structural hierarchy from the nanometer level to the macroscale [1]. In particular the organization on the nanoscale, within the cell wall, is still under debate [2, 3]. The cell wall has been studied with numerous high resolution techniques like Transmission Electron Microscopy, TEM [4, 5], Scanning Near-field Optical Microscopy (SNOM) [6], or Atomic Force Microscopy (AFM) [3, 7–10]. A comparison of these studies shows that the observed nanostructural organization of the cell wall is strongly influenced by the used technique, and the sample preparation. Most of the previously cited studies have focused on the structural arrangement of the S_2 layer and its lamellar organization, but also the other layers

like the S_1 and S_3 are of importance for the mechanical behavior of the entire cell wall composite. TEM measurements on stained samples elucidated a transition zone between the S_1 and S_2 layers, called S_{12}, where the lignin concentration decreases [11–14]. It is well known that in case of intra-wall failure, cell walls rupture predominately between the S_1 and the S_2-layer [14], however mechanical characterization of this mechanically highly relevant zone is missing.

State of the art imaging of secondary cell walls by AFM is in Resonant Contact mode [15, 16] and more common Amplitude Controlled mode (AC mode, an intermittent contact mode) of embedded cells [7, 17, 18]. Recent technical developments in the field of AFM allow to conduct a mechanical characterization in addition to topography studies, without sample embedding. Such multichannel AFM studies have first been applied for the characterization of primary walls [19, 20], and little has been done on secondary cell walls yet. Peakforce Quantitative

*Correspondence: iburgert@ethz.ch
[1] Wood Materials Science, Institute for Building Materials, ETH Zürich, Stefano-Franscini-Platz 3, 8093 Zurich, Switzerland
Full list of author information is available at the end of the article

Nanomechanics (PFQNM) has been used to image bamboo fiber cell walls [21] and to characterize the shape of cellulose microfibrils in primary plant cell walls in water using a very sharp probe with a curvature below 5 nm [19]. Adhesion force mapping was conducted to measure the inactivation of a freshly cut cell wall surface, and to study the influence of surface roughness and tip geometry [22, 23]. Recently, Arnould et al. [24] gave a proof of concept of high resolution force mapping on embedded flax fibers by studying the mechanical gradients using nanoindentation as a reference and additionally Muraille et al. [10] applied the protocol on poplar fiber cell walls, however without providing specific structural details.

The purpose of the present paper is to systematically explore the feasibility of using the Quantitative Imaging mode (QI™ mode), a force spectroscopy mode, which records a force–distance (FD) curve in every pixel, as a tool to characterize the wood cell wall with a specific focus on the transition from the middle lamella to the S_2 layer on a cell wall cross-section of spruce. To be able to compare the structures imaged in QI™ mode to literature data, the cell wall was also scanned in AC mode. Thereby, structural information was revealed with high resolution, helping to better understand the underlying structure–property relationships of wood cell walls.

Methods
Spruce cube preparation
The cross-section of an air dried spruce cube ($5 \times 5 \times 5$ mm^3) was prepared by means of a two-step polishing process. The plane of the sectioning was oriented perpendicular to the longitudinal fiber axis of the wood. A microtome with a steel knife (RM2255, Leica) was used to smoothen the surface under wet conditions. Following the protocol of Keplinger et al. [6], an ultramicrotome (Ultracut, Reichert-Jung) equipped with a Diatome Histo diamond knife was used to polish. The typical root-mean-square roughness of a 1×1 μm^2 cross-section was less than 1.4 nm. The microfibril angle (MFA) was measured on a spruce sample, cut out of the same strip of wood to gain samples from longitudinally matched positions, by wide angle X-ray scattering on the longitudinal-transverse plane. A MFA of 6° was recorded, which points to a mature wood probe.

Quantitative imaging AFM
AFM imaging was performed using a NanoWizard 4 (JPK Instruments AG) in QI™ mode under controlled climatic conditions (temperature 20 °C, humidity 65%). As a cantilever, a non-contact cantilever (NCHR, Nano World, resonance frequency 320 kHz) with a silicon probe was used. The cantilever was calibrated with the contact-free method for a beam shaped cantilever by giving the

environmental conditions and cantilever dimensions [25]. The force constant was calibrated to be 30 N m^{-1} (± 1 N m^{-1}, n = 5) and the deflection sensitivity was determined to be 24 nm V^{-1} (± 1 nm V^{-1}, n = 5). The setpoint of the measurement was defined according to the cantilever stiffness (60 nN) and the z-length (50 nm) and the pixel time (12 ms) were set fixed, to ensure a similar velocity for each measurement. This resulted in an extend rate of 62.5 kHz (the extend rate controls the speed of the movement on the extend part of the FD curve), and an extend speed of 10.42 μm s^{-1}.

The software extension Advanced QI™ mode was used to have full access to all FD curves. The mapping resolution was chosen to be 256×256 pixels and the scan size was set between 10×10 μm^2 and 1×1 μm^2 (theoretical resolution limit 3.9 nm). The AFM was operated in z-closed loop, therefore the nonlinearity and hysteresis of the piezo were corrected during the movement (channel: height measured) and FD curves had a constant speed. The baseline was adjusted to correct for any changes of the setpoint.

Prior and after a cell scan, the tip resolution at 0° and 90° scan direction was checked using a test specimen (Product No. 628-AFM tip and resolution test specimen, Pelcro, TED PELLA INC.). The tip radius was fitted according to the stiffness of polystyrene (2 GPa) after scanning a polymer test specimen (PS-LDPE-GS, Veeco Metrology Group) (Fig. 1).

The data was analyzed in the JPK image processing software (JPK Instruments AG). A line fit was applied to correct the height measure image. To avoid characteristic shadows around high objects, the areas with elevated features were excluded from the line fit by a region of interest. As a second step possible default lines were replaced with the average between the adjacent lines. Force curve batch processing was performed in the following order on the extend curve: (1) Calibration of V-deflection (sensitivity and spring constant). (2) Smoothing of force data. (3) Baseline subtraction for offset and tilt by defining the fit range from 100% to the snap in, normally around 10%. (4) Contact point determination: Calculates vertical tip position, corrects the height signal for the cantilever deflection. (5) The Young's Modulus can be calculated from the slope of the force curve by applying a contact mechanics model. As suggested by the suppliers, in PFQNM (Bruker) the Young's Modulus is determined with the Derjaguin–Müller–Toporov (DMT) model of elastic contact on the retract curve [26], whereas in QI™ the extend curve is fitted to avoid any influence from plastic deformation. The Young's Modulus was calculated with the DMT model, assuming a Poisson's ratio of 0.4 (adapted from Gibson and Ashby [27]), and the respective radius of the cantilever. The quality of the fit

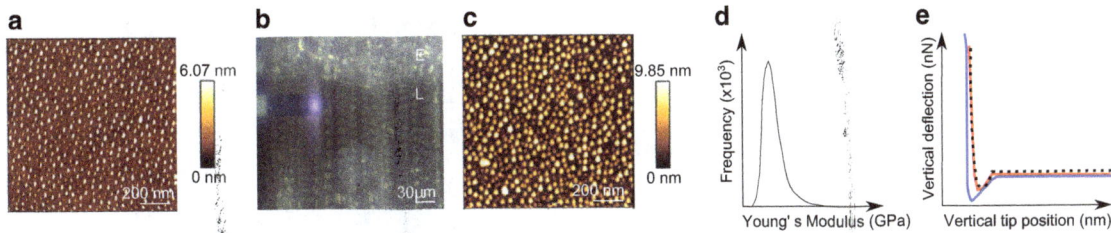

Fig. 1 Basic flow chart of the measurement routine. **a** A tip and resolution test specimen is scanned. **b** The wood sample is scanned, *E* earlywood, *L* latewood. **c** The tip and resolution test specimen is scanned again. **d** A stiffness test specimen is scanned and the tip radius is fitted to the stiffness of the polystyrene. **e** The Young's Modulus can be calculated for the wood sample by fitting (*dotted black line*) the *trace curve* (*red*) with the Derjaguin–Müller–Toporov model (*blue, retrace curve*)

was inspected visually for a typical FD curve. Due to the high amount of curves generated within one image, the fit cannot be optimized for each curve, but the weight of a false processed curve is therefore limited. The results were plotted in a histogram revealing the distribution of the data.

To remove cutting artefacts, the image data can be corrected with a two-dimensional fast Fourier transform (FFT) using Gwyddion 2.47.

AC mode AFM

In AC mode the mapping resolution was increased to 512×512 pixels, taking the equivalent time as the QI™ image. The full z-range of 15 μm was used. For the calibration of the cantilever the setpoint amplitude was selected to be around 70%. The approach was performed with constant velocity and baseline update at the starting point. The gain parameters were optimized in respect to the offset of the trace and retrace line in the oscilloscope. The channels were post-processed according to the height measured image described for the QI™ mode.

Results and discussion
Imaging cell walls in QI™ mode

By using the QI™ mode it is possible to scan over a whole unembedded cell through the lumen, because for FD curves the cantilever only moves the z-length that was set, in this case 200 nm (Fig. 2). Although the resolution of such a large scan (18 μm × 31 μm) is reduced in x-direction to 35 nm and in y-direction to 60 nm, stiffness changes between the stiff S_2 and the softer middle lamella and S_1 layer regions can clearly be seen in the Young's Modulus image (Fig. 2b). Furthermore, the quality of the cut could be judged based on the visible inclination at the lumen/cell wall interface in the height image (Fig. 2a, black arrow).

The middle lamella reveals no clear transition to the adjacent primary wall. Therefore, the middle lamella and

both adjacent primary walls are termed compound middle lamella, CML. The height images of the CML appear different depending on the scanning direction, which can be perpendicular, or in line to the cell wall layers (Fig. 3). This is a typical artefact in AFM imaging when scanning oriented structures [28], and needs to be considered when comparing different positions in the cell wall. Interestingly, the corresponding Young's Modulus images are not affected by the scanning direction.

Comparison of QI™- and AC mode

In QI™ mode the algorithm of the tip motion measures a FD curve in every pixel, with a defined setpoint [29]. Thereby, besides topological information the sample can be characterized mechanically with a high spatial resolution at high speed. There is no xy movement during the FD curve recording, which ensures a measurement under constant velocity. In AC mode the scanning cantilever is oscillating at a kilohertz range frequency, and as an additional channel the phase image can be displayed. The lock-in-amplifier measures a phase shift between the drive signal and the cantilever movement in dependency of the tip-sample interaction, including mechanical information, adhesion, and dissipation of cantilever energy [30]. In order to compare the imaging mode QI™ to the state of the art applied AC mode, we choose to scan with both modes the area of transition from the compound middle lamella to the S_2.

QI™ mode

Figure 4a shows an overview image of a cell corner and Fig. 4b–d display a zoom into the transition zone from the middle lamella to the S_2-layer. The vertical straight lines in the images of the S_2 are cutting artefacts that result from imperfections of the diamond knife. The CML is buildup of an isotropic structure consisting mainly of lignin, that is assumed to organize in a self-assembly process [31, 32]. After the CML follows a narrow zone of

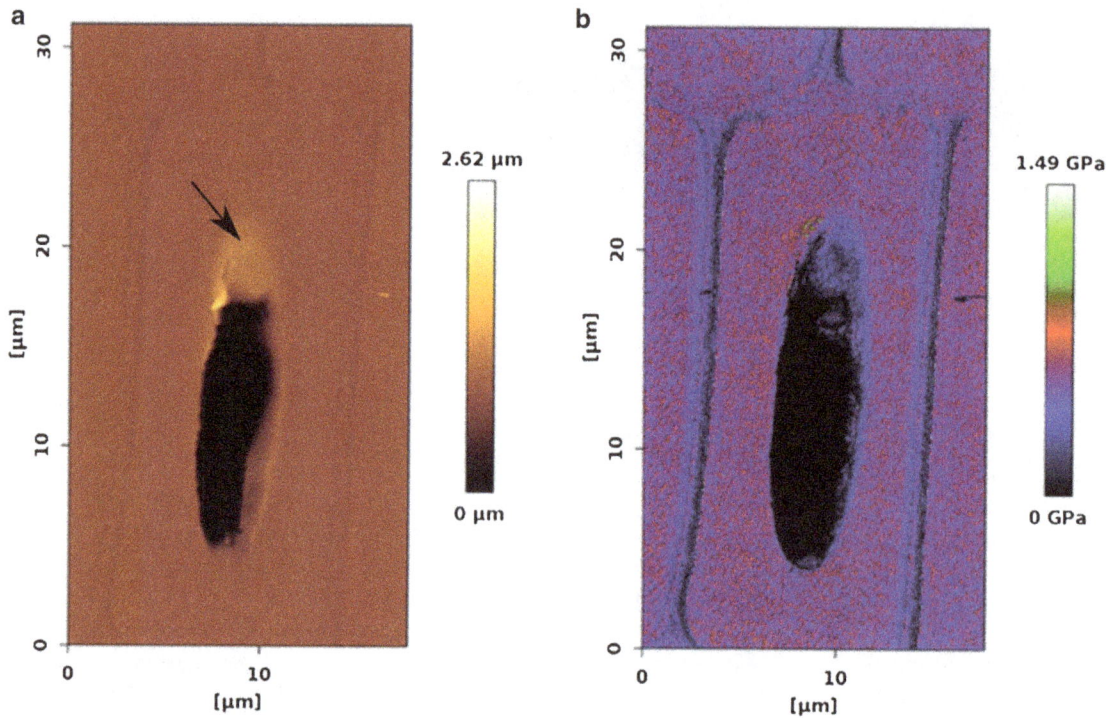

Fig. 2 a Height image and corresponding, **b** Young's Modulus image of a whole latewood cell (512 × 512 pixels). The radial direction runs from *left* to *right*. The *black arrow* point to an inclination at the lumen/cell wall interface

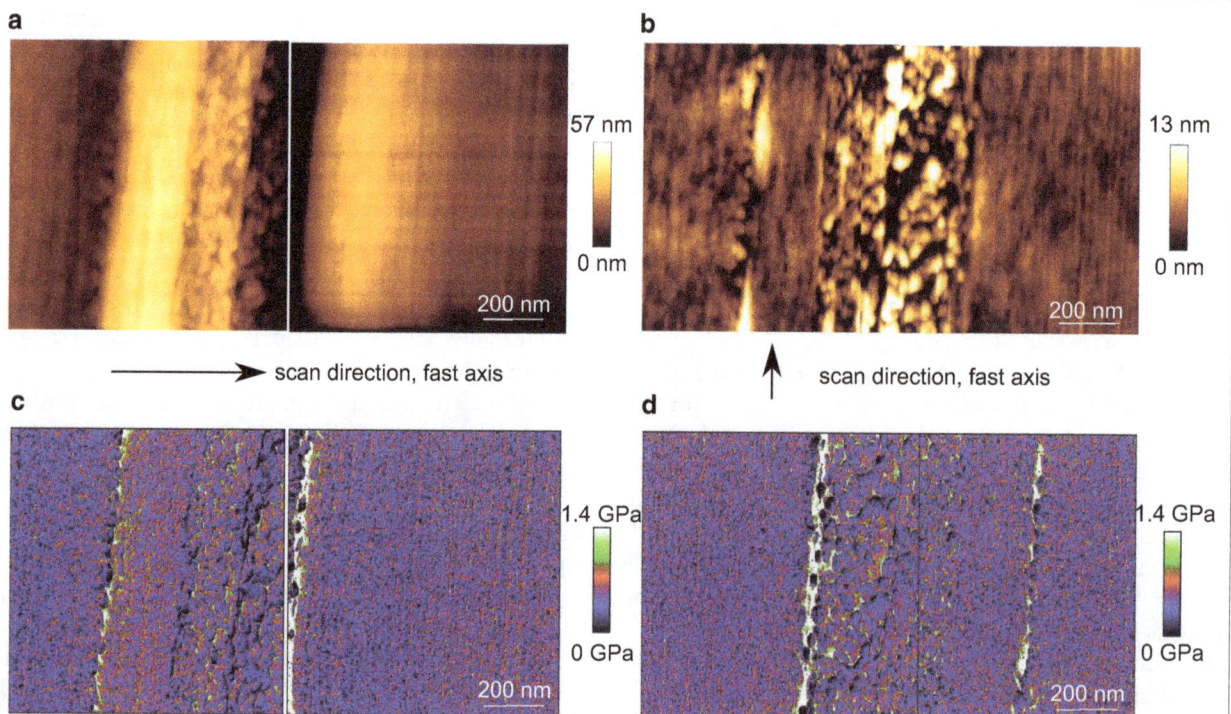

Fig. 3 Height- and Young's Modulus image of a compound middle lamella scanned at two perpendicular directions. **a**, **c** 0° scan direction perpendicular to the structures and **b**, **d** 90° scan direction in line with the structures. The cutting direction is perpendicular to the compound middle lamella

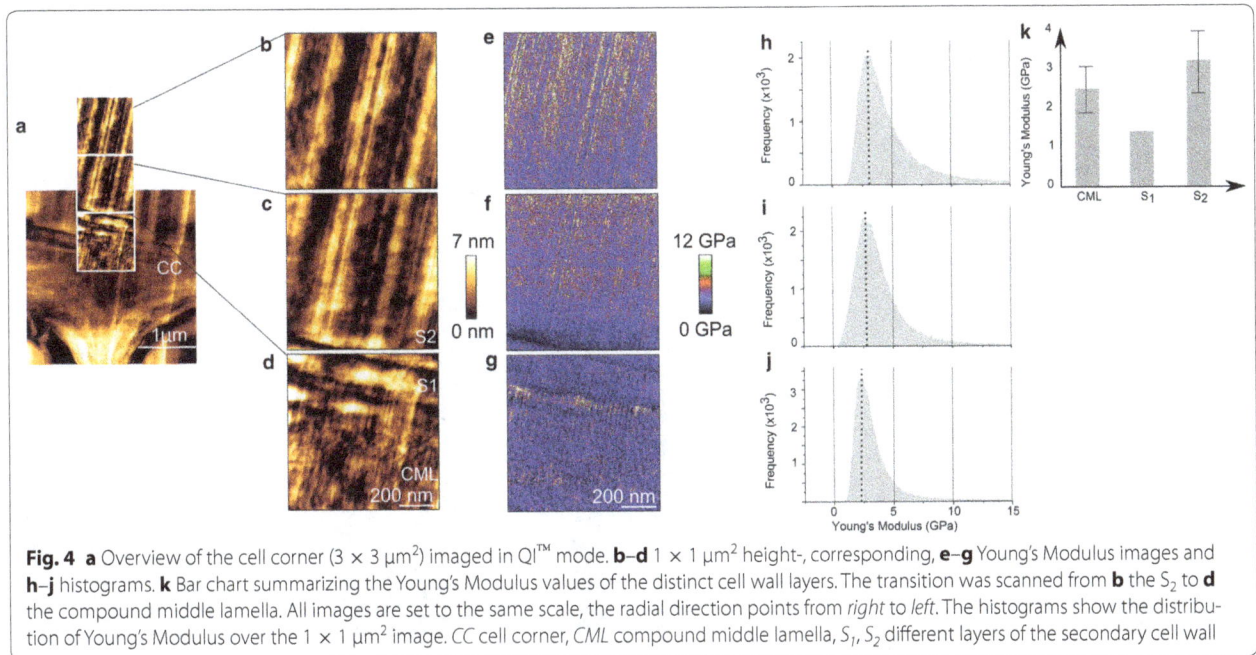

Fig. 4 a Overview of the cell corner (3 × 3 µm²) imaged in QI™ mode. **b–d** 1 × 1 µm² height-, corresponding, **e–g** Young's Modulus images and **h–j** histograms. **k** Bar chart summarizing the Young's Modulus values of the distinct cell wall layers. The transition was scanned from **b** the S₂ to **d** the compound middle lamella. All images are set to the same scale, the radial direction points from *right* to *left*. The histograms show the distribution of Young's Modulus over the 1 × 1 µm² image. *CC* cell corner, *CML* compound middle lamella, *S₁, S₂* different layers of the secondary cell wall

approximately 100 nm, most presumably the S_1. While the S_1 has a comparable denser structure, the S_2 appears as a woven network. The typical lamella structure of the S_2 cannot be clearly visualized because, (1) cutting artefacts may overly the lamella, (2) the scanned area might be too close to the CML, and/or (3) microfibrils are strictly parallel aligned [2].

The Young's Modulus of the S_1 was determined by selecting a rectangle of maximum size inside the cell wall layer and calculating the root mean square value (Fig. 4g). Due to its large MFA, typically 70–90°, the S_1 possesses a lower stiffness than the S_2 (Fig. 4e) [12]. Surprisingly, the S_1 stiffness is with 1.3 GPa lower than the CML (data obtained from a region of interest comprising approximately 5625 FD curves). This might be explained by the different textures of the loose CML and the denser S_1 layer, leading to artificially higher peak values in the CML. This indicates that due to the specific surface- cantilever tip interactions the structural patterns of the cell wall layers affect the stiffness values and that the stiffness ratios between the cell wall layers can be altered.

In the corresponding stiffness images and histograms it can be seen, how the Young's Modulus changes from a narrow Gaussian distribution mean value of around 2.5 ± 0.6 GPa in the CML (Fig. 4j) to a broader distribution around 3.2 ± 0.8 GPa in the S_2 layer (Fig. 4h, for a comparison the values are summarized in Fig. 4k, data obtained from one image comprising approximately 65536 FD curves). The histogram distribution of the S_2 might be larger, because of being more heterogeneous in

terms of biopolymer composition compared to the middle lamella, which is mainly composed of lignin [33]. The above-mentioned transition zone between the S_1 and S_2, called S_{12}, where the lignin concentration decreases [11, 12], can be visualized in the QI™ mode over a length of 2 µm by an increase in Young's Modulus from the S_1 to the S_2 (Fig. 4f,i). Depending on the depth information measured by AFM, the increase in Young's Modulus can arise from (1) the stiff cellulose microfibrils oriented with a small MFA, (2) the change in the surface structure due to a change in MFA, and/or (3) the change in composition of the different cell wall layers. Further experiments on composite model systems, with defined MFA on a similar length scale than wood, are required to make a certain statement about the parameters influencing the measured Young's Modulus. Certainly, very low stiffness values for the wood cell wall, respectively its individual layers were measured in comparison to mechanical data obtained by tensile tests [34–36] and nanoindentation [37, 38]. The entirely different loading conditions and test geometries in micro- and macroscopic tests do not allow for a direct comparison with the AFM data. Since the cell wall stiffness values obtained by nanoindentation and those derived from micro- and macroscopic tests are in the same range, a focus is laid on a comparison with values measured by nanoindentation, which was recently discussed by Arnould and Arinero [39].

The contact mechanism of the probe of a nanoindenter considerably differs from an AFM tip in terms of geometry and indentation depth, which results in different

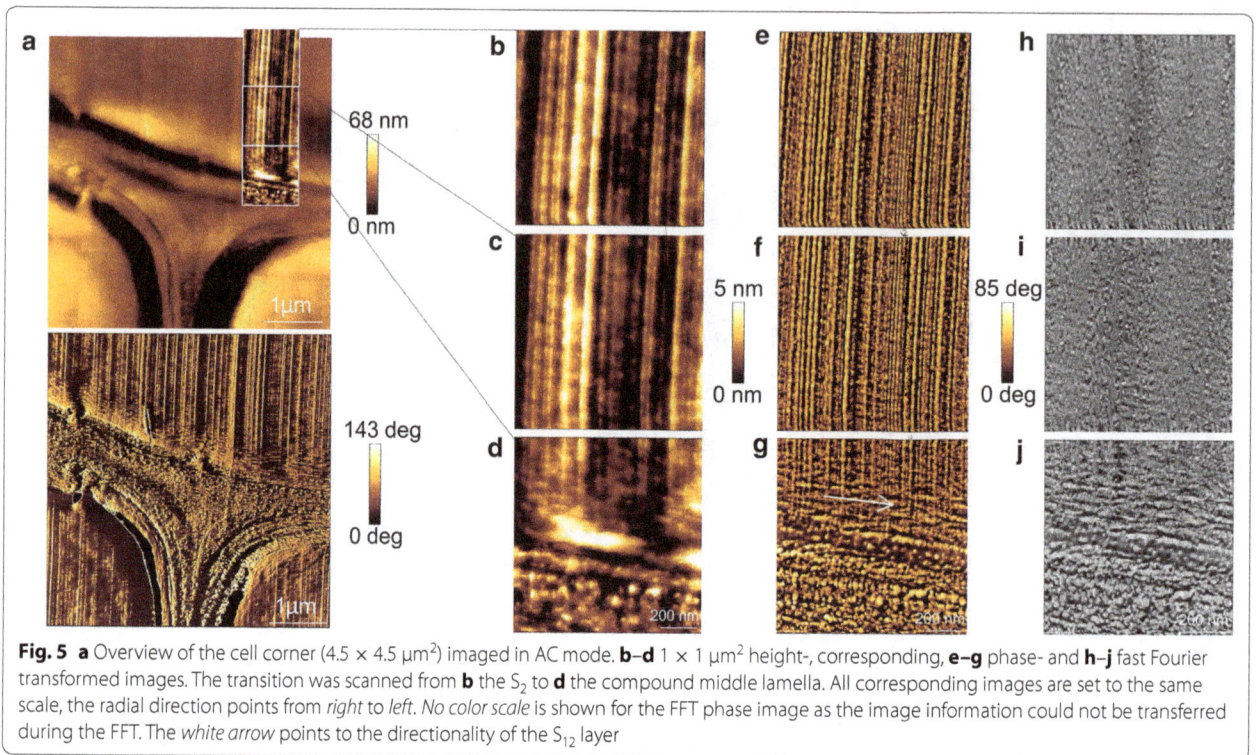

Fig. 5 a Overview of the cell corner (4.5 × 4.5 µm^2) imaged in AC mode. **b–d** 1 × 1 µm^2 height-, corresponding, **e–g** phase- and **h–j** fast Fourier transformed images. The transition was scanned from **b** the S$_2$ to **d** the compound middle lamella. All corresponding images are set to the same scale, the radial direction points from *right* to *left*. *No color scale* is shown for the FFT phase image as the image information could not be transferred during the FFT. The *white arrow* points to the directionality of the S$_{12}$ layer

interactions with the wood surface, because of the anisotropy of wood [39, 40]. Comparing stiffness values obtained by different AFM modes, Arnould et al. [24] applied PFQNM on a flax fiber with a setpoint of 200 nN on the retrace curve. For the S$_1$ and CML they calculated a Young's Modulus of 7 GPa. For the S$_2$ they obtained values from 13 to 18 GPa, which is comparable to the values they measured with a nanoindenter. Muraille et al. [10] applied PFQNM on a transverse section of poplar plant fiber cell wall with a setpoint of 600 nN. The indentation moduli of the CML, S$_1$ and S$_2$ were averaged 17, 21, and 26 GPa. In our measurement the penetration depth is very small as at a setpoint of 60 nN, the penetration is around 5 nm. Thereby, only surface properties are determined and no plastic deformation takes place. The setpoints mentioned in the two studies before are three to ten times higher than in our measurement, therefore it is assumed that also the penetration depth was larger. As wood is a viscoelastic material, the scan speed of the measurements also needs to be taken into account, when comparing the obtained values. In QI$^{™}$ mode the supplier suggests to use the trace curve, whereas in PFQNM it is suggested to use the retrace curve, for the fitting routine of the contact model. In our measurements, the slope of the retrace curve was too steep to be fitted. The trace curves could be fitted with the Hertz- [41] and the DMT model [26], with the DMT model giving higher stiffness

values. The roughness of the microtome polished surface was very small. For the CR-AFM and the PFQNM measurement of Arnould et al. [24] the Young's Modulus increases from the CML to the S$_2$ by a factor of two. Although the values in our measurements were smaller, the same tendency was found.

AC mode

To compare the QI$^{™}$ mode with the AC mode, Fig. 5 displays the transition from the CML to the S$_2$ imaged in AC mode. The height images in the two modes are comparable (Fig. 5b–d). The phase image shows lignin in the CML as ellipsoid structures and the adjacent S$_1$ layer possess a low phase signal (Fig. 5e–g). In phase contrast images, typically darker areas can be correlated with regions of lower stiffness [30]. By applying a FFT the directed cutting artefacts can be selectively removed from the phase image, thereby the structural change from the S$_1$ layer to the S$_2$ layer becomes more obvious (Fig. 5h–j). The transition zone S$_{12}$ shows a parallel orientation to the CML (white arrow), whereas the S$_2$ has a granular structure. The transition zone visualized in the phase image is smaller than the one from the Young's Modulus image, as the Modulus image is more sensitive for detecting changes in mechanical properties.

Fahlén and Salmén [17] visualized individual cellulose aggregates ranging from 10 to 30 nm in the S$_2$ layer by

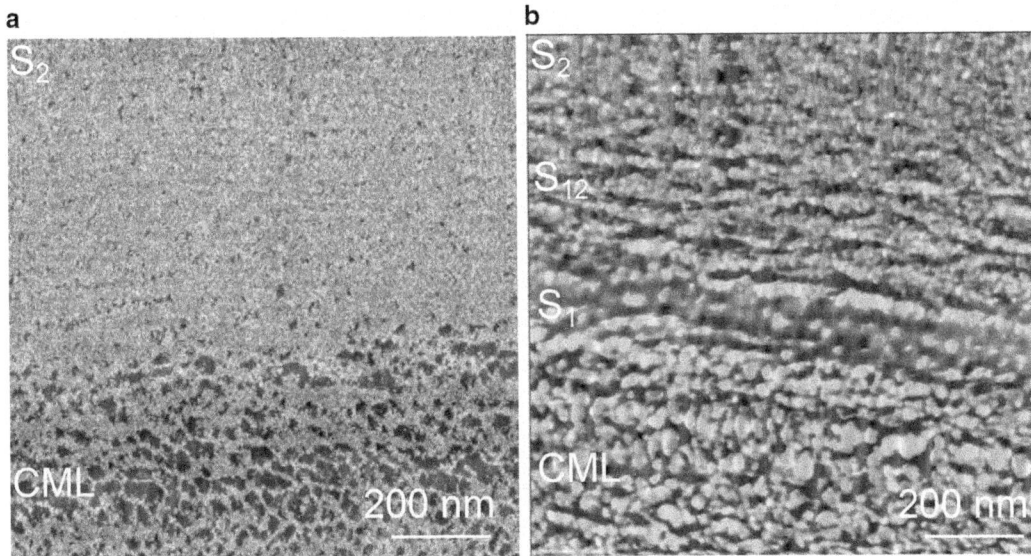

Fig. 6 Comparison of the fast Fourier transform (FFT) image scanned by **a** QI™ mode and **b** AC mode (enlargement of Fig. 5j). *No color scale* is shown as the image information could not be transferred during the FFT. *CML* compound middle lamella, S_1, S_{12}, S_2 different layers of the secondary cell wall layer

phase imaging oriented with a regularity that was interpreted as an additional lignin pattern. Here, we fitted the structures in the FFT of the S_2 and they lay in the same size range compared to the one observed by Fahlén and Salmén [17]. The lignin structures of the CML had an ellipsoid shape ranging from 30 to 60 nm diameter. Due to the influence of the cutting artefacts a clear lamellar structure in the S_2 could not be detected.

Comparison fast Fourier transform

In Fig. 6 two enlarged FFT corrected images of the transition between the CML and S_2 scanned in QI™- (Fig. 6a) and AC mode (Fig. 6b) can be seen. After FFT processing the variability of the Young's Modulus values could be reduced to two colors (the lighter, the stiffer), thereby the gradient information is lost. One can only distinguish the CML from the S_2. The CML appears as a porous network, whereas the S_2 seems to be dense. The phase image has a twice as high resolution, therefore it appears sharper. The CML gives the impression of a negative of the Young's Modulus image and the S_2 has a granular appearance. The phase shift is proportional to the stiffness, but phase contrast interpretation is not as straight forward like the analysis of FD curves, due to the contributions from contact area, viscoelastic properties, and capillary forces [30].

Conclusions

In conclusion, QI™ mode gives the opportunity to mechanically characterize the different secondary cell wall layers on the nanometer level by obtaining

FD curves that can be analyzed with a mechanical model. Although too low stiffness values are measured, it outperforms mechanical characterization by nanoindentation in terms of resolution, and provides a more distinct image of stiffness distribution than what can be obtained by phase contrast imaging in AC mode.

The imaging mode QI™ is on the one hand very robust, as it can scan over whole unembedded cells, and on the other hand it is sensitive enough to detect small changes, as can be seen from the visualization of the transition zone S_{12}. The nanostructure of wooden cell walls is just at the beginning to be characterized mechanically. Comparative studies are needed to unravel the influence of never dried wood, different cell types, or different species on the mechanics. Further insights into the cell wall assembly can be used to understand the localization of chemical modifications within wood beyond the resolution limit of commonly used techniques [42], or to set up reliable computational models on the microscale [43–45].

Abbreviations
AFM: Atomic Force Microscopy; CR-AFM: contact resonance AFM; DMT: Derjaguin–Müller–Toporov; FD curve: force–distance curve; FFT: fast Fourier transform; JKR: Johnson–Kendall–Roberts; MFA: microfibril angle; PFQNM: Peakforce Quantitative Nanomechanics; QI™: Quantitative Imaging.

Authors' contributions
KC designed the experiment and analyzed data. KC, TK and IB co-wrote the paper. All authors discussed results and commented on the manuscript. All authors read and approved the final manuscript.

Author details
[1] Wood Materials Science, Institute for Building Materials, ETH Zürich, Stefano-Franscini-Platz 3, 8093 Zurich, Switzerland. [2] Applied Wood Materials, Empa-Swiss Federal Laboratories for Materials Science and Technology, Überlandstrasse 129, 8600 Dübendorf, Switzerland.

Acknowledgements
The authors thank Thomas Schnider for cutting the wood samples. Valuable discussion with John Berg is acknowledged.

Competing interests
The authors declare that they have no competing interests.

References
1. Fratzl P, Weinkamer R. Nature's hierarchical materials. Prog Mater Sci. 2007;52:1263–334.
2. Donaldson LA. A three-dimensional computer model of the tracheid cell wall as a tool for interpretation of wood cell wall ultrastructure. IAWA J. 2001;22:213–33.
3. Zimmermann T, Thommen V, Reimann P, Hug HJ. Ultrastructural appearance of embedded and polished wood cell walls as revealed by atomic force microscopy. J Struct Biol. 2006;156:363–9.
4. Ruel K, Barnould F, Goring DAI. Lamellation in the S2 layer of softwood tracheids as demonstrated by scanning transmission electron microscopy. Wood Sci Technol. 1978;12:287–91.
5. Kerr AJ, Goring DAI. The ultrastructural arrangement of the wood cell wall. Cellul Chem Technol. 1975;9:563–73.
6. Keplinger T, Konnerth J, Aguié-Béghin V, Rüggeberg M, Gierlinger N, Burgert I. A zoom into the nanoscale texture of secondary cell walls. Plant Methods. 2014;10:1.
7. Fahlén J, Salmén L. On the lamellar structure of the tracheid cell wall. Plant Biol. 2002;4:339–45.
8. Schwarze FWMR, Engels J. Cavity formation and the exposure of peculiar structures in the secondary wall (S2) of tracheids and fibres by wood degrading Basidiomycetes. Holzforschung. 1998;52:117–23.
9. Zimmermann T, Eckstein JSD. Rasterelektronenmikroskopische Untersuchung an Zugbruchflächen von Fichtenholz. Holz Roh Werkst. 1994;52:223–9.
10. Muraille L, Aguié-Béghin V, Chabbert B, Molinari M. Bioinspired lignocellulosic films to understand the mechanical properties of lignified plant cell walls at nanoscale. Sci Rep. 2017;7:1–11.
11. Fromm J, Rockel B, Lautner S, Windeisen E, Wanner G. Lignin distribution in wood cell walls determined by TEM and backscattered SEM techniques. J Struct Biol. 2003;143:77–84.
12. Brändström J, Bardage SL, Daniel G, Nilsson T. The structural organisation of the S1 cell wall layer of norway spruce tracheids. IAWA J. 2003;24:27–40.
13. Reza M, Ruokolainen J, Vuorinen T. Out-of-plane orientation of cellulose elementary fibrils on spruce tracheid wall based on imaging with high-resolution transmission electron microscopy. Planta. 2014;240:565–73.
14. Donaldson LA. Cell wall fracture properties in relation to lignin distribution and cell dimensions among three genetic groups of radiate pine. Wood Sci Technol. 1995;29:51–63.
15. Clair B, Arinero R, Lévèque G, Ramonda M, Thibaut B. Imaging the mechanical properties of wood cell wall layers by atomic force modulation microscopy. IAWA J. 2003;24:223–30.
16. Nair SS, Wang S, Hurley DC. Nanoscale characterization of natural fibers and their composites using contact-resonance force microscopy. Compos A Appl Sci Manuf. 2010;41:624–31.
17. Fahlén J, Salmén L. Cross-sectional structure of the secondary wall of wood fibers as affected by processing. J Mater Sci. 2003;38:119–26.
18. Fahlén J, Salmén L. Pore and matrix distribution in the fiber wall revealed by atomic force microscopy and image analysis. Biomacromol. 2005;6:433–8.
19. Zhang T, Zheng Y, Cosgrove DJ. Spatial organization of cellulose microfibrils and matrix polysaccharides in primary plant cell walls as imaged by multichannel atomic force microscopy. Plant J. 2016;85:179–92.
20. Peaucelle A, Braybrook SA, Le Guillou L, Bron E, Kuhlemeier C, Hofte H. Pectin-induced changes in cell wall mechanics underlie organ initiation in Arabidopsis. Curr Biol. 2011;21:1720–6.
21. Ren D, Wang H, Yu Z, Wang H, Yu Y. Mechanical imaging of bamboo fiber cell walls and their composites by means of peakforce quantitative nanomechanics (PQNM) technique. Holzforschung. 2015;69:975–84.
22. Frybort S, Obersriebnig M, Muller U, Gindl-Altmutter W, Konnerth J. Variability in surface polarity of wood by means of AFM adhesion force mapping. Colloid Surf A. 2014;457:82–7.
23. Jin X, Kasal B. Adhesion force mapping on wood by atomic force microscopy: influence of surface roughness and tip geometry. R Soc Open Sci. 2016;3:160248.
24. Arnould O, Siniscalco D, Bourmaud A, Le Duigou A, Baley C. Better insight into the nano-mechanical properties of flax fibre cell walls. Ind Crop Prod. 2017;97:224–8.
25. Sader JE, Chon JWM, Mulvaney P. Calibration of rectangular atomic force microscope cantilevers. Rev Sci Instrum. 1999;70:3967–9.
26. Derjaguin BV, Muller VM, Toporov YP. Effect of contact deformations on adhesion of particles. J Colloid Interface Sci. 1975;53:314–26.
27. Gibson LJ, Ashby MF. Cellular solids. Structure and properties. Cambridge: Cambridge University Press; 2001.
28. Tsukruk VV, Bliznyuk VN, Visser D, Campbell AL, Bunning TJ, Adams WW. Electrostatic deposition of polyionic monolayers on charged surfaces. Macromolecules. 1997;30:6615–25.
29. Chopinet L, Formosa C, Rols MP, Duval RE, Dague E. Imaging living cells surface and quantifying its properties at high resolution using AFM in QITM mode. Micron. 2013;48:26–33.
30. Tsukruk VV, Singamani S. Scanning probe microscopy of soft matter: fundamentals and practice. 1st ed. London: Wiley; 2012.
31. Salmén L. Wood morphology and properties from molecular perspectives. Ann For Sci. 2015;72:679–84.
32. Salmén L, Olsson A-M, Stevanic JS, Simonovic J, Radotic K. Structural organisation of the wood polymers in the wood fibre structure. Bioresources. 2012;7:512–32.
33. Fengel D, Wegener G. Wood chemistry, ultrastructure, reactions. Berlin: Walter de Gruyter; 1984.
34. Gindl W, Schöberl T. The significance of the elastic modulus of wood cell walls obtained from nanoindentation measurements. Compos A Appl Sci Manuf. 2004;35:1345–9.
35. Page DH, El-Hosseiny F, Winkler K. Behaviour of single wood fibres under axial tensile strain. Nature. 1971;3:229–52.
36. Burgert I, Keckes J, Frühmann K, Fratzl P, Tschegg SE. A comparison of two techniques for wood fibre isolation- evaluation by tensile tests on single fibres with different microfibril angle. Plant Biol. 2002;4:9–12.
37. Jäger A, Hofstetter K, Buksnowitz C, Gindl-Altmutter W, Konnerth J. Identification of stiffness tensor components of wood cell walls by means of nanoindentation. Compos A Appl Sci Manuf. 2011;42:2101–9.
38. Wimmer R, Lucas BN, Tsui TY, Oliver WC. Longitudinal hardness and Young's modulus of spruce tracheid secondary walls using nanoindentation technique. Wood Sci Technol. 1997;31:131–41.
39. Arnould O, Arinero R. Towards a better understanding of wood cell wall characterisation with contact resonance atomic force microscopy. Compos A Appl Sci Manuf. 2015;74:69–76.
40. Gindl W, Gupta HS, Schöberl T, Lichtenegger HC, Fratzl P. Mechanical properties of spruce wood cell walls by nanoindentation. Appl Phys A. 2004;79:2069–73.
41. Sneddon N. The relation between load and penetration in the axisymmetric boussinesq problem for a punch of arbitrary profile. Int J Eng Sci. 1985;3:47–57.
42. Gierlinger N, Keplinger T, Harrington M. Imaging of plant cell walls by confocal Raman microscopy. Nat Protoc. 2012;7:1694–708.
43. Barber NF, Meylan BA. The anisotropic shrinkage of wood: A theoretical model. Holzforschung. 1964;18:146–56.
44. Mishnaevsky L, Qing H. Micromechanical modelling of mechanical behaviour and strength of wood: state-of-the-art review. Comp Mater Sci. 2008;44:363–70.
45. Yamamoto H, Kojima Y. Properties of cell wall constituents in relation to longitudinal elasticity of wood. Wood Sci Technol. 2002;36:55–74.

PERMISSIONS

The contributors of this book come from diverse backgrounds, making this book a truly international effort. This book will bring forth new frontiers with its revolutionizing research information and detailed analysis of the nascent developments around the world.

We would like to thank all the contributing authors for lending their expertise to make the book truly unique. They have played a crucial role in the development of this book. Without their invaluable contributions this book wouldn't have been possible. They have made vital efforts to compile up to date information on the varied aspects of this subject to make this book a valuable addition to the collection of many professionals and students.

This book was conceptualized with the vision of imparting up-to-date information and advanced data in this field. To ensure the same, a matchless editorial board was set up. Every individual on the board went through rigorous rounds of assessment to prove their worth. After which they invested a large part of their time researching and compiling the most relevant data for our readers.

The editorial board has been involved in producing this book since its inception. They have spent rigorous hours researching and exploring the diverse topics which have resulted in the successful publishing of this book. They have passed on their knowledge of decades through this book. To expedite this challenging task, the publisher supported the team at every step. A small team of assistant editors was also appointed to further simplify the editing procedure and attain best results for the readers.

Apart from the editorial board, the designing team has also invested a significant amount of their time in understanding the subject and creating the most relevant covers. They scrutinized every image to scout for the most suitable representation of the subject and create an appropriate cover for the book.

The publishing team has been an ardent support to the editorial, designing and production team. Their endless efforts to recruit the best for this project, has resulted in the accomplishment of this book. They are a veteran in the field of academics and their pool of knowledge is as vast as their experience in printing. Their expertise and guidance has proved useful at every step. Their uncompromising quality standards have made this book an exceptional effort. Their encouragement from time to time has been an inspiration for everyone.

The publisher and the editorial board hope that this book will prove to be a valuable piece of knowledge for researchers, students, practitioners and scholars across the globe.

LIST OF CONTRIBUTORS

Xiong Xiong, Lejun Yu, Ni Jiang and Qian Liu
Britton Chance Center for Biomedical Photonics, Wuhan National Laboratory for Optoelectronics, Huazhong University of Science and Technology, 1037 Luoyu Rd., Wuhan 430074, People's Republic of China

Lizhong Xiong, Kede Liu and Meng Liu
National Key Laboratory of Crop Genetic Improvement and National Center of Plant Gene Research, Huazhong Agricultural University, Wuhan 430070, People's Republic of China

Wanneng Yang
National Key Laboratory of Crop Genetic Improvement and National Center of Plant Gene Research, Huazhong Agricultural University, Wuhan 430070, People's Republic of China
College of Engineering, Huazhong Agricultural University, Wuhan 430070, People's Republic of China

Di Wu
College of Engineering, Huazhong Agricultural University, Wuhan 430070, People's Republic of China

Guoxing Chen
MOA Key Laboratory of Crop Ecophysiology and Farming System in the Middle Reaches of the Yangtze River, Huazhong Agricultural University, Wuhan 430070, People's Republic of China

Fryni Drizou, Neil S. Graham and Rumiana V. Ray
Division of Plant and Crop Sciences, School of Biosciences, University of Nottingham, Sutton Bonington Campus, Loughborough, Leicestershire, UK

Toby J. A. Bruce
School of Life Sciences, Keele University, Keele, Staffordshire, UK

Daisuke Urano, Kang-Ling Liao and Tyson L. Hedrick
Department of Biology, The University of North Carolina at Chapel Hill, Coker Hall, CB#3280, Chapel Hill, NC 27599-3280, USA

Ying Liang
Department of Biology, The University of North Carolina at Chapel Hill, Coker Hall, CB#3280, Chapel Hill, NC 27599-3280, USA
College of Natural Resources and Environment, Northwest A&F University, Yangling 712100, Shaanxi, China

Alan M. Jones
Department of Biology, The University of North Carolina at Chapel Hill, Coker Hall, CB#3280, Chapel Hill, NC 27599-3280, USA
Department of Pharmacology, University of North Carolina at Chapel Hill, Chapel Hill, NC 27599-3280, USA

Yajun Gao
College of Natural Resources and Environment, Northwest A&F University, Yangling 712100, Shaanxi, China

Shouyang Liu, Fred Baret and Xiuliang Jin
INRA, UMR-EMMAH, UMT-CAPTE, UAPV, 228 Route de l'aérodrome CS 40509, 84914 Avignon, France

Denis Allard
UMR BioSP, INRA, UAPV, 84914 Avignon, France

Bruno Andrieu
UMR ECOSYS, INRA, AgroParisTech, Université Paris-Saclay, 78850 Thiverval-Grignon, France

Philippe Burger
UMR AGIR, INRA, INPT, 31326 Toulouse, France

Matthieu Hemmerlé and Alexis Comar
Hi-Phen, 84914 Avignon, France

Chu Zhang, Xuping Feng, Fei Liu and Yong He
College of Biosystems Engineering and Food Science, Zhejiang University, 866 Yuhangtang Road, Xihu District, Hangzhou 310058, China

Jian Wang and Weijun Zhou
College of Agriculture and Biotechnology, Zhejiang University, Hangzhou 310058, China

Marvin Lüpke
Ecoclimatology, Technische Universität München, Hans-Carl-von-Carlowitz-Platz 2, 85354 Freising, Germany

Annette Menze
Ecoclimatology, Technische Universität München, Hans-Carl-von-Carlowitz-Platz 2, 85354 Freising, Germany
TUM Institute for Advanced Study, Lichtenbergstraße 2 a, 85748 Garching, Germany

Michael Leuchner
Ecoclimatology, Technische Universität München, Hans-Carl-von-Carlowitz-Platz 2, 85354 Freising, Germany
Springer Science+Business Media B.V., Van Godewijckstraat 30, 3311 GX Dordrecht, The Netherlands

Rainer Steinbrecher
Department of Atmospheric Environmental Research (IMK-IFU), Institute of Meteorology and Climate Research, Karlsruhe Institute of Technology (KIT), Kreuzeckbahnstraße 19, 82467 Garmisch Partenkirchen, Germany

Lorenzo Baldacci and Luca Masini
NEST, CNR Istituto Nanoscienze and Scuola Normale Superiore, Piazza San Silvestro 12, 56127 Pisa, Italy

Mario Pagano and Paolo Storchi
Consiglio per la ricerca in agricoltura e l'analisi dell'economia agraria, Centro di ricerca per la Viticoltura e l'Enologia, Viale Santa Margherita 80, 52100 Arezzo, Italy

Giorgio Carelli
Dipartimento di Fisica, Università di Pisa, Largo Pontecorvo 3, 56127 Pisa, Italy

Alessandra Toncelli and Alessandro Tredicucci
NEST, CNR Istituto Nanoscienze and Dipartimento di Fisica, Università di Pisa, Largo Pontecorvo 3, 56127 Pisa, Italy

Kyle Benzle and Katrina Cornish
Department of Horticulture and Crop Science, The Ohio State University, 1680 Madison Avenue, Wooster, OH, USA

Sunil Kumar, Frederik G. Görlitz and Paul M. W. French
Photonics Group, Department of Physics, Imperial College London, London SW7 2AZ, UK

Elizabeth Noble
Photonics Group, Department of Physics, Imperial College London, London SW7 2AZ, UK
Department of Chemistry, Imperial College London, London SW7 2AZ, UK
Institute of Chemical Biology, Imperial College London, London SW7 2AZ, UK

Chris Dunsby
Photonics Group, Department of Physics, Imperial College London, London SW7 2AZ, UK
Centre for Pathology, Imperial College London, London SW7 2AZ, UK

Chris Stain
Syngenta, Jealott's Hill International Research Centre, Bracknell, Berkshire RG42 6EY, UK

André C. Velásquez and Kinya Nomura
DOE Plant Research Laboratory, Michigan State University, East Lansing, MI 48824, USA

Sheng Yang He
DOE Plant Research Laboratory, Michigan State University, East Lansing, MI 48824, USA
Department of Plant Biology, Michigan State University, East Lansing, MI 48824, USA

Plant Resilience Institute, Michigan State University, East Lansing, MI 48824, USA

Howard Hughes Medical Institute, Gordon and Betty Moore Foundation, Michigan State University, East Lansing, MI 48824, USA

Max D. Cooper and Brantley R. Herrin
Department of Pathology and Laboratory Medicine, Emory University, Atlanta, GA 30322, USA

Rui Meng and Ying Sun
Computer, Electrical and Mathematical Science and Engineering Division, King Abdullah University of Science and Technology, Thuwal 23955-6900, Saudi Arabia

Stephanie Saade and Mark Tester
Biological and Environmental Science and Engineering Division, King Abdullah University of Science and Technology, Thuwal 23955-6900, Saudi Arabia

Bettina Berger
Australian Plant Phenomics Facility, The Plant Accelerator, University of Adelaide, Urrbrae, South Australia 5064, Australia

Chris Brien
Australian Plant Phenomics Facility, The Plant Accelerator, University of Adelaide, Urrbrae, South Australia 5064, Australia
Phenomics and Bioinformatics Research Centre, University of South Australia, Adelaide, South Australia 5001, Australia

Sebastian Kurtek
Department of Statistics, The Ohio State University, Columbus, OH, USA

Klaus Pillen
Institute of Agricultural and Nutritional Sciences, Martin Luther University Halle-Wittenberg, Betty-Heimann-Str. 3, 06120 Halle, Germany

F. M. Jiménez-Brenes, F. López-Granados, A. I. de Castro and J. Torres-Sánchez
Institute for Sustainable Agriculture, CSIC, 14004 Córdoba, Spain.

N. Serrano
Institute of Agricultural Research and Training (IFAPA), 14004 Córdoba, Spain

J. M. Peña
Institute of Agricultural Sciences, CSIC, 28006 Madrid, Spain

Ignacio López-Ribera and Carlos M. Vicient
Centre for Research in Agricultural Genomics (CRAG) CSIC-IRTA-UAB-UB, Campus UAB Bellaterra, 08193 Barcelona, Spain

A. Clarissa A. Negrini, F. Azzahra Ahmad Rashid, Lucy Hayes, Yuzhen Fan, You Zhang and Owen K. Atkin
ARC Centre of Excellence in Plant Energy Biology, Research School of Biology, Building 134, The Australian National University, Canberra, ACT 2601, Australia

Andrew P. Scafaro
ARC Centre of Excellence in Plant Energy Biology, Research School of Biology, Building 134, The Australian National University, Canberra, ACT 2601, Australia
Bayer CropScience SA-NV, Technologiepark 38, 9052 Gent (Zwijnaarde), Belgium

Brendan O'Leary
ARC Centre of Excellence in Plant Energy Biology, Research School of Biology, Building 134, The Australian National University, Canberra, ACT 2601, Australia
Australian Research Council Centre of Excellence in Plant Energy Biology, University of Western Australia, 35 Stirling Highway, Crawley, WA 6009, Australia

Vincent Chochois and Murray R. Badger
ARC Centre of Excellence for Translational Photosynthesis, Building 134, The Australian National University, Canberra, ACT 2601, Australia

A. Harvey Millar
Australian Research Council Centre of Excellence in Plant Energy Biology, University of Western Australia, 35 Stirling Highway, Crawley, WA 6009, Australia

Yi-Chin Tseng
Department of Physics, National Taiwan University, No. 1, Section 4, Roosevelt Rd, Da'an District, Taipei City 10617, Taiwan

Shi-Wei Chu
Department of Physics, National Taiwan University, No. 1, Section 4, Roosevelt Rd, Da'an District, Taipei City 10617, Taiwan
Molecular Imaging Center, National Taiwan University, No. 81, Changxing Street, Da'an District, Taipei 10672, Taiwan

Hsiang Sing Naik, Alec Lofquist, Soumik Sarkar, David Ackerman and Baskar Ganapathysubramanian
Department of Mechanical Engineering, Iowa State University, Ames, IA 50011, USA

Jiaoping Zhang, Teshale Assefa, Arti Singh and Asheesh K. Singh
Department of Agronomy, Iowa State University, Ames, IA 50011, USA

Dinesh C. Joshi
ICAR- Indian Grassland and Fodder Research Institute, Jhansi, Uttar Pradesh 284003, India

Vijaya Singh, Erik van Oosterom and Graeme Hammer
Queensland Alliance for Agriculture and Food Innovation, The University of Queensland, Brisbane, QLD 4072, Australia

Colleen Hunt and Emma Mace
Department of Agriculture and Fisheries, Hermitage Research Facility, Warwick, QLD 4370, Australia

Richard Sulman
Biosystems Engineering, 323 Margaret Street, Toowoomba, QLD 4350, Australia

David Jordan
Queensland Alliance for Agriculture and Food Innovation, Hermitage Research Facility, The University of Queensland, Warwick, QLD 4370, Australia

Philip-Edouard Shay and C. Peter Constabel
Department of Biology & Centre for Forest Biology, University of Victoria, P.O. Box 3020, STN CSC, Victoria, BC V8W 3N5, Canada

J. A. Trofymow
Department of Biology & Centre for Forest Biology, University of Victoria, P.O. Box 3020, STN CSC, Victoria, BC V8W 3N5, Canada
Pacific Forestry Centre, Canadian Forest Service, Natural Resources Canada, Victoria, BC, Canada

Kirstin Casdorff, Tobias Keplinger and Ingo Burgert
Wood Materials Science, Institute for Building Materials, ETH Zürich, Stefano-Franscini-Platz 3, 8093 Zurich, Switzerland
Applied Wood Materials, Empa-Swiss Federal Laboratories for Materials Science and Technology, Überlandstrasse 129, 8600 Dübendorf, Switzerland

Index

www.ingramcontent.com/pod-product-compliance
Lightning Source LLC
Chambersburg PA
CBHW082057190326
41458CB00010B/3513